Impact of Environmental Factors and Management Practices on Bee Health

Impact of Environmental Factors and Management Practices on Bee Health

Guest Editors

Ivana Tlak Gajger
Franco Mutinelli

MDPI

Basel • Beijing • Wuhan • Barcelona • Belgrade • Novi Sad • Cluj • Manchester

Guest Editors
Ivana Tlak Gajger
Department for Biology and Pathology of Fish and Bees
Faculty of Veterinary Medicine University of Zagreb
Zagreb
Croatia

Franco Mutinelli
National Reference Laboratory for Honey Bee Health
Istituto Zooprofilattico Sperimentale delle Venezie
Legnaro
Italy

Editorial Office
MDPI AG
Grosspeteranlage 5
4052 Basel, Switzerland

This is a reprint of the Special Issue, published open access by the journal *Insects* (ISSN 2075-4450), freely accessible at: https://www.mdpi.com/journal/insects/special_issues/Management_Bee.

For citation purposes, cite each article independently as indicated on the article page online and as indicated below:

Lastname, A.A.; Lastname, B.B. Article Title. *Journal Name* **Year**, *Volume Number*, Page Range.

ISBN 978-3-7258-3277-4 (Hbk)
ISBN 978-3-7258-3278-1 (PDF)
https://doi.org/10.3390/books978-3-7258-3278-1

Contents

About the Editors

Ivana Tlak Gajger

Ivana Tlak Gajger is an accomplished professor at the Faculty of Veterinary Medicine University of Zagreb. She graduated in 2005, and in 2010 she was awarded a Ph.D. in dissertation defence. Since 2006, she has been employed at the Department for Biology and Pathology of Fish and Bees, Faculty of Veterinary Medicine, where she is still working as a full professor. Besides activities in scientific and research work, she is active as a lecturer in the field of biology and pathology of beneficial insects. She has been the supervisor of 47 graduate and 7 doctoral students. She is the head of the postgraduate master study program Honeybee Health Protection and the establisher of the educational-archive station for beekeeping. Also, she is head of accredited (according to HRN EN ISO/IEC 17025), official Croatian National Reference Laboratory for Honeybee Diseases – APISlab. She works in close collaboration with beekeepers and other stakeholders through various science-based activities to identify opportunities to simultaneously promote beekeeping, healthy food production, and natural biodiversity and improve the implementation of honeybee health protection measures. Her goal is to find smart solutions for real beekeepers' problems that ultimately inform policymakers, promote education and good apiary–veterinary–environmental practices, and help primary producers. In recognition of her work, in 2022, she received the Award of the Croatian Academy of Sciences and Arts for the highest scientific achievements in the Republic of Croatia in the field of medical sciences. In 2023, she was elected as a full member of the Collegium of Veterinary Sciences of the Croatian Academy of Medical Sciences. According to research by Stanford University in the USA, she is among the 2% most influential scientists in 2022 and 2023.

Franco Mutinelli

Franco Mutinelli graduated in Veterinary Medicine in 1983 from the University of Bologna, followed by a diploma from the European College of Veterinary Pathologists in 1998 and an executive master's degree in management for healthcare companies in 2006 (EMMAS V, Bocconi University, Milano). He directs the SCS3—Specialist Diagnostics, Histopathology and Sanitary Entomology (2003)—and the Functional Animal Experimental and Welfare Department (2013) of Istituto Zooprofilattico Sperimentale delle Venezie. He is also the Director of the National Reference Centre for Beekeeping (2003) and the National Reference Laboratory for Bee Diseases (2018), as well as the Head of the Regional Centre for Beekeeping in Veneto (1995). His expertise covers the field of diagnosis and control of bee diseases, the safety of hive production, environmental monitoring, and regulations. He carries out studies and applied research in the areas described above within projects of the Ministries of Health, Agriculture, and Environment, EFSA, EU (COST Action, Eurostars, DISCONTOOLS). He has several collaborations in national and international research projects. He participated as a speaker and tutor in the Better Training for Safer Food (BTSF) events "Animal Health of Bees" (2010–2017), Live Animals "Import Control of Bees and Bumblebees" (2017–2025), and "Companion Animals Movement" (2017–2024), as well as in TAIEX events (2018–2025) related to the protection and improvement of the health of honey bees and the production of beehives in Egypt, North Macedonia, Jordan, and Cyprus (Turkish Cypriot Community). He also participates as an expert in the program Green Line Trade-Honey Sampling in Nicosia, Cyprus. He is a contract professor of honey bee health at the School of Veterinary Medicine and Agriculture of the University of Padua. He is the author and co-author of several scientific articles in international journals and Guest Editor of Special Issues on honey bee health.

insects

Editorial

Impact of Environmental Factors and Management Practices on Bee Health

Ivana Tlak Gajger [1,*] and Franco Mutinelli [2]

1 NRL for Honeybee Diseases, Faculty of Veterinary Medicine, University of Zagreb, Heinzelova 55, 10000 Zagreb, Croatia
2 NRL for Honey Bee Health, Istituto Zooprofilattico Sperimentale delle Venezie, Viale dell'Università, 10, 35020 Legnaro, Italy; fmutinelli@izsvenezie.it
* Correspondence: itlak@vef.unizg.hr; Tel.: +385-91-2390041

The honey bee is a symbol of *One Health*, a holistic approach to animal, human and environment health, and beekeeping is an economic branch of exceptional importance for public health [1]. However, in recent years, there has been a lot of discussion about the so-called honey bee health crisis. Consequently, increased attention has been paid to its causes, especially new pathogens, pests, and the geographical redistribution of invasive species, as well as the efficacy of biosecurity measures and practices at the apiary. Here, we outline the different stressors of global importance which are weakening and reducing the number of honeybee and bumblebee colonies [2], as well as *Osmia* spp. solitary bees [3]. Many factors, such as the presence of pathogens, nest destructors, negative environmental drivers, global warming (in particular), agricultural intensification and pesticides, habitat loss, and managing practices, are reported as the main causes of bee depopulation and losses in rearing operations and apiaries. The published articles reflect the complex impacts of environmental factors and management practices on bee health. Herein, we therefore summarize the main focus and results of the twenty-three original articles and one communication that have been published in this Special Issue, the Impact of Environmental Factors and Management Practices on Bee Health. The research has been thematically organized focusing on novel epidemiology studies, diagnostic tools and technologies, monitoring programs, agricultural, beekeeping and veterinary managing practices, biosecurity–control–eradication measures, disinfection methods, and the development of new bee disease control strategies.

The early estimation of infection levels with pathogens and parasites is important to avoid the development of clinical symptoms and further spreading of diseases. The article by Betti et al. [4] described a newly developed multi-scale mathematical model which can investigate the impact of inter-colony interactions on the transmission dynamics of honeybee diseases. The findings indicate that elevated drifting levels significantly enhance the disease spread rate among honey bee colonies. Additionally, the model highlights the necessity for highly efficient worker bees' guarding behavior to effectively mitigate disease transmission. Notably, in densely populated apiaries, the implications of drifting are of greater concern compared to robbing behavior [4]. In recent decades, the ectoparasitic mite *Varroa destructor* has become a major global threat to managed honey bee (*Apis mellifera*) colonies, and drifting is one of the possibilities for its high reinvasions between colonies and neighboring apiaries. So, an integrated pest management approach without using synthetic acaricides against varroosis motivated Qadir et al. [5] to test the effectiveness of formic acid, oxalic acid, and thymol applied at different concentrations and quantities [5]. Also, essential oils and their components are generally known for their acaricidal effects and are used as an alternative to control the population of *V. destructor* instead of synthetic acaricides. The 30 different essential oils were screened by using a glass-vial residual bioassay, and for essential oils with the best selectivity ratio, their main components were

Citation: Tlak Gajger, I.; Mutinelli, F. Impact of Environmental Factors and Management Practices on Bee Health. *Insects* **2024**, *15*, 996. https://doi.org/10.3390/insects15120996

Received: 28 November 2024
Accepted: 13 December 2024
Published: 16 December 2024

detected and quantified by GC-MS/MS, where the most suitable oils were peppermint and manuka, followed by oregano, litsea, carrot, and cinnamon [6]. Bila Dubaić et al. [7] discussed the possibilities and challenges of detecting and effectively monitoring feral and wild honey bees in urban environments, as well as the role of citizen science in these efforts. Their findings will support ongoing initiatives aimed at better understanding and enhancing naturally selected resistance mechanisms against the *V. destructor* mite. This understanding is expected to help mitigate the current threats to and risks of global beekeeping and food production security. Kovačić et al. [8] investigated the effect of queen caging on honey bee colonies' post-treatment and the development and optimal timing of methods applied during honey production during the main summer nectar flow in different Mediterranean countries. The study results showed that caging the queen with subsequent oxalic acid treatment 25 days after caging is an efficient method to control *V. destructor*, while the starting point of queen caging concerning the main summer nectar flow affects honey production [8]. Kolics et al. [9] provided basic information about changes in lithium levels in honey bees and their products following anti-varroa treatment and concluded that lithium treatment left combs (beeswax) lithium-free.

Kušar et al. [10] investigated the relationship between clinical symptoms of American foulbrood and spore counts by quantifying spores in honey and hive debris samples from honeybee colonies with known severity of symptoms. For that purpose, they used a newly developed qPCR assay which enables reliable detection and quantification of *P. larvae* in different in-hive sample types. The proper implementation of control measures and effective final disinfection can significantly reduce the recurrence of visible clinical signs of American foulbrood. Tlak Gajger et al. [11] evaluated ten commercially available disinfectants commonly used in beekeeping, along with some that have proven efficacy in the medicinal and veterinary fields, against different strains of the *P. larvae* bacterium.

Glavinić et al. [12] provided basic information on Nosemosis and reviewed currently known control measures (including beekeeping practices and chemical and natural substances), and they analyzed the application of *Agaricus bisporus* water crude extract on honey bees for the first time. The results showed anti-*Nosema ceranae* and nutrigenomic effects of *A. bisporus* extract when supplementation was preventive or in the moment of infection with *N. ceranae* spores. The stress caused by exposure to thiamethoxam and *Nosema* spp. invasion was mitigated by adjustments in the honey bee colony's dynamics and an increase in the number of worker bees—a behavior known as hormesis. Understanding the factors behind this phenomenon should be included in the prospective risk assessment of plant protection products to enhance the interpretation of field studies and management practices in the future [13].

Mraz et al. [14] studied the prevalence of common honey bee pathogens and viruses in various habitats throughout the Czech Republic. They found that the most prevalent honey bee pathogens belonged to the family Trypanosomatidae, including *Lotmaria passim* and *Crithidia mellificae*, and the most common virus was Deformed Wing Virus, followed by Acute Bee Paralysis Virus. From a location perspective, the highest occurrence of pathogens was observed in urban areas, while fewer pathogens were detected in agroecosystems and the least in national parks. Conversely, the trend for viruses was the opposite.

The Small Hive Beetle (*Aethina tumida* Murray, 1867) is an invasive scavenger of honeybees that is originally endemic in sub-Saharan Africa, and it is regulated internationally to preserve the areas still free from this species. To ensure the reliability of official diagnoses following their introduction, an inter-laboratory comparison was organized to assess the morphological and molecular real-time PCR identification of *A. tumida*. Franco et al. reported the results from 22 official diagnostic laboratories, and the results demonstrated the reliability of the diagnosis, covering the entire analytical process [15].

Pavlović et al. [16] showed differences in the elemental composition of larvae from honey bee colonies with different statuses of chalkbrood disease. Mummies had higher concentrations of macro elements in comparison to typically developed larvae from the same hive, while at the same time, they had much lower concentrations of microelements that have known anti-fungal and antimicrobial activities [16].

Phokasem et al. [17] revealed that the honey bees infected with DWV-A and exposed to thiamethoxam had an increased mortality rate and crippled wings in newly emerged worker bees, and it induced expression of immune gene hymenoptaecin and down-regulation of two apoptosis genes (caspase8-like, caspase9-like genes).

Mejías et al. [18] examined honey and beeswax samples collected in Chile. They performed chemical profiling to assess phenol content using the Folin–Ciocalteu method, measured antioxidant power through the Ferric Reducing Antioxidant Power Assay (FRAP) and evaluated antiradical activity using the 2,2-Diphenyl-1-picrylhydrazyl Assay (DPPH). Additionally, they tested for pesticide residues using HPLC-MS/MS and GC-MS techniques. The findings mapped the suitability index that ranks areas based on their pesticide-free status and biological quality, and those chemical profiles will assist local beekeepers in obtaining international certifications, particularly for the European Union market [18]. Hýbl et al. [19] reported that polyphenols as food supplements improved the food consumption and longevity of honey bees intoxicated by the pesticide thiacloprid, and the expression level of genes encoding detoxification enzymes was quantified. Raimets et al. [20] reported the possibility of the migration of tebuconazole from wax to royal jelly with a strong dilution effect from the original contamination source. Also, no residues were detected in queen bee larvae nor newly emerged queens, indicating that the migration of tebuconazole did not directly endanger the queen but might affect the homeostasis of developing worker bees. Cabezas et al. [21] evaluated the toxicity of six insecticides on buff-tailed bumblebee workers (*Bombus terrestris*): imidacloprid, thiacloprid, deltamethrin, esfenvalerate, sulfoxaflor and a microbial insecticide based on *Bacillus thuringiensis* toxins, which are present in genetically modified (Bt) maize. The results indicated that some currently used insecticides are more acutely toxic to *B. terrestris* than certain neonicotinoids that have been banned [21].

The results of Vilić et al. [22] showed that RF-EMFs at a frequency of 900 MHz can cause oxidative stress in honey bees, with the larval stage being more sensitive than the pupal stage, and there was no linear relationship between the electric field level and effect in any of the developmental stages.

The first survey on the loss rates of honey bee colonies was reported in 2022/2023 in Ethiopia using COLOSS monitoring survey tools, applying a *face-to-face* interview questionnaire [23].

The spider web and barometer tools enable a comprehensive assessment of the implementation status of biosecurity measures, actions taken to protect the environment where stingless bees thrive, the quality and efficiency of nest management techniques, and the monitoring of the health status in Meliponiculture [24].

Understanding how human activities, such as urbanization and agricultural intensification, affect landscapes and insect pollinators is essential for maintaining a healthy pollination system. In this context, Donkersley et al. [25] reported preliminary results indicating that the foraging and nutrition of *A. cerana japonica* may not be negatively impacted by urban land use. They emphasize the need for future comparative studies between *A. cerana* japonica and *A. mellifera*, as these could reveal resilience in pollinators foraging within their native range [25].

Despite specific limitations, Bontšutšnaja et al. [26] saw strong correlations between microscopy and DNA metabarcoding data used for quantification analyses and the botanical origin of bumble bee-collected pollen for *B. terrestris* species. Also, they concluded that the spring vegetation in southern Estonia can support the development of bumblebee colonies, regardless of the specific landscape structure. Additionally, despite the lack of a dominant natural food source, other bumblebee species—particularly generalists—should also be able to find ample forage during their early development phase.

Elfar et al. [27] examined the proteomic variations and biological activities of bee hemolymphs from four localities in Egypt with differing food diversities. The results showed that worker bees fed only sucrose had the lowest protein concentrations and weak biological activities, while those with access to diverse natural foods had the highest protein levels and significant biological activities [27].

In conclusion, the published articles explore complex factors affecting bee health, emphasizing the global crisis in honeybee and pollinator populations. Key insights include the following:

1. Stressors on Bee Populations: factors such as pathogens, pests (e.g., *V. destructor*), pesticides, habitat loss, and climate change contribute to the decline of bee colonies, including honeybees, bumblebees, and solitary bees.
2. Innovative Control Measures: studies highlight effective methods like queen caging combined with oxalic acid treatments for Varroosis control, essential oils as alternatives to synthetic acaricides, and advancements in diagnostic tools like qPCR assays for diseases such as American foulbrood.
3. Impact of Environmental and Agricultural Practices: urbanization, pesticide exposure, and agricultural intensification affect bees differently. For instance, urban areas showed higher pathogen prevalence, while rural areas had more pesticide-related impacts.
4. Chemical and Biological Research: research identified effective natural remedies like *Agaricus bisporus* extract for bee health and studied pesticide residues' effects on honey and beeswax.
5. Behavioral and Genetic Studies: studies on disease transmission behaviors, genetic responses to stressors, and the influence of food diversity reveal how bees adapt to environmental challenges and human activities.
6. Global and Local Initiatives: efforts like citizen science for monitoring wild bees, the development of biosecurity measures, and comparative studies on bee species in diverse habitats aim to mitigate risks and improve resilience.

The compilation of twenty-three original articles and a communication underscores the urgency of interdisciplinary approaches to safeguard bee health and ecosystem sustainability.

Acknowledgments: We want to thank the authors for contributing their articles and all reviewers for their helpful recommendations in this successful Topical Collection—Impact of Environmental Factors and Management Practices on Bee Health.

Conflicts of Interest: The authors declare no conflicts of interest.

References

1. Tlak Gajger, I.; Meana Mañes, A.; Formato, G.; Mortarino, M.; Toporcak, J. Veterinarians and beekeeping: What roles, expectations and future perspectives?—A review paper. *Vet. Arh.* **2021**, *91*, 437–443. [CrossRef]
2. Tlak Gajger, I.; Šimenc, L.; Toplak, I. The First Detection and Genetic Characterization of Four Different Honeybee Viruses in Wild Bumblebees from Croatia. *Pathogens* **2021**, *10*, 808. [CrossRef] [PubMed]
3. Tlak Gajger, I.; Laklija, I.; Jurković, M.; Košćević, A.; Dar, S.A.; Ševar, M. The Impact of Different Biotopes and Management Practices on the Burden of Parasites in Artificial Nests of *Osmia* spp. (Megachilidae) Bees. *Diversity* **2022**, *14*, 226. [CrossRef]
4. Betti, M.; Shaw, K. A Multi-Scale Model of Disease Transfer in Honey Bee Colonies. *Insects* **2021**, *12*, 700. [CrossRef]
5. Qadir, Z.A.; Idrees, A.; Mahmood, R.; Sarwar, G.; Bakar, M.A.; Ahmad, S.; Raza, M.M.; Li, J. Effectiveness of Different Soft Acaricides against Honey Bee Ectoparasitic Mite *Varroa destructor* (Acari: Varroidae). *Insects* **2021**, *12*, 1032. [CrossRef]
6. Hýbl, M.; Bohatá, A.; Rádsetoulalová, I.; Kopecký, M.; Hoštičková, I.; Vaníčková, A.; Mráz, P. Evaluating the Efficacy of 30 Different Essential Oils against *Varroa destructor* and Honey Bee Workers (*Apis mellifera*). *Insects* **2021**, *12*, 1045. [CrossRef]
7. Bila Dubaić, J.; Simonović, S.; Plećaš, M.; Stanisavljević, L.; Davidović, S.; Tanasković, M.; Ćetković, A. Unprecedented Density and Persistence of Feral Honey Bees in Urban Environments of a Large SE-European City (Belgrade, Serbia). *Insects* **2021**, *12*, 1127. [CrossRef]
8. Kovačić, M.; Uzunov, A.; Tlak Gajger, I.; Pietropaoli, M.; Soroker, V.; Adjlane, N.; Benko, V.; Charistos, L.; Dall'Olio, R.; Formato, G.; et al. Honey vs. Mite—A Trade-Off Strategy by Applying Summer Brood Interruption for *Varroa destructor* Control in the Mediterranean Region. *Insects* **2023**, *14*, 751. [CrossRef] [PubMed]
9. Kolics, É.; Sajtos, Z.; Mátyás, K.; Szepesi, K.; Solti, I.; Németh, G.; Taller, J.; Baranyai, E.; Specziár, A.; Kolics, B. Changes in Lithium Levels in Bees and Their Products Following Anti-*Varroa* Treatment. *Insects* **2021**, *12*, 579. [CrossRef] [PubMed]
10. Kušar, D.; Papić, B.; Zajc, U.; Zdovc, I.; Golob, M.; Žvokelj, L.; Knific, T.; Avberšek, J.; Ocepek, M.; Pislak Ocepek, M. Novel TaqMan PCR Assay for the Quantification of *Paenibacillus larvae* Spores in Bee-Related Samples. *Insects* **2021**, *12*, 1034. [CrossRef]

11. Tlak Gajger, I.; Tomljanović, Z.; Mutinelli, F.; Granato, A.; Vlainić, J. Effects of Disinfectants on Bacterium *Paenibacillus larvae* in Laboratory Conditions. *Insects* **2024**, *15*, 268. [CrossRef] [PubMed]
12. Glavinic, U.; Rajkovic, M.; Vunduk, J.; Vejnovic, B.; Stevanovic, J.; Milenkovic, I.; Stanimirovic, Z. Effects of *Agaricus bisporus* Mushroom Extract on Honey Bees Infected with *Nosema ceranae*. *Insects* **2021**, *12*, 915. [CrossRef] [PubMed]
13. Alonso-Prados, E.; González-Porto, A.V.; García-Villarubia, C.; López-Pérez, J.A.; Valverde, S.; Bernal, J.; Martín-Hernández, R.; Higes, M. Effects of Thiamethoxam-Dressed Oilseed Rape Seeds and *Nosema ceranae* on Colonies of *Apis mellifera iberiensis*, L. under Field Conditions of Central Spain. Is Hormesis Playing a Role? *Insects* **2022**, *13*, 371. [CrossRef]
14. Mráz, P.; Hýbl, M.; Kopecký, M.; Bohatá, A.; Hoštičková, I.; Šipoš, J.; Vočadlová, K.; Čurn, V. Screening of Honey Bee Pathogens in the Czech Republic and Their Prevalence in Various Habitats. *Insects* **2021**, *12*, 1051. [CrossRef] [PubMed]
15. Franco, S.; Cougoule, N.; Tison, A.; Del Cont, A.; Gastaldi, C.; Consortium, I.; Duquesne, V. Reliability of Morphological and PCR Methods for the Official Diagnosis of *Aethina tumida* (Coleoptera: Nitidulidae): A European Inter-Laboratory Comparison. *Insects* **2022**, *13*, 33. [CrossRef]
16. Pavlović, R.; Brodschneider, R.; Goessler, W.; Stanisavljević, L.; Vujčić, Z.; Zarić, N.M. Micronutrient Deficiency May Be Associated with the Onset of Chalkbrood Disease in Honey Bees. *Insects* **2024**, *15*, 269. [CrossRef] [PubMed]
17. Phokasem, P.; Mookhploy, W.; Krongdang, S.; Sinpoo, C.; Chantawannakul, P. Interaction between Thiamethoxam and Deformed Wing Virus Type A on Wing Characteristics and Expression of Immune and Apoptosis Genes in *Apis mellifera*. *Insects* **2022**, *13*, 515. [CrossRef] [PubMed]
18. Mejías, E.; Gómez, C.; Garrido, T. Suitable Areas for Apiculture Expansion Determined by Antioxidant Power, Chemical Profiles, and Pesticide Residues in *Caldcluvia paniculata* Honey and Beeswax Samples. *Insects* **2022**, *13*, 31. [CrossRef] [PubMed]
19. Hýbl, M.; Mráz, P.; Šipoš, J.; Hoštičková, I.; Bohatá, A.; Čurn, V.; Kopec, T. Polyphenols as Food Supplement Improved Food Consumption and Longevity of Honey Bees (*Apis mellifera*) Intoxicated by Pesticide Thiacloprid. *Insects* **2021**, *12*, 572. [CrossRef]
20. Raimets, R.; Naudi, S.; Mänd, M.; Bartkevičs, V.; Smagghe, G.; Karise, R. Translocation of Tebuconazole between Bee Matrices and Its Potential Threat on Honey Bee (*Apis mellifera* Linnaeus) Queens. *Insects* **2022**, *13*, 45. [CrossRef] [PubMed]
21. Cabezas, G.; Farinós, G.P. Sensitivity of Buff-Tailed Bumblebee (*Bombus terrestris* L.) to Insecticides with Different Mode of Action. *Insects* **2022**, *13*, 184. [CrossRef] [PubMed]
22. Vilić, M.; Žura Žaja, I.; Tkalec, M.; Tucak, P.; Malarić, K.; Popara, N.; Žura, N.; Pašić, S.; Tlak Gajger, I. Oxidative Stress Response of Honey Bee Colonies (*Apis mellifera* L.) during Long-Term Exposure at a Frequency of 900 MHz under Field Conditions. *Insects* **2024**, *15*, 372. [CrossRef]
23. Hailu, T.G.; Atsbeha, A.T.; Wakjira, K.; Gray, A. High Rates of Honey Bee Colony Losses and Regional Variability in Ethiopia Based on the Standardised COLOSS 2023 Survey. *Insects* **2024**, *15*, 376. [CrossRef] [PubMed]
24. Ocaña-Cabrera, J.S.; Martin-Solano, S.; Saegerman, C. Development of Tools to Understand the Relationship between Good Management Practices and Nest Losses in Meliponiculture: A Pilot Study in Latin American Countries. *Insects* **2024**, *15*, 715. [CrossRef]
25. Donkersley, P.; Covell, L.; Ota, T. Japanese Honeybees (*Apis cerana japonica* Radoszkowski, 1877) May Be Resilient to Land Use Change. *Insects* **2021**, *12*, 685. [CrossRef] [PubMed]
26. Bontšutšnaja, A.; Karise, R.; Mänd, M.; Smagghe, G. Bumble Bee Foraged Pollen Analyses in Spring Time in Southern Estonia Shows Abundant Food Sources. *Insects* **2021**, *12*, 922. [CrossRef] [PubMed]
27. Elfar, S.A.; Bahgat, I.M.; Shebl, M.A.; Lihoreau, M.; Tawfik, M.M. Intraspecific Variability in Proteomic Profiles and Biological Activities of the Honey Bee Hemolymph. *Insects* **2023**, *14*, 365. [CrossRef] [PubMed]

insects

MDPI

Article

A Multi-Scale Model of Disease Transfer in Honey Bee Colonies

Matthew Betti [1,*] and Karalyne Shaw [2]

1 Mount Allison University, Sackville, NB E4L 1E2, Canada
2 Saint Mary's University, Halifax, NS B3H 3C3, Canada; kshaw@mta.ca
* Correspondence: mbetti@mta.ca

Simple Summary: Inter-colony disease spread is a impediment to a healthy apiary. A multi-scale mathematical model is built to explore the effects of inter-colony behaviour on the spread of disease. We model different scenarios corresponding to different behaviours exhibited by honey bees. We show that a colony can use certain behaviours to lower the impact of disease on itself, and show that other behaviours are only relevant under specific conditions. The model can be extended to explore an entire apiary or modified to explore the evolutionary underpinnings of these behaviours.

Abstract: Inter-colony disease transfer poses a serious hurdle to successfully managing healthy honeybee colonies. In this study, we build a multi-scale model of two interacting honey bee colonies. The model considers the effects of forager and drone drift, guarding behaviour, and resource robbing of dying colonies on the spread of disease between colonies. Our results show that when drifting is high, disease can spread rapidly between colonies, that guarding behaviour needs to be particularly efficient to be effective, and that for dense apiaries drifting is of greater concern than robbing. We show that while disease can put an individual colony at greater risk, drifting can help less the burden of disease in a colony. We posit some evolutionary questions that come from this study that can be addressed with this model.

Keywords: honey bee; disease transfer; drift; robbing

Citation: Betti, M.; Shaw, K. A Multi-Scale Model of Disease Transfer in Honey Bee Colonies. *Insects* **2021**, *12*, 700. https://doi.org/10.3390/insects12080700

Academic Editors: Franco Mutinelli and Ivana Tlak Gajger

Received: 18 June 2021
Accepted: 1 August 2021
Published: 4 August 2021

1. Introduction

Pathogens and disease are now agreed to play an important role in global honeybee colony failure [1]. Honeybee colonies seem to be robust against many individual stressors, but coupled stressors can cause effects greater than the sum of their parts [2]. As honeybees continue to be exposed to stressors such as climate change [3] and sub-lethal pesticide exposure [4], it is important to minimise the risk imposed by disease in order to preserve a colony.

Honeybees are exposed to a host of viral and microbial parasites. The most studied of these are the Varroa mite [5] which can transmit lethal viruses such as deformed wing virus (DWV) [6] and acute paralysis virus (APV) [7] and nosemosis [8] caused by either the spores *Nosema ceranae* or *Nosema apis*. There are other less widely studied viruses and diseases which plague colonies such as cloudy wing virus [9], the newly discovered Lake Sinai virus [10].

Honeybees exhibit behaviours that help combat infection within a colony. Bees will often implement hygienic behaviours [11–13] to prevent disease from spreading within a colony. These behaviours consist of removing infected or non-viable brood to stop the spread of infection within a colony.

To prevent disease between colonies, honey bees have a different set of behaviours. One such behaviour is guarding [14], which is when bees stationed at a hive entrance will deny entry to immunocompromised honey bees, bees of another colony, or anything deemed a threat to colony health.

In contrast, some honeybee behaviours facilitate the transfer of infection between colonies. One such behaviour is known as *drift*. Foraging bees can sometimes get confused and return from flight to the wrong hive [15]; if a bee is infected and drifts, they can spread infection to a neighbouring hive [16]. Drifting is worse when hives are visually similar and in close proximity [17]. Honeybees are also known to pillage dying colonies for resources [18]. This is distinct from drift as it is an active behaviour rather than a passive one, and is less dependent on distance given the range of foraging bees, but can still be a transmission route for disease [19].

There are other mechanisms by which disease may pass from one honeybee colony to another. Spillover events between other pollinator species or wild bee populations may cause novel pathogens to infect managed honeybee colonies [20] or vice versa [21] as well as through the environment such as shared foraging patches [20,22].

Diseases in honeybee colonies have been studied mathematically in both the general case [23–26] and in specific cases of prominent infections [27–29]. Recent focus has turned to the study of disease transmission between colonies. A simple model by Muhammad and Eberl illustrates the role of drift using impulsive differential equations [30]. Their results show that drift can allow for the rapid spread of disease through an apiary that drastically reduces the total bee population. In 2019, Bartlett et al. developed a model for networked colonies using an SIR framework to show how colony placement can affect pathogen spread [31].

Infection in honeybee colonies is extremely varied and is confounded by complex social hierarchy and behaviour both within and between hives. This necessitates the use of multi-scale models to aid the understanding of how honeybee behaviour at the individual and colony level effects the transmission of disease. While many diseases have specific, unique routes of transmission, many can be approximated as horizontal transmission between bees. For instance, Varroa mites act as a vector for many viral infections [7,32], but the mite population is dependent on the bee population. As mites often jump from bee to bee [7,33] while feeding, this aspect can be approximated by horizontal transmission. In fact, the mites themselves which weaken bees and lead to health problems [32] can be interpreted as a type of pathogen affecting honey bees. Infections caused by the microsporidian *Nosema* can likewise be approximated by horizontal, bee-to-bee transmission [23], as spores are transmitted through feeding, cleaning, defecating, and transferring pollen and nectar [34]. For this reason, and to increase the tractability of our study, we focus on horizontal transmission between bees on the intra-colony scale.

In the current study, we create a multi-scale model which couples two colonies using the single colony framework introduced in [23], which itself is an extension of the dynamical model of a honey bee colony introduced in [35]. We keep the disease dynamics as general as possible to draw conclusions that can be applied to a wide variety of diseases which affect colonies. The inter-colony behaviours that we will focus on are those pertaining to drift, security, and robbing. We analyse the model using numerical simulations to show the effects of explicit forager and drone drift on the spread of disease as well as under what conditions colony robbing becomes a relevant factor.

2. Model

This model extends the model in [23] to incorporate a multi-hive system and inter-colony interactions. In this model, we model two colonies in order to highlight how various behaviours change disease spread between colonies.

The two colonies will be labelled colony 1 and colony 2. Each hive is made up of three classes of bees: the hive bees, H, the foragers, F and the drones D. We ignore brood dynamics in this model, as the goal is to explore a generalised, mass-action infection at the inter-colony level, which is mostly driven by adult bees. The general hive dynamics are adapted from the works in [35,36], and the generalised intra-colony infection is adapted from in [23].

We assume that hive bees (female bees whose duties are contained entirely within the hive) are contained within the hive and are safeguarded against death (in other words, healthy bees are far more likely to become foragers than to die of natural causes). The dynamics of healthy (i.e., susceptible) hive bees are governed by the equation

$$\frac{\mathrm{d}_1 H_S}{\mathrm{d}t} = p_1 L\mathcal{S}_1({}_1H, {}_1f) - R_1({}_1H, {}_1F)_1 H_S - {}_1H_S \Upsilon_H({}_1H_I, {}_1F_I, {}_1D_I) \tag{1}$$

$$\frac{\mathrm{d}_2 H_S}{\mathrm{d}t} = p_1 L\mathcal{S}_2({}_2H, {}_1f) - R_2({}_2H, {}_1F)_2 H_S - {}_2H_S \Upsilon_H({}_1H_I, {}_1F_I, {}_1D_I) \tag{2}$$

where the subscripts 1 or 2 determine which colony the bees are from, S stands for susceptible bees, and I indicates infected bees. The parameter p_1 is the proportion of eggs laid by the queen of colony 1 which are fertilised (and thus result in female bees), and L is the number of eggs laid per day on average by the queen.

The function $\Upsilon_H({}_1H_I, {}_1F_I, {}_1D_I)$ is an infection transmission function between infected bees and susceptible bees. It is given as

$$\Upsilon_H({}_1H_I, {}_1F_I, {}_1D_I) = \Big(\beta_{HH1}H_I + \beta_{HF}({}_1F_I + {}_2^1F_I) + \beta_{HD1}D_I\Big) \tag{3}$$

where β_{HH} is the interaction term between hive bees, β_{HF} is the interactions between foragers and hive bees, and β_{DH} is the interactions between drones and hive bees. The subscript on Υ denotes the susceptible population being interacted with. Therefore, there are other interactions not shown here, namely, β_{DF}, β_{DD}, and β_{FF}. In practice, we take an average β value and the term reduces to

$$\Upsilon({}_1H_I, {}_1F_I, {}_1D_I) = \beta\Big({}_1H_I + {}_1F_I + {}_2^1F_I + {}_1D_I\Big) \tag{4}$$

These two governing equations, Equations (1) and (2), do not interact with each other as we assume hive bees of one colony are not mixing with other colonies directly.

The two functions $\mathcal{S}(H, f)$ and $R(H, F)$ are survival of brood and recruitment to foraging, respectively, where ${}_1H = ({}_1H_S, {}_1H_I)$ and ${}_1F = ({}_1F_S, {}_1F_I)$ and similarly for colony 2. These terms were first introduced in [35,36]. We have adapted and modified these terms to the following:

$$\mathcal{S}_1(H, f) = \left(\frac{H_S + H_I}{k_1 + H_S + H_I}\right)\left(\frac{f}{b_1 + f}\right) \tag{5}$$

and

$$R_1(H, F) = \alpha_1\left(1 - \sigma_1 \frac{F_S + F_I}{H_S + H_I + F_S + F_I}\right). \tag{6}$$

The parameters k_1 and b_1 are the amount of hive bees and food required to ensure 50% survival of the brood, parameter α_1 is the rate of recruitment to foraging in the absence of foragers, and $1/\sigma_1$ is the optimal proportion of foragers in a colony. We have similar equations for colony 2.

Infected hive bees are created when an infected bee interacts with a susceptible hive bee. These infected hive bees are then susceptible to infection-related death at rate d_H.

$$\frac{\mathrm{d}_1 H_I}{\mathrm{d}t} = {}_1H_S \Upsilon_H({}_1H_I, {}_1F_I) - R_1({}_1H, {}_1F)_1 H_I - d_{H1} H_I \tag{7}$$

$$\frac{\mathrm{d}_2 H_I}{\mathrm{d}t} = {}_2H_S \Upsilon_H({}_2H_I, {}_2F_I) - R_2({}_2H, {}_2F)_2 H_I - d_{H2} H_I \tag{8}$$

Healthy (i.e., susceptible) foragers, F_S, are created through recruitment of hive bees and are susceptible to natural death at rate μ, as well we incorporate a drifting term which allows foragers to visit other colonies. We denote these "drifted foragers" with two left indices.

For example, ${}_2^1F_S$ is a healthy forager from colony 2 who is currently in colony 1. This separation has multiple advantages; it allows for asymmetry in the model (we can allow migrant foragers an increased probability of returning to their proper colony), it allows us to account for their average age before drifting (if we were to add them to ${}_2F_S$ they would then live on average μ_2 days from the day they drifted), and we can track migrant foragers and validate against experimental results. Given this, there are two equations per colony for healthy foragers:

$$\frac{\mathrm{d}_1F_S}{\mathrm{d}t} = R({}_1H, {}_1F){}_1H_S - \mu_{1}{}_1F_S - d_{12}{}_1F_S + r_1{}_2^1F_S - {}_1F_S\Upsilon({}_1H_I, {}_1F_I) \tag{9}$$

$$\frac{\mathrm{d}_1^2F_S}{\mathrm{d}t} = d_{21}{}_2F_S - r_2{}_1^2F_S - {}_1^2F_S\Upsilon({}_1H_I, {}_1F_I) - \frac{\mu_1}{\mu_1 + r_1}{}_1^2F_S \tag{10}$$

$$\frac{\mathrm{d}_2F_S}{\mathrm{d}t} = R({}_1H, {}_1F){}_2H_S - \mu_{2}{}_1F_S - d_{21}{}_1F_S + r_2{}_1^2F_S - {}_2F_S\Upsilon({}_2H_I, {}_2F_I) \tag{11}$$

$$\frac{\mathrm{d}_2^1F_S}{\mathrm{d}t} = d_{21}{}_2F_S - r_1{}_2^1F_S - {}_2^1F_S\Upsilon({}_2H_I, {}_2F_I) - \frac{\mu_2}{\mu_2 + r_2}{}_2^1F_S \tag{12}$$

where d_{12} is the probability of drift from colony 1 to colony 2, d_{21} is the probability of drift from colony 2 to colony 1, and μ is the rate of natural death of foragers.

Infected foragers have similar dynamics to healthy foragers, but there are two routes of creation (infected foragers can be recruited to foraging from infected hive bees, or healthy foragers that become infected). As well, infected foragers will die either through natural death or by death due to infection, d_F. Equations (13)–(16) show the governing behaviour of infected foragers. The notable changes from the susceptible forager equations are highlighted in blue.

$$\frac{\mathrm{d}_1F_I}{\mathrm{d}t} = R({}_1H, {}_1F){}_1H_I - (\mu_1{+}d_F){}_1F_S - d_{12}{}_1F_S + r_1{}_2^1F_S{+}{}_1F_S\Upsilon({}_1H_I, {}_1F_I) \tag{13}$$

$$\frac{\mathrm{d}_1^2F_S}{\mathrm{d}t} = d_{21}{}_2F_S - r_2{}_1^2F_S{+}{}_1^2F_S\Upsilon({}_1H_I, {}_1F_I) - \frac{\mu_1{+}d}{\mu_1 + r_1{+}d}{}_1^2F_S \tag{14}$$

$$\frac{\mathrm{d}_2F_I}{\mathrm{d}t} = R({}_1H, {}_1F){}_2H_I - (\mu_2{+}d_F){}_1F_S - d_{21}{}_1F_S + r_2{}_1^2F_S{+}{}_2F_S\Upsilon({}_2H_I, {}_2F_I) \tag{15}$$

$$\frac{\mathrm{d}_2^1F_S}{\mathrm{d}t} = d_{21}{}_2F_S - r_1{}_2^1F_S{+}{}_2^1F_S\Upsilon({}_2H_I, {}_2F_I) - \frac{\mu_2{+}d}{\mu_2{+}d + r_2}{}_2^1F_S \tag{16}$$

Drones are also a potential vector of infection between colonies. Drone behaviours are incredibly simple, they are born of unfertilised eggs, and tend to be long-lived. Drones live on average 90 days, but at certain times of the year death can be accelerated through mating in spring/early summer [37], or through the mass eviction of drones in autumn by worker bees [38]. Similar to foragers, drones may drift between hives [19]. Our governing equations for healthy drones in colony 1 are given as

$$\frac{\mathrm{d}_1D_S}{\mathrm{d}t} = (1 - p_1)L\mathcal{S}_1(H, f) - {}_1D_S\Upsilon({}_1H_I, {}_1F_I) - \mu_{D1}D_S \tag{17}$$

$$\frac{\mathrm{d}_1^2D_S}{\mathrm{d}t} = \hat{d}_{21}{}_2D_S - \hat{r}_2{}_1^2D_S - {}_1^2D_S\Upsilon({}_1H_I, {}_1F_I) - \frac{\mu_{D1}}{\mu_{D1} + \hat{r}_2}{}_2^1F_S \tag{18}$$

with similar equations for colony 2 and infected equations build in the same way as those for hive bees and foragers.

3. Results

The model is run with parameters given in Table 1. For simplicity we set all parameters equal for both colonies, and set $\beta = \beta_{HH} = \beta_{FH} = \beta_{HD} = \beta_{HF}$, $d_{12} = d_{21} = r_1 = r_2$ and $\hat{d}_{12} = \hat{d}_{21} = \hat{r}_1 = \hat{r}_2$.

Table 1. Parameter values and source references.

Parameter	Definition	Value	Ref.
L	maximum rate of egg laying	2000 eggs/day	[35]
w	number of hive bees for 50% egg survival	5000 bees	[35]
b	mass of food stored for 50% egg survival	500 g	[36]
k	increased food requirements for drone brood survival	variable	
α	maximum rate of recruitment	25%/day	[35]
$\frac{1}{\sigma}$	maximum fraction of colony that can be foraging	1/3	[35]
μ	natural death rate of foragers (summer)	0.14	[35]
μ_D	natural death rate of drones	1/90	[39]
c	food gathered per day per forager	0.1 g/day/bee	[40]
γ_H	daily food requirement per adult hive bee	0.008 g/day/bee	[41]
γ_F	daily food requirement per adult forager	0.008 g/day/bee	[41]
γ_D	daily food requirement per adult drone	0.016 g/day/bee	calculated
γ_{BH}	daily food requirement per worker brood	0.016 g/day/bee	[42]
γ_{BD}	daily food requirement per drone brood	0.049 g/day/bee	[42]
p	proportion of worker eggs	0.95	[43]
β	rate of transmission of infection	5×10^{-4}	[23]
d_{12}	drift rate of foragers	0.1	[44]
$\hat{d}_{12}$	drift rate of drones	$3d_{12}$	[15]

The healthy dynamics of two colonies are shown in Figure 1. Each colony is started with 10,000 hive bees, no foragers, and no drones. This roughly replicates post-winter conditions. Within 1–2 months, each colony is strong and stable. Without infection, and with two identical colonies, we see that the fraction of drifted foragers in a colony, under the given parameter set is roughly 12%; shown in Figure 2. These results are consistent with biological observations: see in [44] for drift values, and in [45] for colony size.

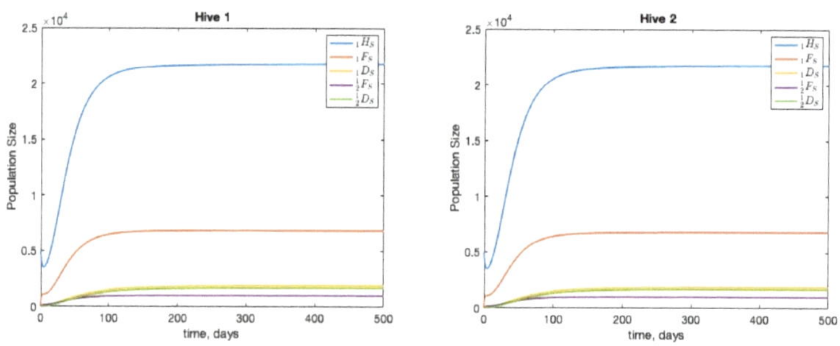

(a) **(b)**

Figure 1. The dynamics of two healthy colonies with interactions through forager and drone drifting for colony 1 (**a**) and colony 2 (**b**). The model is started with initial conditions ${}_1H_S(0) = {}_2H_S(0) =$ 10,000, and all other compartments with 0 individuals. Food is considered abundant and infinite. As identical parameter sets are used for both colonies, the dynamics of both colonies are the same.

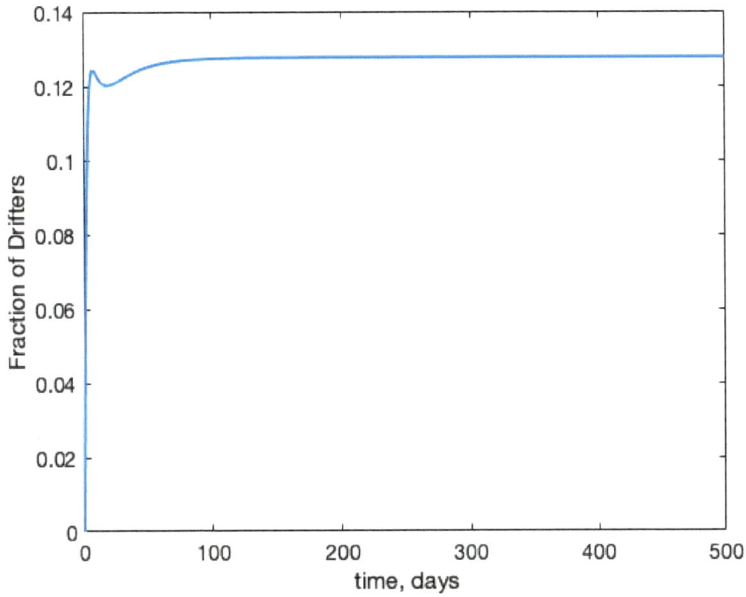

Figure 2. With the drifting rate set at 10%, we see that on average 12% of the foragers in a colony are foreign. These results are consistent with experimental findings.

In Figure 3, we first run the model to reach the disease-free equilibrium and then introduce infection. This is to simulate infection being introduced in mid-summer, when colonies are at full strength. In this case, we see that infection travels between colonies quickly, and despite the colonies being identical, the peak infected population in the second colony is slightly higher than in the source colony. In panel (b) of Figure 3, we see the proportion of bees that are infected in each colony. This is important as a proxy for future risk to the colony as the stress of infection can amplify the effects of other stressors.

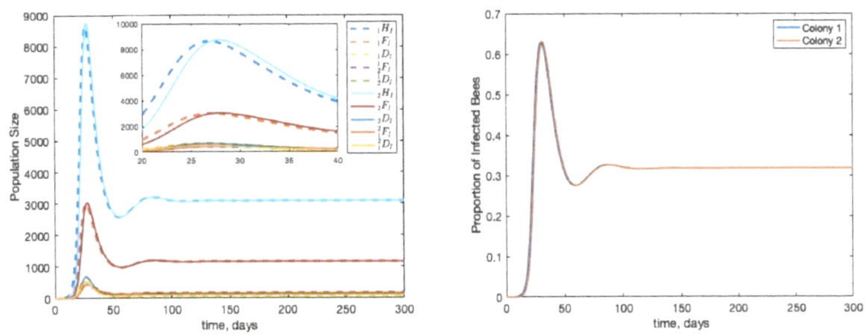

(**a**) (**b**)

Figure 3. Beginning from equilibrium, we introduce one infected forager into colony 1 (${}_1F_I(0) = 1$). We see that infection can travel quickly between two colonies with little delay. Note that the infection in the second colony peaks slightly higher than in the source colony. Panel (**a**) shows the population of infected bees over time and panel (**b**) shows the proportion of bees in each colony that are infected over time.

3.1. Scenario: Drone Drift vs. Forager Drift

We can modify the model by looking at the impact of forager drift versus drone drift on infections within a two colony system. When compared to Figure 3, Figure 4 shows that drift allows the stress of infection to be spread across both colonies. Figure 4 shows the case when there is no drift, in other words, $d_{12} = d_{21} = r_1 = r_2 = 0$ and $\hat{d}_{12} = \hat{d}_{21} = \hat{r}_1 = \hat{r}_2 = 0$. We see that when the infection is limited to the source colony, the peak number of infected bees is much higher than when bees are able to drift between colonies. We also see that the proportion of bees infected in colony 1 is higher, by roughly 10%.

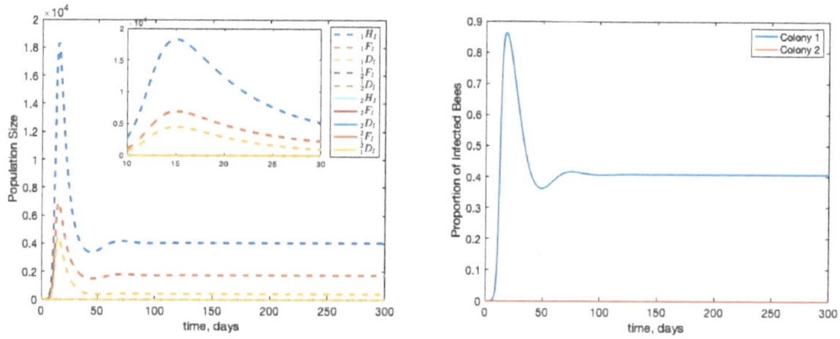

(**a**) (**b**)

Figure 4. Beginning from equilibrium, we introduce one infected forager into colony 1 (${}_1F_I(0) = 1$). We see that infection is more severe and poses greater long-term risk when bees are not able to drift between colonies. Subfigure (**a**) shows the infected population in the hives, subfigure (**b**) shows the percentage of infected bees in each colony.

In Figure 5, we show the impact of forager drift (subfigure (a)) in the absence of drone drift and the impact of drone drift (subfigure (b)) in the absence of forager drift. We see that despite their increased rate of drift, if foragers are not allowed to drift, both colonies see an increase in infection, and the infection peaks roughly 10 days earlier than if both drones and foragers are able to drift. We see that thee same qualitative result is true if only foragers are able to drift, but it is much less pronounced quantitatively.

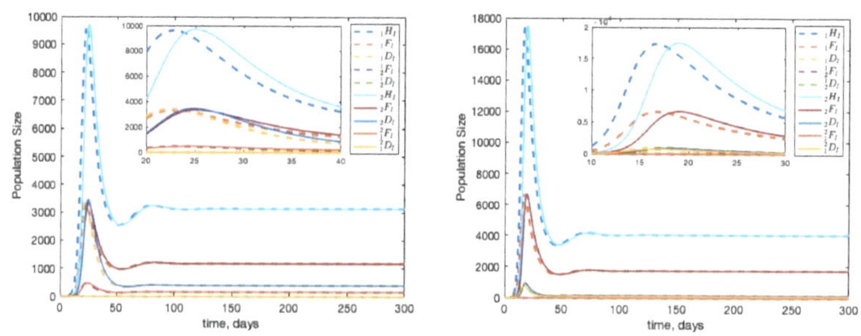

(a) (b)

Figure 5. *Cont.*

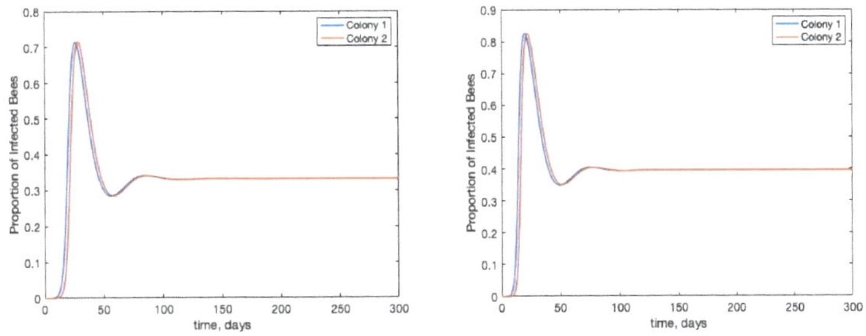

(c) (d)

Figure 5. Beginning from equilibrium, we introduce one infected forager into colony 1 (${}_1F_I(0) = 1$). Subfigure (**a**) shows the impact of only forager drift on the system of two colonies; $\hat{d}_{12} = \hat{d}_{21} = \hat{r}_1 = \hat{r}_2 = 0$ and all other parameters as in Table 1. Subfigure (**b**) shows the impact of only drone drift; $d_{12} = d_{21} = r_1 = r_2 = 0$ and all other parameters as in Table 1. Comparing panels (**c**,**d**), we see that forager drift is mainly responsible for alleviating infection pressure in the colony as when forager drift is removed, the situation in colony 1 is very similar to when there is no drift at all.

3.2. Scenario: Filtering Infected Bees

Another behaviour some honey bee colonies can show is to disallow drifting of compromised bees [46]. In Figure 6, we simulate colony 2 discriminating against the drift of infected bees from colony 1, but colony 1 does not discriminate at all. We see that a perfect filter can save colony 2 in much the same way as no drift; as the guarding bees at colony 2 will often kill the trespassing bees [14], we also see a healthier colony 1 because of the behaviour of hive 2. The figure also shows the case where only 50% of infected drifting bees are stopped and we see that this is not enough to prevent infection in colony 2.

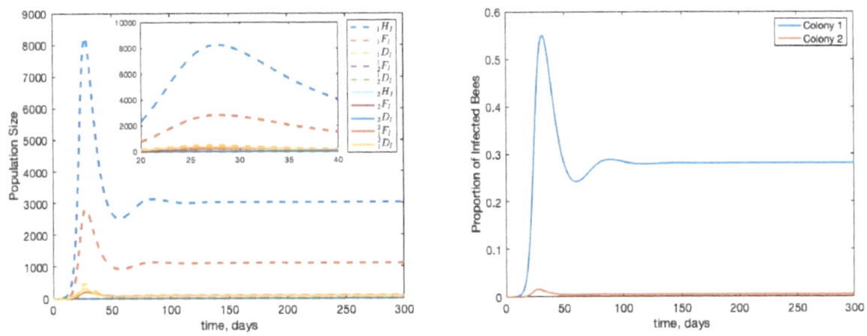

(a) (b)

Figure 6. *Cont.*

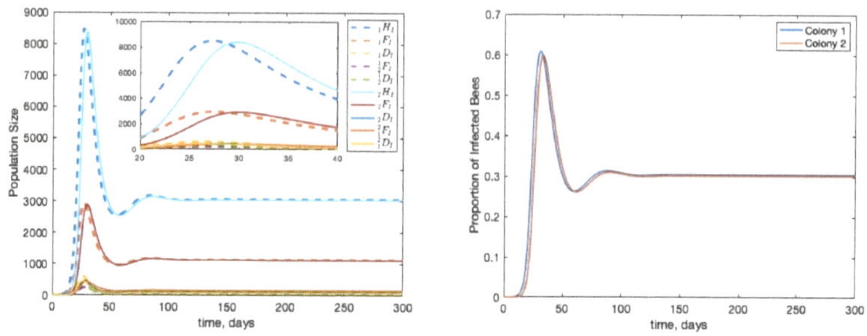

(c) (d)

Figure 6. Beginning from equilibrium, we introduce one infected forager into colony 1 (${}_1F_I(0) = 1$). In this scenario, we allow healthy foragers and drones to drift between colonies, but infected bees are detected and are removed. In panels (**a**,**b**), we assume there is a perfect filter for infected hive bees; in panels (**c**,**d**), we assume this guarding behaviour is 50% effective.

3.3. Scenario: Robbing Behaviour

Another behaviour that bees show is robbing dying colonies for resources [19]. In this situation, healthy foragers from colony 2 will target a dying colony for its resources, often picking up infection along the with the resources [19]. We simulate this situation by inserting an infected forager into colony 1, and running the simulation until the population of colony 1 is below 6000 bees, we then insert an infected forager into colony 2. This is to simulate robbing from a dying colony. When we do this without any drift, we see in Figure 7 that the the colonies independently as expected. Note for these results we set $\beta = 2 \times 10^{-3}$ to ensure colony 1 collapses. As mentioned above, ignoring drift can be interpreted as simulating wild colonies that are often far enough apart that drifting is not an issue [17].

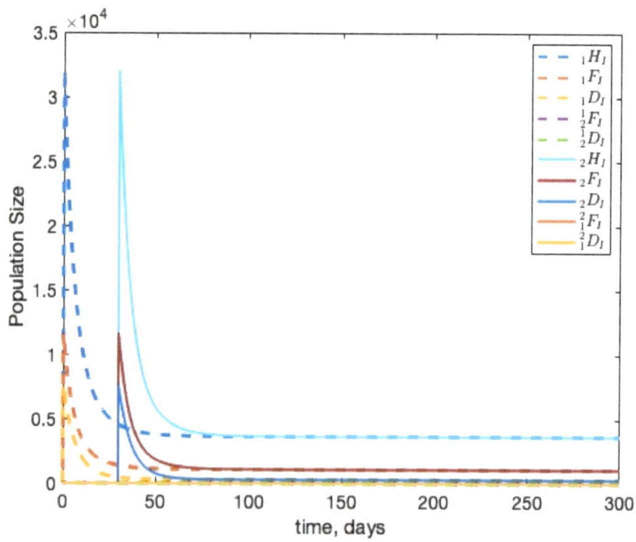

Figure 7. Beginning from equilibrium, we introduce one infected forager into colony 1 (${}_1F_I(0) = 1$). In this scenario, we remove drift but allow colony 2 to rob colony 1 of resources as the colony begins to die. We see that in this case the colonies behave independently.

When we combine the effects of drift and robbing, we see that infection spread by drift dominates. This is visualised in Figure 8.

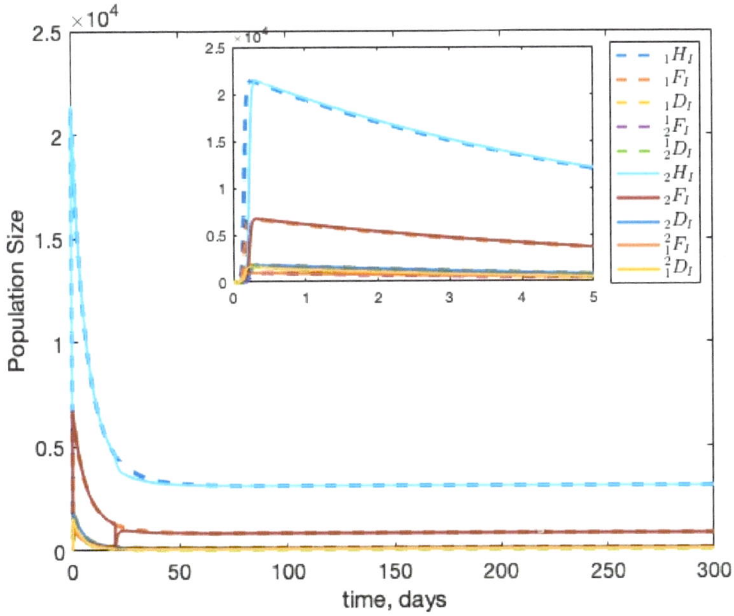

Figure 8. Beginning from equilibrium, we introduce one infected forager into colony 1 (${}_1F_I(0) = 1$). In this scenario, we allow drifting between colonies and allow colony 2 to rob colony 1 of resources as the colony begins to die. We see that in this case the effects of robbing are washed out.

3.4. Basic Reproduction Number

The model is too complex for the use of analytical techniques to determine a basic reproduction number to get a sense of severity of disease. We can however estimate the basic reproduction number from our numerical simulations. Using the heuristic that the proportion of infected individuals, p, at the endemic equilibrium is related by

$$p = 1 - \frac{1}{R_0}$$

we can show that our two values of β in our study correspond to a reproduction number of $R_0 \approx 1.67$ when $\beta = 5 \times 10^{-4}$, and $R_0 \approx 20$ when $\beta = 2 \times 10^{-3}$, in a single colony. These values are in the same range as those used in [23,31].

3.5. Varying Drift & Reproduction Number

As drift is dependent on distance between the colonies, we examine the effects of varying drift on colony behaviour. In Figure 9, we see that lowering drift by distancing colonies can slow spread of infection between colonies by upwards of 10 days. In panel (a), when $\beta = 5 \times 10^{-4}$, we see that as drift rate increases (i.e., distance between colonies decreases), there is a rapid decrease in the time it takes infection to spread from one colony to another. We see in panel (b) that when $\beta = 2 \times 10^{-3}$, the drift coefficient plays little role in spread and drift need only be present.

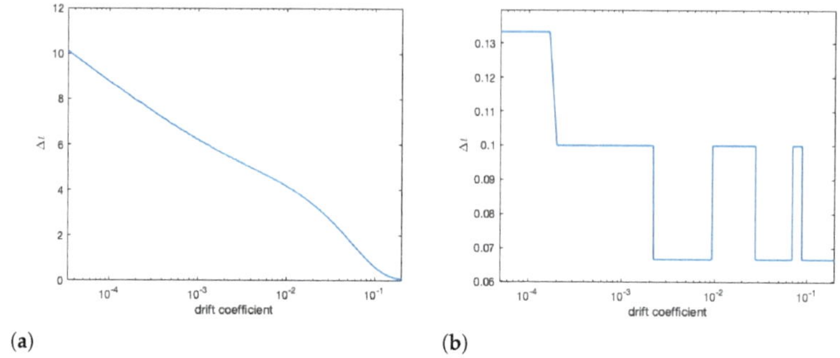

Figure 9. With all other parameters being equal, we change the drift coefficient and measure the time between peak infections in colony 1 and colony 2. We can see for low β (panel (**a**)) that increasing the drift coefficient (i.e., decreasing the distance between colonies) exponentially decreases the time it take an infection to reach colony 2. When β is high (panel (**b**)), we see that there is less importance on drift, it only need be present.

In Figure 10, we vary β and measure the time between peak infections in colony 1 and colony 2. We see that increased infectiousness between bees does not have as large an impact on inter-colony transmission as drift. In panel (a) we set the forager drift coefficients to 0.1 and in panel (b) we set the forager drift coefficient to 1×10^{-4} for comparison so show that the insensitivity to β holds.

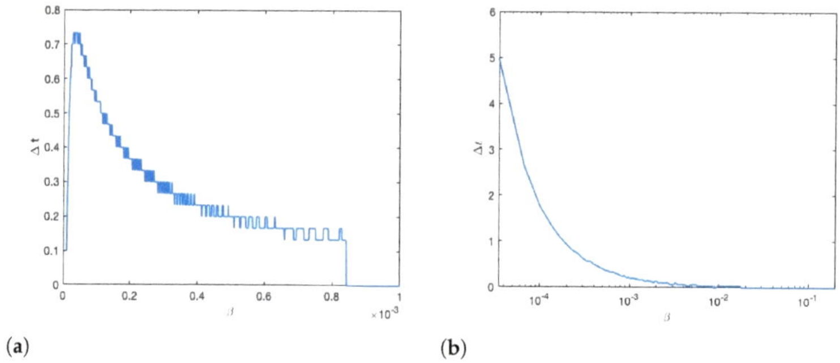

Figure 10. With all other parameters being equal, we change β and measure the time between peak infections in colony 1 and colony 2. We can see that increasing β exhibits exponentially decreases in the time it take an infection to reach colony 2, but the change is quite small. In panel (**a**), we use a drift coefficient of 0.1, and in panel (**b**), we use a drift coefficient of 1×10^{-4}.

4. Discussion

In this study, we explore the dynamics of disease between two coupled honey bee colonies. We use a system of two colonies so that we can clearly highlight the effects of infection in one colony on another. We tune the drift parameters so that the number of expatriated bees from each colony is roughly 12% at any given time. This is in line with observation [17], although drift as high as 80% has been observed [47].

When exploring the interplay between drift and disease spread, we specifically choose disease parameters such that R_0 is low. This allows our results to highlight the qualitative changes induced by our different behavioural scenarios. As we can see in Figures 7 and 8, a larger R_0 allows an infection to spread incredibly rapidly within a colony. Studying under a lower R_0 helps highlight the changes induced by changing behaviours.

Primarily, we notice that when bees are able to drift between colonies, the severity of disease in *both* colonies is reduced. In edge cases, with many colonies, this may help colonies survive in the face of infection instead of dying out. We can expect that if we added more colonies to this model, we may see a further reduction in severity. This may help explain why many diseases that affect colonies are sublethal [48–50] as the population of colonies "share" the stress of infection.

We use the proportion of infected bees in surviving infected colonies as a proxy for future risk from other points of stress. We posit that colonies with high levels of infection will be more susceptible to failure from other stresses. In this sense, we see that drifting can lower future risk in the meta-population by reducing the infection load in each individual colony. This is a particularly interesting result as it seems counterintuitive. That being said, drifting is not the only avenue by which one can reduce stress within an infected colony. Certain types of infections may be better treated by queen replacement [51], better access to forage, etc. There may be cases when it is beneficial to have two working colonies with 60% of the bees healthy versus one failed colony and one healthy colony (for an average of 50% of the bees in the meta-population).

To contrast the above point, the fact that approximately 30% of bees within a colony are infected put the colony at increased risk from other stressors [2,52] and seasonal effects [23,24,53]. In this sense, from the perspective of the healthy colony, it is beneficial to prevent infection. This could be one avenue by which colonies developed guarding behaviour. Current research suggests that some viruses are adapting to circumvent guarding behaviour [54], which may lead to further evolutionary pressure against drifting, causing honey bees to become more territorial.

Our results also show that robbing is mostly a cause for disease spread between colonies when drifting is rare, and in this scenario, honey bee colonies mostly act independently. With commercial colonies often being identical and kept within close proximity, the model predicts that disease spread through drift occurs faster and more frequently than spread through resource robbing. This suggests that disease mitigation in colonies should first focus on reducing drift, which would allow beekeepers more time to isolate dying colonies before they can be robbed.

Finally, we see that extremely infectious pathogens (high β) or extremely dense colonies will results in rapid transmission between colonies (Figures 9 and 10 regardless of the state of the other parameter.

There are also many ways to extend the single colony model to account for different behaviours as well. By including a queen compartment with its own life cycle, one could study the effects of queen replacement on colony infection. The parameters of the model are also set to reflect commercial beekeeping efforts in Canada/United States using *Apis mellifera*. The model could be reparametrized for hobbyist beekeepers, *Apis cerana*, or displaced/imported bees to a new environment. This is dependent on available data for each of these cases. Some of these studies would be better suited to an agent-based simulation such as those in [33,55].

Moreover, extensions of the multi-scale framework to include other pollinators are also interesting. Combining multiple pollinators in a spatially explicit model may lead to insights into how and when wild pollinators are at greatest risk of disease spillover [56–58] from managed honeybee colonies. Such a spatially heterogeneous model will also allow for the study of infections spread through foraging such as in the case of *Nosema* whose spores can be transmitted via bees foraging at the same flower.

From an evolutionary perspective, we see that drifting and closer colonies may have been relevant in reducing overall stress. This could have been beneficial to the meta-population when environmental stresses (i.e., climate change and pesticide use) were less imposing on a colony's overall health. This would require further evolutionary studies. On the same note, it is theorised that robbing behaviour is an adaptation caused by human influence on honeybee colonies [18]. Adapting this mathematical model to an evolutionary framework can help address this question.

Author Contributions: M.B. designed the model and wrote the manuscript. K.S. introduced modifications to the model. M.B. and K.S. performed the analysis of the model. All authors have read and agreed to the published version of the manuscript.

Funding: This research was funded by NSERC, grant number 2019-05679.

Institutional Review Board Statement: Not applicable.

Informed Consent Statement: Not applicable.

Data Availability Statement: Code available upon request.

Conflicts of Interest: The authors declare no conflict of interest.

References

1. Owen, R. Role of human action in the spread of honey bee (Hymenoptera: Apidae) pathogens. *J. Econ. Entomol.* **2017**, *110*, 797–801. [CrossRef] [PubMed]
2. Coulon, M.; Dalmon, A.; Di Prisco, G.; Prado, A.; Arban, F.; Dubois, E.; Ribière-Chabert, M.; Alaux, C.; Thiéry, R.; Le Conte, Y. Interactions between thiamethoxam and Deformed wing virus can drastically impair flight behavior of honey bees. *Front. Microbiol.* **2020**, *11*, 766. [CrossRef] [PubMed]
3. Flores, J.M.; Gil-Lebrero, S.; Gámiz, V.; Rodríguez, M.I.; Ortiz, M.A.; Quiles, F.J. Effect of the climate change on honey bee colonies in a temperate Mediterranean zone assessed through remote hive weight monitoring system in conjunction with exhaustive colonies assessment. *Sci. Total Environ.* **2019**, *653*, 1111–1119. [CrossRef] [PubMed]
4. Colin, T.; Meikle, W.G.; Paten, A.M.; Barron, A.B. Long-term dynamics of honey bee colonies following exposure to chemical stress. *Sci. Total Environ.* **2019**, *677*, 660–670. [CrossRef]
5. De Jong, D. Mites: Varroa and other parasites of brood. In *Honey Bee Pests, Predators and Diseases*; Cornell University Press: Ithaca, NY, USA, 1997; Volume 1, pp. 279–327.
6. Wilfert, L.; Long, G.; Leggett, H.; Schmid-Hempel, P.; Butlin, R.; Martin, S.; Boots, M. Deformed wing virus is a recent global epidemic in honeybees driven by Varroa mites. *Science* **2016**, *351*, 594–597. [CrossRef] [PubMed]
7. Di Prisco, G.; Pennacchio, F.; Caprio, E.; Boncristiani, H.F., Jr.; Evans, J.D.; Chen, Y. *Varroa destructor* is an effective vector of Israeli acute paralysis virus in the honeybee, *Apis mellifera*. *J. Gen. Virol.* **2011**, *92*, 151–155. [CrossRef] [PubMed]
8. Paxton, R.J.; Klee, J.; Korpela, S.; Fries, I. *Nosema ceranae* has infected *Apis mellifera* in Europe since at least 1998 and may be more virulent than *Nosema apis*. *Apidologie* **2007**, *38*, 558–565. [CrossRef]
9. Carreck, N.L.; Ball, B.V.; Martin, S.J. The epidemiology of cloudy wing virus infections in honey bee colonies in the UK. *J. Apic. Res.* **2010**, *49*, 66–71. [CrossRef]
10. Ravoet, J.; De Smet, L.; Wenseleers, T.; De Graaf, D.C. Genome sequence heterogeneity of Lake Sinai Virus found in honey bees and Orf1/RdRP-based polymorphisms in a single host. *Virus Res.* **2015**, *201*, 67–72. [CrossRef]
11. Spivak, M.; Gilliam, M. Hygienic behaviour of honey bees and its application for control of brood diseases and Varroa: Part I. Hygienic behaviour and resistance to American foulbrood. *Bee World* **1998**, *79*, 124–134. [CrossRef]
12. Spivak, M. Honey bee hygienic behavior and defense against Varroa jacobsoni. *Apidologie* **1996**, *27*, 245–260. [CrossRef]
13. Fefferman, N.H.; Traniello, J.F.; Rosengaus, R.B.; Calleri, D.V., II. Disease prevention and resistance in social insects: modeling the survival consequences of immunity, hygienic behavior, and colony organization. *Behav. Ecol. Sociobiol.* **2007**, *61*, 565–577. [CrossRef]
14. Moore, A.J.; Breed, M.D.; Moor, M.J. The guard honey bee: ontogeny and behavioural variability of workers performing a specialized task. *Anim. Behav.* **1987**, *35*, 1159–1167. [CrossRef]
15. Free, J. The drifting of honey-bees. *J. Agric. Sci.* **1958**, *51*, 294–306. [CrossRef]
16. Goodwin, R.; Taylor, M.; Mcbrydie, H.; Cox, H. Drift of Varroa destructor-infested worker honey bees to neighbouring colonies. *J. Apic. Res. Bee World* **2006**, *45*, 155. [CrossRef]
17. Dynes, T.L.; Berry, J.A.; Delaplane, K.S.; Brosi, B.J.; de Roode, J.C. Reduced density and visually complex apiaries reduce parasite load and promote honey production and overwintering survival in honey bees. *PLoS ONE* **2019**, *14*, e0216286. [CrossRef] [PubMed]
18. Rittschof, C.C.; Nieh, J.C. Honey robbing: Could human alterations to the environment change a rare foraging tactic into a maladaptive behavior? *Curr. Opin. Insect Sci.* **2021**, *45*, 84–90. [CrossRef] [PubMed]
19. Peck, D.T.; Seeley, T.D. Mite bombs or robber lures? The roles of drifting and robbing in Varroa destructor transmission from collapsing honey bee colonies to their neighbors. *PLoS ONE* **2019**, *14*, e0218392. [CrossRef]
20. Graystock, P.; Goulson, D.; Hughes, W.O. Parasites in bloom: flowers aid dispersal and transmission of pollinator parasites within and between bee species. *Proc. R. Soc. B Biol. Sci.* **2015**, *282*, 20151371. [CrossRef] [PubMed]
21. Purkiss, T.; Lach, L. Pathogen spillover from Apis mellifera to a stingless bee. *Proc. R. Soc. B* **2019**, *286*, 20191071. [CrossRef] [PubMed]
22. Koch, H.; Brown, M.J.; Stevenson, P.C. The role of disease in bee foraging ecology. *Curr. Opin. Insect Sci.* **2017**, *21*, 60–67. [CrossRef] [PubMed]

23. Betti, M.I.; Wahl, L.M.; Zamir, M. Effects of Infection on Honey Bee Population Dynamics: A Model. *PLoS ONE* **2014**, *9*, e110237. [CrossRef]
24. Betti, M.I.; Wahl, L.M.; Zamir, M. Reproduction Number and Asymptotic Stability for a Model with Continuous Age Structure: An Application to Honey Bee Dynamics. *Bull. Math. Biol.* **2016**, submitted.
25. Betti, M.; Wahl, L.; Zamir, M. Age structure is critical to the population dynamics and survival of honeybee colonies. *R. Soc. Open Sci.* **2016**, *3*, 160444. [CrossRef] [PubMed]
26. Muhammad, N.; Eberl, H.J. Two routes of transmission for Nosema infections in a honeybee population model with polyethism and time-periodic parameters can lead to drastically different qualitative model behavior. *Commun. Nonlinear Sci. Numer. Simul.* **2020**, *84*, 105207. [CrossRef]
27. Ratti, V.; Kevan, P.G.; Eberl, H.J. A mathematical model for population dynamics in honeybee colonies infested with *Varroa destructor* and the Acute Bee Paralysis Virus. *Can. Appl. Math. Q.* **2013**, *21*, 63–93.
28. Eberl, H.J.; Frederick, M.R.; Kevan, P.G. Importance of brood maintenance terms in simple models of the honeybee—*Varroa destructor*—Acute Bee Paralysis Virus complex. *Electron. J. Differ. Equ.* **2010**, *2010*, 85–98.
29. Petric, A.; Guzman-Novoa, E.; Eberl, H.J. A mathematical model for the interplay of Nosema infection and forager losses in honey bee colonies. *J. Biol. Dyn.* **2017**, *11*, 348–378. [CrossRef]
30. Muhammad, N.; Eberl, H.J. A simple model of between-hive transmission of Nosemosis. In *International Conference on Applied Mathematics, Modeling and Computational Science*; Springer: Berlin/Heidelberg, Germany, 2017; pp. 385–395.
31. Bartlett, L.J.; Rozins, C.; Brosi, B.J.; Delaplane, K.S.; de Roode, J.C.; White, A.; Wilfert, L.; Boots, M. Industrial bees: The impact of apicultural intensification on local disease prevalence. *J. Appl. Ecol.* **2019**, *56*, 2195–2205. [CrossRef]
32. Le Conte, Y.; Ellis, M.; Ritter, W. Varroa mites and honey bee health: can Varroa explain part of the colony losses? *Apidologie* **2010**, *41*, 353–363. [CrossRef]
33. Betti, M.; LeClair, J.; Wahl, L.M.; Zamir, M. Bee++: An Object-Oriented, Agent-Based Simulator for Honey Bee Colonies. *Insects* **2017**, *8*, 31. [CrossRef]
34. Fries, I. *Nosema apis*—A parasite in the honey bee colony. *Bee World* **1993**, *74*, 5–19. [CrossRef]
35. Khoury, D.S.; Myerscough, M.R.; Barron, A.B. A Quantitative Model of Honey Bee Colony Population Dynamics. *PLoS ONE* **2011**, *6*, e18491. [CrossRef]
36. Khoury, D.S.; Barron, A.B.; Myerscough, M.R. Modelling Food and Population Dynamics in Honey Bee Colonies. *PLoS ONE* **2013**, *8*, e59084. [CrossRef]
37. Heidinger, I.M.M.; Meixner, M.D.; Berg, S.; Büchler, R. Observation of the mating behavior of honey bee (*Apis mellifera* L.) queens using radio-frequency identification (RFID): Factors influencing the duration and frequency of nuptial flights. *Insects* **2014**, *5*, 513–527. [CrossRef]
38. Seeley, T.; Mikheyev, A. Reproductive decisions by honey bee colonies: tuning investment in male production in relation to success in energy acquisition. *Insectes Sociaux* **2003**, *50*, 134–138. [CrossRef]
39. Fukuda, H.; Ohtani, T. Survival and life span of drone honeybees. *Popul. Ecol.* **1977**, *19*, 51–68. [CrossRef]
40. Russell, S.; Barron, A.B.; Harris, D. Dynamic modelling of honey bee (*Apis mellifera*) colony growth and failure. *Ecol. Model.* **2013**, *265*, 158–169. [CrossRef]
41. Brodschneider, R.; Crailsheim, K. Nutrition and health in honey bees. *Apidologie* **2010**, *41*, 278–294. [CrossRef]
42. Hrassnigg, N.; Crailsheim, K. Differences in drone and worker physiology in honeybees (*Apis mellifera*). *Apidologie* **2005**, *36*, 255–277. [CrossRef]
43. Jay, S. Seasonal development of honeybee colonies started from package bees. *J. Apic. Res.* **1974**, *13*, 149–152. [CrossRef]
44. Neumann, P.; Moritz, R.F.; Mautz, D. Colony evaluation is not affected by drifting of drone and worker honeybees (*Apis mellifera* L.) at a performance testing apiary. *Apidologie* **2000**, *31*, 67–79. [CrossRef]
45. Free, J.; Racey, P. The effect of the size of honeybee colonies on food consumption, brood rearing and the longevity of the bees during winter. *Entomol. Exp. Appl.* **1968**, *11*, 241–249. [CrossRef]
46. Downs, S.G.; Ratnieks, F.L. Adaptive shifts in honey bee (*Apis mellifera* L.) guarding behavior support predictions of the acceptance threshold model. *Behav. Ecol.* **2000**, *11*, 326–333. [CrossRef]
47. Pfeiffer, K.; Crailsheim, K. Drifting of honeybees. *Insectes Sociaux* **1998**, *45*, 151–167. [CrossRef]
48. Wolf, S.; Nicholls, E.; Reynolds, A.M.; Wells, P.; Lim, K.S.; Paxton, R.J.; Osborne, J.L. Optimal search patterns in honeybee orientation flights are robust against emerging infectious diseases. *Sci. Rep.* **2016**, *6*, 1–10. [CrossRef] [PubMed]
49. Milbrath, M.O.; van Tran, T.; Huang, W.F.; Solter, L.F.; Tarpy, D.R.; Lawrence, F.; Huang, Z.Y. Comparative virulence and competition between Nosema apis and Nosema ceranae in honey bees (*Apis mellifera*). *J. Invertebr. Pathol.* **2015**, *125*, 9–15. [CrossRef]
50. Mattila, H.; Otis, G. Effects of pollen availability and Nosema infection during the spring on division of labor and survival of worker honey bees (Hymenoptera: Apidae). *Environ. Entomol.* **2006**, *35*, 708–717. [CrossRef]
51. Botías, C.; Martín-Hernández, R.; Días, J.; García-Palencia, P.; Matabuena, M.; Juarranz, Á.; Barrios, L.; Meana, A.; Nanetti, A.; Higes, M. The effect of induced queen replacement on Nosema spp. infection in honey bee (*Apis mellifera* iberiensis) colonies. *Environ. Microbiol.* **2012**, *14*, 845–859. [CrossRef]
52. Liess, M.; Foit, K.; Knillmann, S.; Schäfer, R.B.; Liess, H.D. Predicting the synergy of multiple stress effects. *Sci. Rep.* **2016**, *6*, 1–8. [CrossRef]

53. Comper, J.R.; Eberl, H.J. Mathematical modelling of population and food storage dynamics in a honey bee colony infected with Nosema ceranae. *Heliyon* **2020**, *6*, e04599. [CrossRef] [PubMed]
54. Geffre, A.C.; Gernat, T.; Harwood, G.P.; Jones, B.M.; Gysi, D.M.; Hamilton, A.R.; Bonning, B.C.; Toth, A.L.; Robinson, G.E.; Dolezal, A.G. Honey bee virus causes context-dependent changes in host social behavior. *Proc. Natl. Acad. Sci. USA* **2020**, *117*, 10406–10413. [CrossRef]
55. Bromenshenk, J.J.; DeGrandi-Hoffman, G. PC BEEPOP: A Microcomputer Model of the Population Dynamics and Economics of Honey Bee Colonies. *Am. Entomol.* **1993**, *39*, 231–237. [CrossRef]
56. Fürst, M.; McMahon, D.P.; Osborne, J.; Paxton, R.; Brown, M. Disease associations between honeybees and bumblebees as a threat to wild pollinators. *Nature* **2014**, *506*, 364–366. [CrossRef]
57. Manley, R.; Boots, M.; Wilfert, L. Emerging viral disease risk to pollinating insects: ecological, evolutionary and anthropogenic factors. *J. Appl. Ecol.* **2015**, *52*, 331–340. [CrossRef]
58. Dalmon, A.; Diévart, V.; Thomasson, M.; Fouque, R.; Vaissière, B.E.; Guilbaud, L.; Le Conte, Y.; Henry, M. Possible spillover of pathogens between bee communities foraging on the same floral resource. *Insects* **2021**, *12*, 122. [CrossRef]

insects

MDPI

Article

Effectiveness of Different Soft Acaricides against Honey Bee Ectoparasitic Mite *Varroa destructor* (Acari: Varroidae)

Ziyad Abdul Qadir [1,2,†], Atif Idrees [3,†], Rashid Mahmood [1], Ghulam Sarwar [1], Muhammad Abu Bakar [4], Saboor Ahmad [5], Muhammad Mohsin Raza [6] and Jun Li [3,*]

1 Honeybee Research Institute, National Agricultural Research Centre, Park Road, Islamabad 45500, Pakistan; zaqadir@parc.gov.pk (Z.A.Q.); directorhbri@parc.gov.pk (R.M.); sarwar1970@parc.gov.pk (G.S.)
2 Department of Entomology and Wildlife Ecology, University of Delaware, Newark, DE 19716, USA
3 Guangdong Key Laboratory of Animal Conservation and Resource Utilization, Guangdong Public Laboratory of Wild Animal Conservation and Utilization, Institute of Zoology, Guangdong Academy of Sciences, Guangzhou 510260, China; atif_entomologist@yahoo.com
4 Department of Entomology, University College of Agriculture, University of Sargodha, Sargodha 40100, Pakistan; mabubakar_26@yahoo.com
5 Institute of Apicultural Research/Key Laboratory of Pollinating Insect Biology, Ministry of Agriculture, Chinese Academy of Agricultural Sciences, Beijing 100081, China; 2018y90100172@caas.cn
6 The Joint Centre for Excellence in Environmental Intelligence, University of Exeter, Exeter EX4 4QF, UK; m.r.raza@exeter.ac.uk
* Correspondence: junl@giabr.gd.cn
† Both authors contributed equally to this work.

Citation: Qadir, Z.A.; Idrees, A.; Mahmood, R.; Sarwar, G.; Bakar, M.A.; Ahmad, S.; Raza, M.M.; Li, J. Effectiveness of Different Soft Acaricides against Honey Bee Ectoparasitic Mite *Varroa destructor* (Acari: Varroidae). *Insects* **2021**, *12*, 1032. https://doi.org/10.3390/insects12111032

Academic Editors: Ivana Tlak Gajger, Franco Mutinelli and Lukasz L. Stelinski

Received: 15 October 2021
Accepted: 11 November 2021
Published: 17 November 2021

Simple Summary: Over the past few decades, the ectoparasitic mite *Varroa destructor* has been a significant threat to managed honey bee (*Apis mellifera*) colonies worldwide. Many control methods, including application of synthetic acaricides, have been adopted to control the infestation of varroa mites in honey bee colonies. Synthetic acaricides such as coumaphos and fluvalinate are only effective in reducing susceptible mites. Besides, synthetic acaricides pose multiple threats to honey bee colonies and the environment, necessitating their alteration with non-synthetic options. Naturally occurring compounds are considered an essential alternative control measure for varroa mites. Natural acaricides are derived from plants that contain essential oils or organic acids. The current study investigated the efficacy of formic acid, oxalic acid, and thymol in the control of Varroa mites. These soft acaricides were applied at various concentrations/quantities. Formic acid, oxalic acid, and thymol were all effective at lowering mite population levels. Formic acid, oxalic acid, and thymol can be used in an integrated management plan to control varroa mite populations. This scientific-based information can be shared with the beekeeping community of Pakistan and elsewhere, which will be helpful in managing this parasite that often affects honey bee productivity.

Abstract: Honey bees (*Apis mellifera*) are essential for their products—honey, royal jelly, pollen, propolis and beeswax. They are also indispensable because they support ecosystems with their pollination services. However, the production and functions of honey bees are hindered by the arthropod pest *Varroa destructor*, which attacks bees through its feeding activities. Efforts to control varroa mites have been made through the development of various synthetic pesticide groups, but have had limited success because the mites developed resistance and some of these pesticides are harmful to bees. Branded pesticides are rarely used in Pakistan, as beekeepers utilize acaricides from unknown sources. There is a need to create awareness of available naturally occurring acaricides that may serve as an alternative to synthetic acaricides. Although some naturally occurring compounds are considered toxic to the environment, the soft acaricides oxalic acid, thymol, and formic acid 65% are usually safe for honey bee colonies and beekeepers, when handled appropriately. The current study investigated the effectiveness of formic acid (10, 15, and 20 mL/hive), oxalic acid (4.2, 3.2, and 2.1%/hive), and thymol (6, 4, and 2 g/hive) in controlling mite infestation. The results indicated that all treatments significantly reduced the mite population ($p < 0.05$). The average efficacies of oxalic acid at 3.2% (94.84% ± 0.34) and 4.2% (92.68% ± 0.37) were significantly higher than those of the other treatments. The lowest efficacy was recorded in formic acid 65% at 10 mL (54.13%). Overall, the results indicated that soft acaricides—such as oxalic acid at 3.2% and 4.2% concentrations—are very effective at controlling varroa mites and can be used in broodless conditions without side effects.

Keywords: honey bee; formic acid; oxalic acid; thymol; *Varroa destructor*

1. Introduction

Honey bees, *Apis mellifera* (L.) (Hymenoptera: Apidae), play an essential role in agriculture by producing commodities, such as honey, propolis, royal jelly, bee pollen, bee venom, and beeswax [1]. Additionally, it is the most crucial eusocial insect, having benefited humankind for medicinal and nutritional purposes for thousands of years [2]. In 1977, grand-scale beekeeping experiments were initiated at the Honey bee Research Institute, under the National Agricultural Research Centre, Islamabad, Pakistan [3]. *Apis mellifera* was introduced in Pakistan to build an industry capable of producing substantial surplus honey for export as a cash crop [4]. In 2010, Pakistani beekeeping was being practiced by 27,000 families that were rearing ~400,000 beehives of *A. mellifera* and were financially benefitting from this profession [4,5]. Pakistan has also exported 4179.953 metric tons of honey valued at U.S. $9.8 million in 2020 [6].

Different insect pests and microorganisms attack honey bees around the world, including in Pakistan [7]. The acarine mite *Varroa destructor* is one of the most serious problems for *A. mellifera* [8]. It has affected both the honey bee industry and honey bee research activities since the dawn of apiculture [9]. *Varroa destructor* is considered the primary cause of the global decline in honey bees, resulting in huge colony losses [10,11].

The mite attacks the occidental bees in Pakistan [3]. After the introduction and successful domestication of the occidental bee, the mite became a serious pest of the bee and destroyed a considerable number of colonies within a few years [12]. In Pakistan, in 2002, two ectoparasitic mites—*V. destructor* and *Tropilaelaps clareae*—reduced honey production by up to 50% [13]. Previously, it was thought that varroa mites fed on the hemolymph of adult honey bees, sealed brood, and larvae, but recently, researchers have found that mites directly feed on body fat tissues and cellular components of immature and mature bees [14].

It is certain that, if appropriate control measures are not practiced to control mites on *A. mellifera* colonies, more damage (such as complete colony loss) can be anticipated [15,16]. In Pakistan, there exist neither honey bee regulatory bodies, nor well-regulated pesticides [17]. Thus, no checks and balances exist on the usage of pesticides, as some beekeepers often use acaricides such as formic acid from unknown sources while others rarely use branded soft acaricides from scientific stores. Environmentally safe options are needed to control various insect pests [18–22]. Different hard acaricides have been used to control this mite but have resulted in increasing mite resistance against these hard acaricides, and ultimately a reduction in their effectiveness [23]. For example, 24 years ago, *V. destructor* developed resistance to fluvalinate [24]. In addition, certain mite populations have established cross-resistance to different pesticide groups, i.e., formamidine (amitraz), organophosphates (coumaphos), and pyrethroids (acrinathin and flumethrin) [25–29]. These pesticides also pose a threat to bees by harming them when the bees are exposed to multiple compounds stored in wax. Accumulation of these hard acaricides in wax can create resistance in mites that are present in sealed cells, making the mites challenging to control [9]. Thus, these acaricides pose a threat to the beekeeping industry and pollute honey and other honey bee products [9]. Multiple soft acaricides are thought to play a critical role in the management of varroa mites. Soft acaricides are natural compounds of plant origin that contain essential oils or organic acids with pesticidal properties [30,31]. They are low-environmental-impact acaricides believed to be harmless to human health when handled appropriately [32]. Hard acaricides on the other hand usually contain synthetic and high-environmental-impact constituents proven to be effective against mites, but can on the other hand affect honey bees and other hive products [9].

Formic acid, lactic acid, oxalic acid, and thymol are organic acids that are used to control varroa and characterize the basis of natural compounds [9]. These natural compounds have various effects against mites. Formic acid interferes with the respiratory system of both adult mites and sealed brood cells [33]. In addition, most of these soft acaricides are water soluble and volatile, have a low risk of accumulation in bee products, and have lower residue levels [9]. Furthermore, they are natural constituents of honey [9]. Thus, they are less likely to contaminate and affect both bees and honey bee products [34–37]. Repeated treatments have a low probability of developing resistance [9]. These soft acaricides are season dependent; colony condition (brood or broodless) is also taken into account before applying these soft acaricides [9]. Similar to thymol, it was reported by Baggio et al. [38] to not use powdered thymol on weak colonies at high temperatures (>27–30 °C). Oxalic acid (OA) is highly effective in treating colonies without brood [39,40].

Varroa mites are becoming a serious threat to the beekeeping industry worldwide. This study aimed to manage varroa mites by using different soft acaricides (formic acid, oxalic acid, and thymol) in respective colonies; changes in honey production were also measured. This study also evaluated whether these chemicals are environmentally friendly, nonresistant, safe for human health, and safe for bee health. The objective of the study was to determine the effectiveness of soft acaricides both at the group level and among different concentration/quantity levels. Similarly, we compared the honey yields of all treatment groups at both the group and different concentration/quantity levels.

2. Materials and Methods

2.1. Study Site

The present study was carried out at the apiary of the Honeybee Research Institute, National Agricultural Research Centre, Islamabad, Pakistan, during the winter season (December to January). In December, the maximum and minimum temperature was 19.3, and 4.13 °C, respectively; the wind speed was 10.56 Km/day; 18.19 mm rainfall, and the average relative humidity was 76%. In January, the maximum and minimum temperature was 17.2 and 3.85 °C; the wind speed was 15.88 Km/day; rainfall was 41.64 mm, and the average relative humidity was 82%. This study was conducted on honey bee *A. mellifera* colonies that were naturally infested with varroa mites.

2.2. Honey Bee Colonies

Before initializing the experiments, 90 honey bee colonies of *A. mellifera* were selected to record the natural varroa mite fall. All of the colonies were housed in standard Langstroth boxes. A total of 45 colonies were screened out of the experiment to have colonies with the following characteristics: a honey bee population roughly equal to that of the other colonies, queen right and possess a mite infestation with an economic threshold level of >10 adult mites [8]. Furthermore, all experimental colonies were fully developed and productive. These were monitored prior to experimentation and had an average of 8 ± 2 brood combs in their brood chambers [41]. Five colonies were in each treatment group.

2.3. Soft Acaricide Treatments

Three soft acaricides, each at three concentration levels or quantities, were used in the study. The acaricides were produced by BDH laboratory supplies, England. Both inter- and intra-comparisons were conducted among the nine treatment groups. Soft acaricides used in the experiments were as follow:

2.3.1. Formic Acid (AnalaR 98/100% 'Safe-Break')

Formic acid (65%) at different quantities (10, 15, and 20 mL/hive) was used in this experimental study. Formic acid treatments were applied by pouring 10, 15, and 20 m of 65% formic acid on a piece of cardboard (7.5″ × 5.5″) placed inside the wire meshed tray inserted above the bottom board from the backside of the hive [42]. Formic acid solution was applied weekly to all colonies over the course of one month for four treatments.

2.3.2. Oxalic Acid (AnalaR)

Three different oxalic acid concentrations (4.2%, 3.2%, and 2.1%) mixed with sugar syrup were applied as treatments. To attain 4.2%, 3.2%, and 2.1% oxalic acid solutions, 100, 75, and 50 g of oxalic acid dehydrate was mixed with 1 L of sugar water (1:1) [43]. Treatments were applied only to frame spaces that contained bees. The 5-mL mixture was trickled directly on the adult bees in between two frames using a syringe as recommended (i.e., ~50 mL/colony) [39,40]. All groups received four doses of oxalic acid solution after a one-week interval.

2.3.3. Thymol (GPR)

Three different thymol concentrations (6, 4, and 2 g) were applied as treatments. Finely ground thymol was placed in petri dishes (80 mm) on top of the brood frame chambers under the top covers of the honey bee colonies [44]. Each treatment was applied after a one-week interval.

2.4. Mite Collection

To assess the adult mite population, mite collection trays were sandwiched between bottom boards from the backside of the beehive with wire screens installed above them to prevent the bees from removing the dead (or live) mites along with the debris [42,44]. Apart from checking for mite infestation levels, mite collection trays also improve treatment effectiveness. Dead fallen mites were examined after 7, 14, 21, and 28 days by counting dead fallen mites on the mite collection trays. To calculate the effectiveness of the applied treatments, Manhao^{TM}, Sichuan Pengshan Wangshi Animal Health Co., Ltd, (Chengdu, China) (fluvalinate) strips were applied to all of the treated colonies to knock down the remaining mites and evaluate the total mite population. Manhao^{TM} (fluvalinate) strips were inserted into each hive immediately following the fourth week of each soft acaricide treatment. Manhao^{TM} (fluvalinate) strips were removed after 28 days, and all the dropped mites were counted on the mite collection trays [45]. Effectiveness of each soft acaricides treatment was calculated separately by using the following efficacy formula [46,47].

$$Efficacy\ (\%) = \frac{Number\ of\ mites\ fallen\ during\ treatment\ with\ each\ soft\ chemical}{Total\ number\ of\ fallen\ mites\ during\ soft\ chemical\ and\ Apistan(r)\ treatment} \times 100$$

2.5. Honey Yield

Honey was harvested after the experiment with the help of a manually operated honey harvester, and the honey yield of treated colonies was recorded. Honey production was measured by taking the weight of each hive body used for honey collection before and after the honey extraction process. The weight difference was considered as the amount of harvestable honey [48].

2.6. Statistical Analysis

We used linear mixed effect models to evaluate the impact of each treatment (3 groups) and their concentrations/quantities (9 treatments) on mites in R software version 4.0.2 (R CoreTeam 2019, Vienna, Austria). Each response variable was separately analyzed. R packages including *"lme4"* [49] and *"lmerTest"* [50] were used for fitting the mixed models. In the case of efficacy and honey yield data, treatments were treated as fixed effects, while replications were treated as random effects. Model quality was evaluated based on scaled residuals simulated from the fitted model provided by the *SimulateResiduals* function of the DHARMa R package [51].

Mite data were not normally distributed; thus, we took the natural log to fulfill the normality and homoscedasticity assumptions. Means of significant interactions were compared using the Tukey HSD test at the 5% level of significance using the emmeans package [52]. For unique lettering in the means comparison, we used the Scott Knott clustering method provided in the ScottKnott package of R [53]. Finally, bar charts of significant means were plotted using the ggplot2 package [54] in R.

3. Results

3.1. Treatment Effectiveness among Major Groups (Formic Acid 65%, Oxalic Acid, and Thymol)

The results revealed that the three major treatments, formic acid 65%, oxalic acid, and thymol, were significantly different in efficacy (%) ($p < 0.05$) (Figure 1). Colonies treated with oxalic acid had the highest efficiency (90.48%), followed by thymol and formic acid 65% (76.74%) (Figure 1).

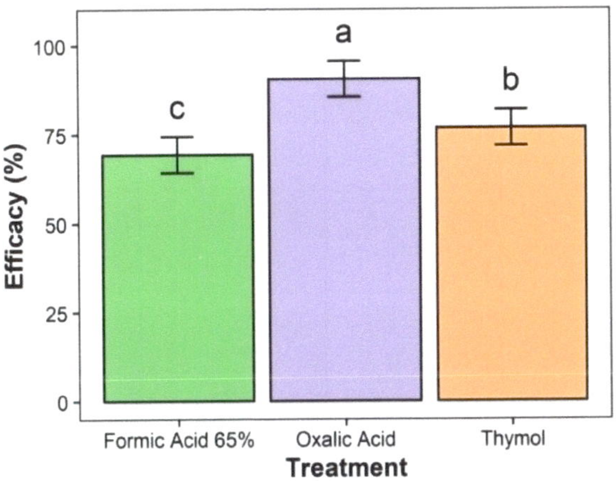

Figure 1. Efficacy of different treatment groups. Bars with the same letters are not significantly different ($p > 0.05$); the error bars represent the confidence interval (95%).

3.2. Treatment Effectiveness within All Concentrations/Quantities

In terms of concentrations/quantities, honey bee colonies treated with thymol at 4 g, oxalic acid 3.2%, and formic acid 20 mL exhibited the highest efficacy within their respective groups. (Figure 2). While the lowest efficacy was recorded against 2 g of thymol, oxalic acid at 2.10%, and 10-mL formic acid (Figure 2).

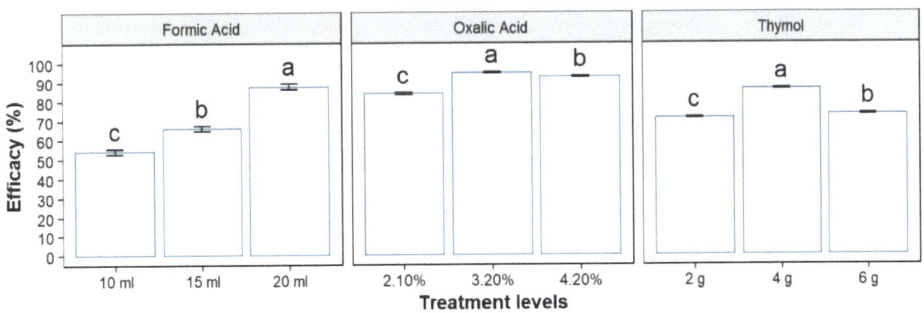

Figure 2. Efficacy of treatments against *Varroa destructor*. Treatments with the same letters are not significantly different; error bars signify confidence intervals (95%).

All concentrations/quantities within the groups were significantly different in terms of efficacy (Figure 2). Maximum effectiveness was observed in colonies treated with oxalic acid at 3.2% (94.84%), and the lowest efficacy was observed in honey bee colonies treated with formic acid at 65% of 10 mL (54.13%) (Figure 2).

3.3. Honey Yield

All Concentrations/Quantities

The highest honey yield was recorded in honey bee colonies treated with oxalic acid (3.20% and 4.20%) at 16.8 kg and 13 kg, respectively (Figure 3). The honey yield was lowest in honey bee colonies treated with 4 g of thymol and 2 g of thymol at 6 kg and 5 kg, respectively (Figure 3).

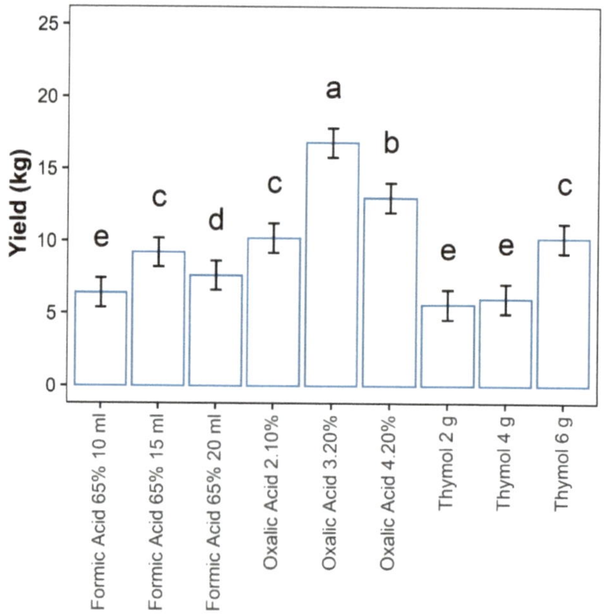

Figure 3. Effect of soft acaricides concentration/quantity on honey yield. Treatments with the same letters are not significantly different; error bars signify confidence intervals (95%).

4. Discussion

Mites are economically important pests in honey bee colonies. Soft acaricide treatments are thought to be effective against the ectoparasite mite *V. destructor*. The current study investigated the effectiveness of soft acaricides against *V. destructor*. Our results revealed that oxalic acid is effective against *V. destructor* followed by thymol and formic acid 65% during winter conditions. Acaricides have different effects against *V. destructor* depending on the mode of application, nature of acaricide formulation and environmental conditions [55,56]. Oxalic acid is a winter treatment, thus it is most efficacious at lower temperatures, especially when the brood is not present [57]. Meanwhile, formic acid and thymol are also dependent of different climatic and beekeeping conditions, which cause temperature-dependent effects [9]. Comparison of these soft acaricides needs to be evaluated in various environmental conditions for a more balanced analysis of their efficiency.

In terms of concentration, 3.2% oxalic acid was most effective in reducing *V. destructor* populations, indicating that 3.2% oxalic acid is the optimum concentration for controlling mites. Our results are consistent with Mahmood et al. [44] and Papežíková et al. [55], who reported that 3.2% oxalic acid is a reliable soft acaricide for controlling honey bee mites. Doses higher than 3.2% oxalic acid have also been found to be effective against mites [55]. Generally, the oxalic concentration registered for varroa control in European countries varies between 3.5% and 4.2%, with 4.2% widely used [55]. Our study could not establish why higher doses of oxalic acid (4.2%) did not transform into higher mite mortalities. This was likely because of varying temperatures, which caused a different treatment effect. No bee mortality, queen mortality, or superseding behavior was observed after completing the treatments, which matches the findings of other authors [44].

Our findings revealed that using 4 g of thymol to control *V. destructor* was more effective than 6 or 2 g of thymol. Conversely, it is expected that higher concentrations (in this case, 6 g) would be more effective in controlling varroa mites; this was not the case in the current study. Thymol is most effective during between 20 and 30 °C with effectiveness lost below 15 °C [58]. This may also partly explain the unclear results from the use of thymol as our study, was conducted during winter season. Thymol is a significant component of different commercially available products and is effective in managing ectoparasitic honey bee mites in *A. mellifera* colonies [59–61]. Thymol is a potential agent that has shown promising results in controlling ectoparasitic honey bee mites and has no negative impact on bee health [37,62].

It is expected that effectiveness is directly proportional to the dose. However, in the current study the expected dose-response relationship was not observed. Our results could not clearly substantiate the explanations for such observations. The need to explore the reasons for the lack of a dose-response relationship in future studies is critical. Understanding the cause for the lack of a dose-response relationship will not only further our scientific knowledge but it will help provide more useful information to beekeepers.

Our results are in accordance with Rashid et al. [63], revealing that formic acid is least effective in controlling *V. destructor* when tested against different groups. The reduced efficacy of formic acid may be due to the distance between formic acid volatilization and the honey bee-containing combs as the acaricide was only applied on cardboards inserted in the hive. In our study, the cardboard with formic acid application was placed inside the wire meshed tray inserted above the bottom board; this may also affect its efficacy [64,65]. The time of the year and temperature may also have affected its efficacy [65–68]. Our results contrast those of Mahmood et al. [42], who reported that a formic acid 20-mL concentration is very effective at controlling varroa mites in the winter season after Sider (*Ziziphus mauritiana*) honey harvest. Furthermore, Giusti et al. [69] reported that formic acid had no side effects on larvae, adult bees, and queens and showed an average efficacy greater than 95%. Our findings revealed that 20 mL of 65% formic acid was effective at controlling the honey bee ectoparasitic mite *V. destructor*.

The differences in soft acaricide efficacy in our study could be due to differences in the original number of mites infesting the bee colonies. Our study did not estimate the starting number of mites per colony; thus, it was difficult to conclusively state that the reduction in mite numbers was attributed to soft acaricide treatment. Additionally, studies conducted previously did not provide an estimate of the initial mite population [44,46,70]. Besides, organic acaricides such as those used in the current study may have certain advantages after repeated usage although their efficacy may be inconsistent when compared to synthetic forms [71].

The harvested honey results are in line with Mahmood et al. [47], who reported that honey yield from different treatment groups was significantly different, with the highest honey yield, obtained from oxalic acid, being 3.2%. However, comparison of honey production results across treatments needs to be accompanied by analysis of the colony strength to obtain a more reliable estimate of honey yield. Additionally, evaluating queen bee performance, climate and pasture conditions would further support the honey yield results [55,56]. Although our results indicate differences in the honey yield across the treatments, the contribution of soft acaricides to honey yield production cannot be conclusive as our experimental methods did not take in to account the honey bee populations in the hive after completion of the treatment.

The findings of the present study showed that the soft acaricides—formic acid, oxalic acid, and thymol—are effective natural products against varroa mite populations. The fact that no effects were observed on the honey bee colonies is an indication that the products are safe for the environment. However, further studies are needed to investigate the effectiveness and action mechanisms of soft acaricides against *V. destructor* and their impact on honey bee colony health (queen longevity, effect on worker, and adult brood longevity). Field experiments need to be conducted in varying environmental conditions as the effects of soft acaricides and mite infestation are environmental dependent [58,72]. Controlled experiments involving a known mite population against various concentrations need to be conducted to estimate the lethal concentration (LD_{50}). Residual analysis studies to assess residual presence in honey bees, bee wax, and other honey bee products will further confirm the safety of the soft acaricides.

5. Conclusions

The soft acaricides (oxalic acid and thymol) used in this experiment at different concentrations were very effective at reducing the damage from the ectoparasitic mite *V. destructor* and controlling its populations in *A. mellifera* colonies without showing any harm to the bees. While different quantities of formic acid 65% showed promising results, these can be included in an integrated mite control program. Beekeepers should use the recommended dosage and registered soft acaricides available in the market while following proper application methods to prevent resistance development in mites. Product use in accordance with the manufacturers directions is essential, as improper use may cause risk to persons or property.

Author Contributions: Conceptualization, Z.A.Q. and R.M.; methodology, R.M.; software, Z.A.Q. and M.M.R.; validation, Z.A.Q., R.M. and G.S.; formal analysis, Z.A.Q. and M.M.R.; investigation, Z.A.Q.; resources, R.M., G.S. and J.L; data curation, M.M.R.; writing—original draft preparation, Z.A.Q.; writing—review and editing, Z.A.Q., A.I., S.A. and M.A.B.; visualization, A.I., M.A.B. and S.A.; supervision, Z.A.Q.; project administration, Z.A.Q. and R.M.; funding acquisition, J.L. All authors have read and agreed to the published version of the manuscript.

Funding: This work is supported by Guangdong Academy of Sciences, GDAS Special Project of Science and Technology Development (2020GDASYL-20200301003).

Institutional Review Board Statement: Not applicable.

Informed Consent Statement: Not applicable.

Data Availability Statement: Not applicable.

Acknowledgments: We acknowledge Qurban Ali (Senior Scientific Assistant), Aman Ullah (Scientific Assistant), and Zafar Iqbal (Bee Attendant) for help with collecting mite collection trays.

Conflicts of Interest: The authors declare no conflict of interest.

References

1. Tiwari, R.; Dhami, M.; Mathur, V.; Bisht, B. Efficacy of Animal Origin Products and Ajwain Powder against Honey Bee Diseases in *Apis mellifera* (Linnaeus) Colonies in Uttarakhand-A Novel Eco-Friendly Approach. *JANS* **2014**, *6*, 68–75. [CrossRef]
2. Bakar, M.A.; Aqueel, M.A.; Raza, A.B.M.; Mahmood, R.; Qadir, Z.A.; Arshad, M.; Sohail, M. Evaluation of Essential Oils for the Management of Parasitic Bee Mites, *Varroa destructor* (Acari: Varroidae) In Vitro. *PJAR* **2019**, *32*, 566–571. [CrossRef]
3. Muzaffar, N.; Wells, H. Beekeeping in Pakistan. *Am. Bee J.* **2002**, *142*, 339–342.
4. PARC. *Annual Report 2010–2011*; Pakistan Agricultural Research Council: Islamabad, Pakistan, 2011; p. 98.
5. Mahmood, R.; Wagchoure, E.S.; Raja, S.; Sarwar, G. Control of *Varroa destructor* Using Oxalic Acid, Formic Acid and Bayvarol Strip in *Apis mellifera* (Hymenoptera: Apidae) Colonies. *Pak. J. Zool.* **2012**, *44*, 1473–1476.
6. Comtrade, U. Download Trade Data | UN Comtrade: International Trade Statistics. Available online: https://comtrade.un.org/data/ (accessed on 5 October 2021).
7. Chantawannakul, P.; de Guzman, L.I.; Li, J.; Williams, G.R. Parasites, Pathogens, and Pests of Honeybees in Asia. *Apidologie* **2016**, *47*, 301–324. [CrossRef]
8. Sajid, Z.N.; Aziz, M.A.; Bodlah, I.; Rana, R.M.; Ghramh, H.A.; Khan, K.A. Efficacy Assessment of Soft and Hard Acaricides against *Varroa destructor* Mite Infesting Honey Bee (*Apis mellifera*) Colonies, through Sugar Roll Method. *Saudi J. Biol. Sci.* **2020**, *27*, 53–59. [CrossRef]

9. Rosenkranz, P.; Aumeier, P.; Ziegelmann, B. Biology and Control of *Varroa destructor*. *J. Invertebr. Pathol.* **2010**, *103*, S96–S119. [CrossRef]
10. Alattal, Y.; Al-Ghamdi, A.; Single, A.; Ansari, M.J.; Alkathiri, H. Fertility and Reproductive Rate of Varroa Mite, *Varroa destructor*, in Native and Exotic Honeybee, *Apis mellifera* L., Colonies under Saudi Arabia Conditions. *Saudi J. Biol. Sci.* **2017**, *24*, 992–995. [CrossRef]
11. De Jong, D. Mites: Varroa and Other Parasites of Brood. In *Honey Bee Pests, Predators, and Diseases*, 2nd ed.; Morse, R.A., Flottum, K., Eds.; A.I. Root Co.: Medina, OH, USA, 1997; pp. 279–327.
12. Ahmad, R. Honeybee parasitic mites and their control in Pakistan. In *Progressive Farming*; Pakistan Agricultural Research Council-Publications: Islamabad, Pakistan, 1988; Volume 8, pp. 34–36.
13. Wagchoure-Camphor, E.S.; Martin, S.J. Beekeeping in Pakistan: A Bright Future in a Troubled Land. *Am. Bee J.* **2008**, *148*, 726–728.
14. Ramsey, S.D.; Ochoa, R.; Bauchan, G.; Gulbronson, C.; Mowery, J.D.; Cohen, A.; Lim, D.; Joklik, J.; Cicero, J.M.; Ellis, J.D.; et al. *Varroa destructor* Feeds Primarily on Honey Bee Fat Body Tissue and Not Hemolymph. *Proc. Natl. Acad. Sci. USA* **2019**, *116*, 1792–1801. [CrossRef] [PubMed]
15. Boecking, O.; Genersch, E. Varroosis–the Ongoing Crisis in Bee Keeping. *J. Verbr. Lebensm.* **2008**, *3*, 221–228. [CrossRef]
16. Rademacher, E.; Harz, M. Oxalic Acid for the Control of Varroosis in Honey Bee Colonies–A Review. *Apidologie* **2006**, *37*, 98–120. [CrossRef]
17. Khan, M.S.; Shah, M.M.; Mahmood, Q.; Hassan, A.; Akbar, K. Assessment of Pesticide Residues on Selected Vegetables of Pakistan. *J. Chem. Soc. Pak.* **2011**, *33*, 816–821.
18. Idrees, A.; Qasim, M.; Ali, H.; Qadir, Z.A.; Idrees, A.; Bashir, M.H. Acaricidal Potential of Some Botanicals against the Stored Grain Mites, *Rhizoglyphus tritici*. *J. Entomol. Zool. Stud.* **2016**, *4*, 611–617.
19. Idrees, A.; Zhangh, H.; Luo, M.; Thu, M.; Cai, P.; Islam, W.; Hussain, M.; Chen, J.; JI, Q. Protein Baits, Volatile Compounds and Irradiation Influence the Expression Profiles of Odorant Binding Protein Genes in *Bactrocera dorsalis* (Diptera: Tephritidae). *Appl. Ecol. Env. Res.* **2017**, *15*, 1883–1899. [CrossRef]
20. Cai, P.; Gu, X.; Yao, M.; Zhang, H.; Huang, J.; Idress, A.; Ji, Q.; Chen, J.; Yang, J. The Optimal Age and Radiation Dose for *Bactrocera dorsalis* (Hendel) (Diptera: Tephritidae) Eggs as Hosts for Mass-Reared *Fopius arisanus* (Sonan) (Hymenoptera: Braconidae). *Biol. Control* **2017**, *108*, 89–97. [CrossRef]
21. Gu, X.; Cai, P.; Yang, Y.; Yang, Q.; Yao, M.; Idrees, A.; Ji, Q.; Yang, J.; Chen, J. The Response of Four Braconid Parasitoid Species to Methyl Eugenol: Optimization of a Biocontrol Tactic to Suppress *Bactrocera dorsalis*. *Biol. Control* **2018**, *122*, 101–108. [CrossRef]
22. Luo, M.; Zhang, H.; Du, Y.; Idrees, A.; He, L.; Chen, J.; Ji, Q. Molecular Identification of Cultivable Bacteria in the Gut of Adult *Bactrocera tau* (Walker) and Their Trapping Effect: Trapping Effect of Cultivable Bacteria from *Bactrocera tau*. *Pest. Manag. Sci.* **2018**, *74*, 2842–2850. [CrossRef]
23. Loglio, G.; Plebani, G. Valutazione Dell'efficacia Dell'Apistan. *Apic.-Mod.* **1992**, *83*, 95–98.
24. Milani, N. Possible Presence of Fluvalinate-Resistant Strains of *Varroa jacobsoni* in Northern Italy. In *New perspectives on Varroa, Proceedings of the International Meeting, Prague, Czech Republic, November 1993*; Matheson, A., Ed.; International Bee Research Association: Cardiff, UK, 1994; p. 87. Available online: https://hal.archives-ouvertes.fr/hal-00891580/document (accessed on 24 September 2021).
25. Elzen, P.J.; Eischen, F.A.; Baxter, J.R.; Elzen, G.W.; Wilson, W.T. Detection of Resistance in US *Varroa jacobsoni* Oud. (Mesostigmata: Varroidae) to the Acaricide Fluvalinate. *Apidologie* **1999**, *30*, 13–17. [CrossRef]
26. Elzen, P.J.; Baxter, J.R.; Spivak, M.; Wilson, W.T. Amitraz Resistance in Varroa: New Discovery in North America. *Am. Bee J.* **1999**, *139*, 362.
27. Lodesani, M.; Colombo, M.; Spreafico, M. Ineffectiveness of Apistan Treatment against the Mite *Varroa jacobsoni* Oud in Several Districts of Lombardy (Italy). *Apidologie* **1995**, *26*, 67–72. [CrossRef]
28. Milani, N. The Resistance of *Varroa jacobsoni* Oud. to Acaricides. *Apidologie* **1999**, *30*, 229–234. [CrossRef]
29. Trouiller, J. Monitoring *Varroa jacobsoni* Resistance to Pyrethroids in Western Europe. *Apidologie* **1998**, *29*, 537–546. [CrossRef]
30. Gashout, H.A.; Guzman-Novoa, E.; Goodwin, P.H. Synthetic and Natural Acaricides Impair Hygienic and Foraging Behaviors of Honey Bees. *Apidologie* **2020**, *51*, 1155–1165. [CrossRef]
31. Flamini, G. Acaricides of Natural Origin, Personal Experiences and Review of Literature (1990-2001). *Stud. Nat. Prod. Chem.* **2003**, *28*, 381–451. [CrossRef]
32. TECA-FAO. Varroosis: A Parasitic Disease of the Brood and Adult Honeybees. Available online: https://www.fao.org/teca/en/technologies/8416 (accessed on 6 November 2021).
33. Islam, N.; Amjad, M.; Ehsan-ul-Haq, S.E.; Naz, F. Management of *Varroa destructor* by essential oils and formic acid in *Apis mellifera* Linn. Colonies. *J. Entomol. Zool. Stud.* **2016**, *4*, 97–104.
34. Fries, I. Treatment of Sealed Honey-Bee Brood with Formic-Acid for Control of *Varroa jacobsoni*. *Am. Bee J.* **1991**, *131*, 313–314.
35. Bogdanov, S.; Imdorf, A.; Kilchenmann, V. Residues in Wax and Honey after Apilife VAR Treatment. *Apidologie* **1998**, 513–524. [CrossRef]
36. Bogdanov, S.; Charrière, J.-D.; Imdorf, A.; Kilchenmann, V.; Fluri, P. Determination of Residues in Honey after Treatments with Formic and Oxalic Acid under Field Conditions. *Apidologie* **2002**, *33*, 399–409. [CrossRef]
37. Floris, I.; Satta, A.; Cabras, P.; Garau, V.L.; Angioni, A. Comparison Between Two Thymol Formulations in the Control of *Varroa destructor*: Effectiveness, Persistence, and Residues. *J. Econ. Entomol.* **2004**, *97*, 5. [CrossRef]

38. Baggio, A.; Arculeo, P.; Nanetti, A.; Marinelli, E.; Mutinelli, F. Field Trials with Different Thymol-Based Products for the Control of Varroosis. *Am. Bee J.* **2004**, *144*, 395–400.
39. Imdorf, A.; Charrière, J.-D.; Bachofen, B. Efficency Checking of the *Varroa jacobsoni* Control Methods by Means of Oxalic Acid. *Apiacta* **1997**, *3*, 89–91.
40. Brødsgaard, C.J.; Jansen, S.E.; Hansen, C.W.; Hansen, H. Spring Treatment with Oxalic Acid in Honey Bee Colonies as Varroa Control. *DIAS Rep. Hortic.* **1999**, *6*, 16.
41. Gregorc, A.; Adamczyk, J.; Kapun, S.; Planinc, I. Integrated Varroa Control in Honey Bee (*Apis mellifera carnica*) Colonies with or without Brood. *J. Apic. Res.* **2016**, *55*, 253–258. [CrossRef]
42. Mahmood, R.; Wagchoure, E.S.; Raja, S.; Sarwar, G.; Aslam, M. Effect of Thymol and Formic Acid Against Ectoparasitic Brood Mite *Tropilaelaps clareae* in *Apis mellifera* Colonies. *Pak. J. Zool.* **2011**, *43*, 91–95.
43. Prandin, L.; Dainese, N.; Girardi, B.; Damolin, O.; Piro, R.; Mutinelli, F. A Scientific Note on Long-Term Stability of a Home-Made Oxalic Acid Water Sugar Solution for Controlling Varroosis. *Apidologie* **2001**, *32*, 451–452. [CrossRef]
44. Mahmood, R.; Wagchoure, E.S.; ul Mohsin, A.; Raja, S.; Sarwar, G. Control of Ectoparasitic Mites in Honeybee (*Apis mellifera* L.) Colonies by Using Thymol and Oxalic Acid. *Pak. J. Zool.* **2012**, *44*, 985–989.
45. Bakar, M.A.; Aqueel, M.A.; Raza, A.B.M.; Arshad, M.; Mahmood, R.; Qadir, Z.A. Comparative Efficacy of Five Commercial Synthetic Acaricides against *Varroa destructor* (Anderson and Trueman) in *Apis mellifera* L. Colonies. *PJZ* **2018**, *50*, 857–861. [CrossRef]
46. Mahmood, R.; Asad, S.; Ahmad, W.; Sarwar, G.; Rafique, M.K.; Islam, N.; Qadir, Z.A.; Abiden, Z.U. Efficacy of Screen Bottom Board Tray with and without Soft Chemicals for Controlling *Varroa destructor* in Honeybee Colonies. *PJZ* **2017**, *49*, 9–13. [CrossRef]
47. Mahmood, R.; Abu Bakar, M.; Raza, M.F.; Qadir, Z.A.; Yahya, M. Efficacy of Naturally Occurring Chemicals for the Integrated Control of *Varroa destructor* (Anderson and Trueman) in Honeybee Colonies. *PJZ* **2021**, *53*, 1173–1176. [CrossRef]
48. Rashid, M.; Wagchoure, E.S.; Mohsin, A.U.; Raja, S.; Sarwar, G. Control of Ectoparasitic Mite *Varroa destructor* in Honeybee (*Apis mellifera* L.) Colonies by Using Different Concentrations of Oxalic Acid. *J. Anim. Plant Sci.* **2012**, *22*, 72–76.
49. Bates, D.; Mächler, M.; Bolker, B.; Walker, S. Fitting Linear Mixed-Effects Models Using Lme4. *arXiv* **2014**, arXiv:1406.5823.
50. Kuznetsova, A.; Brockhoff, P.B.; Christensen, R.H.B. LmerTest Package: Tests in Linear Mixed Effects Models. *J. Stat. Soft.* **2017**, *82*, 1–26. [CrossRef]
51. Hartig, F. DHARMa: Residual Diagnostics for Hierarchical (Multi-Level/Mixed) Regression Models. R Package Version 0.1. 0 2019. Available online: https://cran.r-project.org/web/packages/DHARMa/index.html (accessed on 20 October 2020).
52. Russell, L. Emmeans: Estimated Marginal Means, Aka Least-Squares Means. R Package Version 1.4. 2 2019. Available online: https://cran.r-project.org/web/packages/emmeans/index.html (accessed on 15 August 2020).
53. Jelihovschi, E.; Faria, J.C.; Allaman, I.B. ScottKnott: A Package for Performing the Scott-Knott Clustering Algorithm in R. *Tend. Mat. Apl. Comput.* **2014**, *15*, 3. [CrossRef]
54. Wickham, H. *Ggplot2-Elegant Graphics for Data Analysis*; Springer International Publishing: Cham, Switzerland, 2016.
55. Papežíková, I.; Palíková, M.; Kremserová, S.; Zachová, A.; Peterová, H.; Babák, V.; Navrátil, S. Effect of Oxalic Acid on the Mite *Varroa destructor* and Its Host the Honey Bee *Apis mellifera*. *J. Apic. Res.* **2017**, *56*, 400–408. [CrossRef]
56. Toomemaa, K. The Synergistic Effect of Weak Oxalic Acid and Thymol Aqueous Solutions on *Varroa* mites and Honey Bees. *J. Apic. Res.* **2019**, *58*, 37–52. [CrossRef]
57. Barbero, R.; Panella, F.; Bonizzoni, L. Ácido Oxálico y El Tratamiento de Limpieza Radical de Otoño-Invierno. *Vida Apícola* **1997**, *85*, 8–13.
58. Imdorf, A.; Kilchenmann, V.; Bogdanov, S.; Bachofen, B.; Beretta, C. Toxizität von Thymol, Campher, Menthol und Eucalyptol auf *Varroa jacobsoni* Oud und *Apis mellifera* L im Labortest. *Apidologie* **1995**, *26*, 27–31. [CrossRef]
59. Calderone, N.W.; Wilson, W.T.; Spivak, M. Plant Extracts Used for Control of the Parasitic Mites *Varroa jacobsoni* (Acari: Varroidae) and *Acarapis woodi* (Acari: Tarsonemidae) in Colonies of *Apis mellifera* (Hymenoptera: Apidae). *J. Econ. Entomol.* **1997**, *90*, 1080–1086. [CrossRef]
60. Imdorf, A.; Charrière, J.-D.; Kilchenmann, V.; Bogdanov, S.; Fluri, P. Alternative Strategy in Central Europe for the Control of Varroa destructor in Honey Bee Colonies. *APIACTA* **2003**, *38*, 258–278.
61. Imdorf, A.; Calderone, N.W. Use of Essential Oils for the Control of *Varroa jacobsoni* Oud. in Honey Bee Colonies. *Apidologie* **1999**, *30*, 209–228. [CrossRef]
62. Melathopoulos, A.P.; Gates, J. Comparison of Two Thymol-Based Acaricides, Apilifevar and Apiguard, for the Control of Varroa Mites. *Am. Bee J.* **2003**, *143*, 489–493.
63. Rashid, M.; Asad, S.; Sarwar, G.; Ahmad, W. Control of *Varroa destructor* Mite by Using Oxalic Acid, Formic Acid in Honey Bee *Apis mellifera* L. Colonies in Pakistan. *World Appl. Sci. J.* **2013**, *26*, 5.
64. Fries, I. Short-Interval Treatments with Formic Acid for Control of *Varroa jacobsoni* in Honey Bee (*Apis mellifera*) Colonies in Cold Climates. *Swed. J. Agric. Res.* **1989**, *19*, 213–216.
65. Underwood, R.M.; Currie, R.W. The Effects of Temperature and Dose of Formic Acid on Treatment Efficacy against *Varroa destructor* (Acari: Varroidae), a Parasite of *Apis mellifera* (Hymenoptera: Apidae). *Exp. Appl. Acarol.* **2003**, *29*, 303–313. [CrossRef] [PubMed]
66. Skinner, J.A.; Parkman, J.P.; Studer, M.D. Evaluation of Honey Bee Miticides, Including Temporal and Thermal Effects on Formic Acid Gel Vapours, in the Central South-Eastern USA. *J. Apic. Res.* **2001**, *40*, 81–89. [CrossRef]

67. Maul, V.; Petersen, N.; Wisssen, W. Field Trials Using Formic Acid as a Varroatosis Therapy. *Allg. Dtsch. Imkerztg.* **1980**, *14*, 155–157.
68. Eischen, F.A. Trials (and Tribulations) with Formic Acid for Varroa Control. *Am. Bee J.* **1998**, *138*, 734–735.
69. Giusti, M.; Sabelli, C.; Di Donato, A.; Lamberti, D.; Paturzo, C.E.; Polignano, V.; Lazzari, R.; Felicioli, A. Efficacy and Safety of Varterminator, a New Formic Acid Medicine against the *Varroa* mite. *J. Apic. Res.* **2017**, *56*, 162–167. [CrossRef]
70. Genc, F.; Aksoy, A. Some of the Correlations between the Colony Development and Honey Production on the Honeybee (*Apis mellifera* L.) Colonies. *Apiacta* **1993**, *28*, 33–41.
71. Tlak Gajger, I.; Svečnjak, L.; Bubalo, D.; Žorat, T. Control of *Varroa destructor* Mite Infestations at Experimental Apiaries Situated in Croatia. *Diversity* **2020**, *12*, 12. [CrossRef]
72. Pietropaoli, M.; Tlak Gajger, I.; Costa, C.; Gerula, D.; Wilde, J.; Adjlane, N.; Aldea-Sánchez, P.; Smodiš Škerl, M.I.; Bubnič, J.; Formato, G. Evaluation of Two Commonly Used Field Tests to Assess *Varroa destructor* Infestation on Honey Bee (*Apis mellifera*) Colonies. *Appl. Sci.* **2021**, *11*, 4458. [CrossRef]

insects

MDPI

Article

Evaluating the Efficacy of 30 Different Essential Oils against *Varroa destructor* and Honey Bee Workers (*Apis mellifera*)

Marian Hýbl [1], Andrea Bohatá [1], Iva Rádsetoulalová [2], Marek Kopecký [1], Irena Hoštičková [1], Alena Vaníčková [3] and Petr Mráz [1,*]

[1] Faculty of Agriculture, University of South Bohemia in Ceske Budejovice, Studentska 1668, 370 05 Ceske Budejovice, Czech Republic; mario.eko@seznam.cz (M.H.); bohata@zf.jcu.cz (A.B.); mkopecky@zf.jcu.cz (M.K.); jelini00@zf.jcu.cz (I.H.)

[2] Faculty of AgriSciences, Mendel University in Brno, Zemedelska 1, 613 00 Brno, Czech Republic; radsetoulalova.iva@gmail.com

[3] Aromaterapeutická KH a.s., Kšice 11, 349 01 Stříbro, Czech Republic; vanickova@karelhadek.eu

* Correspondence: mrazpe01@zf.jcu.cz

Simple Summary: Worldwide, mass losses of honey bee colonies are being observed more frequently due to Varroa mite infestation. Therefore, varroosis is considered a major problem in beekeeping participating to a large extent in colony collapse disorder. Except for direct damage of bees and suppressing their immune system caused by parasitism, Varroa mites transfer viral particles straight to bee hemolymph which can have a fatal impact. To control the mite population, several acaricidal treatments are used. Commonly used treatments are synthetic acaricides with a high risk of developing Varroa resistance population and contamination of bee products by acaricidal residues. Other commonly used treatments are organic acids, which are increasingly associated with damage of brood, adult bees, and premature deaths of queens. Therefore, in this study, we evaluated the varroacidal effect of 30 individual essential oils. The toxicity of the most effective oils selected by screening was subsequently tested on Varroa mites and adult honey bee workers simultaneously. In addition, the main components of these essential oils were specified. Several essential oils were proven to be effective against the adult female of Varroa mites and at the same dose safe for adult honey bee workers under laboratory conditions, especially manuka, peppermint, oregano, litsea, and cinnamon.

Citation: Hýbl, M.; Bohatá, A.; Rádsetoulalová, I.; Kopecký, M.; Hoštičková, I.; Vaníčková, A.; Mráz, P. Evaluating the Efficacy of 30 Different Essential Oils against *Varroa destructor* and Honey Bee Workers (*Apis mellifera*). *Insects* **2021**, *12*, 1045. https://doi.org/10.3390/insects12111045

Academic Editors: Ivana Tlak Gajger and Franco Mutinelli

Received: 2 November 2021
Accepted: 19 November 2021
Published: 21 November 2021

Abstract: Essential oils and their components are generally known for their acaricidal effects and are used as an alternative to control the population of the *Varroa destructor* instead of synthetic acaricides. However, for many essential oils, the exact acaricidal effect against Varroa mites, as well as the effect against honey bees, is not known. In this study, 30 different essential oils were screened by using a glass-vial residual bioassay. Essential oils showing varroacidal efficacy > 70% were tested by the complete exposure assay. A total of five bees and five mites were placed in the Petri dishes in five replications for each concentration of essential oil. Mite and bee mortality rates were assessed after 4, 24, 48, and 72 h. The LC_{50} values and selectivity ratio (SR) were calculated. For essential oils with the best selectivity ratio, their main components were detected and quantified by GC-MS/MS. The results suggest that the most suitable oils are peppermint and manuka (SR > 9), followed by oregano, litsea (SR > 5), carrot, and cinnamon (SR > 4). Additionally, these oils showed a trend of the increased value of selective ratio over time. All these oils seem to be better than thymol (SR < 3.2), which is commonly used in beekeeping practice. However, the possible use of these essential oils has yet to be verified in beekeeping practice.

Keywords: acaricidal effect; complete exposure bioassay; honey bee; screening; Varroa mite

1. Introduction

The main threat for beekeeping is a varroosis caused by the obscure ectoparasitic mite *Varroa destructor* Anderson and Trueman (Acari: Varroidae) [1,2]. The mite feeds on the fat body of bees [3] and thus reduces the weight and fitness of newly emerging adult bees, affects cuticle properties [4] and suppresses the immune response system [5]. In addition, *V. destructor* acts as a vector of viruses [6], including deformed wing virus, Kashmir bee virus, and Israeli acute paralysis virus [7–11]. These viruses are transmitted in large doses directly to the hemolymph of the bee brood and adult honey bees [5]. Infected individuals weaken, their lifespan is shorter, and the infection can lead through visible damaged bodies and wings [12,13] to the colony collapse at the final stage [1,6]. For these reasons, and also due to its almost worldwide distribution [1], *V. destructor* is associated with colony collapse disorder (CCD) [14,15].

Reproduction of the *V. destructor* mite is closely related and synchronized with the development of the bee brood [16]. Adult mated female mites enter the bee colony attached to worker and drone bees, usually hidden under the sternites of bees, and then enter brood cells only several hours before capping. Varroa mites can find the adult honey bee workers and bee brood before capping based on chemical communication [1]. In colonies highly infested (>7%) with *V. destructor* [1], the bee population is significantly reduced, and eventually, the entire colony crashes unless the mite population is treated [17]. Colonies in temperate areas must therefore be treated several times in a year against *V. destructor* to keep mite populations at acceptable levels [18].

For a long time, the use of synthetic chemicals has been considered the most effective way to control *V. destructor* [19], especially pyrethroids and organophosphates [20]. Except for their declining efficiency due to emerging resistance against *V. destructor* [21–23], excessive use of these compounds has, in many cases, also led to contamination of bee products [24–26], especially honey and beeswax [24]. This could endanger the health of bees and humans with potential sublethal doses of pesticide residue mixtures [27,28]. As a result, the idea of finding new and safer ways to control the parasite is spreading. Thus, natural products offer a very desirable alternative to synthetic products. Interest in these substances is still growing because they are generally cheap and have lower health risks for humans and bees [29].

In response, beekeepers are showing a growing interest in treatments that work on physical intolerance rather than enzyme degradation, as is the case with synthetic acaricides, to which resistance develops. Therefore, natural chemicals such as organic acids, essential oils, and their derivatives are increasingly used [30,31]. However, several studies suggest that the use of organic acids against Varroa may be harmful to bees. For example, damage and removal of open and capped brood are most commonly observed [32,33]. In addition, permanent damage to the digestive and excretory organs and glands of bees was described [34,35], as well as damage to the queen or often even premature death [36,37], or a decrease in the pH of honey during the following season [38].

Another possible way to reduce Varroa mites is essential oils (EO) [39]. According to The Commission of the European Pharmacopoeia, EOs are odorous products, usually with a complex composition, obtained from a botanically defined plant raw material by steam distillation, dry distillation, or a suitable mechanical process without heating. They are usually separated from the aqueous phase by a physical process that does not significantly affect their composition. EOs are lipophilic and may contain over 100 different plant secondary metabolites (terpenoids and phenylpropanoids, monoterpenes, sesquiterpenes, aldehydes, alcohols, etc.) [40]. Among natural substances, essential oils represent one of the most promising alternatives to synthetic chemicals [41–48], with minimal side effects [49–52]. The effectiveness of EOs against *V. destructor* is comparable to organic acids, but the application of EOs causes a lesser degree of stress in bees than the application of organic acids [29].

In addition to acaricidal effects, the application of EOs into hives often also causes antimicrobial effects, which can lead to an overall improvement in the health status of honey bee colonies [53]. Most research suggests that essential oils may be a useful alternative to maintaining a low level of mite infestation in hives [39,54–57]. Adamczyk et al. [58] concluded that the presence of residues of essential oil components in honey samples does not pose a hygiene risk or a risk to human health.

Despite the promising acaricidal effects of various EOs found in vitro [54,55,57], only a fraction of them has been tested under beehive conditions [39]. This could be the reason why EOs have not yet been included in many commercial formulations, with the exception of some cases [53].

The aim of the study was therefore to determine the acaricidal effect of a large number of selected EOs against Varroa mites, as well as their effect on honey bees in vitro, which select the most promising essential oils for the in vivo experiments. In addition, the most promising EOs were quantified for their major components.

2. Materials and Methods

2.1. Biological Material and Essential Oils

V. destructor mites and honey bees (*Apis mellifera*) used in this study were obtained from the experimental apiary of the Faculty of Agriculture, the University of South Bohemia in České Budějovice, (Czech Republic). To rear mites, 4 honey bee colonies were infested by Varroa mites and untreated for over 12 months. From the infested beehives, the bees were collected in a mesh container by sweeping from the brood frames and subsequently exposed to CO_2. After anesthesia of the bees, the vessel was closed and shaken until mites fell over the mesh bottom [59]. Thus, a sufficient number of adult vital female mites were collected. Mites showing signs of defect, newly molded, or poorly mobile were eliminated.

A total of 30 essential oils (EO) were obtained from company 1. Aromaterapeutická KH a.s. (Czech Republic). The list of EOs, their abbreviations, Latin names, and part of used plants are given in Table 1.

Table 1. The list of essential oils, abbreviations, Latin names, and part of the used plants.

English Name	Abbreviation	Latin Name	Part of Plant
Black pepper	PEP	*Piper nigrum*	berry
Blue chamomile	BCH	*Matricaria chamomilla*	flower
Carrot	CAT	*Daucus carota*	seeds
Cinnamon	CIN	*Cinnamomum zeylanicum*	bark
Clove Bud	CB	*Eugenia caryophyllata*	leaves, buds, and twigs
Copaiba	COP	*Copaifera reticulata*	resin
Coriander	COR	*Coriandrum sativum*	seeds
Fennel	FEN	*Foeniculum vulgare*	seeds
Ginger	GIN	*Zingiber officinale*	rhizome
Green cardamom	CAR	*Elettaria cardamomum*	seeds
Laurel	LAU	*Laurus nobilis*	leaves
Lavender	LAV	*Lavandula angustifolia*	flowering herb
Litsea	LIT	*Litsea cubeba*	fruits
Mace	MAC	*Myristica fragrans*	flower
Manuka	MAN	*Leptospermum scoparium*	leaves and twigs
Maroc chamomile	MCH	*Ormenis multicaulis*	herb
Nutmeg	NUT	*Myristica fragrans*	seeds
Oregano	ORG	*Origanum vulgare*	herb
Pelargonium	PEL	*Pelargonium graveolens*	leaves and flowers
Peppermint	PPM	*Mentha piperita*	herb
Ravensara	RAV	*Ravensara aromatica*	leaves and twigs
Roman chamomile	RCH	*Anthemis nobilis*	flower
Rosemary	ROS	*Rosmarinus officinalis*	herb
Sage	SAG	*Salvia officinalis*	leaves
Savory	SAV	*Satureja montana*	herb
Spearmint	SPM	*Mentha spicata crispa*	flowering herb

Table 1. *Cont.*

English Name	Abbreviation	Latin Name	Part of Plant
Thyme	TYM	*Thymus vulgaris*	herb
Turmeric	TUR	*Curcuma longa*	root
Wild thyme	WTYM	*Thymus serpyllum*	herb
Wormwood	WW	*Artemisa absinthium*	herb

2.2. Screening of Essential Oils for Their Acaricidal Activity

To evaluate EO acute toxicity on *V. destructor*, a glass-vial residual bioassay was used [60]. Each tested product was diluted in acetone (0.375 μL EO/500 μL acetone). This solution was pipetted into a 10 mL glass vial. Glass vials were rolled on their side until the acetone evaporated and EOs created a cohesive film. Then, 5 vital female adult mites were placed in each glass vial using a fine brush. The glass vials were sealed and placed in a dark room at 25 °C and 65% RH. For each treatment, including acetone as a negative control and thymol (THM) as a positive control; 5 repetitions were provided (each repetition in an individual glass vial).

The mortality rates of Varroa mites were evaluated 2 and 4 h after the treatment, and the efficacy of tested EOs was determined [55]. The mites were transferred to a white pad and encouraged to move with the brush. Mites that did not move even after repeated brushing were considered dead.

2.3. Complete Exposure Bioassay

EOs showing >70% mite mortality in the screening test were subjected to further testing in the complete exposure method [61]. Dosages of EOs were prepared based on the mortality of previous experiments with honey bees (data not included). A selected amount of EOs was diluted in 0.5 mL of acetone. This solution was pipetted on the bottom of the Petri dish and subsequently covered with filter paper (Whatman 1). After evaporation of the solvent, five vital adult honey bee workers were placed in each Petri dish, together with five vital female adult Varroa mites. Positive control (thymol) and negative control (acetone only) were included. Altogether, 5 replicates were established for each treatment (each repetition in an individual Petri dish). Immediately after the establishment, the Petri dishes were transferred to an incubator (28 °C ± 0.5). Honey bee and mite mortality were assessed after 4, 24, 48, and 72 h. The values of LC_{50} and selectivity ratio (SR) were calculated. SR is a ratio between mite and bee toxicity, and it was determined according to the following formula: $SR = LC_{50\ A.\ mellifera}/LC_{50\ V.\ destructor}$.

2.4. Assessment of the Main Components of the Examined EOs

Samples of essential oils were analyzed diluted 1:10,000 in hexane by GC MS/MS system consisting of TriPlus autosampler, Trace GC Ultra gas chromatograph equipped with a TG-5MS fused silica capillary column, 30 m × 0.25 mm × 0.25 μm and coupled to a mass spectrometer TSQ Quantum XLS all from Thermo Fischer Scientific, Cleveland, OH, USA. Helium was used as a carrier gas at 1.0 mL/min. A total of 1 μL of the sample was injected into the SSL injector in the splitless mode set at 280 °C. The oven temperature was programmed as follows: start at 40 °C and held for 5 min, then increased to 150 °C at a rate of 3 °C/min and held for 0.5 min, then increased to 250 °C at a rate of 10 °C/min, then increased to 290 °C at a rate of 25 °C, and finally maintained at 290 °C for 10 min. The temperature of the transfer line was held at 250 °C, and the ion source was operating at 200 °C. TIC mode was performed on Q1 at 70 eV of ionization energy and mass range 50–450 m/z. To exclude congestion of detector the scanning was performed after 6 min of injection. The data were processed in Thermo Xcalibur 3.0.63 (Thermo Fisher, Waltham, MA, USA). Component identification was made based on comparison with the NIST Mass Spectral Search Program library v 2.0 f (Thermo Fisher). The quantification was achieved based on Q3 SIM mode focused on fragmentation ions of desired compounds and also via an external calibration curve. The Thujone (Sigma Aldrich, St. Louis, MO, USA) was used as an internal and also external standard.

2.5. Statistical Analyses

Statistical analyses of the screening of essential oils, including graphical outputs, were processed in STATISTICA (version 14, TIBCO Software Inc., Palo Alto, CA, USA, 2021), specifically, the analysis of variance procedure ANOVA, preceded by a normality test. Statistical significance was tested at a level of significance = 0.05.

Probit analyses were calculated in XLSTAT (Addinsoft, 2016) incorporating natural mortality into the analyses. The concentration of essential oils was transformed logarithmically. LD_{50} with 95% confidence intervals ($p < 0.05$) were fitted.

The in vitro effect of each active substance on mortality of both Varroa mite and honey bees was analyzed by the test of hypothesis for two samples representing independent binomial experiments, and the acaricidal effects of active substances were subsequently evaluated (GenStat 17). Significant differences among substances were stated where $p \leq 0.05$.

3. Results

3.1. Screening of Essential Oils for Their Acaricidal Activity

All 30 EOs were screened for acaricidal effect in glass vials (Figure 1). Based on these results, the EOs were divided into three categories according to their efficacy. A total of 11 EOs showed a high acaricidal efficacy (>70%) and were further tested on Petri dishes (complete exposure assay) simultaneously with honey bees and mites. These were MAN, TYM, WTYM, ORG, SAV, CIN, CB, PPM, CAT, PEL, LIT, and THM as a positive control. The category of moderately effective oils (30–50%) includes ROS, RAV, TUR, RCH, LAV, CAR, PEP, and GIN. The last category with an efficiency of less than 30% includes NUT, FEN, MAC, BCH, MCH, COP, SAG, SPM, COR, LAU, and WW. Oils showing less than 70% efficacy were further tested.

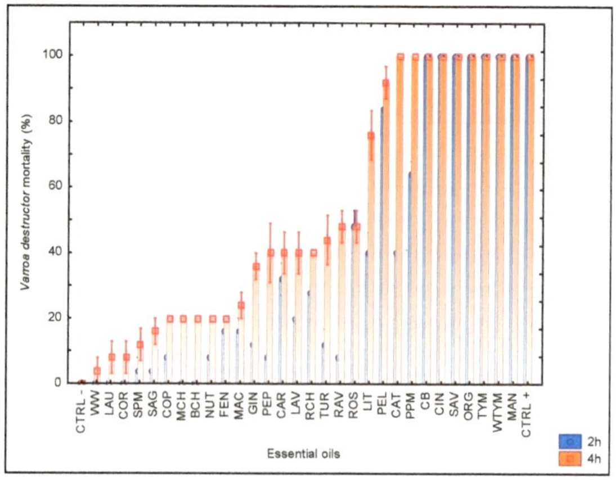

Figure 1. Mortality rates of *Varroa destructor* in glass vial bioassay after 2 and 4 h of EO exposition. The error bars denote standard deviation. Full name of each abbreviation is shown in Table 1.

3.2. Complete Exposure Bioassay

The complete exposure bioassay reveals that EO from MAN showed by far the lowest LC_{50} value against Varroa mites both after 4 and 72 h of exposure. EOs from TYM, ORG, and control THM also had a low LC_{50} value after 72 h of exposure. A moderate LC_{50} value after 72 h showed PPM, SAV, WTYM, CB, and CIN. EO from PEL and CAT showed a relatively high value. The lowest LC_{50} value for bees after 72 h of exposure had EOs from

MAN, TYM, and control THM, slightly high values had CB and ORG. In contrast, bees were most tolerant of EOs from CAT, PPM, LIT, and PEL (Table 2).

Table 2. Complete exposure bioassay. LC_{50} (µL) of essential oils on *V. destructor* and *A. mellifera* and their selectivity ratio in a monitored period. Green highlighting means low value of selectivity ratio (<3), yellow highlighting means moderate value of selectivity ratio (3–5), and red highlighting means high value of selectivity ratio (>5).

EO	Species	4 h			24 h			48 h			72 h		
		LC_{50}	95% CL		LC_{50}	95% CL		LC_{50}	95% CL		LC_{50}	95% CL	
THM	*V. destructor*	1.505	1.180	1.937	0.834	0.629	1.052	0.660	0.475	0.846	0.660	0.475	0.846
	A. mellifera	6.181	5.074	7.847	4.090	3.189	6.759	2.427	2.097	2.871	2.112	1.940	2.320
	Selectivity ratio	4.107			4.903			3.675			3.198		
CAT	*V. destructor*	10.449	6.806	34.882	4.167	2.457	6.630	3.276	1.930	4.590	2.539	1.187	3.653
	A. mellifera	18.607	13.845	64.136	13.048	9.588	27.137	11.557	8.855	19.447	11.557	8.855	19.447
	Selectivity ratio	1.781			3.131			3.527			4.552		
PPM	*V. destructor*	8.121	6.159	13.576	2.512	1.430	3.578	1.732	0.499	2.806	1.066	0.011	2.197
	A. mellifera	12.951	11.259	14.994	10.759	9.483	12.156	10.285	9.109	11.568	10.285	9.109	11.568
	Selectivity ratio	1.595			4.283			5.939			9.651		
SAV	*V. destructor*	3.825	3.165	4.918	2.008	1.323	2.754	1.459	0.626	2.075	1.364	0.417	1.996
	A. mellifera	11.657	9.247	16.335	5.786	4.607	7.218	5.275	4.273	6.467	4.621	3.884	5.897
	Selectivity ratio	3.048			2.881			3.615			3.386		
WTYM	*V. destructor*	8.185	5.218	22.390	2.549	1.495	8.207	2.013	0.926	7.327	1.861	0.825	5.487
	A. mellifera	9.074	8.106	10.780	7.517	6.606	8.265	6.512	5.958	7.494	6.250	5.603	6.897
	Selectivity ratio	1.109			2.949			3.236			3.358		
ORG	*V. destructor*	3.517	2.339	7.322	0.879	0.638	1.302	0.577	0.280	0.924	0.577	0.280	0.924
	A. mellifera	6.982	6.136	7.889	3.362	2.997	3.803	3.362	2.997	3.803	3.362	2.997	3.803
	Selectivity ratio	1.985			3.827			5.830			5.830		
PEL	*V. destructor*	2.798	0.113	4.758	2.291	0.247	3.825	2.402	0.804	3.532	2.272	0.788	3.311
	A. mellifera	17.122	13.427	27.935	12.401	10.159	17.209	9.479	8.132	11.201	9.479	8.132	11.201
	Selectivity ratio	6.120			5.413			3.945			4.171		
MAN	*V. destructor*	1.262	0.848	3.192	1.029	0.558	2.880	0.265	0.020	0.540	0.158	0.011	0.497
	A. mellifera	1.975	1.662	2.681	1.415	1.218	1.666	1.472	1.277	1.720	1.472	1.277	1.720
	Selectivity ratio	1.565			1.375			5.551			9.333		
LIT	*V. destructor*	4.801	3.522	7.436	2.716	1.322	4.311	2.116	0.243	3.761	1.807	0.243	2.989
	A. mellifera	11.660	9.524	15.222	11.590	8.994	20.096	9.207	7.721	12.115	9.678	7.255	18.278
	Selectivity ratio	2.429			4.267			4.352			5.354		
TYM	*V. destructor*	1.279	0.985	1.613	0.678	0.314	0.940	0.678	0.314	0.940	0.587	0.202	0.851
	A. mellifera	8.759	6.684	14.553	3.887	3.295	4.837	3.113	2.696	3.763	2.677	2.418	2.982
	Selectivity ratio	6.848			5.731			4.590			4.557		
CB	*V. destructor*	2.337	1.829	2.962	1.690	1.237	2.143	1.490	1.207	1.776	1.490	1.207	1.776
	A. mellifera	5.965	4.620	10.868	4.860	4.023	6.546	3.305	2.790	4.179	3.305	2.790	4.179
	Selectivity ratio	2.553			2.875			2.218			2.218		
CIN	*V. destructor*	4.321	3.163	5.979	2.820	1.577	4.002	2.529	1.370	3.590	1.543	0.829	2.484
	A. mellifera	10.635	9.559	11.972	7.488	6.680	8.408	7.488	6.680	8.408	7.007	5.835	8.664
	Selectivity ratio	2.461			2.655			2.960			4.542		

The selectivity ratio was calculated based on the LC_{50} values. The estimated LC_{50} values, including standard deviation obtained at each observation time and selectivity ratio for every treatment, are shown in Table 2. By far, the highest value of the selective ratio was reached after 72 h of exposure to EO from PPM (SR = 9.65) and MAN (SR = 9.33). ORG (SR = 5.83) and LIT (SR = 5.35) also reached high values at 72 h. All these four oils had an increasing SR value over time. In contrast, TYM and PEL oils had the highest SR value after 4 h of exposure (SR = 6.85; SR = 6.12). A significant decrease in this value was observed in the following measurements. Moderately high SR values were observed after 72 h of exposure in CAT, SAV, and WTYM, which showed an increasing tendency of SR value in time (SR = 4.55; SR = 3.39; SR = 3.36). A moderate-to-high SR value was also observed in THM (positive control). After 4 h of experiment, THM showed even one of the highest SR values (SR = 4.11), however, with a declining trend of SR values in time. A constantly low value of SR was observed with CB, as in each measurement SR was less than 3. Similarly, CIN also had a low value of SR, and with the exception of the last measurement after 72 h of exposure, the level of SR increased significantly (SR = 4.54).

The main components and their quantity of the most effective EOs were assessed (Table 3). The most frequent substances were carvacrol and p-cymene.

Table 3. Composition of the most effective essential oils and their constituents' quantity (>5%).

EO	Main Components and Their Quantity (%)				
Carrot	Ceratol 30.28	α-Pinen 15.462	Sabinen 10.22	β- Caryophyllen 8.31	β-bisabolen 5.63
Peppermint	Limonen 38.02	Menthol 16.41	α-Pinen 15.92	β-Pinen 11.46	Menthon 5.65
Savory	Carvacrol 41.67	ɣ-Terpinen 35.82	p-Cymen 11.73	-	-
Wild thyme	Thymol 16.33	Carvacrol 15.38	p-Cymen 15.01	Geraniol 10.62	ɣ-Terpinen 10.30
Oregano	Carvacrol 73.50	p-Cymen 6.97	ɣ-Terpinen 6.02	-	-
Pelargonium	Citronellol 33.51	Geraniol 15.36	Citronellylformiat 7.81	Isomenthon 5.61	10-epi-g-Eudesmol 5.37
Manuka	Calamenene 17.92	Leptospermone 16.02	Flaveson 5.95	α-Selinene 4.62	-
Litsea	Citral A 39.03	Citral B 29.35	Limonen 13.74	-	-
Thyme	Thymol 40.96	p-Cymen 16.76	-	-	-
Clove Bud	Eugenol 86.62	β-Caryophyllen 10.21	-	-	-
Cinnamon	trans-Cinnamaldehyde 77.69	Eugenol 7.50	-	-	-

4. Discussion

Investigation of the acaricidal activity of essential oils is a major concern of many scientific studies. However, large-scale screening of a number of EOs is rare, and most of the effort is devoted to an individual or a small number of selected oils, such as thyme, clove bud, or oregano [55,56,62]. In this study, the acaricidal effect of 30 EOs on *V. destructor* mites was assessed by the glass vials bioassay (Figure 1), which represents a simple and quick way to determine the effectiveness of individual EOs [60].

Thymol, as a derivate of thyme, was included in the screening as a positive control, as it is commonly used in beekeeping practice as an acaricide [63]. However, thymol could have some negative effects on bees, including toxicity on bee brood, metabolic disorders, changes in bee's behaviors, etc. [64–71].

In the experiment, after 4 h of exposure, all EOs showed either the same or higher acaricidal effect than after 2 h. Based on the results of mortality after 4 h of exposure, the individual EOs were divided into three categories according to their effectiveness: highly effective, moderately effective, and minimally effective. The oils in the highly effective group, including MAN, WTYM, TYM, ORG, SAV, CIN, CB, PPM, CAR, PEL, and LIT, were further tested. Almost all oils in this group were able to kill 100% of mites after 2 h, with the exception of PPM, CAR, PEL, and LIT. The EOs from the moderately effective group have still the potential to participate in the mite control; however, a higher dose or applying a certain mixture showing a stronger synergistic effect would be needed. From the group of moderately effective EOs, the best acaricidal activity belonged to ROS, RAV, and TUR. The oils from the minimally effective group showed a very low varroacidal effect, and therefore, they were not suitable for further testing. Especially WW, LAU, and COR appear to be ineffective.

The 11 EOs from the highly effective group were further tested in order to determine the most suitable EOs for the best potential use in beekeeping practice. In addition to mite toxicity, the bee tolerance was necessary to be evaluated. Therefore, the method of complete exposure assay [61] was chosen, which allows the evaluation of selectivity ratio (SR), the most telling data for this purpose, in addition to LC for mites and bees [57].

In the complete exposure bioassay, after 4 h of exposure, only MAN and TYM showed a higher level of mite toxicity than THM (control). After 72 h of exposure at the end of the experiment, MAN, TYM, and ORG showed higher mite toxicity. The higher degree of toxicity of the above-mentioned EOs, compared with THM, is probably due to the content of other active substances (carvacrol, p-cymene, calamenene, leptospermone), which can additionally act synergistically [56,72]. While the varroacidal effect has already been described for TYM and ORG [55,56,62], for MAN, it has not been described yet. However, its antimicrobial and also acaricidal effects against other mite species (*Dermatophagoides* and *Tyrophagus*) have been observed [73,74]. Regarding bee toxicity, only EOs from CB and MAN were more toxic than THM after 4 h of exposure. After 72 h, at the end of the experiment, a higher degree of toxicity was observed only in EO from MAN. Thus, the results indicate higher toxicity of THM to Varroa mites but also to honey bees [55,57].

The ratio between mite and bee toxicity is defined as selectivity ratio (SR) values. At the beginning of the experiment, after 4 h of exposure, THM showed an SR value of 4.107, which was better than most EOs tested. Higher SR value was observed only at PEL (6.120) and TYM (6.848). However, with the duration of exposure, the SR value of THM decreased. After 72 h of exposure, the value was only 3.198. This can be explained by a decrease in mite toxicity, an increase in bee toxicity, or a combination of both in time. [64,75]. A similar trend was observed for PEL and TYM. In both EOs, the SR value also decreased with the duration of exposure; however, in both EOs, the SR value was always higher, compared with THM. This declining trend in the SR value with increasing exposure time for TYM is consistent with the results of Damiani et al. [62] and is probably due to the high thymol content that is characteristic of thyme [76]. This declining trend in the SR value indicates the potential unsuitability of EOs with these properties, and these EOs need to be subjected to further testing.

Stable to slightly fluctuating development of SR values depending on the duration of exposure was observed at SAV and CB. The initial values at the beginning of the measurement were very similar to the values at the end of the experiment and do not change significantly during the experiment. However, the SR value was significantly lower in CB than in THM, which is in accordance with the results of Damiani et al. [62], and in the case of SAV, the SR values are similar to THM. In the other tested EOs, an opposite trend was observed, and the SR increased with the time of exposure.

The best SR value after 72 h was determined at EOs from PPM (SR = 9.651) and MAN (SR = 9.333), followed by ORG (SR = 5.830) and LIT (SR = 5.354). From the results of Nazer and Al-Abbadi [77], it seems EO from PPM is more suitable to control varroosis than THM in vivo. The same conclusion can be drawn from the results of Damiani et al. [62] in the case of ORG in in vitro conditions. There is still a lack of varroacidal data from MAN and LIT in the literature; however, a strong antimicrobial effect against *Clostridium*, *Bifidobacterium*, *Escherichia*, *Staphylococcus*, *Lactobacillus*, and an acaricidal effect against *Dermatophagoides* and *Tyrophagus* is known for both EOs [73,74,78].

A very good result after 72 h was also observed at EOs from CIN (SR = 4.542) and CAT (SR = 4.552). CIN is proposed as a suitable option for reducing the population of *V. destructor*. In addition, CIN has a strong repellent effect on *V. destructor* mites and is also gentle on bees [39]. The suitability of CAT for further testing in the beehive conditions is also proved by its strong inhibitory effect against *Ascosphera apis* and *Paenibacillus larvae* [79,80]. A slightly lower SR value, but still higher than THM, was observed in EO from WTYM (SR = 3.358).

Since the chemical composition of EOs is influenced by many factors (geographical origin, part of the plant, agrotechnics, genotype, extraction technology, etc.), it is necessary to know their composition to interpret the effect of individual EOs [78].

According to SR of EOs from PPM and MAN, they seem to be the most promising oils against *V. destructor*. The most represented substances in PPM were limonene, menthol, and α-pinene. Limonene has been shown to be effective in reducing the population of *V. destructor* at a colony level [81] and has strong antimicrobial effects [82]. Varoacidal [55] and antimicrobial effects have also been reported for menthol [83], whereas α-pinene is known for its inhibitory effects on bacteria [84]. In addition, it can also be produced in larger quantities by genetically modified bacteria [85]. In the case of MAN, calamenene and leptospermone were the most abundant constituents. Celemonene-containing oils show high antimicrobial and fungicidal activity and are effective against a wide range of pathogens, including methicillin-resistant *Staphylococcus aureus* (MRSA) strains, and also have high antioxidant activity [86]. Leptospermone is known for its bactericidal, antiviral, and acaricidal effects [73,74].

Other EOs with suitable results were LIT and ORG. The main components of LIT were citral (A and B) and limonene. The findings of Liu et al. [87] agree with ours that citral is the main component of litsea and has a strong aroma and strong antimicrobial effects [88] against both, gram-positive and gram-negativebacteria [78]. At ORG, carvacrol was absolutely dominant and is known for its significant acaricidal and antimicrobial effects. In addition, it also has anti-inflammatory and antimutagenic, and antigenotoxic effects [89]. CIN and CAT also showed a significant acaricidal effect. The main component of CIN was cinnamaldehyde, to which Conti et al. [39] attributed the main varroacidal effects in cinnamon EO. It also has antibacterial effects [90]. Ceratol and α-pinen were predominant in CAT. The last EO with better results than THM was WTYM, with an almost balanced representation of thymol, carvacrol, and p-cymene.

5. Conclusions

The results based on selectivity ratio (SR) value for individual EOs showed that potential best EOs for Varroa control are PPM and MAN, followed by ORG and LIT. Other suitable candidates seem to be CAT, SAV, WTYM, and CIN. All these oils showed better SR values at the end of the experiment than THM (control group), which is used in beekeeping practice. Additionally, these oils showed a trend of an increased value of the selective ratio.

Thymol showed very good SR at the beginning of the experiment, but this value declined with all following measurements. At the end of the experiment, the SR value was lower than the values of most tested essential oils. This trend was also observed in EOs from PEL and TYM.

Except for well-known substances such as thymol, menthol, and carvacrol, other components appear to be potentially interesting for the control of Varroa, especially citral, limonene, calamenene, leptospermone, p-cymene, and cinnamaldehyde, as the main compounds of the most effective EOs.

Author Contributions: Conceptualization, M.H., M.K. and P.M.; data curation, A.B., I.R. and I.H.; investigation, M.K., I.H., A.V. and P.M.; methodology, M.H.; supervision, A.B. and A.V.; writing—original draft preparation, M.H. and P.M. All authors have read and agreed to the published version of the manuscript.

Funding: This research received no external funding.

Institutional Review Board Statement: Not applicable.

Informed Consent Statement: Not applicable.

Data Availability Statement: The study did not report any data.

Conflicts of Interest: The authors declare no conflict of interest.

References

1. Rosenkranz, P.; Aumeier, P.; Ziegelmann, B. Biology and control of *Varroa destructor*. *J. Invertebr. Pathol.* **2010**, *103*, S96–S119. [CrossRef]
2. Sammataro, D.; Gerson, U.; Needham, G. Parasitic mites of honey bees: Life history, implications, and impact. *Annu. Rev. Entomol.* **2000**, *45*, 519–548. [CrossRef] [PubMed]
3. Ramsey, S.D.; Ochoa, R.; Bauchan, G.; Gulbronson, C.; Mowery, J.D.; Cohen, A.; Lim, D.; Joklik, J.; Cicero, J.M.; Ellis, J.D. *Varroa destructor* feeds primarily on honey bee fat body tissue and not hemolymph. *Proc. Natl. Acad. Sci. USA* **2019**, *116*, 1792–1801. [CrossRef]
4. Bowen-Walker, P.L.; Gunn, A. The effect of the ectoparasitic mite, *Varroa destructor* on adult worker honeybee (*Apis mellifera*) emergence weights, water, protein, carbohydrate, and lipid levels. *Entomol. Exp. Appl.* **2001**, *101*, 207–217. [CrossRef]
5. Yang, X.; Cox-Foster, D.L. Impact of an ectoparasite on the immunity and pathology of an invertebrate: Evidence for host immunosuppression and viral amplification. *Proc. Natl. Acad. Sci. USA* **2005**, *102*, 7470–7475. [CrossRef] [PubMed]

6. Le Conte, Y.; Ellis, M.; Ritter, W. Varroa mites and honey bee health: Can varroa explain part of the colony losses? *Apidologie* **2010**, *41*, 353–363. [CrossRef]
7. Bowen-Walker, P.; Martin, S.; Gunn, A. The transmission of deformed wing virus between honeybees (*Apis mellifera* L.) by the ectoparasitic mite *Varroa jacobsoni* Oud. *J. Invertebr. Pathol.* **1999**, *73*, 101–106. [CrossRef]
8. Chen, Y.; Zhao, Y.; Hammond, J.; Hsu, H.; Evans, J.; Feldlaufer, M. Multiple virus infections in the honey bee and genome divergence of honey bee viruses. *J. Invertebr. Pathol.* **2004**, *87*, 84–93. [CrossRef]
9. Di Prisco, G.; Pennacchio, F.; Caprio, E.; Boncristiani, H.F., Jr.; Evans, J.D.; Chen, Y. *Varroa destructor* is an effective vector of Israeli acute paralysis virus in the honeybee, *Apis mellifera*. *J. Gen. Virol.* **2011**, *92*, 151–155. [CrossRef]
10. Martin, S.J.; Highfield, A.C.; Brettell, L.; Villalobos, E.M.; Budge, G.E.; Powell, M.; Nikaido, S.; Schroeder, D.C. Global honey bee viral landscape altered by a parasitic mite. *Science* **2012**, *336*, 1304–1306. [CrossRef]
11. Shen, M.; Cui, L.; Ostiguy, N.; Cox-Foster, D. Intricate transmission routes and interactions between picorna-like viruses (Kashmir Bee Virus and Sacbrood Virus) with the honeybee host and the parasitic varroa mite. *J. Gen. Virol.* **2005**, *86*, 2281–2289. [CrossRef] [PubMed]
12. Genersch, E. Honey bee pathology: Current threats to honey bees and beekeeping. *Appl. Microbiol. Biotechnol.* **2010**, *87*, 87–97. [CrossRef]
13. Meixner, M.D. A historical review of managed honey bee populations in Europe and the United States and the factors That may affect them. *J. Invertebr. Pathol.* **2010**, *103*, S80–S95.
14. Annoscia, D.; Del Piccolo, F.; Nazzi, F. How does the mite *Varroa destructor* kill the honeybee *Apis mellifera*? Alteration of cuticular hydrcarbons and water loss in infested honeybees. *J. Insect Physiol.* **2012**, *58*, 1548–1555. [CrossRef]
15. Potts, S.G.; Biesmeijer, J.C.; Kremen, C.; Neumann, P.; Schweiger, O.; Kunin, W.E. Global pollinator declines: Trends, impacts and drivers. *Trends Ecol. Evol.* **2010**, *25*, 345–353. [CrossRef] [PubMed]
16. Martin, S.J. Ontogenesis of the mite *Varroa jacobsoni* Oud. in worker brood of the honeybee *Apis mellifera* L. under natural conditions. *Exp. Appl. Acarol.* **1994**, *18*, 87–100. [CrossRef]
17. Murilhas, A.M. *Varroa destructor* infestation impact on *Apis mellifera* Carnica capped worker brood production, bee population and honey storage in a Mediterranean climate. *Apidologie* **2002**, *33*, 271–281. [CrossRef]
18. Delaplane, K.S.; Hood, W.M. Effects of delayed acaricide treatment in honey bee colonies parasitized by *Varroa jacobsoni* and a late-season treatment threshold for the South-Eastern USA. *J. Apic. Res.* **1997**, *36*, 125–132. [CrossRef]
19. Meikle, W.G.; Sammataro, D.; Neumann, P.; Pflugfelder, J. Challenges for developing pathogen-based biopesticides against *Varroa destructor* (Mesostigmata: Varroidae). *Apidologie* **2012**, *43*, 501–514. [CrossRef]
20. Calderone, N.W.; Shimanuki, H.; Allen-Wardell, G. An in vitro evaluation of botanical compounds for the control of the honeybee pathogens *Bacillus larvae* and *Ascosphaera apis*, and the secondary invader *B. Alvei*. *J. Essent. Oil Res.* **1994**, *6*, 279–287. [CrossRef]
21. Cakmak, I.; Aydin, L.; Gulegen, E.; Wells, H. Varroa (*Varroa destructor*) and tracheal mite (*Acarapis woodi*) incidence in the Republic of Turkey. *J. Apic. Res.* **2003**, *42*, 57–60. [CrossRef]
22. Price, K.L.; Lummis, S.C. An atypical residue in the pore of *Varroa destructor* GABA-activated RDL receptors affects picrotoxin block and thymol modulation. *Insect Biochem. Mol. Biol.* **2014**, *55*, 19–25. [CrossRef] [PubMed]
23. Rinkevich, F.D. Detection of amitraz resistance and reduced treatment efficacy in the varroa mite, *Varroa destructor*, within commercial beekeeping operations. *PLoS ONE* **2020**, *15*, e0227264. [CrossRef]
24. Wallner, K. Varroacides and their residues in bee products. *Apidologie* **1999**, *30*, 235–248. [CrossRef]
25. Kanga, L.H.; Jones, W.A.; Gracia, C. Efficacy of strips coated with *Metarhizium anisopliae* for control of *Varroa destructor* (Acari: Varroidae) in honey bee colonies in Texas and Florida. *Exp. Appl. Acarol.* **2006**, *40*, 249–258. [CrossRef] [PubMed]
26. Milani, N. The resistance of *Varroa jacobsoni* Oud. to acaricides. *Apidologie* **1999**, *30*, 229–234. [CrossRef]
27. Akyol, E.; Özkök, D. The use of organic acids for varroa (*Varroa destructor*) control. *Uludag Bee J.* **2005**, *5*, 167–174.
28. Aydın, L. Control of honey bee diseases and pets in autumn. *Uludag Bee J.* **2005**, *5*, 159–161.
29. Isman, M.B. Plant essential oils for pest and disease management. *Crop Prot.* **2000**, *19*, 603–608. [CrossRef]
30. Emsen, B.; Dodoloğlu, A.; Genç, F. Natural compounds for the control of *Varroa destructor* in honey bees (*Apis mellifera* L.) and an economic assessment of these products. *Mellifera* **2010**, *10*, 32–36.
31. Bahreini, R. Comparison of two methods of applying oxalic acid for control of varroa. *J. Apic. Res.* **2003**, *42*, 82–83. [CrossRef]
32. Hatjina, F.; Haristos, L. Indirect effects of oxalic acid administered by trickling method on honey bee brood. *J. Apic. Res.* **2005**, *44*, 172–174. [CrossRef]
33. Gregorc, A.; Škerl, M.I.S. Toxicological and immunohistochemical testing of honeybees after oxalic acid and rotenone treatments. *Apidologie* **2007**, *38*, 296–305. [CrossRef]
34. Hernández, R.M.; Pascual, M.H.; Pérez, J.; del Nozal Nalda, M.J.; Meana, A. Short term negative effect of oxalic acid in "*Apis mellifera* Iberiensis". *Span. J. Agric. Res.* **2007**, *5*, 474–480. [CrossRef]
35. Silva-Zacarin, E.C.; Gregorc, A.; de Moraes, R.L.S. In situ localization of heat-shock proteins and cell death labelling in the salivary gland of acaricide-treated honeybee larvae. *Apidologie* **2006**, *37*, 507–516. [CrossRef]
36. Underwood, R.; Currie, R. Use of formic acid to control varroa and tracheal mites in indoor overwintering facilities. *Am. Bee J.* **2003**, *143*, 323.
37. Higes, M.; Meana, A.; Suárez, M.; Llorente, J. Negative long-term effects on bee colonies treated with oxalic acid against *Varroa jacobsoni* Oud. *Apidologie* **1999**, *30*, 289–292. [CrossRef]

38. Borsuk, G.; Olszewski, K.; Paleolog, J.; Strachecka, A.; Gryzińska, M. The effect of different varroacides on the acidity of winter stores and honey stores. *Ann. UMCS Zootech.* **2012**, *30*, 11–16. [CrossRef]
39. Conti, B.; Bocchino, R.; Cosci, F.; Ascrizzi, R.; Flamini, G.; Bedini, S. Essential oils against *Varroa destructor*: A soft way to fight the parasitic mite of *Apis mellifera*. *J. Apic. Res.* **2020**, *59*, 774–782. [CrossRef]
40. Carson, C.F.; Hammer, K.A. Chemistry and bioactivity of essential oils. *Lipids Essent. Oils Antimicrob. Agents* **2011**, *25*, 203–238.
41. Athanassiou, C.G.; Rani, P.U.; Kavallieratos, N.G. The use of plant extracts for stored product protection. In *Advances in Plant Biopesticides*; Springer: Berlin/Heidelberg, Germany, 2014; pp. 131–147.
42. Bedini, S.; Flamini, G.; Girardi, J.; Cosci, F.; Conti, B. Not just for beer: Evaluation of spent hops (*Humulus lupulus* L.) as a source of eco-friendly repellents for insect pests of stored foods. *J. Pest Sci.* **2015**, *88*, 583–592. [CrossRef]
43. Bedini, S.; Cosci, F.; Tani, C.; Pierattini, E.C.; Venturi, F.; Lucchi, A.; Ioriatti, C.; Ascrizzi, R.; Flamini, G.; Ferroni, G. Essential oils as post-harvest crop protectants against the fruit fly *Drosophila suzukii*: Bioactivity and organoleptic profile. *Insects* **2020**, *11*, 508. [CrossRef]
44. Bedini, S.; Flamini, G.; Cosci, F.; Ascrizzi, R.; Echeverria, M.C.; Guidi, L.; Landi, M.; Lucchi, A.; Conti, B. *Artemisia* spp. essential oils against the disease-carrying blowfly *Calliphora vomitoria*. *Parasites Vectors* **2017**, *10*, 1–10. [CrossRef]
45. Omar, A.; Zayed, M. Bioactivity impact of essential oils *Allium sativum* L. and *Citrus reticulata* L. against stored product insects *Tribolium castaneum* (Herbst) and *Rhyzopertha dominica* (F.). *J. Plant Prot. Pathol.* **2021**, *12*, 465–471.
46. Ebadollahi, A.; Ziaee, M.; Palla, F. Essential oils extracted from different species of the lamiaceae plant family as prospective bioagents against several detrimental pests. *Molecules* **2020**, *25*, 1556. [CrossRef]
47. Bedini, S.; Bougherra, H.H.; Flamini, G.; Cosci, F.; Belhamel, K.; Ascrizzi, R.; Conti, B. Repellency of anethole-and estragole-type fennel essential oils against stored grain pests: The different twins. *Bull. Insectol.* **2016**, *69*, 149–157.
48. Conti, B.; Benelli, G.; Leonardi, M.; Afifi, F.U.; Cervelli, C.; Profeti, R.; Pistelli, L.; Canale, A. Repellent effect of *Salvia dorisiana*, *S. longifolia*, and *S. sclarea* (Lamiaceae) essential oils against the mosquito *Aedes albopictus* Skuse (Diptera: Culicidae). *Parasitol. Res.* **2012**, *111*, 291–299. [CrossRef] [PubMed]
49. Lima, T.C.; da Silva, T.K.M.; Silva, F.L.; Barbosa-Filho, J.M.; Marques, M.O.M.; Santos, R.L.C.; de Holanda Cavalcanti, S.C.; de Sousa, D.P. Larvicidal activity of *Mentha* x *villosa* Hudson essential oil, rotundifolone and derivatives. *Chemosphere* **2014**, *104*, 37–43. [CrossRef]
50. Benelli, G.; Bedini, S.; Cosci, F.; Toniolo, C.; Conti, B.; Nicoletti, M. Larvicidal and ovideterrent properties of neem oil and fractions against the filariasis vector *Aedes albopictus* (Diptera: Culicidae): A bioactivity survey across production sites. *Parasitol. Res.* **2015**, *114*, 227–236. [CrossRef] [PubMed]
51. Regnault-Roger, C.; Vincent, C.; Arnason, J.T. Essential oils in insect control: Low-risk products in a high-stakes world. *Annu. Rev. Entomol.* **2012**, *57*, 405–424. [CrossRef]
52. Rajendran, S.; Sriranjini, V. Plant products as fumigants for stored-product insect control. *J. Stored Prod. Res.* **2008**, *44*, 126–135. [CrossRef]
53. Eguaras, M.J.; Fuselli, S.; Gende, L.; Fritz, R.; Ruffinengo, S.R.; Clemente, G.; Gonzalez, A.; Bailac, P.N.; Ponzi, M.I. An in vitro evaluation of *Tagetes minuta* essential oil for the control of the honeybee pathogens *Paenibacillus larvae* and *Ascosphaera apis*, and the parasitic mite *Varroa destructor*. *J. Essent. Oil Res.* **2005**, *17*, 336–340. [CrossRef]
54. Imdorf, A.; Bogdanov, S.; Ochoa, R.I.; Calderone, N.W. Use of essential oils for the control of *Varroa jacobsoni* Oud. in honey bee colonies. *Apidologie* **1999**, *30*, 209–228. [CrossRef]
55. Gashout, H.A.; Guzmán-Novoa, E. Acute toxicity of essential oils and other natural compounds to the parasitic mite, *Varroa destructor*, and to larval and adult worker honey bees (*Apis mellifera* L.). *J. Apic. Res.* **2009**, *48*, 263–269. [CrossRef]
56. Ramzi, H.; Ismaili, M.R.; Aberchane, M.; Zaanoun, S. Chemical characterization and acaricidal activity of *Thymus satureioides* C. & B. and *Origanum elongatum* E. & M. (*Lamiaceae*) essential oils against *Varroa destructor* Anderson & Trueman (Acari: Varroidae). *Ind. Crops Prod.* **2017**, *108*, 201–207.
57. Lindberg, C.M.; Melathopoulos, A.P.; Winston, M.L. Laboratory evaluation of miticides to csontrol *Varroa jacobsoni* (Acari: Varroidae), a honey bee (Hymenoptera: Apidae) parasite. *J. Econ. Entomol.* **2000**, *93*, 189–198. [CrossRef]
58. Adamczyk, S.; Lázaro, R.; Pérez-Arquillué, C.; Conchello, P.; Herrera, A. Evaluation of residues of essential oil components in honey after different anti-varroa treatments. *J. Agric. Food Chem.* **2005**, *53*, 10085–10090. [CrossRef] [PubMed]
59. Ariana, A.; Ebadi, R.; Tahmasebi, G. Laboratory evaluation of some plant essences to control *Varroa destructor* (Acari: Varroidae). *Exp. Appl. Acarol.* **2002**, *27*, 319–327. [CrossRef]
60. Aliano, N.P.; Ellis, M.D.; Siegfried, B.D. Acute contact toxicity of oxalic acid to *Varroa destructor* (Acari: Varroidae) and their *Apis mellifera* (Hymenoptera: Apidae) hosts in laboratory bioassays. *J. Econ. Entomol.* **2006**, *99*, 1579–1582. [CrossRef] [PubMed]
61. Ruffinengo, S.; Eguaras, M.; Floris, I.; Faverin, C.; Bailac, P.; Ponzi, M. LD50 and repellent effects of essential oils from Argentinian wild plant species on *Varroa destructor*. *J. Econ. Entomol.* **2005**, *98*, 651–655. [CrossRef] [PubMed]
62. Damiani, N.; Gende, L.B.; Bailac, P.; Marcangeli, J.A.; Eguaras, M.J. Acaricidal and insecticidal activity of essential oils on *Varroa destructor* (Acari: Varroidae) and *Apis mellifera* (Hymenoptera: Apidae). *Parasitol. Res.* **2009**, *106*, 145–152. [CrossRef] [PubMed]
63. Tihelka, E. Effects of synthetic and organic acaricides on honey bee health: A review. *Slov. Veter.-Res.* **2018**, *55*, 114–140. [CrossRef]
64. Charpentier, G.; Vidau, C.; Ferdy, J.; Tabart, J.; Vetillard, A. Lethal and sub-lethal effects of thymol on honeybee (*Apis mellifera*) larvae reared in vitro. *Pest Manag. Sci.* **2014**, *70*, 140–147. [CrossRef] [PubMed]

65. Boncristiani, H.; Underwood, R.; Schwarz, R.; Evans, J.D.; Pettis, J. Direct effect of acaricides on pathogen loads and gene expression levels in honey bees *Apis mellifera*. *J. Insect Physiol.* **2012**, *58*, 613–620. [CrossRef]
66. Bogdanov, S.; Imdorf, A.; Kilchenmann, V. Residues in wax and honey after Apilife VAR® treatment. *Apidologie* **1998**, *29*, 513–524. [CrossRef]
67. Floris, I.; Satta, A.; Cabras, P.; Garau, V.L.; Angioni, A. Comparison between two thymol formulations in the control of *Varroa destructor*: Effectiveness, persistence, and residues. *J. Econ. Entomol.* **2004**, *97*, 187–191. [CrossRef]
68. Tananaki, C.; Goras, G.; Huggett, N.; Karazafiris, E.; Dimou, M.; Thrasyvoulou, A. Evaluation of the impact of Exomite Pro™ on varroa mite (*Varroa destructor*) populations and honeybee (*Apis mellifera*) colonies: Efficacy, side effects and residues. *Parasitol. Res.* **2014**, *113*, 1251–1259. [CrossRef]
69. Ares, A.M.; Nozal, M.J.; Bernal, J.L.; Bernal, J. Simultaneous determination of carvacrol and thymol in bee pollen by using a simple and efficient solvent extraction method and gas chromatography-mass spectrometry. *J. Pharm. Biomed. Anal.* **2020**, *181*, 113124. [CrossRef]
70. Bergougnoux, M.; Treilhou, M.; Armengaud, C. Exposure to thymol decreased phototactic behaviour in the honeybee (*Apis mellifera*) in laboratory conditions. *Apidologie* **2013**, *44*, 82–89. [CrossRef]
71. Colin, T.; Lim, M.Y.; Quarrell, S.R.; Allen, G.R.; Barron, A.B. Effects of thymol on European honey bee hygienic behaviour. *Apidologie* **2019**, *50*, 141–152. [CrossRef]
72. Gunes, N.; Aydın, L.; Belenli, D.; Hranitz, J.M.; Mengilig, S.; Selova, S. Stress responses of honey bees to organic acid and essential oil treatments against varroa mites. *J. Apic. Res.* **2017**, *56*, 175–181. [CrossRef]
73. Jeong, E.-Y.; Jeon, J.-H.; Kim, H.-W.; Kim, M.-G.; Lee, H.-S. Antimicrobial activity of leptospermone and its derivatives against human intestinal bacteria. *Food Chem.* **2009**, *115*, 1401–1404. [CrossRef]
74. Jeong, E.; Kim, M.; Lee, H. Acaricidal activity of triketone analogues derived from *Leptospermum scoparium* oil against house-dust and stored-food mites. *Pest Manag. Sci.* **2009**, *65*, 327–331. [CrossRef]
75. Mattila, H.; Otis, G.; Daley, J.; Schulz, T. Trials of apiguard, a thymol-based miticide part 2. Non-target effects on honey bees. *Am. Bee J.* **2000**, *140*, 68–70.
76. Zambonelli, A.; D'Aulerio, A.Z.; Severi, A.; Benvenuti, S.; Maggi, L.; Bianchi, A. Chemical composition and fungicidal activity of commercial essential oils of *Thymus vulgaris* L. *J. Essent. Oil Res.* **2004**, *16*, 69–74. [CrossRef]
77. Nazer, I.; Al-Abbadi, A. Control of varroa mite (*Varroa destructor*) on honeybees by aromatic oils and plant materials. *J. Agric. Mar. Sci. [JAMS]* **2003**, *8*, 15–20. [CrossRef]
78. Thielmann, J.; Muranyi, P.; Kazman, P. Screening essential oils for their antimicrobial activities against the foodborne pathogenic bacteria *Escherichia coli* and *Staphylococcus aureus*. *Heliyon* **2019**, *5*, e01860. [CrossRef]
79. Ansari, M.J.; Al-Ghamdi, A.; Usmani, S.; Khan, K.A.; Alqarni, A.S.; Kaur, M.; Al-Waili, N. In vitro evaluation of the effects of some plant essential oils on *Ascosphaera apis*, the causative agent of chalkbrood disease. *Saudi J. Biol. Sci.* **2017**, *24*, 1001–1006. [CrossRef]
80. Ansari, M.J.; Al-Ghamdi, A.; Usmani, S.; Al-Waili, N.; Nuru, A.; Sharma, D.; Khan, K.A.; Kaur, M.; Omer, M. In vitro evaluation of the effects of some plant essential oils on *Paenibacillus larvae*, the causative agent of American foulbrood. *Biotechnol. Biotechnol. Equip.* **2016**, *30*, 49–55. [CrossRef]
81. Choi, Y.; Woo, S.; Hong, I.; Han, S.; Byoun, G.; Thapa, R.; Lee, M. Effects of limonene composition for controlling of the *Varroa destructor* and *Tropilaelaps clareae* in *Apis mellifera* Hives. *Korean J. Apic.* **2012**, *27*, 137–142.
82. van Vuuren, S.; Viljoen, A.M. Antimicrobial activity of limonene enantiomers and 1, 8-cineole alone and in combination. *Flavour Fragr. J.* **2007**, *22*, 540–544. [CrossRef]
83. Pattnaik, S.; Subramanyam, V.R.; Bapaji, M.; Kole, C.R. Antibacterial and antifungal activity of aromatic constituents of Essential oils. *Microbios* **1997**, *89*, 39–46.
84. Leite, A.M.; Lima, E.D.O.; De Souza, E.L.; Diniz, M.D.F.F.M.; Trajano, V.N.; De Medeiros, I.A. Inhibitory effect of beta-pinene, alpha-pinene and eugenol on the growth of potential infectious endocarditis causing gram-positive bacteria. *Rev. Bras. Ciênc. Farm.* **2007**, *43*, 121–126. [CrossRef]
85. Yang, J.; Nie, Q.; Ren, M.; Feng, H.; Jiang, X.; Zheng, Y.; Liu, M.; Zhang, H.; Xian, M. Metabolic engineering of *Escherichia coli* for the biosynthesis of alpha-pinene. *Biotechnol. Biofuels* **2013**, *6*, 1–10. [CrossRef] [PubMed]
86. Azevedo, M.; Chaves, F.; Almeida, C.A.; Bizzo, H.R.; Duarte, R.S.; Campos-Takaki, G.M.; Alviano, C.S.; Alviano, D.S. Antioxidant and antimicrobial activities of 7-hydroxy-calamenene-rich essential oils from *Croton cajucara* Benth. *Molecules* **2013**, *18*, 1128–1137. [CrossRef] [PubMed]
87. Liu, T.-T.; Yang, T.-S. Antimicrobial impact of the components of essential oil of *Litsea cubeba* from Taiwan and antimicrobial Activity of the oil in food systems. *Int. J. Food Microbiol.* **2012**, *156*, 68–75. [CrossRef]
88. Onawunmi, G.O. Evaluation of the antimicrobial activity of citral. *Lett. Appl. Microbiol.* **1989**, *9*, 105–108. [CrossRef]
89. Can Baser, K. Biological and pharmacological activities of carvacrol and carvacrol bearing essential oils. *Curr. Pharm. Des.* **2008**, *14*, 3106–3119. [CrossRef] [PubMed]
90. Doyle, A.A.; Stephens, J.C. A review of cinnamaldehyde and its derivatives as antibacterial agents. *Fitoterapia* **2019**, *139*, 104405. [CrossRef]

insects

Article

Unprecedented Density and Persistence of Feral Honey Bees in Urban Environments of a Large SE-European City (Belgrade, Serbia) †

Jovana Bila Dubaić [1], Slađan Simonović [2], Milan Plećaš [1], Ljubiša Stanisavljević [1,*], Slobodan Davidović [3], Marija Tanasković [3] and Aleksandar Ćetković [1]

1 Faculty of Biology, University of Belgrade, Studentski Trg 16, 11000 Belgrade, Serbia; jovanabila@bio.bg.ac.rs (J.B.D.); mplecas@bio.bg.ac.rs (M.P.); acetkov@bio.bg.ac.rs (A.Ć.)
2 SOS Mobile Team for Rescue and Removal of Honey Bee Swarms and Colonies, Koste Glavinića 12, 11000 Belgrade, Serbia; bioplodplus@ptt.rs
3 Department of Genetics of Populations and Ecogenotoxicology, Institute for Biological Research "Siniša Stanković"—National Institute of the Republic of Serbia, University of Belgrade, Bulevar Despota Stefana 142, 11060 Belgrade, Serbia; slobodan.davidovic@ibiss.bg.ac.rs (S.D.); marija.tanaskovic@ibiss.bg.ac.rs (M.T.)
* Correspondence: ljstanis@bio.bg.ac.rs

† Some segments of this study were presented at the Symposium of Serbian Entomologists in 2019: Bila Dubaić, J.; Simonović, S.; Plećaš, M.; Ćetković, A.; Stanisavljević, L. Mass occurrence and longevity of feral honey bee colonies in the highly urbanized area of Belgrade [in Serbian]. In Book of Abstracts of the XII Symposium of Serbian Entomologists, Niš, Serbia, 25–29 September 2019; p. 45.

Citation: Bila Dubaić, J.; Simonović, S.; Plećaš, M.; Stanisavljević, L.; Davidović, S.; Tanasković, M.; Ćetković, A. Unprecedented Density and Persistence of Feral Honey Bees in Urban Environments of a Large SE-European City (Belgrade, Serbia). *Insects* **2021**, *12*, 1127. https://doi.org/10.3390/insects12121127

Academic Editor: Mariano Higes Pascual

Received: 7 November 2021
Accepted: 30 November 2021
Published: 16 December 2021

Simple Summary: The western honey bee, a pollinator species that is essential for modern agriculture and food production sustainability, is under various anthropogenic pressures. In the last few decades, these have led to serious worldwide problems concerning the health and stability of honey bee colonies. The importance of wild or feral honey bees has only recently been recognized, as these populations are crucial for research on the processes that enable their survival under such pressures. Here, we present a case of an unmanaged free-living population of honey bees that, unlike in other known studies, live in a highly populated urban environment. This extraordinarily dense feral honey bee population, which is not directly associated with managed apiaries, provided us with a large dataset of various life history parameters that will considerably fill in the knowledge gaps on unmanaged colonies. Furthermore, we want to underline the importance of citizen science in the data collection process and suggest it as a suitable approach to study feral honey bees in urban landscapes. We believe that highly populated urban landscapes can support and reinforce feral honey bees and that similar citizen science projects should be set up in other urban areas or other countries.

Abstract: It is assumed that wild honey bees have become largely extinct across Europe since the 1980s, following the introduction of exotic ectoparasitic mite (*Varroa*) and the associated spillover of various pathogens. However, several recent studies reported on unmanaged colonies that survived the *Varroa* mite infestation. Herewith, we present another case of unmanaged, free-living population of honey bees in SE Europe, a rare case of feral bees inhabiting a large and highly populated urban area: Belgrade, the capital of Serbia. We compiled a massive data-set derived from opportunistic citizen science (>1300 records) during the 2011–2017 period and investigated whether these honey bee colonies and the high incidence of swarms could be a result of a stable, self-sustaining feral population (i.e., not of regular inflow of swarms escaping from local managed apiaries), and discussed various explanations for its existence. We also present the possibilities and challenges associated with the detection and effective monitoring of feral/wild honey bees in urban settings, and the role of citizen science in such endeavors. Our results will underpin ongoing initiatives to better understand and support naturally selected resistance mechanisms against the *Varroa* mite, which should contribute to alleviating current threats and risks to global apiculture and food production security.

Keywords: *Apis mellifera*; citizen science; feral honey bees; natural selection; unmanaged honey bees; urban environment

1. Introduction

The western honey bee, *Apis mellifera* Linnaeus, 1758, is recognized as the single most important pollinator species, vital to the success of modern agriculture and the stability of human food production [1–5], let alone the provisioning of honey and other bee-products. There is a worldwide concern regarding the trends of its populations, including various health issues and maintenance of sufficient pollination ecosystem service [3,6–14]. In the last two decades, numerous studies identified several factors that negatively affect honey bee populations and beekeeping [7,15–17]. Among other factors, the exotic ectoparasitic mite *Varroa destructor* (hereafter: *Varroa*) is considered the most significant threat to western honey bee in many parts of the world [18–23], largely contributing to widespread colony losses [24,25]. There is a widely accepted view that unmanaged (wild/feral) honey bee populations were completely eradicated in Europe since the 1980s, following the introduction and spread of *Varroa* and the associated spillover of various pathogens [26,27]. However, several studies reported that both feral and managed colonies can survive for an extended period despite of *Varroa* infestation, and without receiving any treatments, triggering wide scientific and public attention [28–37].

There is a general lack of knowledge regarding feral colonies, since the majority of research is focused only on managed honey bees [38–40]. Given that unmanaged honey bees are not receiving any chemical treatments against *Varroa* or other pests/pathogens, it is important to explain how they manage to survive. Successful cohabitation with *Varroa* is well known in closely related Asian species (*Apis cerana*), which co-evolved with this parasite [23]. Understanding the mechanisms behind these interactions might provide clues for improving the health status and fitness of managed *A. mellifera* colonies and securing perspectives for sustainable beekeeping [41–44]. In that respect, various studies focused on genetic differences between feral and managed colonies [31], or on more efficient hygienic and grooming behaviors [35,45] that might have evolved in colonies left untreated against parasites. Several studies suggested that differences of living conditions between free-living and managed honey bees (Table S5) may generate essentially different susceptibility to *Varroa* mite [36,46,47].

Finding wild/feral honey bees in nature is difficult and time consuming [38,48]. In areas where they exist (e.g., in the USA), feral colonies are scattered, mostly in woodland areas, usually high in the trees, and generally not easy to spot [38,48]. Potentially, urban environments could also be suitable for feral honey bees, due to ample nesting opportunities [49,50], the presence of diverse floral resources throughout the season [51–53], and lesser exposure to pesticide spraying [54]. However, there are only very few records of feral honey bees in urban environments [48–50].

Here, we present another case of an unmanaged, free-living population of honey bees. Unlike the previously reported cases, our study area is a large and highly populated urban environment, the city of Belgrade (capital of Serbia). Until several years ago, local scientists involved in various bee studies were generally unaware of the existence of abundant feral honey bees in and around Belgrade, while local beekeepers were not familiar with the importance of this phenomenon. Therefore, a 'discovery' of a huge data-set documenting numerous thriving colonies living completely without any kind of human interference or treatment came as a surprise. Data were obtained through the long-term activity of one of the authors (S.S.), a beekeeper then affiliated with the Belgrade Beekeepers Society (BBS). Specifically, S.S. was engaged in extensive communication with the Belgrade citizens who reported on honey bee colonies or swarms in their surroundings, mostly asking for their removal. Therefore, the compilation of the core data-set may be regarded as "incidental crowd sourced observations" also known as "opportunistic citizen science" [55]. It is well known that citizens' help (e.g., the implementation of citizen science) can effectively overcome various difficulties in gathering information on biodiversity in urban environments [56,57].

The purpose of this study was to:

- present evidence of the extraordinary frequent incidence of unmanaged honey bee colonies throughout the wider Belgrade area (based on the occurrence data compiled for 2011–2017), and a similarly frequent incidence of swarms (probably many of them from unmanaged colonies);
- assess the basic parameters of the recorded occurrences (the features of documented nesting/swarming sites: type of substrate/microspace, height, neighbourhood or habitat type) and the parameters of the recording and reporting process (features of the compiled data-set, its limitations, peculiarities, and utility);
- analyze the patterns of distribution of free-living honey bee colonies and observed swarms across the different local urban and sub-urban/peri-urban settings of a large city, relative to the distribution of managed hives;
- evaluate the status of unmanaged honey bee occurrences in Belgrade, hypothesizing that they indicate the existence of a large, long-lasting and self-sustaining feral population;
- present the experiences (advantages, problems and perspectives) of the citizen science approach for detecting, assessing the status and monitoring of feral honey bees in urban environments.

2. Materials and Methods

2.1. Case Study Area

The city of Belgrade (44°49′ N 20°27′ E), the capital and the largest city in Serbia, is located at the confluence of two rivers, the Danube and the Sava. The municipal boundary of the city is roughly 35 × 36 square km, while the core administrative–urbanistic unit covers around 776 km^2. Around 1.7 million people live within the total area of Belgrade, while up to 1.2 million people are located in the urban zone [58]. The climate is transitional between the temperate–continental and steppic regimes and its relief is spanning the altitude of 65–506 m.

2.2. Data Source

Data for this survey were gained through the compilation of the citizen requests for removal of honey bee colonies and/or swarms, mostly occurring near their homes or work place in the period 2011–2017. As Belgrade does not have an official utility service for this, all such requests have been redirected to the Belgrade Beekeepers Society (hereafter: BBS), who kept records and provided removal service (in that period operated by S.S.). Most records included the following information: date of request, location—address, category (swarm or colony), nesting/swarming site (type of substrate, etc.), contact information of observer, and intervention status. The intervention status refers to the personal notes of the visited site and of the removed swarm/colony by the beekeeper. He also kept notes on swarms that flew away on their own in which instances no intervention was required. Due to logistic and time constraints, especially in the swarming season, not all reported locations were visited. However, at most visited locations, successful removal or capture of swarms was conducted, which was noted in the original database. Sometimes, the removal of colonies was unsuccessful due to a lack of equipment to reach colonies in high places in the reported location.

2.3. Data Processing

During the studied period of seven years (2011–2017), more than 1700 removal requests were received. We performed several steps of refinement and cleaning of the initial data. Each removal record was first categorized into one of four categories: "bee colony", "bee swarm", "colony or swarm", and "other". The third category, "colony or swarm", was created to account for incomplete reports that could not be resolved—when it was not possible to distinguish between colony and swarm, based on the information from the initial notes. The fourth category, "other", contains reports of other forms of insects (mainly social wasps: yellowjackets and hornets), and they were excluded from the analysis. The next step was to detect duplicated records of swarms and colonies. When two (or more) records of swarms were reported on the same location and on the same date, or in the interval of one to two days, we regarded them as duplicates (i.e., the same occurrence reported by the different persons). Detecting duplicates in the category of honey bee colonies posed more difficulties. Even if the records originate on the same day and at the same location, there is no certainty whether it is one or more colonies involved. On some occasions the same colony is a nuisance to two or more neighboring apartments, and if both inhabitants filed a report, this makes one of the reports a duplicate. Additional difficulty may arise from the following situation: there were several reports of a honey bee colony all noted at the same location but in different years, which may relate to the same long-lasting colony reported by the same (or different) person year after year. To clarify situations like this, as well as to confirm and verify the rest of the data from our database, we carried out a process of validation. The validation process consisted of two steps: the first step was to conduct interviews with citizens who reported swarms/colonies, and the second step was to carry out a field study where possible (which was carried out in 2019) and gather information by observing reported locations (Figure 1).

Figure 1. Some of the free-living honey bee colonies observed in Belgrade during 2019: (**a**) inside the hollow tree, (**b**) in the hollow space within a multi-story building floor, (**c**) inside the steel tubular electric pole, and (**d**) in a cavity of a damaged façade (showing propolis staining around the entrance).

During the interviews, our focus was to address the following: to ask about missing information, to describe what they can remember about swarms/colonies they reported, to inquire on the precise location of the hive/swarm, the height of the occurrence, and if they knew about the longevity of a colony they noticed. Furthermore, our intention was to see what happened after they reported it—are the bees still there? Did the beekeeper remove them? Was there a case of a newly established colony in the same cavity?

Original data rarely contained information about the longevity of reported colonies. During the validation process and interviews, we were particularly interested in this aspect. We were especially focused on reporters from the core of urban area, since occurrences within and nearby the tentative feral zone are considerably more relevant for the purpose of our study.

After each interview, the person was classified as "reliable" or "not reliable", and therefore the information received was equally categorized. This thorough quality check was possible mainly because all interviews were conducted by one person (JBD), who spent considerable time working with this data set, first getting familiar with it, then reviewing each record and detecting duplicates, then mapping data (georeferencing locations), and finally conducting interviews with citizens. This allowed for an informed estimation of whether the person and resulting data were valuable or anecdotal (characterized by insufficient information or chaotic/contradictory statements, obvious lack of understanding of the object reported, etc.). We eliminated unreliable, duplicated, and incorrect data. The remaining records were mapped using the QGIS program [59].

We also mapped the presence of managed apiaries in the wider city area, based on the list of registered hives, personal communication with BBS, and recorded unregistered apiaries that were found during location inspections or other field surveys conducted in the study area.

For an informed analysis and better visualization of compiled data, we used the QGIS visualization tool—Heatmap (Kernel Density Estimation; hereafter: KDE)—to create density maps of reported colonies and swarms [59]. This analysis creates differently shaded surfaces, each shade representing a different density of the respective analyzed units. After a couple of trial runs, we used a search radius of 1000 m and 10 m px size since these parameterizations were found to be the best fitting for our data set.

In order to analyze the relationship between human population density and unmanaged honey bee colony or swarm occurrences, we performed point pattern analysis (Poisson point process model—ppm) [60,61]. We used the Belgrade official map of local communities (the lowest administrative unit for which standardized census data are available) as a spatial layer to define the analysis and population density of each local community in the form of a raster layer as a covariate. The reported occurrences of feral bees were defined as a planar point pattern (ppp). The analysis was done using the spatstat R package v.1.63 [60] within R v3.6.3 [62].

3. Results

For the period of seven successive seasons (2011–2017), we compiled a total of 1745 calls/reports (Table S1), of which 1495 were reporting on honey bees and 250 on other insects. After the removal of duplicated reports (124), our data-set includes 1371 unique reports (Table 1). Of those, for 56 reports we could not retrieve locations, hence 1315 reports were georeferenced (Figure 2; Figure S1). Unique reports comprise 460 colonies, 537 swarms, and 374 ambiguous/unspecified reports (occurrences that could represent either a colony or a swarm, but original evidence lacked details or contained contradictory elements).

Table 1. All uniquely reported observations of unmanaged honey bees (i.e., without duplicated reports of the same units), grouped into: colonies, swarms, and ambiguous or unspecified units (could be either a colony or a swarm). Sums include 56 reports that could not be georeferenced.

Year	Colonies	Swarms	Ambiguous/Unspecified	Total Honey Bees
2011	157	95	-	252
2012	116	78	-	194
2013	34	38	38	110
2014	54	115	109	278
2015	33	56	78	167
2016	44	95	110	249
2017	22	60	39	121
∑	460	537	374	1371

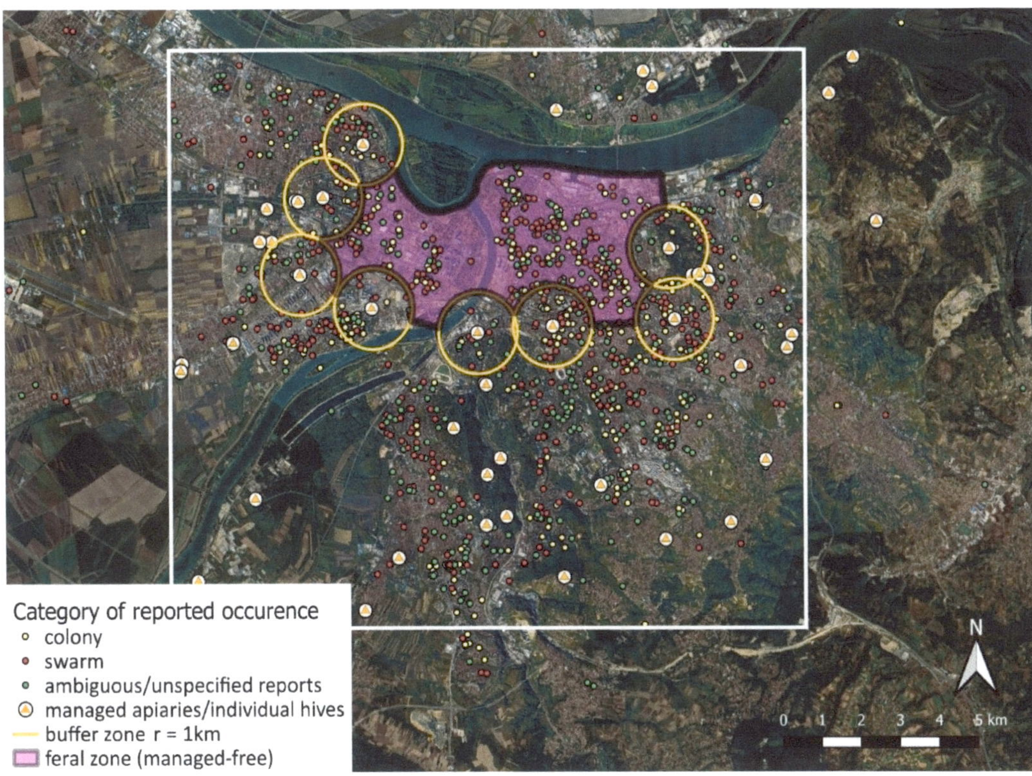

Figure 2. Distribution of georeferenced occurrences of unmanaged (potentially feral) honey bee colonies and swarms of unknown origin in the period of 2011–2017, within the wider Belgrade area (24 more remote locations from our data-set are not shown, being too widely scattered beyond the coverage of this map). Three categories of reported cases are shown combined but with differently colored circles (nesting colonies, swarms, and ambiguous or unspecified reports—the latter could be either a colony or a swarm; the separated distribution maps are available in Figure S1a–c). The locations of known managed apiaries (or individual hives) are also shown, with particular focus on those that surround the urban core area (each shown with a circular 'buffer zone' of r = 1 km). The urban core is presumed to harbor mostly the self-sustained feral bee colonies, and consequently, the swarms produced mostly by them; accordingly, we delimited a tentative 'feral zone' (or 'managed-free zone'). The white rectangle (ca. 16.2 × 14.4 km) delimits the area shown in more detail in further maps. (This map is also available as a high-resolution image, upon request to J.B.D.).

Out of 1371 reports, information about height of the nesting/swarming site was available for 391 (Figure 3; Table S3), while information about the type of nesting/swarming site was known for 1195 reports (Figure 4; Table S4). Of these, around 85% colonies were recorded between 3 and 21 m, while 90% swarms were recorded between 1 and 15 m. Nesting/swarming site type was recorded for 437 of the reported colonies (95%) and for 463 of reported swarms (86%). The most common type of nesting cavity was inside the walls or various building/house façades (182 occurrences, ca. 42%), followed by the wooden shutter window box (134 occurrences, ca. 31%). The most common swarming situation was the hanging from a tree branch (277 occurrences, ca. 60%) (Figure 4).

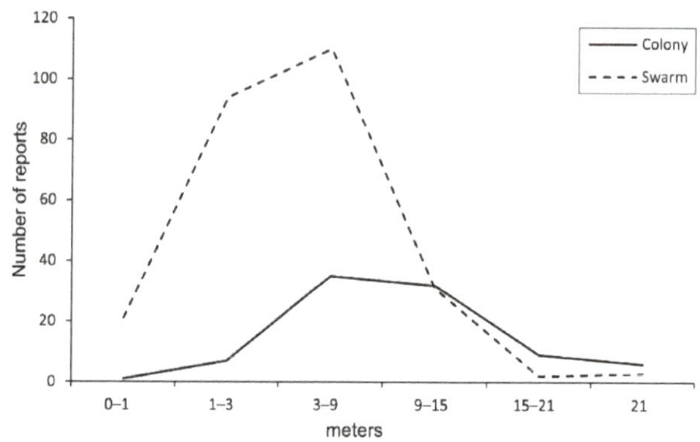

Figure 3. Height distribution of reported nesting/swarming sites of unmanaged honey bees (full line: colonies, dashed line: swarms).

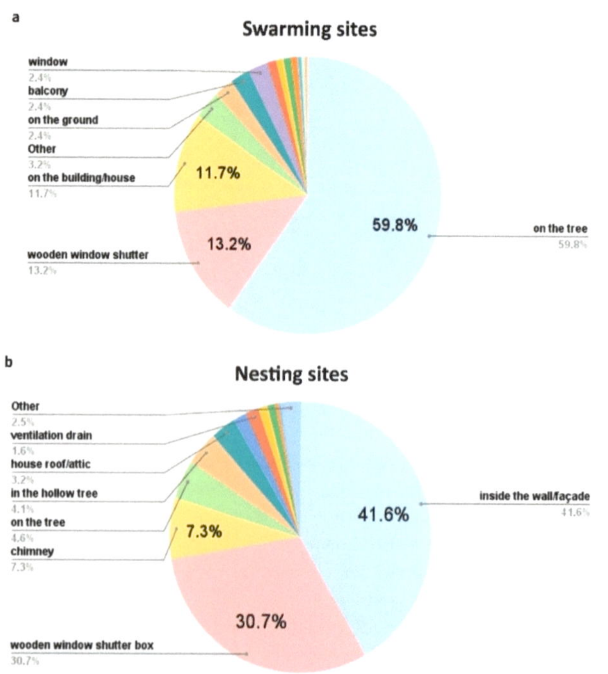

Figure 4. Distribution of records categorized by types of (**a**) swarming and (**b**) nesting sites.

Analysis of the effect of human population density on reporting about unmanaged honey bee occurrences showed a strong positive relationship (Table 2, Figure 5). The overlapping distribution of human population density (by analyzed 'local communities' of Belgrade; data from 2011) and georeferenced occurrences of feral honey bee units is shown in Figure S2.

Table 2. Estimated regression coefficients and their standard errors from a point process model showing the relationship between human population density and reports of unmanaged honey bees.

Variable	Coefficient (SE)	95% CI	z Value	z Test
Intercept	0.842421 (0.046938)	0.7504–0.9344	17.947	<0.05
Population density	0.000128 (0.000004)	0.000121–0.000136	34.384	<0.05

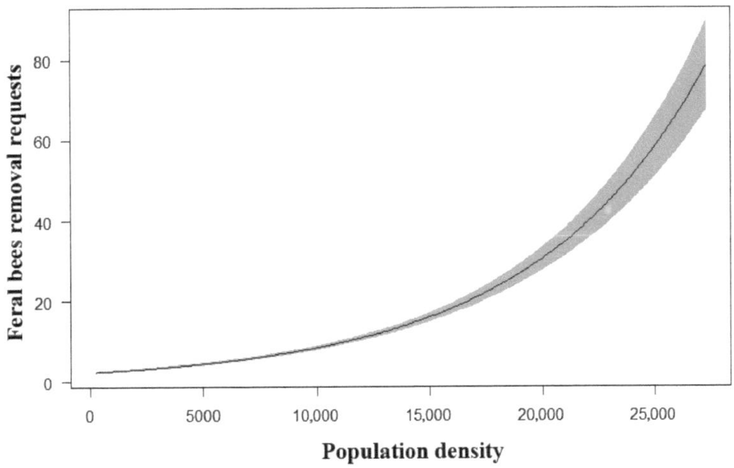

Figure 5. Relationship between human population density and the reports of unmanaged honey bees from the point process model. The gray zone represents a 95% confidence interval.

Based on the presence and distance from managed apiaries (the circular buffer of 1 km), we defined a tentative 'feral' (=managed-free) zone in the core area of the city (Figure 6). Due to specific geographical features, the northern border of feral zone is represented by the Danube River, which is ca. 480–680 m wide in this section, hence represents a barrier to the potential influx of swarms from the north.

We generated three separate KDE heatmaps: for all georeferenced reports of unmanaged honey bees (combined: the sum of all three categories), for the colonies, and for the swarms (Figure 7). Densities are shown as six color grades of the same intensity range, but the respective city areas with same color represent quite different maximal estimated values: ca. 18 for swarms, 27 for colonies, and 44 for all occurrences (the later include also the ambiguous records, which represent >27% of the whole data-set). The distribution of areas with highest density of different categories shows various levels of concordance.

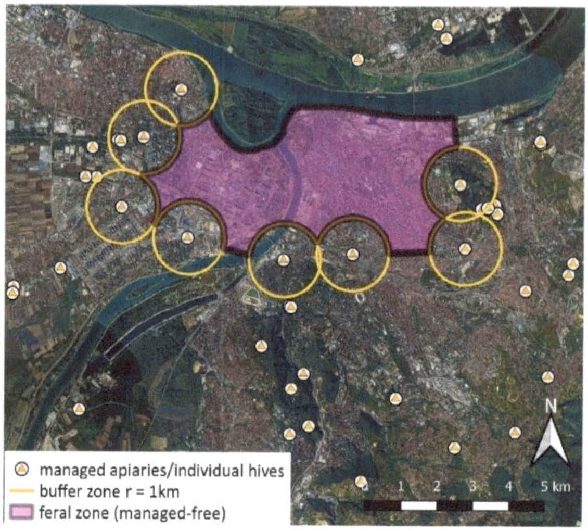

Figure 6. The central part of the study area, most intensively covered by citizens' reports. Close-up view of the core feral/managed-free zone of Belgrade, and the locations of managed apiaries/hives which surround it (with their respective buffer zones).

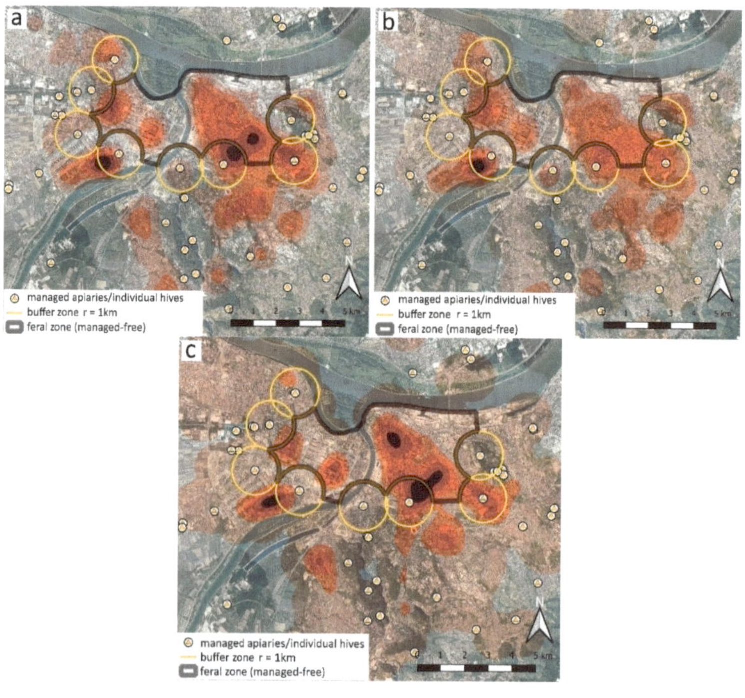

Figure 7. Kernel Density Estimation heatmaps: shading intensity represents the increasing density of respective occurrences (defined through six density zones for the search radius 1 km), based on summed reports for 2011–2017: (**a**) all categories, combined (min. 5.7, max. 43.8, zonal mid-point value range: 6.5–40.1); (**b**) colonies (min. 3.8, max. 26.6, zonal mid-point value range: 4.1–24.4); (**c**) swarms (min. 2.7, max. 18.1, zonal mid-point value range: 2.9–16.6).

We managed to obtain more detailed and sufficiently reliable information for 78 nesting sites with tentative long-lasting continual activity of respective colonies (Table 3, Figure 8). Out of these, 18 colonies were from the interior of the feral zone, and further 22 from the neighboring area (within the buffer zone of ca. 2 km around feral zone, mostly within the respective buffer zones around centrally positioned managed hives/apiaries).

Table 3. Tentative longevity of feral colonies: number of nesting sites with documented honey bee activity potentially extending beyond the single season.

Years	**1–2**	**3**	**4–6**	**7–10**	**>10**	**∑**
Number of Colonies	16	28	19	12	3	78

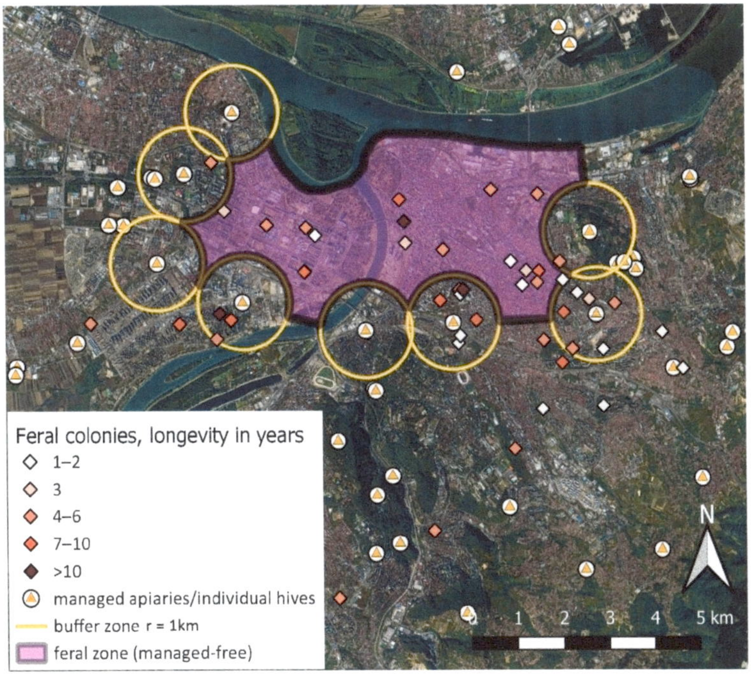

Figure 8. Distribution of feral colonies with documented activity potentially extending beyond the single season: evidence was principally based on repeated observations of the very same nesting space, obtained through reports/interviews with citizens and/or through personal checking by one of the authors (S.S.).

4. Discussion

For the context of our research, it is important to stress that urban beekeeping was almost a non-existent activity in most of the urban core of Belgrade (BBS, personal communication; the situation is slowly changing after 2018). Even at a moderate distance from central urban areas, managed apiaries are still relatively rare and scattered (Figure 6). Therefore, it seems quite unlikely that so many free-living colonies (and consequently most of the swarms) in these central areas are derived from regular swarm inflow from managed hives. Accordingly, we expect that >90% of the occurrences within the tentative feral zone might be essentially feral.

This zone represents the urban core of residential and business activities, characterized by dense urban infrastructure, high human population density, and economic activities, accompanied by appropriate traffic dynamics. In contrast, the periphery of Belgrade is characterized with more widespread presence of beekeepers and apiaries of various types. Many of the reported swarms or unmanaged colonies in that area are likely not truly feral, but recently derived from managed hives. Between the feral zone and the periphery, a type of 'transitional area' may be operatively defined. The transition is related to two principal patterns: type of settlements is gradually changing (from densely urbanized to nearly rural/agricultural), while the incidence of beekeeping is growing (but largely varying in management efficiency). Accordingly, we expect that as many as 50–75% of the reported swarms and colonies in this 'transition zone' might also be feral. Much lower values are expected across the peripheral areas, but also the incidence of unmanaged honey bee reporting was generally more sparse.

The distribution pattern of reported occurrences, presented with HDE heatmaps from three data-subsets, shows reasonable concordance between the most densely inhabited city areas (Figure 7). Obviously, large extent of central urban area and few peripheral tracts are quite densely populated with free-living honey bees, with a few distinctive "cores of incidence" being situated in or nearby our feral (managed-free) zone. Two prominent cores are seemingly related to the two locations with managed apiaries (known, from anecdotal beekeepers sources, for excessive poorly managed swarming in past times). However, it is noteworthy that none of these cores are exactly centered at respective apiary, but rather offsetted in different directions, indicating the likely influence of habitat suitability and attractiveness. Indeed, the 'western feral core' is effectively settled within the particular type of relatively uniform neighbourhood, with multi-storey buildings of peculiar construction and uniform age, known empirically for numerous suitable nesting cavities. The 'eastern feral core' shows a 'tendency of spreading' towards the most urbanized downtown area of Belgrade. This is the largest continuous while also almost evenly populated part of the city; it is most distantly situated from the managed hives around the feral zone. Noteworthy, two of the three cases of longest-living colonies (or alternatively, the longest active nesting site—see also later) are practically centered in these two 'feral cores' (Figure 8). Therefore, it is possible that, even if initially established and 'boosted' from the managed hives, these two cores may now serve—for more than a decade—as 'secondary source' of numerous feral swarms, which continually populate the preferred urban sectors in a self-sustainable manner.

This possibility, that many unmanaged colonies (and respective share of swarms) in the Belgrade urban area represent an essentially feral and persistent population has been further corroborated by the results of a separate genetic study. Within the project SERBHIWE [63], S.D. and M.T. analysed the genetic variability of feral and managed honey bee colonies in the Belgrade area by molecular genetics methods; a brief summary outcomes are presented herewith. They sampled feral bees from 40 colonies selected from within our data-set and used two types of genetic markers [64]: uniparentally inherited mitochondrial DNA (mtDNA) and biparentally inherited microsatellites from autosomal loci. The analysis of the mtDNA $tRNA^{leu}$-*cox2* intergenic region demonstrated a similar pattern of genetic variability for both types of sampled colonies (feral and managed), while the presence of rare haplotypes was detected in the mtDNA gene pool that could be found in only one group, either feral or managed. The analysis of 14 microsatellites loci showed that the feral honey bee colonies possess greater genetic diversity compared to the managed ones, and the assessed relatedness showed that on average, a feral honey bee colony is more related to other feral honey bee colonies than managed ones. These results suggest that swarming from managed apiaries is not the only reason for existence of such a great number of feral honey bee colonies in Belgrade. In other words, the abundant feral population may not be regarded as primarily or predominantly derived from the contemporary managed hives. Overall, molecular genetic analysis suggests the existence of a strong and genetically diverse population of feral honey bees in Belgrade. Hypothetically, this urban population of free-living honey bees may have existed in continuum from the period before the introduction of *Varroa* mite. Greater genetic diversity could have contributed to natural selection for improved tolerance against parasites and pathogens, and hence their capability for successful survival despite these pressures.

Available chemical treatments against *Varroa* mite do not provide a long-term solution (due to the development of resistance to pesticides), while some are even harmful to bees [65,66]. Moreover, any kind of treatment may inhibit natural selection pressures preventing coevolution between parasite and host, thus being counterproductive [33,43,67,68]. In uncontrolled/wild settings natural selection will lead toward disease-resistant bees and less virulent forms of pathogens [69,70]. Generally, wild populations of domesticated plants and animals are important reservoirs of genetic diversity [71] and this applies to wild and feral honey bees as well [27,39,72]. However, beekeepers are generally not willing to support such a course for economic reasons—expected high colony losses in initial period [43,73], but also the possibility of losing some of the desirable traits of managed bees [23]. For decades, breeding programs selected traits like greater honey production, less aggressive temperament, a lower tendency for swarming, high fecundity of queen, etc. [23,46,74,75]. In fact, honey bees that are in balance with new parasite are economically more beneficial to manage in the long run [43,70]. Recently, selection programs started to include disease resistance as one of the wanted traits of honey bees [74,75]. Hence, the spontaneously established *Varroa*-resistant honey bee population, like the one thriving in Belgrade, could be of great importance for future breeding programs in Europe.

Wild honey bees in natural habitats prefer to inhabit cavities in tree trunks [76]. Several studies reported on feral honey bees being opportunistic regarding the cavity types used for nesting in urban environments [49,50]. Our study provided extensive further details about the nesting site features of feral honey bees. The most commonly observed type of cavity was inside the walls or building/house façades (41.6%), followed by the cavities of wooden window shutter box (30.7%). Both types of cavities are closely associated with human living spaces that are heated/insulated, hence energetically beneficial for the colony, especially during the winter [77]. The advantage of certain human-associated cavity types for honey bees has also been emphasized in a recent study from Ireland [78].

Information on the longevity of individual colonies is a direct indicator of *Varroa* resistance level. In our data set, there were several colonies with exceptional duration. However, a seemingly continuous bee activity around a certain cavity does not necessarily imply the continual usage and persistence of the same colony (or its direct descendants). Instead, a colony may die off, but a swarm from another colony may move in, and this change in cavity status (from inactive to active) is called "turnover" [79]. Turnovers are easily missed; hence, the colony may be categorized in a wrong longevity class. Therefore, the longevity of free-living colonies has to be checked frequently and regularly [80]. Based on our experience from Belgrade (from 2019), for reliable longevity estimate, monitoring must be conducted as frequently as 10–14 days. If the long-active bee cavity cannot be monitored appropriately, such situations are better termed 'extended cavity occupancy' than colony longevity.

In addition to presenting a huge data set, this paper also highlights the importance and suitability of the citizen science (CS) approach for studying feral honey bees in urban settings. Finding feral honey bee colonies is difficult and time-consuming [38,48], particularly in the cities. The presence of a nuisance species, e.g., the new pest, will be primarily detected by the members of the public [81]. Honey bees are generally considered as beneficial insects, but they can represent a nuisance if inhabiting undesirable locations—too close to human activities. In addition to 'ordinary' citizens, beekeepers and urban pest control agencies can be a particularly important source of data [50,82]. Certainly, participatory research has its shortcomings, e.g., requires a longer process of validation, as in case of our data-set. Communicating with a substantial number of people can be challenging for several reasons; various problems which we encountered are summarized in Table S2.

Our study falls into the category of incidental citizen science, known also as "opportunistic citizen science" [55,83,84], which is linked with several biases (i.e., reporting, taxonomical, observational and geographical biases). Honey bees are widely known species, distinctive, and easily recognizable; hence, taxonomic bias in our data set is presumably not significant. Monitoring of easily recognizable species, such as honey bees, is generally feasible with CS projects [85], but untrained people often have difficulties distinguishing between swarm and colony. Honey bees are perceived in several different ways by the public [50], ranging from aversion and fear [86], to fondness and generally positive attitude towards them [87,88]. Consequently, people had strong reasons not to ignore, but to report occurrences of honey bees. Therefore, reporting bias is probably not significant to our study. Observational bias also could not be greatly present in our data since all observations of honey bee swarms and colonies were detected in close proximity to homes of people who reported it, not as a result of a search effort. On the other hand, we noticed significant geographical bias—there is evident lack of data of the city's industrial and peripheral areas. Lack of data from industrial, non-populated areas, occurred simply because colonies and swarms are more likely to be noticed in densely populated residential areas of the city (proximity to human daily lives, high frequency of human activity), rather than in unpopulated or sparsely populated parts [48,49]. There were practically no reports of bee colonies or swarms in unpopulated areas because there was no one who could make a report [89]. On the other hand, lack of data from the periphery is due to the people living there would likely deal with the problem themselves. On the contrary, in the urban core, colonies are more likely to be reported because citizens who live there are less likely to have beekeeping experience [90]. The presence of swarms and colonies on the edges of the city, where managed apiaries are frequent, could originate from the swarm escapes from managed hives [76].

The successful existence of a dense, self-sustaining feral honey bee population in Belgrade is further promoted by certain socio-economic circumstances. For example, problems with unregulated jurisdiction (responsibilities) of public services in Belgrade, combined with a shortage of specialised employees and funding, make public utility companies inefficient and inadequate to respond to citizens' requests for removing honey bee colonies/swarms; accordingly, many of them remain untreated. Furthermore, the extended poor state of Serbia's economy (for many decades now) reflects on poor maintenance of many buildings (which makes urban areas rich in suitable cavities). For similar reasons, the majority of people cannot afford to pay for removal to private services (by professional beekeepers), again contributing to many colonies/swarms being left undisturbed. All these circumstances made Belgrade almost the perfect environment for the establishment and persistence of a large free-living honey bee population. There are also less specific favourable aspects, common to many large urban areas [91], but they certainly contributed to the overall good conditions for wild bees: variable and long-lasting floral resources, and decreased exposure to pesticides (relative to dominating agricultural systems, etc.).

In our case, the existence of a strong and self-sustaining feral honey bee population confirms that urban environments can be highly favourable for this species. However, a combination of circumstances which enabled or enhanced its establishment might have been locally or regionally context-specific, in several aspects. Nonetheless, it could be rewarding to investigate, following our experiences, if other cities in Europe (particularly larger cities in SE Europe) might also harbor the overlooked free-living honey bees.

Supplementary Materials: The following are available online at https://www.mdpi.com/article/10.3390/insects12121127/s1, Table S1: Overview of all received reports. Table S2: Outcome of citizens/reporter's interviews. Table S3: Summary of reports by height of nesting/swarming sites. Table S4: Summary of reports by nesting/swarming site types. Table S5: Comparison of 'lifestyle' features: key differences between feral/wild and managed honey bee colonies. Figure S1: Distribution of georeferenced unmanaged bee colonies and swarms: (a) nesting colonies, (b) swarms, (c) ambiguous or unspecified reports. Figure S2: Distribution of human population density of Belgrade by 'local communities' units, overlayed with distribution of feral honey bee units.

Author Contributions: Conceptualization, A.Ć. and L.S.; methodology, A.Ć., J.B.D., M.P., M.T. and S.D.; formal analysis, M.P., S.D. and J.B.D.; investigation, J.B.D. and S.S.; data curation, S.S., A.Ć. and J.B.D.; writing—original draft preparation, J.B.D.; writing—review and editing, A.Ć., L.S., M.P. and M.T.; visualization, J.B.D. All authors have read and agreed to the published version of the manuscript.

Funding: This research was funded by the Ministry of Education, Science, and Technological Development of the Republic of Serbia, Grant Nos. III43001 and 451-03-68/2020-14/200178 for J.B.D., M.P., L.S., A.Ć., Grant No. 451-03-9/2021-14/200007 for M.T. and S.D. and Science Fund of the Republic of Serbia, PROMIS, Grant No. 6066205, SERBHIWE for M.T., L.S. and S.D.

Institutional Review Board Statement: Not applicable.

Informed Consent Statement: Not applicable.

Data Availability Statement: Data used in the submitted manuscript can be made available to a limited degree after a reasonable request to the corresponding author. Personal data of citizen scientists/reporters or the location–exact addresses will not be available due to violation of privacy.

Acknowledgments: We express our gratitude to all the citizens who provided data for our study, especially those who helped us during the field survey validation process. We also thank our colleague Milena Vučković (VRVis Zentrum für Virtual Reality und Visualisierung Forschungs GmbH) who advised us at the very beginning of our data processing and analyzing. Special thanks go to Bernard Vaissière (INRA, Avignon, France), who firstly pointed us to the peculiarity of this phenomenon in Belgrade, hence giving impetus to this study.

Conflicts of Interest: The authors declare no conflict of interest. The funders had no role in the design of the study; in the collection, analyses, or interpretation of data; in the writing of the manuscript, or in the decision to publish the results.

References

1. Chaplin-Kramer, R.; Tuxen-Bettman, K.; Kremen, C. Value of Wildland Habitat for Supplying Pollination Services to Californian Agriculture. *Rangelands* **2011**, *33*, 33–41. [CrossRef]
2. Delaplane, K.S.; Mayer, D.F. *Crop Pollination by Bees*; CABI Publishing: New York, NY, USA, 2000.
3. De la Rúa, P.; Jaffé, R.; Dall'Olio, R.; Muñoz, I.; Serrano, J. Biodiversity, Conservation and Current Threats to European Honeybees. *Apidologie* **2009**, *40*, 263–284. [CrossRef]
4. Gaines-Day, H.R.; Gratton, C. Crop Yield is Correlated with Honey Bee Hive Density but not in High-Woodland Landscapes. *Agric. Ecosyst. Environ.* **2016**, *218*, 53–57. [CrossRef]
5. Belsky, J.; Joshi, N.K. Impact of Biotic and Abiotic Stressors on Managed and Feral Bees. *Insects* **2019**, *10*, 223. [CrossRef]
6. Aebi, A.; Neumann, P. Endosymbionts and Honey Bee Colony Losses? *Trends Ecol. Evol.* **2011**, *26*, 494. [CrossRef]
7. Ellis, J.D.; Evans, J.D.; Pettis, J. Colony Losses, Managed Colony Population Decline, and Colony Collapse Disorder in the United States. *J. Apic. Res.* **2010**, *49*, 134–136. [CrossRef]
8. Higes, M.; Meana, A.; Bartolome, C.; Botías, C.; Martin-Hernandez, R. *Nosema ceranae* (Microsporidia), A Controversial 21st Century Honey Bee Pathogen. *Environ. Microbiol. Rep.* **2013**, *5*, 17–29. [CrossRef]
9. Maini, S.; Medrzycki, P.; Porrini, C. The Puzzle of Honey Bee Losses: A Brief Review. *Bull. Insectology* **2010**, *63*, 153–160.
10. Meixner, M.D.; Kryger, P.; Costa, C. Effects of Genotype, Environment, and Their Interactions on Honey Bee Health in Europe. *Curr. Opin. Insect.* **2015**, *10*, 177–184. [CrossRef]
11. Moritz, R.F.A.; de Miranda, J.; Fries, I.; Le Conte, Y.; Neumann, P.; Paxton, R.J. Research Strategies to Improve Honeybee Health in Europe. *Apidologie* **2010**, *41*, 227–242. [CrossRef]
12. Naug, D. Nutritional Stress due to Habitat Loss may Explain Recent Honeybee Colony Collapses. *Biol. Conserv.* **2009**, *142*, 2369–2372. [CrossRef]
13. Paudel, Y.P.; Mackereth, R.; Hanley, R.; Qin, W. Honey Bees (*Apis mellifera* L.) and Pollination Issues: Current Status, Impacts and Potential Drivers of Decline. *J. Agric. Sci.* **2015**, *7*, 93–109. [CrossRef]
14. Stanimirović, Z.; Glavinić, U.; Ristanić, M.; Aleksić, N.; Jovanović, N.; Vejnović, B.; Stevanović, J. Looking for the Causes of and Solutions to the Issue of Honey Bee Colony Losses. *Acta Vet.* **2019**, *69*, 1–31. [CrossRef]
15. van Engelsdorp, D.; Evans, J.D.; Saegerman, C.; Mullin, C.; Haubruge, E.; Nguyen, B.K.; Frazier, M.; Frazier, J.; Cox-Foster, D.; Chen, Y.; et al. Colony Collapse Disorder: A Descriptive Study. *PLoS ONE* **2009**, *4*, e6481. [CrossRef]
16. Sánchez-Bayo, F.; Wyckhuys, K.A.G. Worldwide Decline of the Entomofauna: A Review of its Drivers. *Biol. Conserv.* **2019**, *232*, 8–27. [CrossRef]
17. Goulson, D.; Nicholls, E.; Botías, C.; Rotheray, E.L. Bee Declines Driven by Combined Stress from Parasites, Pesticides, and Lack of Flowers. *Science* **2015**, *347*, 1255957. [CrossRef]

18. Fries, I.; Camazine, S.; Sneyd, J. Population Dynamics of *Varroa jacobsoni*: A Model and a Review. *Bee World* **1994**, *75*, 5–28. [CrossRef]
19. Le Conte, Y.; Ellis, M.; Ritter, W. *Varroa* Mites and Honey Bee Health: Can *Varroa* Explain Part of the Colony Losses? *Apidologie* **2010**, *41*, 353–363. [CrossRef]
20. Sammataro, D.; Gerson, U.; Needham, G. Parasitic Mites of Honey Bees: Life History, Implications, and Impact. *Annu. Rev. Entomol.* **2000**, *45*, 519–548. [CrossRef]
21. Solignac, M.; Cornuet, J.-M.; Vautrin, D.; Le Conte, Y.; Anderson, D.; Evans, J.; Cros-Arteil, S.; Navajas, M. The Invasive Korea and Japan Types of *Varroa destructor*, Ectoparasitic Mites of the Western Honeybee (*Apis mellifera*), are Two Partly Isolated Clones. *Proc. Royal Soc. B* **2005**, *272*, 411–419. [CrossRef] [PubMed]
22. Potts, S.G.; Roberts, S.P.M.; Dean, R.; Marris, G.; Brown, M.; Jones, R.; Neumann, P.; Settele, J. Declines of Managed Honey Bees and Beekeepers in Europe. *J. Apic. Res.* **2010**, *49*, 15–22. [CrossRef]
23. Rosenkranz, P.; Aumeier, P.; Ziegelmann, B. Biology and Control of *Varroa destructor*. *J. Invertebr. Pathol.* **2010**, *103*, 96–119. [CrossRef]
24. Guzmán-Novoa, E.; Eccles, L.; Calvete, Y.; McGowan, J.; Kelly, P.G.; Correa-Benítez, A. *Varroa destructor* is the Main Culprit for the Death and Reduced Populations of Overwintered Honey Bee (*Apis mellifera*) Colonies in Ontario, Canada. *Apidologie* **2010**, *41*, 443–450. [CrossRef]
25. van Dooremalen, C.; Gerritsen, L.; Cornelissen, B.; van der Steen, J.J.M.; van Langevelde, F.; Blacquière, T. Winter Survival of Individual Honey Bees and Honey Bee Colonies Depends On Level of *Varroa destructor* infestation. *PLoS ONE* **2012**, *7*, e36285. [CrossRef]
26. Büchler, R.; Berg, S.; Le Conte, Y. Breeding for Resistance to *Varroa destructor* in Europe. *Apidologie* **2010**, *41*, 393–408. [CrossRef]
27. Oleksa, A.; Gawroński, R.; Tofilski, A. Rural Avenues as a Refuge for Feral Honey Bee Population. *J. Insect Conserv.* **2013**, *17*, 465–472. [CrossRef]
28. Le Conte, Y.; de Vaublanc, G.; Crauser, D.; Jeanne, F.; Rousselle, J.-C.; Bécard, J.-M. Honey Bee Colonies that Have Survived *Varroa destructor*. *Apidologie* **2007**, *38*, 566–572. [CrossRef]
29. Seeley, T.D. Honey Bees of the Arnot Forest: A Population of Feral Colonies Persisting with *Varroa destructor* in the Northeastern United States. *Apidologie* **2007**, *38*, 19–29. [CrossRef]
30. Seeley, T.D. Life-History Traits of Wild Honey Bee Colonies Living in Forests Around Ithaca, NY, USA. *Apidologie* **2017**, *48*, 743–754. [CrossRef]
31. Seeley, T.D.; Tarpy, D.R.; Griffin, S.R.; Carcione, A.; Delaney, D.A. A Survivor Population of Wild Colonies of European Honeybees in the Northeastern United States: Investigating Its Genetic Structure. *Apidologie* **2015**, *46*, 654–666. [CrossRef]
32. Rinderer, T.E.; De Guzman, L.I.; Delatte, G.T.; Stelzer, J.A.; Lancaster, V.A.; Kuznetsov, V.; Beaman, L.; Watts, R.; Harris, J.W. Resistance to the Parasitic Mite *Varroa destructor* in Honey Bees from Far-Eastern Russia. *Apidologie* **2001**, *32*, 381–394. [CrossRef]
33. Fries, I.; Imdorf, A.; Rosenkranz, P. Survival of Mite Infested (*Varroa destructor*) Honey Bee (*Apis mellifera*) Colonies in a Nordic Climate. *Apidologie* **2006**, *37*, 564–570. [CrossRef]
34. De Jong, D.; Soares, A.E.E. An Isolated Population of Italian Bees that Has Survived *Varroa jacobsoni* Infestation without Treatment for over 12 Years. *Am. Bee J.* **1997**, *137*, 742–745.
35. Locke, B.; Fries, I. Characteristics of Honey Bee Colonies (*Apis mellifera*) in Sweden Surviving *Varroa destructor* Infestation. *Apidologie* **2011**, *42*, 533–542. [CrossRef]
36. Kohl, P.L.; Rutschmann, B. The Neglected Bee Trees: European Beech Forests as a Home for Feral Honey Bee Colonies. *PeerJ* **2018**, *6*, e4602. [CrossRef]
37. Locke, B.; Le Conte, Y.; Crauser, D.; Fries, I. Host Adaptations Reduce the Reproductive Success of *Varroa destructor* in Two Distinct European Honey Bee Populations. *Ecol. Evol.* **2012**, *2*, 1144–1150. [CrossRef]
38. Moritz, R.F.A.; Kraus, F.B.; Kryger, P.; Crewe, R.M. The Size of Wild Honeybee Populations (*Apis mellifera*) and Its Implications for the Conservation of Honeybees. *J. Insect Conserv.* **2007**, *11*, 391–397. [CrossRef]
39. Alaux, C.; Le Conte, Y.; Decourtye, A. Pitting Wild Bees against Managed Honey Bees in Their Native Range, a Losing Strategy for the Conservation of Honey Bee Biodiversity. *Front. Ecol. Evol.* **2019**, *7*, 60. [CrossRef]
40. Requier, F.; Garnery, L.; Kohl, P.L.; Njovu, H.K.; Pirk, C.W.W.; Crewe, R.M.; Steffan-Dewenter, I. The Conservation of Native Honey Bees is Crucial. *Trends Ecol. Evol.* **2019**, *34*, 789–798. [CrossRef] [PubMed]
41. Moro, A.; Blacquière, T.; Panziera, D.; Dietemann, V.; Neumann, P. Host-Parasite Co-Evolution in Real-Time: Changes in Honey Bee Resistance Mechanisms and Mite Reproductive Strategies. *Insects* **2021**, *12*, 120. [CrossRef]
42. Kurze, C.; Routtu, J.; Moritz, R.F. Parasite Resistance and Tolerance in Honeybees at the Individual and Social Level. *Zoology* **2016**, *119*, 290–297. [CrossRef]
43. Blacquière, T.; Panziera, D. A Plea for Use of Honey Bees' Natural Resilience in Beekeeping. *Bee World* **2018**, *95*, 34–38. [CrossRef]
44. Moro, A.; Beaurepaire, A.; Dall'Olio, R.; Rogenstein, S.; Blacquière, T.; Dahle, B.; de Miranda, J.R.; Dietemann, V.; Locke, B.; Licón Luna, R.M.; et al. Using Citizen Science to scout Honey Bee Colonies that Naturally Survive *Varroa destructor* Infestations. *Insects* **2021**, *12*, 536. [CrossRef]
45. Panziera, D.; van Langevelde, F.; Blacquière, T. *Varroa* Sensitive Hygiene Contributes to Naturally Selected Varroa Resistance in Honey Bees. *J. Apic. Res.* **2017**, *56*, 635–642. [CrossRef]

46. Powell, J. Learning from Wild Bees and Tree Beekeeping [The Beekeepers Quarterly, Issue 123]. Available online: https://www.naturalbeekeepingtrust.org/learning-from-wild-bees-trees (accessed on 30 October 2021).
47. Loftus, J.C.; Smith, M.L.; Seeley, T.D. How Honey Bee Colonies Survive in the Wild: Testing the Importance of Small Nests and Frequent Swarming. *PLoS ONE* **2016**, *11*, e0150362. [CrossRef]
48. Thompson, C.E. The Health and Status of the Feral Honeybee (*Apis mellifera* sp) and *Apis mellifera mellifera* Population of the UK. Ph.D. Thesis, University of Leeds, Leeds, UK, November 2012.
49. Baum, K.; Tchakerian, M.D.; Thoenes, S.C.; Coulson, R.N. Africanized Honey Bees in Urban Environments: A Spatio-Temporal Analysis. *Landsc. Urban Plan.* **2008**, *85*, 123–132. [CrossRef]
50. Baum, K.A.; Tchakerian, M.D.; Birt, A.G.; Coulson, R.N. Honey Bee Ecology from an Urban Landscape Perspective. The Spatial Ecology of Feral Honey Bees. In *Silico Bees*; Devillers, J., Ed.; CRC Press, Taylor & Francis Group: Boca Raton, FL, USA, 2014; Chapter 6; pp. 109–134.
51. Wojcik, V.A.; Frankie, G.W.; Thorp, R.W.; Hernandez, J.L. Seasonality in Bees and Their Floral Resource Plants at a Constructed Urban Bee Habitat in Berkeley, California. *J. Kans. Entomol.* **2008**, *81*, 15–28. [CrossRef]
52. Wenzel, A.; Grass, I.; Belavadi, V.V.; Tscharntke, T. How Urbanization is Driving Pollinator Diversity and Pollination—A Systematic Review. *Biol. Conserv.* **2020**, *241*, 108321. [CrossRef]
53. Geslin, B.; Le Féon, V.; Folschweiller, M.; Flacher, F.; Carmignac, D.; Motard, E.; Perret, S.; Dajoz, I. The Proportion of Impervious Surfaces at the Landscape Scale Structures Wild Bee Assemblages in a Densely Populated Region. *Ecol. Entomol.* **2016**, *6*, 6599–6615. [CrossRef]
54. Lawrence, T.J.; Culbert, E.M.; Felsot, A.S.; Hebert, V.R.; Sheppard, W.S. Survey and Risk Assessment of *Apis mellifera* (Hymenoptera: Apidae) Exposure to Neonicotinoid Pesticides in Urban, Rural, and Agricultural Settings. *J. Econ. Entomol.* **2016**, *109*, 520–528. [CrossRef] [PubMed]
55. Prudic, K.L.; Oliver, J.C.; Brown, B.V.; Long, E.C. Comparisons of Citizen Science Data-Gathering Approaches to Evaluate Urban Butterfly Diversity. *Insects* **2018**, *9*, 186. [CrossRef]
56. Dickinson, J.L.; Zuckerberg, B.; Bonter, D.N. Citizen Science as an Ecological Research Tool: Challenges and Benefits. *Annu. Rev. Ecol. Evol. Syst.* **2010**, *41*, 149–172. [CrossRef]
57. Li, E.; Parker, S.S.; Pauly, G.B.; Randall, J.M.; Brown, B.V.; Cohen, B.S. An Urban Biodiversity Assessment Framework that Combines an Urban Habitat Classification Scheme and Citizen Science Data. *Front. Ecol. Evol.* **2019**, *7*, 277. [CrossRef]
58. Statistical Office of the Republic of Serbia. Available online: https://www.stat.gov.rs/en-US (accessed on 1 October 2021).
59. QGIS Development Team, QGIS Geographic Information System. Open Source Geospatial Foundation. Available online: http://qgis.org (accessed on 1 October 2021).
60. Baddeley, A.; Rubak, E.; Turner, R. *Spatial Point Patterns. Methodology and Applications with R*; CRC Press, Taylor & Francis Group: Boca Raton, FL, USA, 2016.
61. Baddeley, A.; Turner, R. Spatstat: An R Package for Analyzing Spatial Point Patterns. *J. Stat. Softw.* **2005**, *12*, 1–42. [CrossRef]
62. R Core Team: A Language and Environment for Statistical Computing, R Foundation for Statistical Computing. Available online: https://www.r-project.org/ (accessed on 1 October 2021).
63. Project SERBHIWE-Honey Bees of Serbia, Wild vs. Managed Colonies through the Eyes of Population Geneticists. Available online: http://www.ibiss.bg.ac.rs/index.php/en/present/national/item/1623-serbhiwe-honey-bees-of-serbia-wild-vs-managed-colonies-through-the-eyes-of-population-geneticists (accessed on 1 September 2021).
64. Patenković, A.; Tanasković, M.; Erić, P.; Erić, K.; Mihailović, M.; Tanasić, V.; Stanisavljević, L.; Davidović, S. Urban Ecosystem Drives Genetic Diversity in Feral Honey Bees' (*Apis mellifera*) Colonies. manuscript in preparation.
65. Spivak, M.; Reuter, G.S. *Varroa destructor* Infestation in Untreated Honey Bee (Hymenoptera: Apidae) Colonies Selected for Hygienic Behavior. *J. Econ. Entomol.* **2001**, *94*, 326–331. [CrossRef]
66. Burley, L.M.; Fell, R.D.; Saacke, R.G. Survival of Honey Bee (Hymenoptera: Apidae) Spermatozoa Incubated at Room Temperature from Drones Exposed to Miticides. *J. Econ. Entomol.* **2008**, *101*, 1081–1087. [CrossRef]
67. Fries, I.; Hansen, H.; Imdorf, A.; Rosenkranz, P. Swarming in Honey Bees (*Apis mellifera*) and *Varroa destructor* Population Development in Sweden. *Apidologie* **2003**, *34*, 389–397. [CrossRef]
68. Fries, I.; Camazine, S. Implications of Horizontal and Vertical Pathogen Transmission for Honey Bee Epidemiology. *Apidologie* **2001**, *32*, 199–214. [CrossRef]
69. van Alphen, J.J.M.; Fernhout, B.J. Natural Selection, Selective Breeding, and the Evolution of Resistance of Honeybees (*Apis mellifera*) against *Varroa*. *Zool. Lett.* **2020**, *6*, 6. [CrossRef] [PubMed]
70. Blacquière, T.; Boot, W.; Calis, J.; Moro, A.; Neumann, P.; Panziera, D. Darwinian Black Box Selection for Resistance to Settled Invasive *Varroa destructor* Parasites in Honey Bees. *Biol. Invasions* **2019**, *21*, 2519–2528. [CrossRef]
71. Rosenthal, J.S. A Review of the Role of Protected Areas in Conserving Global Domestic Animal Diversity. *Anim. Genet. Resour. Inf.* **2010**, *47*, 101–113. [CrossRef]
72. López-Uribe, M.M.; Soro, A.; Jha, S. Conservation Genetics of Bees: Advances in the Application of Molecular Tools to Guide Bee Pollinator Conservation. *Conserv. Genet.* **2017**, *18*, 501–506. [CrossRef]
73. Broeckx, B.J.G.; De Smet, L.; Blacquiere, T.; Maebe, K.; Khalenkow, M.; Van Poucke, M.; Dahle, B.; Neumann, P.; Bach Nguyen, K.; Smagghe, G.; et al. Honey Bee Predisposition of Resistance to Ubiquitous Mite Infestations. *Sci. Rep.* **2019**, *9*, 7794. [CrossRef]
74. Uzunov, A.; Brascamp, E.W.; Büchler, R. The Basic Concept of Honey Bee Breeding Programs. *Bee World* **2017**, *94*, 84–87. [CrossRef]

75. Costa, C.; Büchler, R.; Berg, S.; Bienkowska, M.; Bouga, M.; Bubalo, D.; Charistos, L.; Le Conte, Y.; Drazic, M.; Dyrba, W.; et al. A Europe-Wide Experiment for Assessing the Impact of Genotype-Environment Interactions on the Vitality and Performance of Honey Bee Colonies: Experimental Design and Trait Evaluation. *J. Apic. Sci.* **2012**, *56*, 147–158. [CrossRef]
76. Requier, F.; Paillet, Y.; Laroche, F.; Rutschmann, B.; Zhang, J.; Lombardi, F.; Svoboda, M.; Steffan-Dewenter, I. Contribution of European Forests to Safeguard Wild Honeybee Populations. *Conserv. Lett.* **2019**, *13*, e12693. [CrossRef]
77. Mitchell, D. Ratios of Colony Mass to Thermal Conductance of Tree And Man-Made Nest Enclosures of *Apis mellifera*: Implications for Survival, Clustering, Humidity Regulation and *Varroa destructor*. *Int. J. Biometeorol.* **2016**, *60*, 629–638. [CrossRef] [PubMed]
78. Browne, K.A.; Hassett, J.; Geary, M.; Moore, E.; Henriques, D.; Soland-Reckeweg, G.; Ferrari, R.; Mac Loughlin, E.; O'Brien, E.; O'Driscoll, S.; et al. Investigation of Free-Living Honey Bee Colonies in Ireland. *J. Apic. Res.* **2020**, *60*, 229–240. [CrossRef]
79. Baum, K.A.; Rubink, W.L.; Pinto, A.M.; Coulson, R.N. Spatial and Temporal Distribution and Nest Site Characteristics of Feral Honey Bee (Hymenoptera: Apidae) Colonies in a Coastal Prairie Landscape. *Environ. Entomol.* **2005**, *34*, 610–618. [CrossRef]
80. Collins, A. Feral Friend or Foe? *Nat. Bee Husb.* **2017**, *4*, 4–7.
81. Geoghegan, H.; Dyke, A.; Pateman, R.; West, S.; Everett, G. *Understanding Motivations for Citizen Science. Final Report on Behalf of UKEOF*; University of Reading, Stockholm Environment Institute (University of York) and University of the West of England: Wilthsire, UK, 2016.
82. Saunders, M.E.; Goodwin, E.K.; Santos, K.C.B.S.; Sonter, C.A.; Rader, R. Cavity Occupancy by Wild Honey Bees: Need for Evidence of Ecological Impacts. *Front. Ecol. Environ.* **2021**, *19*, 349–354. [CrossRef]
83. van Strien, A.J.; van Swaay, C.A.M.; Termaat, T.; Devictor, V. Opportunistic Citizen Science Data of Animal Species Produce Reliable Estimates of Distribution Trends If Analysed with Occupancy Models. *J. Appl. Ecol.* **2013**, *50*, 1450–1458. [CrossRef]
84. Jackson, M.M.; Gergel, S.E.; Martin, K. Citizen Science and Field Survey Observations Provide Comparable Results for Mapping Vancouver Island White-Tailed Ptarmigan (*Lagopus leucura saxatilis*) Distributions. *Biol. Conserv.* **2015**, *181*, 162–172. [CrossRef]
85. Zapponi, L.; Cini, A.; Bardiani, M.; Hardersen, S.; Maura, M.; Maurizi, E.; Redolfi De Zan, L.; Audisio, P.; Bologna, M.A.; Carpaneto, G.M.; et al. Citizen Science Data as an Efficient Tool for Mapping Protected Saproxylic Beetles. *Biol. Conserv.* **2016**, *208*, 139–145. [CrossRef]
86. Olkowski, H.; Olkowski, W. Entomophobia in the Urban Ecosystem, Some Observations and Suggestions. *Ann. Entomol. Soc.* **1976**, *22*, 313–318. [CrossRef]
87. Schönfelder, M.L.; Bogner, F.X. Individual Perception Of Bees: Between Perceived Danger and Willingness to Protect. *PLoS ONE* **2017**, *12*, e0180168. [CrossRef]
88. Sumner, S.; Law, G.; Cini, A. Why We Love Bees and Hate Wasps. *Ecol. Entomol.* **2018**, *43*, 836–845. [CrossRef]
89. Dennis, R.L.H.; Thomas, C.D. Bias in Butterfly Distribution Maps: The Influence of Hot Spots and Recorder's Home Range. *J. Insect Conserv.* **2000**, *4*, 73–77. [CrossRef]
90. Thompson, C.E.; Budge, G.E.; Biesmeijer, J.C. Feral Bees in the UK: The Real Story. *Bee Craft* **2010**, *92*, 22–24.
91. Wastian, L.; Unterweger, P.A.; Betz, O. Influence of the Reduction of Urban Lawn Mowing on Wild Bee Diversity (Hymenoptera, Apoidea). *J. Hymenopt. Res.* **2016**, *49*, 51–63. [CrossRef]

insects

MDPI

Article

Honey vs. Mite—A Trade-Off Strategy by Applying Summer Brood Interruption for *Varroa destructor* Control in the Mediterranean Region

Marin Kovačić [1,†], Aleksandar Uzunov [2,3,†], Ivana Tlak Gajger [4,*], Marco Pietropaoli [5], Victoria Soroker [6], Noureddine Adjlane [7], Valerija Benko [4], Leonidas Charistos [8], Raffaele Dall'Olio [9], Giovanni Formato [5], Fani Hatjina [8], Valeria Malagnini [10], Fabrizio Freda [10], Asaf Otmi [6], Zlatko Puškadija [1], Claudio Villar [11] and Ralph Büchler [12]

1 Faculty of Agrobiotechnical Sciences Osijek, J.J. Strossmayer University of Osijek, V. Preloga 1, 31000 Osijek, Croatia; marin.kovacic@fazos.hr (M.K.); zpuskadi@fazos.hr (Z.P.)
2 Faculty of Agricultural Sciences and Food, Ss. Cyril and Methodius University in Skopje, 1000 Skopje, North Macedonia; uzunov@fznh.ukim.edu.mk
3 State Key Laboratory of Resource Insects, Institute of Apicultural Research, Chinese Academy of Agricultural Sciences, Beijing 100193, China
4 Department for Biology and Pathology of Fish and Bees, Faculty of Veterinary Medicine, University of Zagreb, Heinzelova ul. 55, 10000 Zagreb, Croatia; vbenko@vef.unizg.hr
5 Istituto Zooprofilattico Sperimentale del Lazio e della Toscana "M. Aleandri", Via Appia Nuova 1411, 00178 Rome, Italy; marco.pietropaoli@izslt.it (M.P.); giovanni.formato@izslt.it (G.F.)
6 Agricultural Research Organization (ARO), The Volcani Center, 68 HaMacabim Road, Rishon LeZion 7505101, Israel; sorokerv@volcani.agri.gov.il (V.S.); assafot@volcani.agri.gov.il (A.O.)
7 Department of Agronomy, Faculty of Science, University of Boumerdes, 35000 Boumerdes, Algeria; adjlanenoureddine@hotmail.com
8 Department of Apiculture, Institute of Animal Sciences, Ellinikos Georgikos Organismos "DIMITRA", 63 200 Nea Moudania, Greece; leoharistos@gmail.com (L.C.); fhatjina@gmail.com (F.H.)
9 BeeSources, 40132 Bologna, Italy; raffaele.dallolio@gmail.com
10 Fondazione Edmund Mach, Centro Trasferimento Tecnologico, Via E. Mach, 1 San Michele all'Adige, 38098 Trento, Italy; valeria.malagnini@fmach.it (V.M.); fabriziofreda89@gmail.com (F.F.)
11 Consejería de Agricultura de la Junta de Comunidades de Castilla La Mancha, 02600 Albacete, Spain; claudiov@jccm.es
12 Hintergasse 30, 35274 Kirchhain, Germany; ralph.buechler@t-online.de
* Correspondence: itlak@vef.unizg.hr
† Marin Kovacic and Aleksandar Uzunov shared first authorship.

Citation: Kovačić, M.; Uzunov, A.; Tlak Gajger, I.; Pietropaoli, M.; Soroker, V.; Adjlane, N.; Benko, V.; Charistos, L.; Dall'Olio, R.; Formato, G.; et al. Honey vs. Mite—A Trade-Off Strategy by Applying Summer Brood Interruption for *Varroa destructor* Control in the Mediterranean Region. *Insects* **2023**, *14*, 751. https://doi.org/10.3390/insects14090751

Academic Editor: Kirk E. Anderson

Received: 1 August 2023
Revised: 3 September 2023
Accepted: 5 September 2023
Published: 7 September 2023

Simple Summary: Ectoparasitic mite *Varroa destructor* with its associated viruses is a common global threat to the health of honey bee colonies. If colonies are not treated, the vast majority die in a 3-year period. Existing acaricides used for treatment are becoming less effective, and new approaches to honey bee protection are required. A reliable method is to create a broodless condition in a colony by preventing the queen from laying eggs, and after 25 days all mites will be exposed to the treatment with organic acids or essential oils. The focus of our study, performed on 178 colonies in six Mediterranean countries, was to compare different periods of queen caging on honey production, colony development, and the effect of treatment. Queen caging had no negative effect on colony strength before the wintering period, while it affected honey production; colonies in which queens were caged two weeks before the main summer nectar flow produced significantly less honey. However, tested colonies ten weeks after the treatment had significantly lower infestation with *V. destructor* mites. This study shows that caging the queen with subsequent oxalic acid treatment 25 days after caging is an efficient method to control *V. destructor* infestation, while the starting point of queen caging in relation to the main summer nectar flow affects honey production.

Abstract: In this study, we investigated the effect of queen caging on honey bee colonies' post-treatment development and the optimal timing of method application on honey production during the main summer nectar flow. We conducted the study in nine apiaries (N = 9) across six Mediterranean countries, with a total of 178 colonies. The colonies were divided into three test groups: QC1, QC2, and C. The QC1 group involved queens caged for a total of 28 days before the expected harvesting day. In the QC2 group, queens were caged for 28 days, but only 14 days before the expected harvesting day. The C group consisted of queens that were not caged, and the colonies received common local treatments. In both the QC1 and QC2 groups, the colonies were treated with a 4.2% oxalic acid (OA) solution by trickling after the queen release. Our findings revealed no significant adverse effects ($p > 0.05$) on colony strength at the end of the study resulting from queen caging. However, significantly lower amounts of honey were extracted from the QC1 group compared to both the QC2 group ($p = 0.001$) and the C group ($p = 0.009$). Although there were no initial differences in *Varroa destructor* infestation between the groups, ten weeks later, a significantly higher infestation was detected in the C group compared to both the QC1 group ($p < 0.01$) and the QC2 group ($p = 0.003$). Overall, our study demonstrates that queen caging, in combination with the use of OA, is an effective treatment for controlling *V. destructor*. However, the timing of caging plays a crucial role in honey production outcomes.

Keywords: honey bee; *Varroa destructor*; queen caging; honey yield

1. Introduction

The *Varroa destructor* mite is an ectoparasite of the honey bee (*Apis mellifera* L.) and is recognized as the leading cause of worldwide colony losses [1,2]. From the beginning of the invasion, beekeepers prioritized using chemical substances, mainly synthetic acaricides [3]. Even over half a century later, synthetic chemicals are commonly used by many beekeeper operations despite the potential of residues in hive products [4,5] and, more importantly, *V. destructor* mite resistance due to overuse of these chemicals [6–8]. These aspects primarily threaten consumers' safety and sustainable beekeeping management.

In parallel, alternative beekeeping techniques, known as api-biotechnical methods, were developed to counteract *V. destructor* with limited or no use of acaricides. A comprehensive overview of different api-biotechnical methods to prevent and control mite infestation is given in an article by Rosenkranz et al., 2010 [9]. Many of those methods in beekeeping, such as screened bottom boards, trapping of mites in worker or drone brood, and colony arrangement, prevent reinfestation. Methods relying on a brood interruption during the active beekeeping season, followed by oxalic acid treatment, are currently gaining popularity among the beekeeping and research communities [10,11]. The fundamental mechanism behind this approach is trapping and physical removal of the mites in the sealed brood and/or treating the exposed mites (known as the phoretic or dispersal stage [12,13]), during the broodless conditions in the colony. Thus, the methods of brood removal, queen caging, and trapping comb seem best suited to the various beekeeping practices, particularly for the geographical regions with prolonged brood rearing [13].

Among the available acaricides for *V. destructor* control, oxalic acid shows high efficacy [14–19], does not leave residues in beehive products [20,21], and does not lead to resistance phenomena [22]. However, to achieve a high acaricide efficacy, colonies should be in a broodless stage, which in temperate climates may naturally happen only for a short period during the winter or seldom in dry summer season. In the brood's presence, oxalic acid's efficacy is less than 50% [19,23,24].

Several studies have shown that summer brood interruption combined with a subsequent OA application, either via the trickle or sublimation method, is an effective strategy to reduce *V. destructor* infestation [10,11,16,25–28] and virus load [28–30]. Furthermore, no adverse effects on honey production early in the season [31], and colony strength before winter, were detected [32,33].

To create a broodless condition, beekeepers can confine the queen for a defined period [10,13]. By caging a queen for 21 to 25 days (depending on the presence of drone brood), the colony becomes broodless, forcing mites into the dispersal phase when they are susceptible to organic acid treatments, like oxalic acid. Büchler et al. [10] demonstrated high efficacy of the method when 4.2% oxalic acid was applied by trickling after the caging period of 25 days. Previous results of studies combining queen caging and oxalic acid treatment look promising, but it is important to consider the consequences of such a treatment on honey production and honey bee colony development.

Beekeepers from both hobby and commercial sectors are predominantly concerned about queen performance, colony development, and honey production. Therefore, our study investigated both the timing and effect of the queen caging method combined with an oxalic acid treatment on the post-treatment colony development and honey production during the main summer nectar flow.

2. Materials and Methods

The experiment was conducted in six Mediterranean countries in the summer of 2021 (Figure 1). A total of nine test apiaries and 178 honey bee colonies were involved in the study. The study protocol (Supplement File S1) involved selecting full-size colonies of similar comparative strength in each apiary and dividing them into three homogeneous groups.

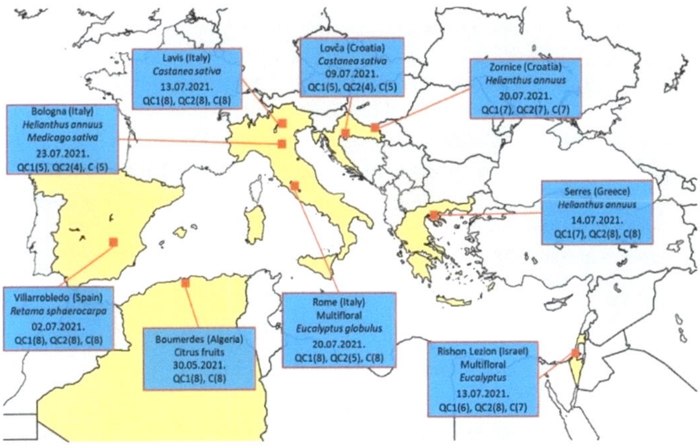

Figure 1. Map with test apiary's location, main summer nectar sources, date of honey extraction (day 0), and size of the groups at each testing apiary.

In the first (QC1) and second (QC2) groups, queens were caged for a total of 28 days in a small-sized cage without the possibility of laying eggs [10] (Figure 2). Briefly, queens from the QC1 group were caged 28 days before "day 0" (day of the expected honey harvest of the main summer nectar flow), while queens from the QC2 group were caged 14 days prior to, and released 14 days after, "day 0". In the control (C) group, queens were not caged. Honey bee colonies in QC1 and QC2 groups were treated, after queen release, on day 0 and day 14, respectively, by trickling 5 mL of oxalic acid 4.2% solution per occupied comb [10], while control colonies were treated using the usual local treatment (such as Apivar, Apiraz, CheckMite, formic acid, and total brood removal).

Colony strength was assessed by counting the number of combs occupied by adult bees and combs with brood, as previously described [34]. The net amount of honey produced by each colony was measured by weighing the honey super before and after extraction. The infestation rate of *V. destructor* on adult bees was determined using either the alcohol/soapy water wash or the powder sugar shake method [35]. The number of *V. destructor* mites per 100 bees was calculated following the method described by Dietemann et al., 2013 [36]. As a general rule, no major colony management techniques/methods that could potentially bias colony development and mite population growth were applied during the test period.

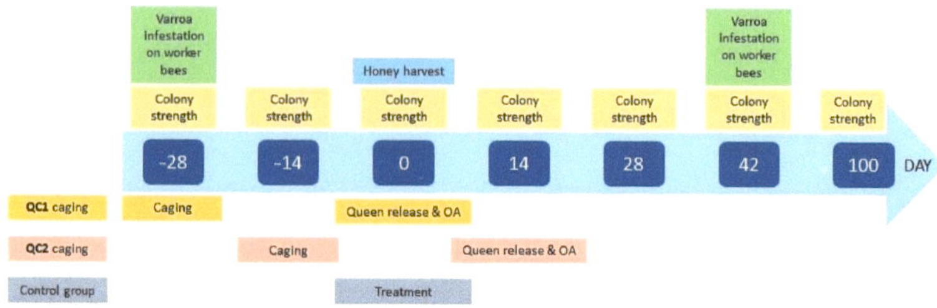

Figure 2. Timeline of the activities during the course of the study. Assessments listed above the timeline were performed on all colonies at the given day. Activities listed below the timeline were performed on a specific group at the given day.

The timeline of the study activities is reported in Figure 2, where the timing of the queens' caging, the frequency of honey bee colony strength estimations, the monitoring of *V. destructor* infestation, and the colony treatments are also shown.

Statistical analysis was performed in the SPSS software package, release 19.0 (SPSS Inc., Chicago, IL, USA). The effect of the fixed factors, namely, location (apiary, N = 9), study group (Q1, Q2, C), and their interaction (N = 16) on honey bee colony strength (number of combs occupied with bees and number of brood combs), were analyzed using a GLM ANOVA model. The same GLM model was applied to analyze honey production (only one measurement) and *V. destructor* infestation at the beginning and at the end of the experiment. Adjusted means between study groups were compared using Bonferroni post hoc analysis. Pearson's correlation (r) analysis was used to calculate the correlation between colony strength and honey production.

3. Results

3.1. Colony Strength

The apiaries involved differed significantly ($p < 0.01$) in colony strength, as assessed by the number of combs occupied with bees and the number of brood combs at the beginning (day −28) and end (day 100) of the study (Tables 1 and 2). However, there were no significant differences in colony strength between the groups at the beginning and end of the study ($p > 0.05$, Figures 3 and 4). On the inspection at "day −14", the QC2 group had a significantly higher number of combs with bees compared to the other two groups ($p < 0.01$), while on the next two measurements (days 0 and 14), the QC1 group had a significantly lower number of combs with bees compared to the other two groups. On "day 28" and "day 42", the colonies from the C group had significantly more combs occupied with bees compared to the other groups, as a consequence of queen caging.

Table 1. GLM analysis on colony strength (number of combs with bees and combs with brood) at the start of the experiment (at day −28), with apiary, group, and their interaction as fixed factors.

	Number of Combs with Bees			Number of Brood Combs		
Source	**df**	**Mean Square**	**F**	**df**	**Mean Square**	**F**
Model	26	1504.739	359.315 **	26	447.303	260.864 **
Apiary	8	570.201	136.158 **	8	80.272	46.814 **
Group	2	1.172	0.411	2	2.017	1.176
Apiary × Group	15	3.020	0.721	15	3.263	1.903 *
Error	152	4.188		152	1.715	
Total	178			178		
	R^2 = 0.984 (adjusted R^2 = 0.981)			R^2 = 0.978 (adjusted R^2 = 0.974)		

** $p < 0.01$; * $p < 0.05$ Bonferroni test.

Table 2. GLM analysis on colony strength (number of combs with bees and combs with brood) at the end of the experiment (at day 100), with apiary, group, and their interaction as fixed factors.

	Number of Combs with Bees			Number of Brood Combs		
Source	**df**	**Mean Square**	**F**	**df**	**Mean Square**	**F**
Model	26	297.319	123.159 **	26	43.323	12.773 **
Apiary	8	84.157	34.976 **	8	29.556	8.714 **
Group	2	0.538	0.224	2	2.786	0.821
Apiary × Group	15	2.978	1.237	15	1.805	0.532
Error	133	2.406		133	3.392	
Total	159			159		
	$R^2 = 0.960$ (adjusted $R^2 = 0.952$)			$R^2 = 0.714$ (adjusted $R^2 = 0.658$)		

** $p < 0.01$; Bonferroni test.

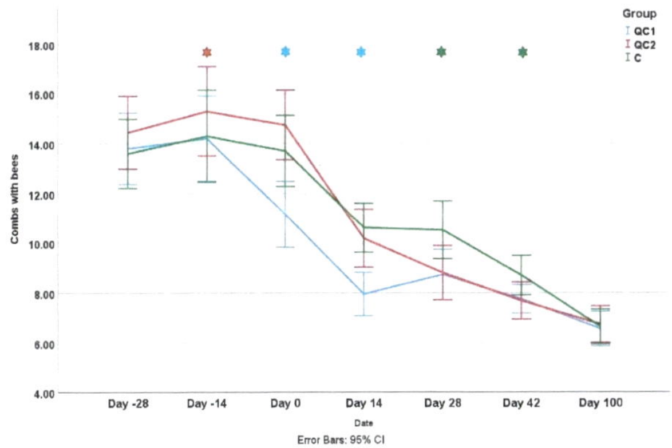

Figure 3. The mean number of combs occupied with bees during the experiment for three different groups. Stars represent significant differences between groups at a certain inspection day and the color of the star shows which group differs.

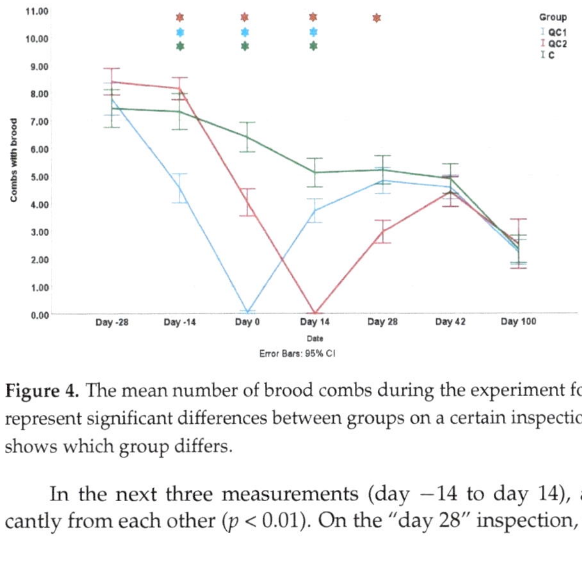

Figure 4. The mean number of brood combs during the experiment for three different groups. Stars represent significant differences between groups on a certain inspection day and the color of the star shows which group differs.

In the next three measurements (day −14 to day 14), all groups differed significantly from each other ($p < 0.01$). On the "day 28" inspection, the QC2 group had significantly fewer brood combs compared to the other two groups ($p < 0.01$), while in the last two inspections, there were no significant differences in the amount of brood.

3.2. Honey Production

The honey yield differed significantly between the groups and apiaries (Table 3). The average amount of extracted honey per colony across all apiaries was 14.353 ± 0.429 kg (mean ± SD). Generally, the highest honey production was recorded in Israel (29.5 ± 7.15 kg), while the lowest was in Greece (8.19 ± 2.44 kg). Significantly lower amounts of honey were extracted from the QC1 group compared to both the QC2 ($p = 0.001$) and C groups ($p = 0.009$, Table 4. However, there was no significant difference between the QC2 and C groups. The initial colony strength had a significant effect on production. Pearson's correlation analysis revealed a significant moderate positive correlation between honey extraction and the number of combs occupied with bees (r = 0.629, $p < 0.01$) as well as the number of brood combs (r = 0.257, $p < 0.01$) on "day −28" at the beginning of the study.

Table 3. GLM analysis for honey production. Apiary, group, and their interaction are set as fixed effects.

Source of Variation	df	Mean Square	F
Model	26	2087.885	67.980 **
Apiary	8	2146.907	69.901 **
Group	2	199.852	6.507 **
Apiary ×Group	15	80.000	2.605 **
Error	151	30.713	
Total	177		

R^2 = 0.921 (adjusted R^2 = 0.908); ** significance < 0.01; (** $p < 0.01$; Bonferroni test).

Table 4. Estimated marginal means of honey extraction (in kg) for different groups.

Group	Mean	Standard Error	95% Confidence Interval	
			Lower Bound	Upper Bound
QC1	12.091 [a]	0.729	10.650	13.532
QC2	16.009 [b]	0.803	14.423	17.596
C	15.142 [b]	0.705	13.749	16.535

[a,b] Different apex letters represent significant differences between the groups (adjustment for multiple comparisons: Bonferroni, $p < 0.05$).

3.3. *V. destructor* Infestation

At the beginning of the field study, the infestation of adult bees with *V. destructor* varied between apiaries but not between groups (Table 5). On "day −28", the infestation of adult bees (mean ± standard error) in the QC1, QC2, and C groups was 1.64 ± 0.19, 1.48 ± 0.22, and 1.52 ± 0.19 mites per 100 bees, respectively. However, on day 42 (following the completion of the control group treatment), there were significant differences in the infestation rate both between apiaries and between the groups. Specifically, the infestation rates were reduced to 0.44 ± 0.13 (QC1), 0.63 ± 0.16 (QC2), and 1.18 ± 0.13 (C) mites per 100 bees, with the C group exhibiting a significantly higher infestation compared to the QC1 group ($p < 0.01$) and QC2 group ($p = 0.003$).

Table 5. GLM analysis on the infestation of colonies with *V. destructor* at the beginning (day −28) and end of the experiment (day 42) with apiary and group as fixed factors.

	Infestation with *V. destructor* on day −28			Infestation with *V. destructor* on day 42		
Source	**df**	**Mean Square**	**F**	**df**	**Mean Square**	**F**
Model	11	61.277	25.277 **	11	15.185	15.186 **
Apiary	8	30.962	12.772 **	8	5.170	5.171 **
Group	2	0.417	0.172	2	8.894	8.894 **
Error	167	2.424		152	1.000	
Total	178			163		
	R^2 = 0.625 (adjusted R^2= 0.600)			R^2 = 0.542 (adjusted R^2= 0.511)		

** $p < 0.01$; Bonferroni test.

4. Discussion

Here, we studied the impact of caging the queen at different times during the main summer nectar flow in combination with an oxalic acid treatment on honey production, *V. destructor* population, and honey bee colony development until winter. Our data support the usage of queen caging to achieve artificial summer brood interruption and the following oxalic acid application as a strategy for efficient *V. destructor* control. The number of adult bees in the autumn is negatively correlated to the *V. destructor* infestation level in the previous summer. Further, the number of bees in spring is negatively correlated to the *V. destructor* infestation levels in the previous October [37,38], so an efficient method of treating against Varroosis during the summer would improve the adult bee population needed for overwintering. We found that the timing of queen caging played an important role in the subsequent productivity of the colonies. Honey bee colonies in which queens were caged at the beginning of the summer nectar flow (QC2 group) produced, on average, the same amount of honey as the control colonies (where queens were not caged), while colonies in the QC1 group produced significantly less (on average 3–4 kg or 20–25%). On the other hand, both caging groups had significantly lower mite infestation at the end of the experiment compared to the control group, thus demonstrating the high efficacy of the caging method. It should be noted that we only measured the harvested honey regardless of the honey stores in the brood chamber. During the period of brood interruption, colonies usually store part of their honey in the brood chamber, which will afterward be used for new brood development and may reduce the need for extra feeding.

The strength of the colonies in different groups at the beginning of the study was equal, as well as at the end, before the winter. Still, there is an obvious positive correlation between colony strength and honey production. This indicates that beekeepers need to closely monitor and maximize honey bee colony strength, particularly the adult bee population, prior to the honey flow. Even if this is a trivial recommendation, one should keep in mind the significant differences between the regions and climates and the recommendations provided by the literature for the particular region. Thus, our results show that the timing is equally as relevant as the method. One of the effects that may be expected following the brood break is that after a few weeks, young bees have low juvenile hormone titers [39] and high protein and vitellogenin concentrations [33], as in long-lived wintering bees, and live significantly longer [1,31]. After the queen is caged, the amount of brood that needs to be fed decreases, so young bees can reach higher longevity and may start foraging earlier [40,41]. This can at least partly explain why the caging groups reached the same wintering colony size as the control colonies, although those had a higher overall brood production.

The starting point in combating *V. destructor* should consider the effect of management strategy on honey production, as this is the hive product of greatest interest for most beekeepers. For instance, brood breaks resulting from swarming negatively affect mite population development [37,42] but also honey production [28,43]. Therefore, our first point of interest was how a different starting time of queen caging in relation to the beginning of the main summer nectar flow would affect the amount of the extracted honey. In our study, honey production was highest in the QC2 and control groups, showing that caging the queens two weeks before the start of the main summer nectar flow (group QC1) is too early. A possible reason for lower nectar intake is that the strength of these colonies dwindled when the summer flow started. In addition, the lower amount of brood pheromones may have a negative impact on nectar intake [44]. Colonies in which queens were caged at the beginning of the nectar flow (QC2) were as productive as colonies from the control group. Decreased number of bees after the honey harvest in caging groups was no longer so important from the beekeepers' aspect because the strength of the colonies at day 100 was equal. However, caution should be taken when using this queen caging method, and adaptation to the local environment is recommended as differences occur in the duration of brooding and nectar flow among the different geographic regions [38]. If there is late summer or fall nectar flow expected, the question is how this would affect possible additional honey harvest, as the tested group of colonies reached the control colonies in strength before winter. In addition, once the queen is released from the cage into a crowded hive, she starts to lay intensively, and the resulting few frames of open brood might lead some foragers to revert to nurse bees [45]. However, we did not measure the strength of colonies from day 42 until day 100, and we did not distinguish when in this period colonies equalized in strength. Similar values were obtained by Kovačić et al. [27], where colonies with caged queens had a 20–35% bee population reduction 28 days after queen release. In the work of Lodesani et al. [31], equalization between caging and control groups happened at least 67 days after queen release, which corresponds to the three weeks after day 42 in our study (three complete brood cycles instead of two). Brood interruption by queen caging in September seems to be late, as it affects the strength of the colonies entering the winter [46]. On the other hand, early spring queen caging is shown to be effective in reducing mite load without a negative effect on honey production and final colony size when caging is performed 9 weeks before the main spring nectar flow [31].

At the beginning of the study, there were no significant differences in the infestation rate of *V. destructor* of the colonies from different groups. However, upon measurement of the infestation rate after the treatment, we found a significantly lower infestation rate in caging groups compared to the control group. The control group of the study was treated as "business as usual" and consisted of different well-known and verified methods by partners. However, in this testing season, brood interruption followed by the OA treatment was shown to be more effective. This confirms the results of the previous study [27], where caging groups also had higher efficacy. In this study, we used 4.2% oxalic acid solution, which is proven to be effective [10], and the correct concentration and dosage of treatment are two of the most important details which should be considered, as lower concentrations when using the trickling method [10] or lower quantity when using sublimation [18,46] will result in lower efficacy. It is essential to highlight that the brood break is also an effective control method for *Tropilaelaps* spp. mite [47], a new possible threat to the European beekeeping industry [48]. From an economic point of view, it is possible to reduce costs since low *V. destructor* infestation leads to lower cost requirements for treatments, higher quality products, vital colonies, higher survival rates, and fewer winter colony losses which, according to Popovska Stojanov et al. [49], has substantial economic negative consequences on the overall beekeeping operation.

It is essential to emphasize the advantages of the tested method from the aspect of food safety as this approach does not compromise honey or other products in the hive as oxalic acid does not leave residues. One of the main challenges for successful *V. destructor* management is to reduce the infestation level in time before the development of long-living winter bees starts. While most registered chemical products may not be applied before the last honey harvest, which is often too late in the season, brood interruption can be started some weeks earlier without adverse effects on honey production and the in-hive products' safety. Given the growing reports of resistance of mites to the active substances of medicines [7,8,50,51] and the negative effects of pesticides residues in wax on drone semen viability [52] and on workers longevity [53], future strategies of colonies protection should mainly focus on biotechnical methods and breeding honey bees with increased resistance against *V. destructor* mites [54–56].

Supplementary Materials: The following supporting information can be downloaded at: https://www.mdpi.com/article/10.3390/insects14090751/s1, Supplement file S1: Coloss_protocol_2021.

Author Contributions: Conceptualization, R.B., M.K. and A.U.; methodology, M.K., A.U., Z.P. and R.B.; investigation, M.K., M.P., V.S., N.A., V.B., L.C., R.D., G.F., F.H., V.M., F.F., A.O., Z.P., I.T.G. and C.V.; writing—original draft preparation, M.K., A.U., M.P. and V.S.; writing—review and editing, M.K., A.U., M.P., V.S., N.A., V.B., L.C., R.D., G.F., F.H., V.M., F.F., A.O., Z.P., I.T.G., C.V. and R.B. All authors have read and agreed to the published version of the manuscript.

Funding: This research received no external funding.

Data Availability Statement: The datasets analyzed during the present study are available upon reasonable request from the corresponding authors.

Acknowledgments: We would like to acknowledge the "COLOSS" honey bee research association, which enabled us to organize meetings and to develop successful and effective cooperation.

Conflicts of Interest: The authors declare no conflict of interest.

References

1. van Dooremalen, C.; Gerritsen, L.; Cornelissen, B.; van der Steen, J.J.M.; van Langevelde, F.; Blacquière, T. Winter Survival of Individual Honey Bees and Honey Bee Colonies Depends on Level of *Varroa destructor* Infestation. *PLoS ONE* **2012**, *7*, e36285. [CrossRef]
2. Thoms, C.A.; Nelson, K.C.; Kubas, A.; Steinhauer, N.; Wilson, M.E.; vanEngelsdorp, D. Beekeeper Stewardship, Colony Loss, and *Varroa destructor* Management. *Ambio* **2019**, *48*, 1209–1218. [CrossRef]
3. De la Rúa, P.; Jaffé, R.; Dall'Olio, R.; Muñoz, I.; Serrano, J. Biodiversity, Conservation and Current Threats to European Honeybees. *Apidologie* **2009**, *40*, 263–284. [CrossRef]
4. Martel, A.C.; Zeggane, S.; Aurières, C.; Drajnudel, P.; Faucon, J.P.; Aubert, M. Acaricide Residues in Honey and Wax after Treatment of Honey Bee Colonies with Apivar® or Asuntol® 50. *Apidologie* **2007**, *38*, 534–544. [CrossRef]
5. Jamal, M.; Aziz, M.A.; Naeem, M.; Iqbal, Z.; Khalid, A.; Siddique, F.; Khan, K.A.; Ghramh, H.A. Detection of Flumethrin Acaricide Residues from Honey and Beeswax Using High Performance Liquid Chromatography (HPLC) Technique. *J. King Saud Univ. Sci.* **2020**, *32*, 2229–2235. [CrossRef]
6. Almecija, G.; Poirot, B.; Cochard, P.; Suppo, C. Inventory of *Varroa destructor* Susceptibility to Amitraz and Tau-Fluvalinate in France. *Exp. Appl. Acarol.* **2020**, *82*, 1–16. [CrossRef]
7. Higes, M.; Martín-Hernández, R.; Hernández-Rodríguez, C.S.; González-Cabrera, J. Assessing the Resistance to Acaricides in *Varroa destructor* from several Spanish locations. *Parasitol. Res.* **2020**, *119*, 3595–3601. [CrossRef] [PubMed]
8. Rinkevich, F.D. Detection of Amitraz Resistance and Reduced Treatment Efficacy in the Varroa Mite, *Varroa destructor*, within Commercial Beekeeping Operations. *PLoS ONE* **2020**, *15*, e0227264. [CrossRef]
9. Rosenkranz, P.; Aumeier, P.; Ziegelmann, B. Biology and Control of *Varroa destructor*. *J. Invertebr. Pathol.* **2010**, *103*, S96–S119. [CrossRef] [PubMed]
10. Büchler, R.; Uzunov, A.; Kovačić, M.; Prešern, J.; Pietropaoli, M.; Hatjina, F.; Pavlov, B.; Charistos, L.; Formato, G.; Galarza, E.; et al. Summer Brood Interruption as Integrated Management Strategy for Effective Varroa Control in Europe. *J. Apic. Res.* **2020**, *59*, 764–773. [CrossRef]
11. Pietropaoli, M.; Giacomelli, A.; Milito, M.; Pizzariello, M.; Gobbi, C.; Scholl, F.; Formato, G. Integrated Pest Management Strategies against *Varroa destructor*: The Use of Oxalic Acid Combined with Innovative Cages to Obtain the Absence of Brood. *Eur. J. Integr. Med.* **2012**, *15*, 93.
12. Traynor, K.S.; Mondet, F.; de Miranda, J.R.; Techer, M.; Kowallik, V.; Oddie, M.A.Y.; Chantawannakul, P.; McAfee, A. *Varroa destructor*: A Complex Parasite, Crippling Honey Bees Worldwide. *Trends Parasitol.* **2020**, *36*, 592–606. [CrossRef]
13. Uzunov, A.; Büchler, R. *Alternative Methods for Varroa Control in Honey Bee Colonie*; Swiss Agency for Development and Cooperation—SDC: Skopje, North Macedonia, 2020.
14. Imdorf, A.; Charrière, J.D.; Bachofen, B. Efficiency Checking of the *Varroa jacobsoni* Control Methods by Means of Oxalic Acid. *Apiacta* **1997**, *32*, 89–91.
15. Gregorc, A.; Planinc, I. The Control of *Varroa destructor* Using Oxalic Acid. *Vet. J.* **2002**, *163*, 306–310. [CrossRef]
16. Nanetti, A.; Büchler, R.; Charrière, J.-D.; Friesd, I.; Helland, S.; Imdorf, A.; Korpela, S.; Kristiansen, P. Oxalic Acid Treatments for *Varroa control* (Review). *Apiacta* **2003**, *38*, 81–87.

17. Al Toufailia, H.; Scandian, L.; Ratnieks, F.L.W. Towards Integrated Control of Varroa: 2) Comparing Application Methods and Doses of Oxalic Acid on the Mortality of Phoretic *Varroa destructor* Mites and Their Honey Bee Hosts. *J. Apic. Res.* **2015**, *54*, 108–120. [CrossRef]
18. Berry, J.A.; Bartlett, L.J.; Bruckner, S.; Baker, C.; Braman, S.K.; Delaplane, K.S.; Williams, G.R. Assessing Repeated Oxalic Acid Vaporization in Honey Bee (Hymenoptera: Apidae) Colonies for Control of the Ectoparasitic Mite *Varroa destructor*. *J. Insect Sci.* **2022**, *22*, 15. [CrossRef] [PubMed]
19. Kulhanek, K.; Hopkins, B.K.; Sheppard, W.S. Comparison of Oxalic Acid Drip and HopGuard for Pre-Winter *Varroa destructor* Control in Honey Bee (*Apis mellifera*) Colonies. *J. Apic. Res.* **2022**. [CrossRef]
20. Bogdanov, S.; Charrière, J.D.; Imdorf, A.; Kilchenmann, V.; Fluri, P. Determination of Residues in Honey after Treatments with Formic and Oxalic Acid under Field Conditions. *Apidologie* **2002**, *33*, 399–409. [CrossRef]
21. Rademacher, E.; Harz, M. Oxalic Acid for the Control of Varroosis in Honey Bee Colonies—A Review. *Apidologie* **2006**, *37*, 98–120. [CrossRef]
22. Maggi, M.D.; Damiani, N.; Ruffinengo, S.R.; Brasesco, M.C.; Szawarski, N.; Mitton, G.; Mariani, F.; Sammataro, D.; Quintana, S.; Eguaras, M.J. The Susceptibility of *Varroa destructor* against Oxalic Acid: A Study Case. *Bull. Insectol.* **2017**, *70*, 39–44.
23. Domatsky, A.N.; Domatskaya, T.F. Effectiveness of Oxalic Acid in Varroatosis in the Apiaries of Tyumen Region, Russia. *Ukr. J. Ecol.* **2018**, *8*, 143–147.
24. Toomemaa, K. The Synergistic Effect of Weak Oxalic Acid and Thymol Aqueous Solutions on Varroa Mites and Honey Bees. *J. Apic. Res.* **2018**, *58*, 37–52. [CrossRef]
25. Gregorc, A.; Adamczyk, J.; Kapun, S.; Planinc, I. Integrated Varroa Control in Honey Bee (*Apis mellifera carnica*) Colonies with or without Brood. *J. Apic. Res.* **2016**, *55*, 253–258. [CrossRef]
26. Gregorc, A.; Alburaki, M.; Werle, C.; Knight, P.R.; Adamczyk, J. Brood Removal or Queen Caging Combined with Oxalic Acid Treatment to Control Varroa Mites (*Varroa destructor*) in Honey Bee Colonies (*Apis mellifera*). *Apidologie* **2017**, *48*, 821–832. [CrossRef]
27. Kovačić, M.; Puškadija, Z.; Charistos, L.; Hatjina, F.; Pietropaoli, M.; Formato, G.; Wilde, J.; Uzunov, A.; Büchler, R. Effect of Timing of Queen Caging on Honey Production. In Proceedings of the Coloss Varroa Task Force Spring 2020 Workshop Proceedings, Pulawy, Poland, 18–21 February 2020.
28. Evans, K.C.; Underwood, R.M.; López-Uribe, M.M. Combined Effects of Oxalic Acid Sublimation and Brood Breaks on Varroa Mite (*Varroa destructor*) and Deformed Wing Virus Levels in Newly Established Honey Bee (*Apis mellifera*) Colonies. *J. Apic. Res.* **2022**, *61*, 197–205. [CrossRef]
29. Büchler, R. Vital Colonies Thanks to Complete Brood Removal. *ADIZ/db/IF* **2009**, 10–12. Available online: brood-removal-web.pdf (accessed on 3 September 2023).
30. Lodesani, M.; Costa, C.; Besana, A.; Dall'Olio, R.; Franceschetti, S.; Tesoriero, D.; Vaccari, G. Impact of Control Strategies for *Varroa destructor* on Colony Survival and Health in Northern and Central Regions of Italy. *J. Apic. Res.* **2014**, *53*, 155–164. [CrossRef]
31. Lodesani, M.; Franceschetti, S.; Dall'Ollio, R. Evaluation of Early Spring Bio-Technical Management Techniques to Control Varroosis in *Apis mellifera*. *Apidologie* **2019**, *50*, 131–140. [CrossRef]
32. Hatjina, F.; Haristos, L. Indirect Effects of Oxalic Acid Administered by Trickling Method on Honey Bee Brood. *J. Apic. Res.* **2005**, *44*, 172–174. [CrossRef]
33. Cabbri, R.; Ferlizza, E.; Nanetti, A.; Monari, E.; Andreani, G.; Galuppi, R.; Isani, G. Biomarkers of Nutritional Status in Honeybee Haemolymph: Effects of Different Biotechnical Approaches for *Varroa destructor* Treatment and Wintering Phase. *Apidologie* **2018**, *49*, 606–618. [CrossRef]
34. Uzunov, A.; Büchler, R.; Costa, C.; Mondet, F.; Bienkowska, M.; Hatjina, F.; Meixner, M.; Andonov, S.; Kovačić, M.; Dall'Olio, R.; et al. *Book of Methods-Performance Testers*; EURBEST Project (AGRI-2017-0346); Bee Institute in Kirchhain: Kirchhain, Germany, 2021.
35. Pietropaoli, M.; Tlak Gajger, I.; Costa, C.; Gerula, D.; Wilde, J.; Adjlane, N.; Sánchez, P.A.; Škerl, M.I.S.; Bubnič, J.; Formato, G. Evaluation of Two Commonly Used Field Tests to Assess *Varroa destructor* Infestation on Honey Bee (*Apis mellifera*) Colonies. *Appl. Sci.* **2021**, *11*, 4458. [CrossRef]
36. Dietemann, V.; Nazzi, F.; Martin, S.J.; Anderson, D.L.; Locke, B.; Delaplane, K.S.; Wauquiez, Q.; Tannahill, C.; Frey, E.; Ziegelmann, B.; et al. Standard Methods for Varroa Research. *J. Apic. Res.* **2013**, *52*, 1–54. [CrossRef]
37. Fries, I.; Imdorf, A.; Rosenkranz, P. Survival of Mite Infested (*Varroa destructor*) Honey Bee (*Apis mellifera*) Colonies in a Nordic Climate. *Apidologie* **2006**, *37*, 564–570. [CrossRef]
38. Hatjina, F.; Costa, C.; Büchler, R.; Uzuno, A.; Drazic, M.; Filipi, J.; Charistos, L.; Ruottinen, L.; Andonov, S.; Meixner, M.D.; et al. Population Dynamics of European Honey Bee Genotypes under Different Environmental Conditions. *J. Apic. Res.* **2014**, *53*, 233–247. [CrossRef]
39. Fluri, P.; Lüscher, M.; Wille, H.; Gerig, L. Changes in Weight of the Pharyngeal Gland and Haemolymph Titres of Juvenile Hormone, Protein and Vitellogenin in Worker Honey Bees. *J. Insect Physiol.* **1982**, *28*, 61–68. [CrossRef]
40. Greenberg, J.K.; Xia, J.; Zhou, X.; Thatcher, S.R.; Gu, X.; Ament, S.A.; Newman, T.C.; Green, P.J.; Zhang, W.; Robinson, G.E.; et al. Behavioral Plasticity in Honey Bees Is Associated with Differences in Brain MicroRNA Transcriptome. *Genes Brain Behav.* **2012**, *11*, 660. [CrossRef] [PubMed]

41. Ma, R.; Rangel, J.; Grozinger, C.M. Honey Bee (*Apis mellifera*) Larval Pheromones May Regulate Gene Expression Related to Foraging Task Specialization. *BMC Genom.* **2019**, *20*, 592. [CrossRef]
42. Fries, I.; Hansen, H.; Imdorf, A.; Rosenkranz, P. Swarming in Honey Bees (*Apis mellifera*) and *Varroa destructor* Population Development in Sweden. *Apidologie* **2003**, *34*, 389–397. [CrossRef]
43. Rangel, J.; Seeley, T.D. Colony Fissioning in Honey Bees: Size and Significance of the Swarm Fraction. *Insectes Sociaux* **2012**, *59*, 453–462. [CrossRef]
44. Lait, C.G.; Borden, J.H.; Kovacs, E.; Moeri, O.E.; Campbell, M.; MacHial, C.M. Treatment with Synthetic Brood Pheromone (SuperBoost) Enhances Honey Production and Improves Overwintering Survival of Package Honey Bee (Hymenoptera: Apidae) Colonies. *J. Econ. Entomol.* **2012**, *105*, 304–312. [CrossRef] [PubMed]
45. Münch, D.; Amdam, G.V. The Curious Case of Aging Plasticity in Honey Bees. *FEBS Lett.* **2010**, *584*, 2496–2503. [CrossRef] [PubMed]
46. Jack, C.J.; Van Santen, E.; Ellis, J.D.; Johnson, R. Evaluating the Efficacy of Oxalic Acid Vaporization and Brood Interruption in Controlling the Honey Bee Pest *Varroa destructor* (Acari: Varroidae). *J. Econ. Entomol.* **2020**, *113*, 582–588. [CrossRef]
47. Roberts, J.M.K.; Schouten, C.N.; Sengere, R.W.; Jave, J.; Lloyd, D. Effectiveness of Control Strategies for *Varroa jacobsoni* and *Tropilaelaps mercedesae* in Papua New Guinea. *Exp. Appl. Acarol.* **2020**, *80*, 399–407. [CrossRef]
48. Chantawannakul, P.; Ramsey, S.; vanEngelsdorp, D.; Khongphinitbunjong, K.; Phokasem, P. Tropilaelaps Mite: An Emerging Threat to European Honey Bee. *Curr. Opin. Insect Sci.* **2018**, *26*, 69–75. [CrossRef]
49. Popovska Stojanov, D.; Dimitrov, L.; Danihlík, J.; Uzunov, A.; Golubovski, M.; Andonov, S.; Brodschneider, R. Direct Economic Impact Assessment of Winter Honeybee Colony Losses in Three European Countries. *Agriculture* **2021**, *11*, 398. [CrossRef]
50. Milani, N. The Resistance of *Varroa jacobsoni* Oud. to Acaricides. *Apidologie* **1999**, *30*, 229–234. [CrossRef]
51. Sammataro, D.; Untalan, P.; Guerrero, F.; Finley, J. The Resistance of Varroa Mites (Acari: Varroidae) to Acaricides and the Presence of Esterase. *Int. J. Acarol.* **2005**, *31*, 67–74. [CrossRef]
52. Fisher, A.; Rangel, J. Exposure to Pesticides during Development Negatively Affects Honey Bee (*Apis mellifera*) Drone Sperm Viability. *PLoS ONE* **2018**, *13*, e0208630. [CrossRef] [PubMed]
53. Wu, J.Y.; Anelli, C.M.; Sheppard, W.S. Sub-Lethal Effects of Pesticide Residues in Brood Comb on Worker Honey Bee (*Apis mellifera*) Development and Longevity. *PLoS ONE* **2011**, *6*, e14720. [CrossRef]
54. Büchler, R.; Berg, S.; Le Conte, Y. Breeding for Resistance to Varroa Destructor in Europe. *Apidologie* **2010**, *41*, 393–408. [CrossRef]
55. Guichard, M.; Dietemann, V.; Neuditschko, M.; Dainat, B. Advances and Perspectives in Selecting Resistance Traits against the Parasitic Mite *Varroa destructor* in Honey Bees. *Genet. Sel. Evol.* **2020**, *52*, 71. [CrossRef] [PubMed]
56. Büchler, R.; Kovačić, M.; Buchegger, M.; Puškadija, Z.; Hoppe, A.; Brascamp, E.W. Evaluation of Traits for the Selection of *Apis mellifera* for Resistance against *Varroa destructor*. *Insects* **2020**, *11*, 618. [CrossRef] [PubMed]

insects

MDPI

Communication

Changes in Lithium Levels in Bees and Their Products Following Anti-*Varroa* Treatment

Éva Kolics [1,2], Zsófi Sajtos [3,4], Kinga Mátyás [1], Kinga Szepesi [1], Izabella Solti [1], Gyöngyi Németh [1], János Taller [1], Edina Baranyai [4], András Specziár [5] and Balázs Kolics [1,2,*]

1 Festetics Bioinnovation Group, Institute of Genetics and Biotechnology, Georgikon Campus, Hungarian University of Agriculture and Life Sciences, H-8360 Keszthely, Hungary; kolicseva@gmail.com (É.K.); petrovicsnemkk@gmail.com (K.M.); szepesikinga66@gmail.com (K.S.); izabella.solti@gmail.com (I.S.); drnemethgyongyi@gmail.com (G.N.); taller.janos@uni-mate.hu (J.T.)
2 Kolics Apiaries, H-8710 Balatonszentgyörgy, Hungary
3 Doctoral School of Chemistry, University of Debrecen, H-4032 Debrecen, Hungary; sajtos.zsofi@science.unideb.hu
4 Atomic Spectrometry Partner Laboratory, Department of Inorganic and Analytical Chemistry, Faculty of Science and Technology, University of Debrecen, H-4032 Debrecen, Hungary; baranyai.edina@science.unideb.hu
5 Balaton Limnological Research Institute, ELKH, H-8237 Tihany, Hungary; specziar.andras@blki.hu
* Correspondence: bkolics@gmail.com; Tel.: +36-302629236

Simple Summary: Varroosis caused by the ectoparasitic mite *Varroa destructor* has been the biggest threat to managed bee colonies over recent decades. Chemicals available to treat the disease imply problems of resistance, inconsistent efficacy, and residues in bee products. Recently, alongside novel compounds to defeat the pest, lithium chloride has been found to be effective. In this study, we found that lithium treatments leave beeswax residue-free. The possibility of decontamination in adult bees, bee bread, and uncapped honey was revealed. On the other hand, ripe honey was found to be affected by lithium administered via feeding. Case studies are necessary to uncover the level of exposition in harvested honey to estimate its potential risk once it becomes a registered veterinary medicine.

Abstract: The biggest threat to beekeeping is varroosis caused by the mite *Varroa destructor*. Chemicals available to treat this fatal disease may present problems of resistance or inconsistent efficacy. Recently, lithium chloride has appeared as a potential alternative. To date, the amount of residue lithium treatments may leave in honeybee products is poorly understood. Honeybees were fed with 25 mM lithiated sugar syrup, which was used in earlier studies. The accumulation and elimination of the lithium were monitored in bees and their products for 22 days. Lithium concentration increased in the entire body of the bees to day 4 post-treatment and then recovered rapidly to the control level. Lithium exposure was found to affect uncapped honey in the short term (<16 days), but ripe (capped) honey measured at the end of the trial remained affected. On the other hand, lithium treatment left beeswax lithium-free. Based on these data, we propose that comprehensive research on harvested honey is needed to decide on the veterinary use of lithium.

Keywords: lithium chloride; beeswax; honey; chemical residues; *Apis mellifera*; *Varroa destructor*

Citation: Kolics, É.; Sajtos, Z.; Mátyás, K.; Szepesi, K.; Solti, I.; Németh, G.; Taller, J.; Baranyai, E.; Specziár, A.; Kolics, B. Changes in Lithium Levels in Bees and Their Products Following Anti-*Varroa* Treatment. *Insects* **2021**, *12*, 579. https://doi.org/10.3390/insects12070579

Academic Editor: Yanping(Judy) Chen

Received: 19 May 2021
Accepted: 21 June 2021
Published: 25 June 2021

1. Introduction

Hive products are associated with nutritional benefits and are value-added products of the human food chain. However, honey, beeswax, and bee bread may be exposed to pesticides as a consequence of anti-*Varroa* treatment.

The overwhelming majority of pollination carried out by honeybees (*Apis mellifera*) is performed by managed bee colonies [1]. The biggest threat to the honeybee worldwide is varroosis, which involves virus diseases transmitted by the ectoparasite *V. destructor* originally parasitizing a closely related species (*A. cerana*). If left untreated, mites can kill an entire colony within one or two years [2,3], but, in areas of a high density of honeybee colonies, it might occur within an apicultural season. Controls currently in practice based on synthetics can be adequate but are restricted to a few chemicals such as amitraz, coumaphos, flumethrin, and fluvalinate, the formulations of which, however, are demonstrated to lead to the risk of development of resistance [4,5]. As a result, they offer a limited possibility of mite eradication in the foreseeable future. With the exception of oxalic acid, one of the most widely used varroacides [6,7], alternatively used essential oils or organic acids in some instances may be inconsistent in efficacy [8]. Alongside novel approaches (RNAi) to treating varroosis, lithium salts were found to be effective in eradicating *V. destructor* in vitro [9]. Although publications concerning the effects of lithium on harnessed bee individuals are available, these remain restricted to physiological studies [10–12].

Concerning other invertebrates (e.g., sea urchins, marine polychaete worms), disturbances in embryonal development were raised in relation to lithium compounds [13]. Interestingly, however, beneficial effects of lithium on longevity were detected in adult individuals of *Drosophila* [14]. For both honeybee adults and brood, adverse effects on the longevity of lithiated sugar syrup administration is reported but restricted mainly to in vitro trials [9,15]. Furthermore, it is of note that a freely moving bee might react differently to aversive compounds. Being an obligatory social organism, a honeybee colony could actively reject aversive substances [12]. Moreover, feeding sugar syrup infused with varroacide is not typically the way of administering an anti-*Varroa* treatment in apicultural practice [16].

Lithium chloride (PubChem CID: 433294) may provide an effective, commercially available, and relatively cheap alternative, and therefore, it may be increasingly applied as an unregulated veterinary medicine [16–19]. Despite its potential to treat varroosis in the short term, only a few studies are dealing with the consequences of lithium treatments on honey and other bee products [15,20]. However, with honey and beeswax being the most remarkable hive products worldwide, human exposure to lithium when it is used needs to be extensively studied in the apiculture-related food chain to evaluate its impacts before it ever becomes a registered veterinary medicine.

We aimed to understand the consequences of anti-*Varroa* treatment using lithium chloride feeding, monitoring the changes of the lithium level in the bees and their most important products.

2. Materials and Methods

2.1. Colony Setup and Samplings of Biological Materials and Apicultural Products

The experiment was started in early October 2018 in Hungary (Keszthely, 46°45′55.6″ N, 17°14′52.6″ E), excluding outer nectar flow. Carniolan (*A. m. carnica*) colony splits were populated into four hives (local type). On the same day (day 6), colonies were transferred away from their flight range into a dark room and kept for five days. Apart from one frame originating from the donor colonies, each hive was equipped with wax foundations (day 3) only. After making sure the colonies were queenright, they were placed outdoors in the evening of day 1. Hives were set at least 3 m away from each other, with geographical landmarks to prevent drifting.

Sampling was initiated on day zero. Before treatment, bees and their products were sampled to measure control lithium concentrations in the four hives. Then, the frame originating from the donor colonies containing the honey and bee bread store from pretreatment was discarded from each hive. The colonies were subsequently fed with one liter of 1:1 sucrose syrup containing 25 mM lithium chloride (126.5 mg kg^{-1} Li+) [9].

Sampling was carried out in a standardized manner to prevent cross-contamination of the hive products as follows. First, bottom boards were cleaned to collect hive debris. Adult bees (25 workers, mixed of age, from each hive in each occasion) were collected from the bee space of the hives to make a pooled sample. To be able to sample beeswax secreted under lithium exposure from the hive, the colony was forced to build brace combs (about

10×10 cm in size). Cells from which the bee bread was collected (2 g from each hive) were marked to prevent their re-sampling (except the pre-treatment control originating from the donor colony). Taken from all combs, 30 mL honey was collected from each hive on each sampling occasion. Samplings were carried out on day 0 (pre-treatment control), and days 1, 4, 8, 16, and 22 (post-treatment) for hive debris (n = 24), bees (n = 24) divided later into three body parts (head, thorax and abdomen, and legs), brace combs (n = 19), bee bread (n = 24), and uncapped (unripe) honey (n = 24).

The experiment was terminated on day 28. Queens were killed to examine their whole body (n = 4). Mature, capped honey was sampled (n = 4). Beeswax was rendered from the combs; during this process, slumgum was collected (n = 4). Furthermore, sediments (n = 4) of the wax cakes and the melting waters (n = 4) in which the wax (n = 4) was processed were collected. Altogether, 139 different samples were collected. An overview of the whole sampling process is visualized in Figure 1.

2.2. Sample Preparation

Samples were stored at −5 °C in plastic tubes before the sample pre-treatment process. Bees' heads were measured separately since pharyngeal glands produce and excrete royal jelly, presenting the food for honeybee larvae. The lithium content of the legs was measured separately.

Bees were separated into the three main parts according to the details above: approx. 75 mg for the head, and the same amount for the legs, and approx. 500 mg for the thorax & abdomen in each sampling. The body parts, as well as the exact known weight of honey and beeswax (0.5 g of each), bee bread, and hive debris (0.1 g of each), were measured on an analytical balance (ES 225SM-DR, Precisa, Dietikon, ZH, USA) into 50 mL glass beakers. Samples were dried at 50 °C to constant weight in an electric drying cabinet.

Dried samples were wet digested in the same vessels by the mixture of 4.0 mL 65% (m/m) HNO_3 (reagent grade, Scharlau, Germany) and 1.0 mL 30% (m/m) H_2O_2 (reagent grade, Merck, Kenilworth, NJ, USA) to evade the cross-contamination from changing glassware. Digested samples were transferred without loss into volume calibrated plastic centrifuge tubes and diluted up to the volume of 10.00 mL with ultrapure water (Synergy UV, Sigma-Aldrich, St. Louis, MO, USA). Solutions were kept at room temperature before further elemental analysis. Each piece of glassware used was decontaminated by immersion in a 1:5 HNO_3: H_2O solution for 24 h and rinsed with deionized water before use.

2.3. Analytical Measurements

The quantitative analysis of the lithium content in the different samples was carried out by microwave plasma atomic emission spectrometry (MP-AES 4200, Agilent Technologies, Santa Clara, CA, USA). The plasma gas was continuously supplied during measurement by a nitrogen generator (4107, Agilent Technologies). The MP-AES instrument operates with a vertical torch alignment together with an axial observation position. As well as sample solutions, standards were introduced by autosampler (SPS, Agilent Technologies) with 30 s of rinsing between each with 0.1M HNO_3 prepared in ultrapure water. The MP-AES operating conditions and measurement parameters are indicated in Table 1. Lithium standard stock solution of 1000 mg L^{-1} (Scharlau, Germany) was used to prepare the 5-point calibration series. The limit of detection (LOD) was defined as 0.3246 μg kg^{-1} at the applied wavelength of 610.365 nm (the measurement parameters are summarized in Table 1).

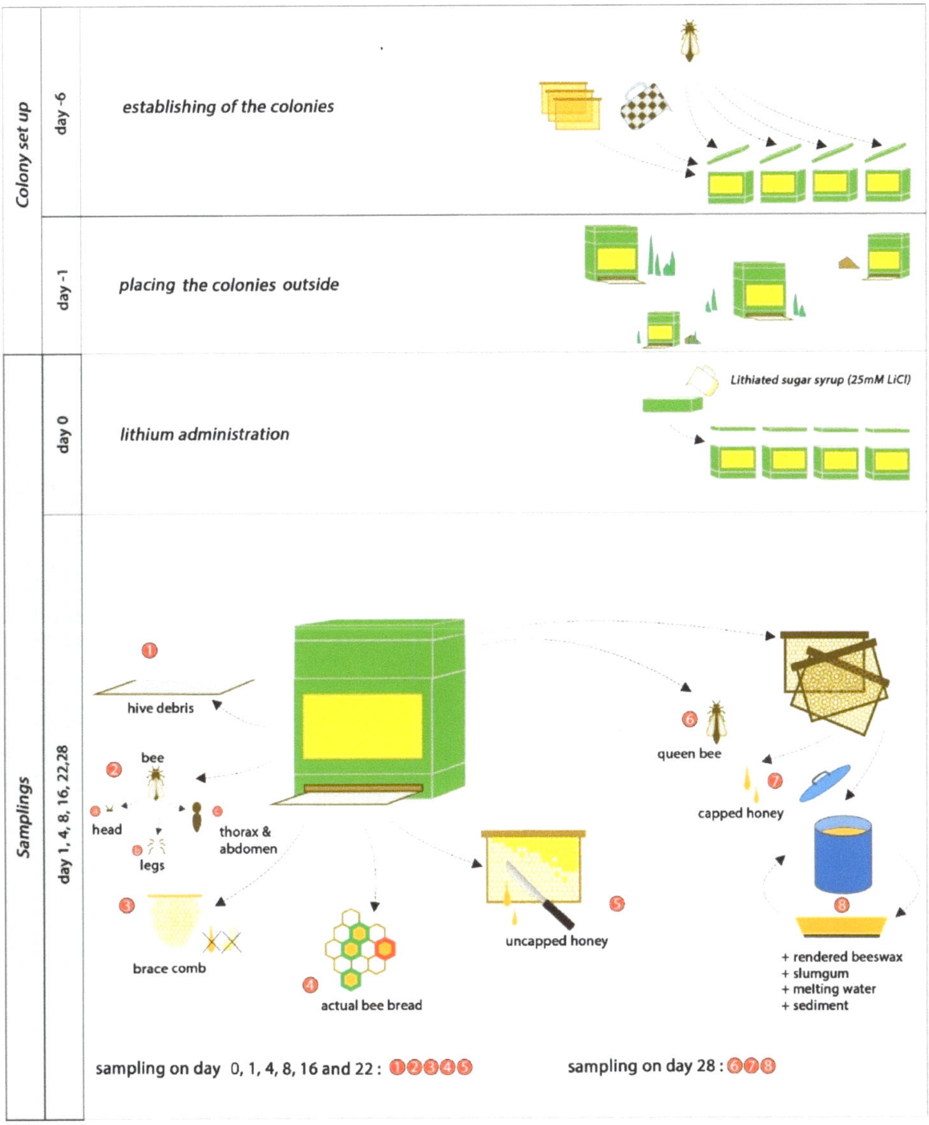

Figure 1. Experimental design and samplings.

Table 1. Analytical measurement parameters.

Replicates Pump Speed	3 15 rpm
Uptake time	15 s
Rinse time	30 s
Stabilization time	15 s
Read time	3 s
Nebulizer pressure	240 kPa
Wavelength	610.365 nm

The following formula calculated the LOD: LOD = (3 × s)/S where s is the standard deviation of 15 blank samples, and S is the specificity (slope of the calibration curve). The results of the elemental analysis were given on a dry mass basis.

2.4. Statistical Analysis

To analyze the effect of LiCl treatment on the Li concentration of bees (head, thorax and abdomen, and leg were analyzed separately), honey, bee bread, beeswax and hive debris (response variables), we used linear mixed models (LMMs) in Statistica 8.0 (http://www.statsoft.hu) (accessed on 19 May 2021). Prior to analysis, Li concentration data were log10 transformed to improve normality. The LMMs included sampling time as a fixed factor representing pre-treatment (control, at day 0) and post-treatment measurements (days 1, 4, 8, 16, and 22). In order to account for repeated measures, the hive was included as a random factor. Differences among means were identified using Tukey HSD post hoc tests when the model fixed effect was significant.

3. Results and Discussion

3.1. Lithium Level Returns to Normal Values in Adult Bees

Feeding lithium syrup at a concentration (25 mM) applied in earlier studies [9,15,20] resulted in an average lithium peak of 130.13 mg kg^{-1} (average of the hives) in bees' bodies (thorax and abdomen), with an absolute maximum value of 167.71 mg kg^{-1} in hive 1, on day 4. Lithium concentration decreased in all body parts of the bees from day 4 post-treatment (Table 2, Figure 2). This pattern may be consistent with the findings of Prešern and colleagues, who revealed that in bee larvae, lithium level started to drop on day 3 post-treatment [15]. By day 22 post-treatment, lithium level showed full recovery to the pre-treatment control level (0.15 mg kg^{-1} on average; Figure 2). Data from the present study indicate that adult bees seem to be able to excrete lithium at the colony level.

Table 2. Effect of lithium in bees and bee products in factor of time. Analysis of log-transformed Li concentration data using linear mixed models revealed a significant effect of time relative to the LiCl treatment (control: pre-treatment (day 0); treated: 1, 4, 8, 16, and 22 days post-treatment) of honeybee colonies on lithium concentration of the bees, the honey, and the bee bread, but not that of the wax and the hive debris. Results of Tukey HSD post hoc tests are shown in Figure 2.

	Main Effects					Overall Model			
	Factor	**Effect type**	**d.f. (Effect, Error)**	**F**	***p***	**$R^2_{adj.}$**	**d.f. (Model, Residual)**	**F**	***p***
Bee head	day	fixed	5, 15	154.4	<0.001	0.971	8, 15	96.8	<0.001
	hive	random	3, 15	0.8	0.529				
Bee thorax and abdomen	day	fixed	5, 15	395.1	<0.001	0.988	8, 15	247.9	<0.001
	hive	random	3, 15	2.5	0.102				
Bee leg	day	fixed	5, 15	156.2	<0.001	0.971	8, 15	97.9	<0.001
	hive	random	3, 15	0.7	0.592				
Honey	day	fixed	5, 15	57.3	<0.001	0.925	8, 15	36.5	<0.001
	hive	random	3, 15	1.7	0.210				
Bee bread	day	fixed	5, 15	6.6	0.002	0.543	8, 15	4.4	0.006
	hive	random	3, 15	0.8	0.512				
Hive debris	day	fixed	5, 15	1.0	0.451	0.000	8, 15	1.0	0.474
	hive	random	3, 15	1.0	0.418				

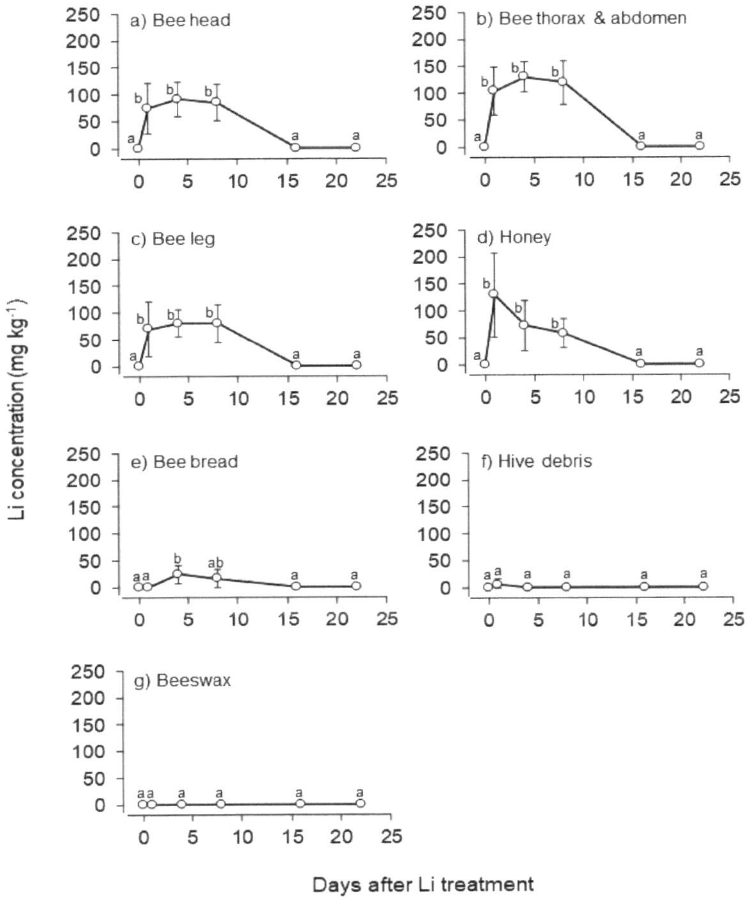

Figure 2. Post-treatment changes in lithium concentration in hive products and bees. Linear mixed model analysis (LMM) revealed a significant increase in Li concentrations (mean ± SD) in bee head (**a**), thorax and abdomen (**b**), and leg (**c**), as well as in honey (**d**) between 1 and 8 days after LiCl treatment, followed by a return of concentrations to the control level after 16 days post-treatment. Only a small increase in Li concentration was revealed in bee bread (**e**), and only on the 4th day post-treatment, and no increase in Li concentration at all was detected in hive debris (**f**) or beeswax (**g**). For LMM statistics, see Table 2. Plotted values not sharing any index letter are statistically different at $p < 0.05$ (Tukey HSD post hoc test). Note that the threshold of detecting Li was 0.0003 mg kg^{-1}.

Residues were found in the entire body of bees, reaching the legs as well, the parts of the body in which lithium may have been transferred via the hemolymph. Therefore, all parts of the bees' bodies may be eligible to estimate changes in the lithium level of the colony. We hypothesize that the Li concentration peak measured in the bees and their offspring might help to predict the timeframe of treatment efficacy for future research in this field.

A sampling of the queens was only possible by killing the individuals. Therefore, the queens were measured only at the termination of the trial (day 28), revealing no detectable lithium. No signs of an attempt to refuse the queens by the worker bees, known as supersedure, were observed in the whole period of the experiment. We conclude that lithium has no detrimental effect on the queen, at least in the short term.

Hive debris usually contains wax particles and may contain bee parts, traces of honey, or pollen in a variable composition. Nonetheless, it revealed no response to LiCl treatment (Table 2, Figure 2). Despite being readily available without opening the hive, the debris does not appear to be suitable for collecting information about the lithium level in the hive.

3.2. Bee Bread Is the Least Affected of Beekeeping Products

Of the bee products in which lithium appeared, the stored, fermented pollen (known as bee bread) was found to be the least exposed to lithium contamination (Table 2).

Bee bread is commercialized for its beneficial nutritional and therapeutic properties. However, collecting it for human food is time-consuming and suffers from limitations [21]. Similar to other samples investigated, lithium peaked on day 4 with a lithium level of 28.11 mg kg^{-1} (Figure 2). Representing four hives but a single time measurement, a similar value (30.75 mg kg^{-1}) was reported for day 4 post-treatment by Prešern and collegues [15]. Bee bread is the primary protein resource that bees utilize, especially for feeding larvae and adults. High lithium exposure may adversely affect the development of the larvae **[15]**. Furthermore, increased mortality in the lithium-treated colonies was recorded. Thus, lithium treatment may have an impact on colony reproduction. Nevertheless, reduced lithium levels were measured in 5-day-old larvae three days after the lithium culmination (on day 7 post-treatment). [22].

Our data support the rapid decrease of LiCl in the bee bread after the peak caused by treatment (Figure 2). Lithium concentration recovery in adult bees and in the diet of larvae enables the brood to be less exposed over time. We propose that the possible adverse effects of lithium might be compensated for or minimized by applying it only in naturally or artificially induced brood-free or brood-poor periods. It should be noted, however, that the veterinary use of LiCl has not been authorized yet. Further research is needed to accurately determine a low-risk timing or a withholding period in queen rearing.

3.3. Lithium Treatment Leaves Beeswax Unaffected

In brace combs, representing the wax secreted directly by the bees, no lithium was detected in any samples (n = 19) at any sampling time (Figure 2). No lithium was detected in any lumps of wax (n = 4) rendered from the old combs of the hives, either. Moreover, in by-products of the wax rendering process from old combs such as the slumgum, sediment, and the melting water, no lithium content has been confirmed. These facts are of great importance as comb wax is commonly recycled in apicultural practice; recycled comb is used to make beeswax foundations and widely distributed to beekeepers.

Our finding is significant also because other commonly used acaricides such as amitraz [23], coumaphos [24], tau fluvalinate [25], flumethrin [26], and thymol [27] affect the beeswax.

3.4. Lithium Levels Decrease during Dehydration, but Residues May Remain in the Ripe Honey

Honey, the most important apicultural product, represents aggregate honey taken from thousands of honey storage sites from capped and uncapped cells.

Uncapped cells initially contain freshly collected unripe nectar, which undergoes the process of dehydration and transposition to cells to be filled into the vicinity of the brood, where they are then capped by the bees. In the present study, uncapped honey served to uncover the kinetics of lithium within the hive, whilst capped honey was separately handled to represent the store to be harvested at the termination of the experiment.

Treatment with lithiated sugar syrup containing 126.5 mg kg^{-1} LiCl (25 mM) affected the uncapped honey considerably in the short term (Table 2, Figure 2). The highest lithium content in the honey was measured on day 1 (Figure 2). Despite the dehydration process of the honey carried out by the bees, the concentration of lithium started to decrease from day 4 post-treatment. Lithium concentration in uncapped honey showed full recovery to the control level (below 0.25 mg kg^{-1}) by day 22. Based on the obtained data, the possibility of decontamination of uncapped honey has been confirmed. This hive product, being the transposed honey stock of the colony, is most affected by lithium. Incoming nectar or sugar syrup is processed and exposed to transposition from being passed from bee to bee several times. It is hypothesized that one possible point of lithium depletion may occur via the bees.

In the capped, ripe honey, a value corresponding to one-fifth of the initial lithium syrup concentration (22.40 mg kg^{-1}) was measured on average at the termination of the experiment (day 28). This amount may be comparable to the natural lithium trace element content measured in honey so far [28]. Honey exerts positive nutritional and health effects if consumed at high doses of 50 to 80 g per daily intake [29]. Considering it as a proposed intake, capped honey from the present trial would equal 1.12–1.79 mg lithium. This amount can be achieved from other alimentary products as a daily lithium intake [30–32]. It is of note that hardly any similar modes of administration are used in which the anti-*Varroa* active ingredient is applied via a large amount of sugar syrup (e.g., feeding), since it will inevitably induce an elevated residue level in the honey store. Although a single treatment is not likely to result in an alarming level in ripe honey, a trickling mode of administration may be preferred once lithium is registered as a veterinary medicine.

4. Conclusions

In this study, the progress of contamination and the possibility of subsequent elimination of lithium in the most important bee products and adult bees were investigated in situ after lithiated sugar syrup feeding. Unlike commonly used varroaicides, lithium treatment left beeswax unexposed as a clear positive property of lithium. On the other hand, it was revealed that lithium could contaminate ripe honey. Despite inducing the pollution of the honey by feeding the bees lithiated syrup, lithium levels remained under the level of commercialized honey, which naturally possess higher lithium content (38–110 mg/kg^{-1} [28,33]. Currently, no maximum residue levels (MRL) exist for lithium, nor it is recognized as veterinary medicine. More extensive research is needed to determine lithium residues under field conditions in harvested honey and bee bread, as well as to determine a waiting time after the Li treatment. Further experiments are necessary to reveal how application methods like trickling [16] would affect the appearance of residues in honey, especially if performed repeatedly against the devastating pest *V. destructor*.

Author Contributions: Conceptualization, É.K. and B.K.; methodology, B.K. and K.M.; software, A.S.; validation, A.S.; investigation, K.M., É.K., B.K., Z.S., and E.B.; resources, B.K. and E.B.; data curation, A.S.; writing—original draft preparation, É.K., B.K., K.M., K.S., E.B., and Z.S.; writing—review and editing, B.K., É.K., K.M., G.N., I.S., and A.S.; visualization, B.K. and A.S.; funding acquisition, J.T. and B.K. All authors have read and agreed to the published version of the manuscript.

Funding: This research and the APC was supported by the Hungarian Government and the European Union, with the co-funding of the European Regional Development Fund in the frame of Széchenyi 2020 Programme GINOP-2.3.2-15-2016-00054 project.

Institutional Review Board Statement: Not applicable.

Data Availability Statement: Not applicable.

Acknowledgments: We would like to thank Mr Imre Bognár for providing the colonies for the experiment. We would like to highly acknowledge Agilent Technologies (Novo-Lab Ltd., Dallas, TX, USA) for providing an MP-AES 4200 instrument for the elemental analysis. This publication was supported by the GINOP-177 2.3.2-15-2016-00054 project.

Conflicts of Interest: The authors declare no conflict of interest. The funders had no role in the design of the study, in the collection, analyses, or interpretation of data, or the writing of the manuscript.

References

1. Genersch, E. American foulbrood in honeybees and its causative agent, *Paenibacillus larvae*. *J. Invertebr. Pathol.* **2010**, *103*, S10–S19. [CrossRef] [PubMed]
2. Spivak, M.; Reuter, G. *A Sustainable Approach to Controlling Honey Bee Diseases and Varroa Mites*; USDA: Washington, DC, USA, 2005; pp. 1–6.
3. Barlow, V.M.; Fell, R.D. *Sampling Methods for Varroa Mites on the Domesticated Honeybee*; Virginia Cooperative Extension: Virginia, VA, USA, 2006; pp. 1–3. Available online: http://hdl.handle.net/10919/50392 (accessed on 19 May 2021).
4. Mozes-Koch, R.; Slabezki, Y.; Efrat, H.; Kalev, H.; Kamer, Y.; Yakobson, B.; Dag, A. First detection in Israel of fluvalinate resistance in the varroa mite using bioassay and biochemical methods. *Exp. Appl. Acarol.* **2000**, *24*, 35–43. [CrossRef] [PubMed]
5. Spreafico, M.; Bernardinelli, I.; Colombo, M.P. First detection of strains of Varroa destructor resistant to coumaphos. Results of laboratory tests and field trials. *Apidologie* **2001**, *32*, 49–55. [CrossRef]
6. Coffey, M.F.; Breen, J. The efficacy and tolerability of Api-Bioxal® as a winter varroacide in a cool temperate climate. *J. Apic. Res.* **2016**, *55*, 65–73. [CrossRef]
7. Al Toufailia, H.; Francis, L.W.R. Towards integrated control of varroa: 5) Monitoring Honey bee brood rearing in winter, and the proportion of varroa in small patches of sealed brood cells. *J. Apic. Res.* **2018**, *57*, 444–451. [CrossRef]
8. Rosenkranz, P.; Pia, A.; Bettina, Z. Biology and control of varroa destructor. *J. Invertebr. Pathol.* **2010**, *103*, S96–S119. [CrossRef] [PubMed]
9. Ziegelmann, B.; Elisabeth, A.; Stefan, H.; Michaela, B.; Stefan, B.; Peter, R. Lithium chloride effectively kills the honey bee parasite varroa destructor by a systemic mode of action. *Sci. Rep.* **2018**, *8*, 683. [CrossRef]
10. Hurst, V.; Philip, C.S.; Geraldine, A.W. Toxins induce "Malaise' behaviour in the Honeybee (Apis Mellifera)". *J. Comp. Physiol. A* **2014**, *200*, 881–890. [CrossRef] [PubMed]
11. Whitehead, A.T. Electrophysiological response of honey bee labial palp contact chemoreceptors to sugars and electrolytes. *Physiol. Entomol.* **1978**, *3*, 241–248. [CrossRef]
12. Ayestaran, A.; Martin, G.; María, G.d.B.S. Toxic but Drank: Gustatory aversive compounds induce post-ingestional malaise in harnessed honeybees. *PLoS ONE* **2010**, *5*, e15000. [CrossRef]
13. Léonard, A.; Hantson, P.; Gerber, G. Mutagenicity, carcinogenicity and teratogenicity of lithium compounds. *Mutat. Res. Genet. Toxicol.* **1995**, *339*, 131–137. [CrossRef]
14. Castillo-Quan, J.I.; Tain, L.S.; Kinghorn, K.J.; Li, L.; Gronke, S.; Hinze, Y.; Blackwell, K.T.; Bjedov, I.; Partridge, L. A triple drug combination targeting components of the nutrient-sensing network maximizes longevity. *Proc. Natl. Acad. Sci. USA* **2019**, *116*, 20817–20819. [CrossRef]
15. Prešern, J.; Kur, U.; Bubnič, J.; Šala, M. Lithium contamination of honeybee products and its accumulation in brood as a consequence of anti-varroa treatment. *Food Chem.* **2020**, *330*, 127334. [CrossRef] [PubMed]
16. Kolics, É.; Specziár, A.; Taller, J.; Mátyás, K.; Kolics, B. Lithium chloride outperformed oxalic acid sublimation in a preliminary experiment for varroa mite control in pre-wintering honey bee colonies. *Acta Vet. Hung.* **2021**, *68*, 370–373. [CrossRef] [PubMed]
17. Stanimirovic, Z.; Uroš, G.; Marko, R.; Nevenka, A.; Nemanja, J.; Branislav, V.; Jevrosima, S. Looking for the causes of and solutions to the issue of honey bee colony losses. *Acta Vet. Beogr.* **2019**, *69*, 1–31. [CrossRef]
18. Ziegelmann, B.; Blumenschein, M.; Rein, C.; Lang, V.; Hannus, S.; Rosenkranz, P. Varroa Treatment of Brood-Free Honey Bee Colonies with Lithium Chloride. In Proceedings of the 46th APIMONDIA—International Apicultural Congress, Montréal, QC, Canada, 8–12 September 2019.
19. Kolics, É.; Kinga, M.; János, T.; András, S.; Balázs, K. Contact effect contribution to the high efficiency of lithium chloride against the mite parasite of the honey bee. *Insects* **2020**, *11*, 333. [CrossRef]
20. Kolics, B.; Sajtos, Z.; Matyas, K.; Kolics, É.; Taller, J.; Baranyai, E. Lithium Chloride—Hazard or Possibility? In Proceedings of the 46th APIMONDIA—International Apicultural Congress: Montréal, QC, Canada, 8–12 September 2019.
21. Khalifa, S.A.; Elashal, M.; Kieliszek, M.; Ghazala, N.E.; Farag, M.A.; Saeed, A.; Xiao, J.; Zou, X.; Khatib, A.; Göransson, U.; et al. Recent insights into chemical and pharmacological studies of bee bread. *Trends Food Sci. Technol.* **2020**, *97*, 300–316. [CrossRef]
22. Prešern, J. Neurostatistical approach to toxicological testing in honeybees. *MethodsX* **2020**, *7*, 101077. [CrossRef]
23. Jiménez, J.J.; José, L.B.; María, J.d.N.; María, T.M. Residues of organic contaminants in beeswax. *Eur. J. Lipid Sci. Technol.* **2005**, *107*, 896–902. [CrossRef]
24. Bajuk, B.P.; Katarina, B.; Tomaž, S.; Luka, M.; Metka, P.; Ocepek, M.Š.; Vlasta, J.; Ayhan Filazi, D.Š.; Silvestra, K. Coumaphos residues in honey, bee brood, and beeswax after varroa treatment. *Apidologie* **2017**, *48*, 588–598. [CrossRef]
25. Wilmart, O.; Anne, L.; Marie-Louise, S.; Wim, R.; Bruno, U.; Dirk, C.D.G.; Walter, S.; Philippe, D.; Pascal, G.; Bach, K.N. Residues in Beeswax: A health risk for the consumer of honey and beeswax? *J. Agric. Food Chem.* **2016**, *64*, 8425–8434. [CrossRef]
26. Bogdanov, S.; Verena, K.; Ueli, B. Determination of acaricide residues in beeswax: Collaborative study. *Apiacta* **2003**, *38*, 235–245.
27. Carayon, J.-L.; Téné, N.; Bonnafé, E.; Alayrangues, J.; Hotier, L.; Armengaud, C.; Treilhou, M. Thymol as an alternative to pesticides: Persistence and effects of Apilife Var on the phototactic behavior of the honeybee Apis mellifera. *Environ. Sci. Pollut. Res.* **2013**, *21*, 4934–4939. [CrossRef] [PubMed]
28. Hernández, O.M.; Fraga, J.M.G.; Jiménez, A.I.; Jiménez, F.; Arias, J.J. Characterization of honey from the Canary Islands: Determination of the Mineral content by atomic absorption spectrophotometry. *Food Chem.* **2005**, *93*, 449–458. [CrossRef]

29. Bogdanov, S.; Jurendic, T.; Sieber, R.; Gallmann, P. Honey for Nutrition and Health: A Review. *J. Am. Coll. Nutr.* **2008**, *27*, 677–689. [CrossRef]
30. Voica, C.; Roba, C.; Iordache, A.M. Lithium levels in food from the romanian market by Inductively Coupled Plasma-Mass Spectrometry (ICP-MS): A pilot study. *Anal. Lett.* **2021**, *54*, 242–254. [CrossRef]
31. Schrauzer, G.N. Lithium: Occurrence, Dietary intakes, nutritional essentiality. *J. Am. Coll. Nutr.* **2002**, *21*, 14–21. [CrossRef]
32. González-Weller, D.; Rubio, C.; Gutiérrez Ángel, J.; González, G.L.; Mesa, J.M.C.; Gironés, C.R.; Ojeda, A.B.; Hardisson, A. Dietary intake of barium, bismuth, chromium, lithium, and strontium in a Spanish population (Canary Islands, Spain). *Food Chem. Toxicol. Int. J. Publ. Br. Ind. Biol. Res. Assoc.* **2013**, *62*, 856–868. [CrossRef]
33. García, J.C.R.; Rodríguez, R.I.; Crecente, R.M.P.; García, J.B.; Martín, S.G.; Latorre, C.H. Preliminary chemometric study on the use of honey as an environmental marker in Galicia (Northwestern Spain). *J. Agric. Food Chem.* **2006**, *54*, 7206–7212. [CrossRef]

insects

MDPI

Article

Novel TaqMan PCR Assay for the Quantification of *Paenibacillus larvae* Spores in Bee-Related Samples

Darja Kušar [1,*,†], Bojan Papić [1,†], Urška Zajc [1], Irena Zdovc [1], Majda Golob [1], Lucija Žvokelj [2], Tanja Knific [3], Jana Avberšek [1], Matjaž Ocepek [1] and Metka Pislak Ocepek [2]

1 Institute of Microbiology and Parasitology, Veterinary Faculty, University of Ljubljana, Gerbičeva 60, 1000 Ljubljana, Slovenia; bojan.papic@vf.uni-lj.si (B.P.); urska.zajc@vf.uni-lj.si (U.Z.); irena.zdovc@vf.uni-lj.si (I.Z.); majda.golob@vf.uni-lj.si (M.G.); jana.avbersek@vf.uni-lj.si (J.A.); matjaz.ocepek@vf.uni-lj.si (M.O.)

2 Institute of Pathology, Wild Animals, Fish and Bees, Veterinary Faculty, University of Ljubljana, Gerbičeva 60, 1000 Ljubljana, Slovenia; lucija.zvokelj@vf.uni-lj.si (L.Ž.); metka.pislakocepek@vf.uni-lj.si (M.P.O.)

3 Institute of Food Safety, Feed and Environment, Veterinary Faculty, University of Ljubljana, Gerbičeva 60, 1000 Ljubljana, Slovenia; tanja.knific@vf.uni-lj.si

* Correspondence: darja.kusar@vf.uni-lj.si

† Both authors contributed equally to the study.

Citation: Kušar, D.; Papić, B.; Zajc, U.; Zdovc, I.; Golob, M.; Žvokelj, L.; Knific, T.; Avberšek, J.; Ocepek, M.; Pislak Ocepek, M. Novel TaqMan PCR Assay for the Quantification of *Paenibacillus larvae* Spores in Bee-Related Samples. *Insects* **2021**, *12*, 1034. https://doi.org/10.3390/insects12111034

Academic Editors: Ivana Tlak Gajger and Franco Mutinelli

Received: 26 October 2021
Accepted: 14 November 2021
Published: 17 November 2021

Simple Summary: American foulbrood (AFB) is the most severe bacterial disease of honeybees, caused by *Paenibacillus larvae*. Larvae become infected by ingesting food contaminated with *P. larvae* spores, which are extremely resistant and can remain infectious for decades. Burning affected colonies is widely used to prevent further spread of the disease. The presence of *P. larvae* spores in bee-related samples is associated with an increased risk of developing clinical symptoms, and spore counts can be used for early detection of at-risk colonies, which should then undergo thorough clinical examination. Because quantification of *P. larvae* spores by plate counting is time-consuming and unreliable, due to poor and inconsistent germination, molecular quantification is more suitable. To overcome the limitations of available quantification methods, we developed a quantitative PCR (qPCR) assay for reliable quantification of *P. larvae* that also performs well at low spore counts. The assay was validated for honey and hive debris samples but can be extended to other sample types. Spore counts in AFB-positive colonies were significantly higher than those in asymptomatic colonies, both for honey and hive debris samples. By comparing plate and qPCR counts, the germination rate of *P. larvae* spores was found to be low and inconsistent.

Abstract: *Paenibacillus larvae* is the causative agent of American foulbrood (AFB), a devastating disease of honeybees. *P. larvae* spore counts in bee-related samples correlate with the presence of AFB symptoms and may, therefore, be used to identify at-risk colonies. Here, we constructed a TaqMan-based real-time PCR (qPCR) assay targeting a single-copy chromosomal metalloproteinase gene for reliable quantification of *P. larvae*. The assay was calibrated using digital PCR (dPCR) to allow absolute quantification of *P. larvae* spores in honey and hive debris samples. The limits of detection and quantification were 8 and 58 spores/g for honey and 188 and 707 spores/mL for hive debris, respectively. To assess the association between AFB clinical symptoms and spore counts, we quantified spores in honey and hive debris samples originating from honeybee colonies with known severity of clinical symptoms. Spore counts in AFB-positive colonies were significantly higher than those in asymptomatic colonies but did not differ significantly with regard to the severity of clinical symptoms. For honey, the average spore germination rate was 0.52% (range = 0.04–6.05%), indicating poor and inconsistent in vitro germination. The newly developed qPCR assay allows reliable detection and quantification of *P. larvae* in honey and hive debris samples but can also be extended to other sample types.

Keywords: American foulbrood (AFB); *Paenibacillus larvae*; real-time or quantitative PCR (qPCR); digital PCR (dPCR); plate counting; spore germination rate; honey; hive debris

1. Introduction

Honeybees (*Apis mellifera*) play an essential role in pollination and biodiversity conservation [1]. Several stress factors affect bee health on national and global scales, leading to significant reductions of their populations [2,3]. Biotic factors include pathogens from both prokaryotic or eukaryotic taxa and pests causing various diseases of bees. On the other hand, bee health is endangered by abiotic stressors, including unfavorable weather conditions, habitat or diet changes, introduction of invasive species, intoxication of beehives with acaricides, and treatment of crops with insecticides.

American foulbrood (AFB) is one of the most widespread and most severe diseases of honeybees and is caused by the spore-forming Gram-positive bacterium *Paenibacillus larvae* [4]. Honeybee larvae become infected by ingesting food contaminated with *P. larvae* spores, which are highly resistant to environmental factors and can remain infectious for several decades [5]. Clinical onset of AFB depends on the virulence and spore count of *P. larvae* in the honeybee colony as well as many intrinsic (genetic) and extrinsic (environmental) factors [6,7]. In Slovenia, the disease is diagnosed when characteristic clinical symptoms are identified in the honeybee colony and the causative agent is confirmed by laboratory examination [8]. Due to its severity, AFB is a statutory notifiable disease in the European Union (Council Directive 92/65/EEC, 1992) and is regulated by national legislation in many countries worldwide, including Slovenia [8]. Elimination of the infected honeybee colonies and contaminated equipment is required, and trade restrictions are enforced in the outbreak area, resulting in significant economic losses to the beekeeping industry. Early detection of the disease is the most effective way to prevent its spread.

Apparently healthy honeybee colonies can harbor *P. larvae* spores [9–11]. Spore counts in honey, hive debris, and adult bees correlate with the severity of symptoms in the clinical stage of AFB and may be useful in identifying colonies in the pre-clinical stage of the disease [10–17]. Thus, quantification of *P. larvae* spores offers a promising prognostic tool for early detection of at-risk apiaries based on increased spore counts in the colonies, which should then undergo thorough clinical examination. For this purpose, a method for reliable quantification of *P. larvae* spores in different samples is needed.

Honey and (winter) hive debris samples provide non-invasive and easily accessible material for *P. larvae* surveillance. Moreover, they may reflect a long-term accumulation of *P. larvae* spores in the hive, whereas spore counts in adult bees mostly reflect the current disease status of the colony [12,15]. Previous studies that used culture-based methods for the detection and quantification of *P. larvae* showed a limited ability of honey to reliably identify diseased colonies, since spores may remain undetected even in symptomatic colonies [9,10]. On the contrary, spore counts in accumulated winter hive debris have shown promise in reflecting disease status, since hive debris allowed identification of higher numbers of *P. larvae*-positive colonies compared with adult bees, regardless of disease status [15], and was also a better predictor of the onset of AFB [11]. In Slovenia, it is common practice to collect honey samples from colonies in areas with an increased risk of AFB to monitor *P. larvae* spore counts by cultivation, whereas hive debris is usually not present during clinical examinations of apiaries in summer. Sampling of adult bees for spore counts has not yet been introduced into our laboratory as of the time of this study.

Two main approaches for the quantification of *P. larvae* spores in bee-related samples are culture- and PCR-based methods. To date, the detection and quantification of *P. larvae* in honey and hive debris have mostly relied on culture-based methods [9,11,17,18]. However, these suffer from certain limitations. First, germination of *P. larvae* spores is markedly affected by the sample pretreatment procedure, *P. larvae* genotype, type of honey, and the choice of culture media [19–22]. Moreover, the method is time-consuming and requires species confirmation [23]. Thus, alternative and state-of-the-art methods for reliable detection and quantification of *P. larvae* are needed to reassess the relationship between spore counts in bee-related samples and the disease status of the corresponding honeybee colonies.

Quantitative real-time PCR (qPCR) offers a time- and cost-effective alternative for quantifying *P. larvae* spores and overcomes the limitations of culture-based methods.

Although several qPCR assays have been developed for the detection and characterization of *P. larvae* in different sample types [21,23–28], only two assays have been optimized for the quantification of *P. larvae* spores in honey and/or hive debris samples [23,26]. Both target the 16S rRNA gene and use SYBR technology, which suffer from important limitations. Moreover, in these two studies, the standard curve for qPCR quantification was based on plate counts [23] or flow cytometry [26]. The latter requires a clean culture (suspension of vegetative cells or spores) and is, therefore, unsuitable for direct quantification of *P. larvae* in complex sample types (e.g., bee-related samples).

Contrary to flow cytometry, digital PCR (dPCR) allows direct quantification of the target in complex sample types and uses the same DNA template as well as amplification and detection chemistry as qPCR. dPCR allows absolute quantification without the need for a standard curve [29] but, to our knowledge, has not yet been used for *P. larvae*. Here, we developed a novel TaqMan-based qPCR assay for the quantification of *P. larvae* which was validated for honey and hive debris samples and calibrated using dPCR to allow absolute quantification. For honey samples, spore counts derived from qPCR were compared with those derived from plate counting, and the correlation between spore counts and the intensity of AFB clinical symptoms in the corresponding honeybee colonies was assessed. The latter was also investigated for hive debris samples.

2. Materials and Methods

2.1. Samples

Sampling of honey and hive debris was performed in 2019–2020. In most cases, only one sample type (honey or hive debris) per apiary was obtained. Both sample types were collected in sterile containers from individual honeybee colonies. Honey samples were collected from the honeycomb cells near the brood within the scope of veterinary clinical examinations of colonies upon clinical suspicion of AFB; they originated from AFB-positive (n = 41) or asymptomatic (n = 26) colonies positioned in AFB-positive apiaries (Table S1). To collect hive debris, sampling boards were placed at the bottom of hives in apiaries with different AFB status. One month after placing the boards, a control examination of colonies was performed. Hive debris samples were collected from newly clinically affected (i.e., AFB-positive or symptomatic) colonies (n = 17) and asymptomatic colonies positioned in AFB-positive apiaries (n = 52). Hive debris was also collected from colonies positioned in asymptomatic apiaries with a history of AFB and/or located within the active AFB zone (n = 25) or apiaries with a complete absence of AFB (n = 13) (Table S1).

2.1.1. Honey

For the quantification of *P. larvae* spores by plate counting and qPCR, 67 honey samples were collected (Table S1). The colonies had AFB clinical symptoms of various severities (0–4; Table S2) and originated from 18 apiaries (1–18); more than one honeybee colony per apiary was sampled for 13 apiaries (Table S1). Honey samples were heated to 45–50 °C, and 50 g was poured into sterile centrifuge containers, there supplemented with 150 mL of dH_2O (50 °C), and mixed to a homogeneous suspension. Samples were centrifuged at 4000× g for 30 min, and the supernatant was discarded, leaving ~1 mL to resuspend the pellet. The resulting suspension was supplemented with sterile 0.9% NaCl (saline) to a final volume of 10 mL and incubated at 80 °C for 10 min to kill the vegetative cells (suspension H1).

For quantification by plate counting, the prepared suspension was diluted by a factor of 10 once more (1 mL of suspension H1 supplemented with 9 mL of sterile saline; suspension H2). A total of 500 µL of suspensions H1 and H2 was inoculated onto Brain heart infusion (BHI; Oxoid, Basingstoke, UK) agar plates supplemented with 5% sheep blood, 1 mg/L thiamine (Fargon Hellas, Trikala, Greece), and 30 mg/L nalidixic acid (Millipore-Sigma by Merck, Burlington, MA, USA) using Drigalski spatulas. Plates were incubated at 37 °C for seven days. Presumptive *P. larvae* colonies were counted, and one colony per plate was confirmed by matrix-assisted laser desorption/ionization time-of-flight (MALDI-TOF) mass spectrometry (Microflex LT system; Bruker Daltonics, Leipzig, Germany) according to the manufacturer's instructions.

For qPCR quantification, 1 mL of suspension H1 was used for DNA extraction, corresponding to 5 g of honey sample (1/10 of the initial sample). For qPCR validation, a single naturally contaminated honey sample (positive by bacteriological examination) was used; three 1 mL aliquots of suspension H1 were collected to represent three biological replicates used for DNA extraction. A honey sample determined as negative by bacteriological examination and showing no qPCR amplification curve in the preliminary analysis was used as a negative template control (NTC).

2.1.2. Hive Debris

P. larvae spores in the hive debris samples were quantified by qPCR; plate counting was not performed, because we had observed poor spore germination for honey samples in our previous laboratory examinations. A total of 107 hive debris samples were quantified (Table S1); pooled samples were obtained from three apiaries without current AFB symptoms in any of the colonies (20, 29, and 30 in Table S1). The colonies had varying severities of AFB clinical symptoms (0–2; Table S2) and originated from 16 apiaries (17–32); for 11 apiaries, more than one honeybee colony per apiary was sampled for individual (non-pooled) quantification (Table S1). Volume (mL) was selected as the basic metric unit for hive debris samples, due to the variation in moisture content. A total of 5 mL of the collected hive debris was transferred to a 50 mL centrifuge tube and supplemented with sterile saline to a final volume of 50 mL. The suspension was vigorously mixed (KS 4000 i control; IKA, Staufen, Germany) at 160 rpm for 2 h and homogenized (MiniMix; Interscience, Saint Nom, France) at the highest speed for 1.5 min (suspension D1).

For qPCR quantification, 1 mL of suspension D1 was used for DNA extraction, corresponding to 0.1 mL of hive debris sample (1/50 of the initial sample). A single naturally contaminated hive debris sample originating from an AFB-positive colony was used for qPCR validation; three 1 mL aliquots of suspension D1 were collected to represent three biological replicates used for DNA extraction. One of the negative hive debris samples originating from an apiary with a complete absence of AFB was used as a NTC.

2.2. DNA Extraction

The extraction of DNA from honey and hive debris samples was performed using a commercial kit (DNA isolation from complex samples, Institute of Metagenomics and Microbial Technologies, Ljubljana, Slovenia; https://www.immt.eu, accessed on 12 November 2021) according to the manufacturer's instructions. The protocol included bead-beating combined with enzymatic and heat-induced lysis between mechanical shearing steps. Briefly, 1 mL of suspensions H1 and D1 was centrifuged at 10,000× *g* for 5 min in 2 mL screw-cap tubes containing 150 mg of glass beads with diameter ≤106 μm (MilliporeSigma by Merck, Burlington, MA, USA). The supernatants were discarded, and lysis buffer (392 μL of destroy buffer D) supplemented with 8 μL of proteinase K (20 mg/mL; MilliporeSigma by Merck, Burlington, MA, USA) was added to the pellets. Twice, samples were subjected to bead-beating (45 s at 6400 rpm) using a tissue homogenizer (MagNA Lyser Instrument; Roche, Basel, Switzerland) and incubation at 56 °C for 15 min. After the third beat-beating, samples were incubated at 100 °C for 10 min and cooled in the refrigerator for 3 min. After centrifugation at 10,000× *g* for 5 min, supernatants were mixed with three times the volume of binding buffer B and loaded onto spin columns for centrifugation at 16,000× *g* for 1 min. After rinsing two times with wash buffer W, DNA was eluted from the spin columns with elution buffer E. For DNA extraction, 1 mL of the prepared sample suspensions (Sections 2.1.1 and 2.1.2) was used, and DNA was eluted to a final volume of 100 μL; these volumes, together with DNA dilutions and DNA-to-PCR-mixture volume ratios, were considered when calculating *P. larvae* spore numbers per sample unit (g for honey and mL for hive debris) from qPCR (Cq values) and dPCR (copies of target per μL of the extracted DNA).

2.3. Design of a Quantitative *P. larvae* TaqMan Assay

We selected the chromosomal metalloproteinase (MP) gene (NCBI genome accession number CP020557.1, locus tag B7C51_23310; [30]) as the target gene, because it is present in a single copy in all *P. larvae* genomes and is highly conserved (≥99.7% identity and 100% coverage) among the currently established *P. larvae* ERIC types [4,31]. Primers (MPF, MPR) and probe (MPP) for its amplification were constructed with the IDT RealTime qPCR Assay Entry tool (https://eu.idtdna.com/scitools/Applications/RealTimePCR/ accessed on 14 March 2019) and analyzed, using the IDT OligoAnalyzer tool (https://eu.idtdna.com/calc/analyzer accessed on 14 March 2019), by applying the qPCR parameter sets. The TaqMan probe was labeled with the reporter 6-FAM at the 5′-end (56-FAM), with a double quencher, ZEN, as an internal quencher, and Iowa Black FQ at the 3′-end (3IABkFQ; Integrated DNA Technologies [IDT], Coralville, IA, USA). The constructed 5′–3′ sequences were GGT AAC TAT TCT GGC AGG AGC for the forward primer (MPF), AAG TTC ACG GTT AGG GTC TTC for the reverse primer (MPR), and [56-FAM] TTG GTA GGA [ZEN] ACG TCA TTG TCC GCA [3IABkFQ] for the probe (MPP).

2.4. In Silico and In Vitro Inclusivity and Exclusivity of *P. larvae* TaqMan Assay

In silico inclusivity was assessed based on nine *P. larvae* complete genomes available in the NCBI Genomes Database (Table S3; accessed on 30 July 2021). In addition, inclusivity was tested based on 40 field *P. larvae* isolates from Slovenia of different ERIC types with available whole-genome sequencing (WGS) data (Table S3). Metalloproteinase gene sequences were identified using BLASTn by applying an identity of ≥90% and a coverage of ≥70% as cut-off values. Gene alignments were performed and visualized using the Geneious aligner implemented in Geneious v11.1.5 (Biomatters, Auckland, New Zealand).

To assess in silico exclusivity, the five most closely related *Paenibacillus* species with available complete or draft genome data were identified based on the highest identity of the 16S rRNA gene using the EzBioCloud identification service [32] (Table S3). BLASTn was used to identify the homologs of the metalloproteinase gene by applying a cut-off of ≥90% identity and ≥70% coverage.

A total of 23 field *P. larvae* isolates of different ERIC types and five *P. larvae* ERIC reference strains were used to assess in vitro inclusivity (Table S4). Two related species (*Bacillus pumilus* and *Paenibacillus alvei*) and five common honeybee pathogens (*Chritidia mellificae, Lotmaria passim, Melissococcus plutonius, Nosema apis,* and *Nosema ceranae*) were used to assess in vitro qPCR exclusivity (Table S4).

2.5. qPCR and dPCR Conditions

The qPCR reaction mix contained 10 μL of 2× master mix (Maxima Probe/ROX qPCR MasterMix; Thermo Fisher Scientific, Waltham, MA, USA), 0.12 μL of passive reference dye ROX (diluted 1:10), 300 nM of both primers and 200 nM of probe, 5 μL of template DNA (undiluted or from a dilution series prepared for method validation; Section 2.6), and PCR-grade water to a final volume of 20 μL. Amplification and detection were performed on the 7500 Fast Real-Time PCR System (Applied Biosystems by Thermo Fisher Scientific, Foster City, CA, USA). The following amplification protocol was used: 50 °C for 2 min, 95 °C for 10 min, and 45 cycles of 95 °C for 15 s and 60 °C for 1 min. Results were expressed in quantification cycle (Cq) values. After validation (Section 2.6), the threshold line was set at 0.1 for all samples. Positive control, NTC, and water no template control (WNTC) were included in each run.

For dPCR, the same TaqMan assay was employed as for qPCR. The reaction mix contained 7.5 μL of 2× master mix (QuantStudio 3D Digital PCR Master Mix v2; Applied Biosystems by Thermo Fisher Scientific, Foster City, CA, USA), 300 nM of both primers and 200 nM of probe, 3 μL of template DNA (appropriately diluted to allow dPCR quantification or selected from a dilution series prepared for qPCR validation), and PCR-grade water to a final volume of 15 μL. The prepared dPCR reactions were loaded onto QuantStudio 3D Digital PCR 20K Chips v2 (consisting of 20,000 reaction wells per chip), and amplification was performed on a ProFlex 2× Flat PCR System (Applied Biosystems by Thermo Fisher Scientific, Foster City, CA, USA) according to the manufacturer's instructions. The following amplification protocol was employed: 96 °C for 10 min, 39 cycles of 60 °C for 2 min and 98 °C for 30 s, and 60 °C for 2 min. After chip imaging and initial analysis with the QuantStudio 3D Digital PCR Instrument, results were analyzed using QuantStudio 3D AnalysisSuite v3.1.6 Cloud Software; after validation (Section 2.6), the quality and fluorescence (FAM) threshold values were set at 0.6 and 2000, respectively. Results were expressed in copies/μL (i.e., number of single-copy metalloproteinase gene targets, corresponding to the number of *P. larvae* spores per μL of the extracted DNA), considering the DNA dilution rate and its volume in the PCR mix. The precision of dPCR, which refers to the ability to discriminate between two measurements with a certain confidence (the lower the precision, the narrower the confidence interval), was calculated by the software. Positive control, NTC, and WNTC were included in each run.

2.6. *Validation of qPCR and Its Calibration Using dPCR*

For the quantification of *P. larvae* spores, linear regression of a standard curve was performed to validate qPCR. Three 1 mL aliquots of the suspensions prepared from a naturally contaminated honey and hive debris sample (see Sections 2.1.1 and 2.1.2) represented three biological replicates for DNA extraction, all of which were analyzed in three technical replicates, resulting in nine qPCR results per dilution; the extracted DNA was diluted in a 5-fold dilution series (for honey, the first two dilutions were 10-fold, followed by a 5-fold dilution series). For the calculation of a linear regression equation, only the data within the linear dynamic range were considered, defined as dilutions with the coefficient of variation (CV) below 33%, since the CV markedly increases below the LOQ [33]. The CV was defined for each dilution as the ratio between the standard deviation of the calculated spore concentration and the average calculated concentration; spore concentrations were calculated from the obtained regression equation. The qPCR amplification efficiency was calculated according to the equation $E = 10^{-1/\text{slope}} - 1$ [33].

Considering the obtained CVs, the limit of quantification (LOQ) and Cq cut-off value were determined both for honey and hive debris. The LOQ was defined from the last serial dilution with CV below 33%, indicating the lower limit of the linear dynamic range. For determination of the Cq cut-off value, the first standard dilution where no amplification was observed in some of the replicates was considered; the highest value was selected from Cq values obtained for this dilution, rounded up to the next half value, and increased by 0.5 to determine the cut-off value [34]. From the LOQ, the limit of detection (LOD) was determined by considering the next dilution in the series, since LOD is 5–10 times lower than LOQ in complex samples [35].

To calibrate qPCR using dPCR, the prepared DNA dilution series was also quantified using dPCR. For dPCR, all three biological replicates were analyzed in one technical replicate. Since dPCR is characterized by a narrower dynamic range [29], only the three most optimal dilutions were considered. The obtained dPCR results were reported as the number of targets per μL of DNA and were used for the calibration of qPCR. For the conversion of Cq values to spores per sample unit, all dilutions of the standard curve from sample preparation to PCR were considered. For field samples, spores per sample unit were calculated from Cq values according to the standard curve equation.

2.7. *Agreement between qPCR and dPCR*

After calibration, the agreement between qPCR and dPCR measurements was assessed using Bland–Altman analysis and Spearman's correlation. For this purpose, additional honey (n = 17) and hive debris (n = 24) samples (obtained independently of the samples shown in Table S1) were quantified using qPCR and dPCR, as described in Section 2.5. For honey samples, the correlation between plate and qPCR counts was also assessed.

For dPCR, a single chip per sample was run, and those with 20–10,000 copies/μL of the analyzed DNA were used for comparison. According to the manufacturer's instructions, the best results are obtained when the concentration of the target sequence per reaction mix is within the optimal range of 200–2000 copies/μL, where precision of at least 10% at a 95% confidence interval can be achieved. Because dPCR measurements were in high agreement with those of qPCR, even outside this dynamic range, the acceptable precision was arbitrarily set at <30%, extending the dynamic range of dPCR to 20–10,000 copies/μL and allowing dPCR/qPCR comparison on a larger number of samples.

All samples with dPCR precision <30% were quantified using qPCR run in triplicate, and the average spore count was considered. For statistical analyses, all spore counts were log-transformed and a *p*-value of <0.05 was considered significant. GraphPad Prism v8.0.2 (GraphPad Software, San Diego, CA, USA) was used to generate Bland–Altman plots and calculate Spearman's correlation.

2.8. Association between Spore Counts and Clinical Symptoms

To assess the association between AFB clinical symptoms and spore counts, a set of 67 honey and 107 hive debris samples originating from honeybee colonies (hives) with known severity of clinical symptoms of AFB were used (Table S1). AFB clinical symptoms were rated on a scale of 0 to 4, with "0" representing no clinical symptoms and "4" representing the highest severity of clinical symptoms (Table S2). Spore counts per sample unit were determined by qPCR (both sample types) and plate counting (honey); Spearman's coefficient was used to assess the correlation between the two methods. Honey samples with counts above the LOQ by both methods were also included in the calculation of spore germination rate, which was defined as the ratio between plate and qPCR counts (expressed as a percentage). For qPCR, LOQ was determined during validation and calibration of the method (Section 2.6). For plate counting, plates with 15–300 colony forming units (CFUs) were considered suitable for enumeration. For the purpose of statistical analyses, all values below the LOQ of plate counting or qPCR were set at 0.5 LOQ value (Table S1).

A Kruskal–Wallis test, followed by Dunn's post hoc multiple comparison test, was used to evaluate the differences in spore counts of *P. larvae* with respect to the severity of clinical symptoms of AFB. The Mann–Whitney test was used to compare spore counts in the asymptomatic colonies with different histories of AFB. All of the above tests were performed using GraphPad Prism v8.0.2 (GraphPad Software, San Diego, CA, USA). The univariate logistic regression in R statistical software v4.0.5 [36] was used to investigate whether *P. larvae* spore counts determined by qPCR (honey and hive debris) or plate counting (honey) have the potential to classify colonies as symptomatic (i.e., clinically affected) or asymptomatic (i.e., clinically unaffected). A *p*-value of <0.05 was considered significant.

3. Results

3.1. In Silico and In Vitro Inclusivity and Exclusivity of P. larvae TaqMan Assay

The assay showed perfect in silico inclusivity with no mismatches observed in the primer/probe binding sites (Figure S1, Table S3). In silico exclusivity was also perfect, since none of the five most closely related *Paenibacillus* species harbored metalloproteinase homologs (Table S3). Moreover, a BLASTn search yielded no hits with identity of ≥90% and coverage of ≥70% in the complete NCBI nr/nt database when the complete metalloproteinase gene was used as a query and the taxon '*Paenibacillus larvae*' was excluded.

After establishing reaction conditions (Section 2.5) and performing validation (Section 2.6), field and reference strains underwent qPCR. None of the tested non-*P. larvae* isolates were qPCR-positive, whereas all *P. larvae* isolates were positive (Table S4); thus, the assay also showed perfect (100%) in vitro inclusivity and exclusivity. Furthermore, the expected qPCR amplicon length (94 bp) was confirmed by separation by QIAxcel capillary electrophoresis (Qiagen, Hilden, Germany) using the QIAxcel DNA High Resolution Kit, QX Alignment Marker 15–1000 bp, QX Size Marker 50–800 bp, and OM500 separation method (data not shown).

3.2. Validation of qPCR and Its Calibration Using dPCR

Standard curves for honey and hive debris were based on average Cq values obtained from nine replicates of the quantitative serial dilutions plotted against the estimated number of *P. larvae* spores per sample unit (g of honey and mL of hive debris) in logarithmic values; absolute quantification by qPCR was possible after its calibration using dPCR, which reported the absolute number of targets (*P. larvae* spores) per µL of DNA extracted from positive honey and hive debris calibration samples (Table 1). Parameters for both qPCR standard curves were within the bounds of expected performance (Figure 1), with amplification efficiencies of 96.0% for honey and 102.1% for hive debris.

Table 1. dPCR quantification of the DNA dilution series prepared from positive honey (H) and hive debris (D) samples used for calibration of qPCR. dPCR counts for the negative control samples (NTC for honey, NTC for hive debris, and WNTC), collected from different dPCR runs, are also given to show the differences in precision and the maximum number of positive wells per chip. The average numbers were calculated from at least four technical dPCR replicates (chips). See Section 2.6 for details.

dPCR	Spores/µL DNA (Range) *	Precision [%]	Avg. No. of Positive Wells/Chip ** (% of All Filled)	Avg. No. of Negative Wells/Chip ** (% of All Filled)
Positive (H)	27,125 (24,767–28,587)	1.3	–	–
Positive (D)	243,000 (154,000–293,000)	0.9	–	–
NTC (H)	1.2 (0.4–2.8)	229.9	3.0 (0.2)	17,149 (94.5)
NTC (D)	0.6 (0.2–1.9)	104.2	3.5 (0.2)	16,971 (94.2)
WNTC	1.0 (0.2–1.6)	137.9	4.2 (0.2)	17,348 (95.1)

NTC, negative template control; WNTC, water no template control. * DNA dilution rate was considered to calculate the number of spores/µL of the extracted DNA in the positive calibration samples. For each technical replicate of the biological replicates, more than one dilution was quantified by dPCR. "Spores" refers to spore equivalents as determined by dPCR, since the assay targets a single-copy metalloproteinase gene. For the negative control samples, the calculated number of spores is a result of false-positive dPCR events, since the samples contained no *P. larvae* spores determined by qPCR or plate counting and precision was far outside the acceptable range to obtain reliable results. ** For the positive calibration samples, the number of positive or negative wells/chip is not reported, since it depends on the analyzed DNA dilution. On average, the number of wells/chip filled (18,517 with a range of 18,107–18,834 for honey; 18,528 with a range of 18,061–19,173 for hive debris) and qualified by quality threshold (17,513 with a range of 16,818–17,888 for honey; 16,935 with a range of 16,164–17,609 for hive debris) was comparable with those of the negative control samples (18,153 filled, with range 17,447–18,771, and 17,187 qualified, with range 16,263–18,125), depending on the applied dPCR technology.

According to the CVs, LOQ values were set at 58 *P. larvae* spores/g honey and 707 *P. larvae* spores/mL hive debris, and LOD values at 8 *P. larvae* spores/g honey and 188 *P. larvae* spores/mL hive debris; from a 5-fold dilution series, the LOD was calculated from the dilution which was the first one higher than the LOQ dilution, generating a LOD approximately five times lower than LOQ. Considering all dilutions during DNA extraction and qPCR mix preparation, LOD values could be translated into 2 and 1 *P. larvae* spores/qPCR reaction for honey and hive debris, respectively, reaching the reported theoretical limit of three target copies per PCR [37]. The Cq cut-off value was set at 38.0 for both sample types.

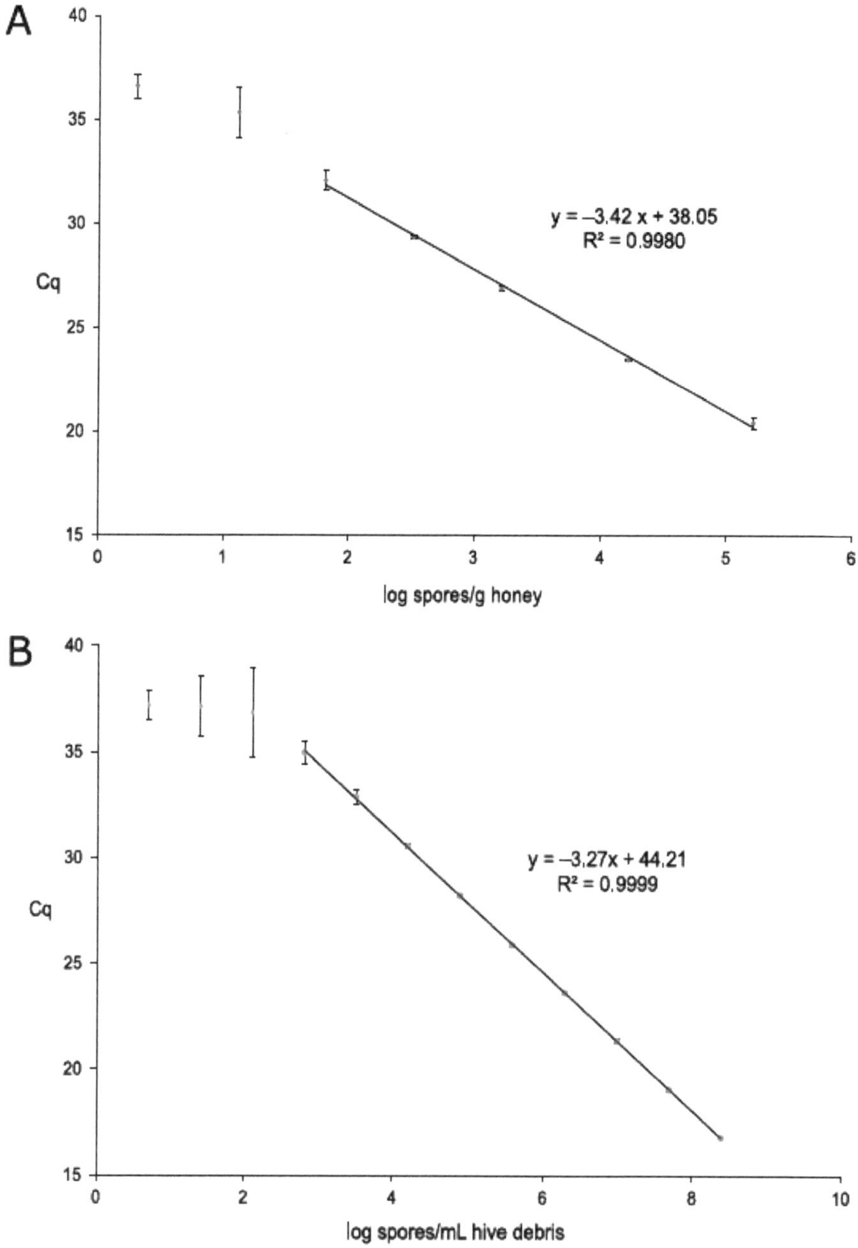

Figure 1. Standard curves for qPCR quantification of *Paenibacillus larvae* in honey (**A**) and hive debris (**B**). Each of the three biological replicates per dilution was measured in three technical replicates, giving a total of nine measurements per dilution. Performance parameters (standard curve equation and R^2) are given. Dilutions outside the linear range (i.e., dilutions where CV was greater than 33%; two for honey and three for hive debris) are also shown.

3.3. Agreement between qPCR and dPCR

Based on the results of dPCR, 14 honey and 12 hive debris samples with 20–10,000 copies/µL were selected for comparison of dPCR and qPCR (Table S5). For honey, 9/14 samples showed a precision of <10% (range = 1.9–5.7%) and 5/14 a reasonable precision of <30% (range = 17.6–27.8%). For hive debris, 8/12 samples showed high precision (range = 2.0–14.5%), and 4/12 showed acceptable precision in the range of 24.2–26.8%. For dPCR, the number of spores ranged from 490 to 190,506 per g of honey and 27,512 to 9,756,100 per mL of hive debris. For qPCR, the number of spores ranged from 534 to 315,604 per g of honey and from 18,249 to 8,313,821 per mL of hive debris; the same samples (H5, H12, D3, and D12) were at the extreme ends of the range both by dPCR and qPCR (highlighted in Table S5).

Bland–Altman analysis showed high agreement between qPCR and dPCR, both for honey (Figure 2A) and hive debris (Figure 2B), indicating interchangeability of both methods and successful calibration of qPCR using dPCR. Although it could be noted that qPCR had slightly higher values than dPCR for honey samples and the opposite was true for hive debris (Table S5), the mean bias value was close to zero for both sample types. Furthermore, the correlation between both methods was highly significant ($p < 0.0001$), with Spearman's $r_s = 0.9912$ for honey, 0.9492 for hive debris, and 0.9682 for both sample types combined.

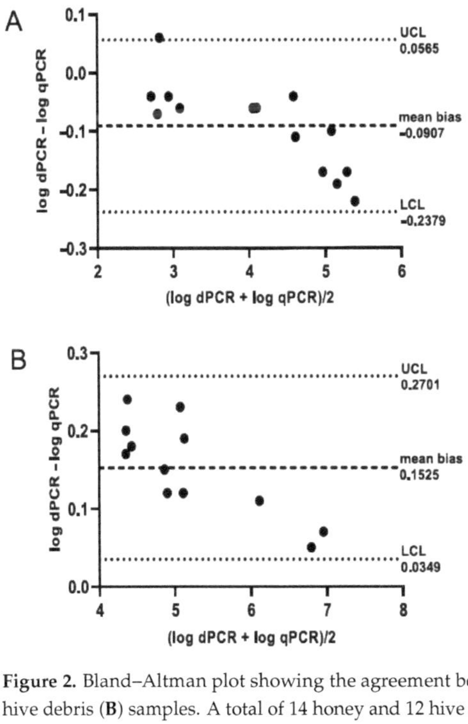

Figure 2. Bland–Altman plot showing the agreement between qPCR and dPCR for honey (**A**) and hive debris (**B**) samples. A total of 14 honey and 12 hive debris samples with spore counts above the limit of quantification by both methods were included in the analysis; see Section 2.7 for details. LCL, lower confidence level; UCL, upper confidence level.

3.4. Association between Spore Counts and Clinical Symptoms

For honey (n = 67) and hive debris (n = 107) samples collected from colonies with varying severity of clinical symptoms (honey: 26 samples with level "0", 14 with "1", 17 with "2", nine with "3", and one with "4", which was regarded as "3" for the purpose of statistical analyses; hive debris: 96 samples with level "0", nine with "1" and eight with

"2"), spore counts were determined by plate counting (honey) and qPCR (honey and hive debris) (Table S1). For apiaries, we indicated if AFB was active (i.e., clinical symptoms were present in at least one colony per apiary; asymptomatic colonies from such AFB-positive apiaries were also included in the sampling) or completely absent (i.e., the apiary was free from AFB, since it contained no clinically affected colonies, was not located in any of the active AFB zones, and had no history of AFB). As expected, qPCR counts in hive debris samples originating from asymptomatic colonies (clinical symptoms designated as "0") were significantly higher in apiaries with a history of AFB than in those with full absence of AFB ($p < 0.0001$, Mann–Whitney test); no such association could be assessed for honey samples, since all originated from AFB-positive apiaries. In addition, spore counts in hive debris samples collected from asymptomatic colonies with a history of AFB were significantly lower than in those from affected colonies with clinical symptoms designated as "1" ($p < 0.0001$, Mann–Whitney test).

For honey samples, a significant positive correlation was observed between plate and qPCR counts (Spearman's $r_s = 0.9185$, $p < 0.0001$), although plate counts were always lower than qPCR counts (Table S1). Median spore counts in honey (Figure 3) and hive debris (Figure 4) samples from clinically affected colonies were significantly different ($p < 0.001$) from those in asymptomatic colonies; for honey samples, this was observed for both quantification methods. Although a trend of increased spore counts with increasing severity of clinical symptoms was observed, median spore counts in AFB-positive samples did not differ significantly with respect to the severity of AFB symptoms, regardless of sample type.

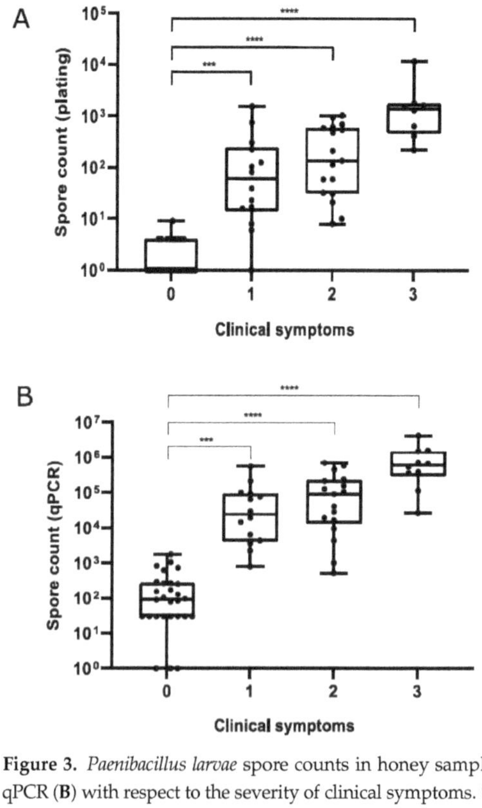

Figure 3. *Paenibacillus larvae* spore counts in honey samples determined by plate counting (**A**) and qPCR (**B**) with respect to the severity of clinical symptoms. Significant differences (Kruskal–Wallis test followed by Dunn's post hoc test) are marked with asterisks. Legend: ***, $p \leq 0.001$; ****, $p \leq 0.0001$.

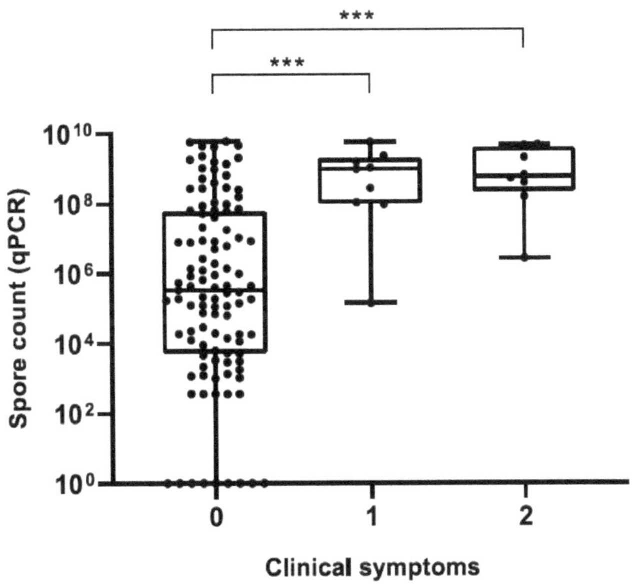

Figure 4. *Paenibacillus larvae* spore counts in hive debris samples determined by qPCR with respect to the severity of clinical Scheme 0. Legend: ***, $p \leq 0.001$.

The univariate logistic regression revealed that plate and qPCR counts in honey samples were associated with the presence of clinical symptoms in the colony (odds ratios: 8.64 and 10.42; confidence intervals: 1.06–4.59 and 1.24–4.45; $p = 0.0090$ and 0.0024, respectively). The same was observed for qPCR counts in hive debris samples (odds ratio: 1.62; confidence interval: 0.29–0.76; $p < 0.0001$). At a probability of 0.5, the threshold to classify colonies as affected (i.e., symptomatic) or unaffected (i.e., asymptomatic) was 1077.21 spores/g honey for qPCR and 6.76 spores/g honey for plate counting (Figure 5). For hive debris, a broader range of qPCR spore counts was observed in the asymptomatic colonies, and a relatively small number of colonies exhibited clinical symptoms, resulting in the logistic regression model not converging (Figure 5). Nevertheless, all the affected colonies harbored >5 log spores/mL debris by qPCR.

For 31/67 honey samples, spore counts were above the LOQ by both methods, enabling calculation of the germination rate (Table S1); it was low and inconsistent (average = 0.52%, range = 0.04–6.05%), making plate counting unsuitable for reliable quantification. Out of 67 honey samples, 66 were positive by qPCR, of which 66 were culture-negative (Table S1), representing a sensitivity of 50/66 (75.8%) for plate counting. All culture-positive samples were also qPCR-positive. All qPCR counts were higher than the plate counts, with an average ratio of qPCR count to plate count of 2.4 logs.

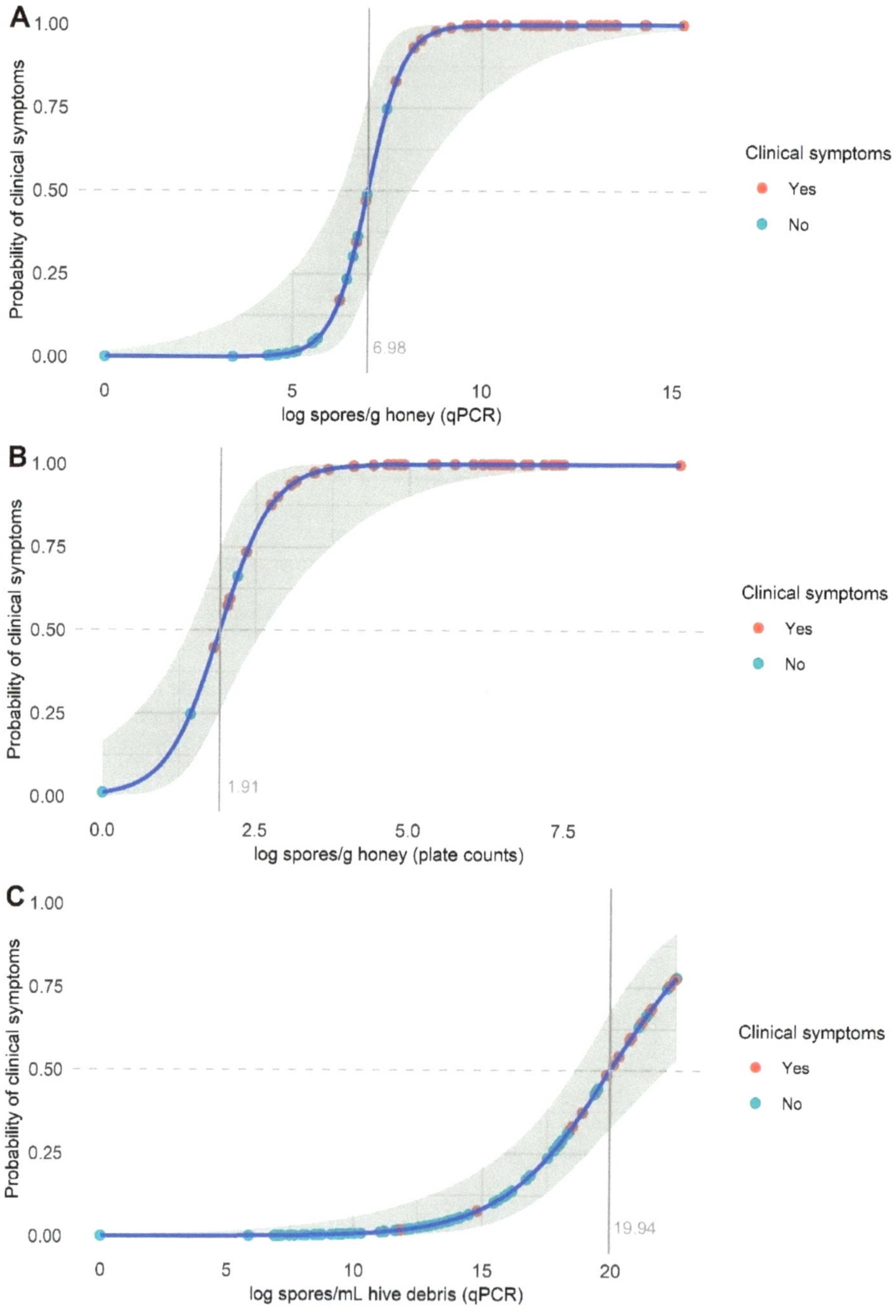

Figure 5. Logistic regression plot (blue line, logistic curve; shaded area, 95% confidence interval) of *Paenibacillus larvae* spore counts in honey (**A**,**B**) and hive debris (**C**) samples. Spore counts were determined by qPCR (**A**,**C**) and plate counting (**B**). The points on the fitted logistic curve show the predicted probability of a colony exhibiting clinical symptoms for each sample.

4. Discussion

This study is the first report of the use of dPCR for the initial calibration of a novel TaqMan probe-based qPCR assay for the quantification of *P. larvae*. Both previously developed qPCR assays optimized for the detection and quantification of *P. larvae* spores in honey and hive debris [23,26] were based on 16S rRNA gene and SYBR technologies, which are known for their limitations. First, the 16S rRNA gene is present in eight copies per *P. larvae* genome, which can exhibit single nucleotide polymorphisms, resulting in preferential amplification of a particular subset of 16S rRNA gene variants [23]. Second, SYBR-based technology is known to be less specific and sensitive compared with TaqMan-based probes, and highly similar 16S rRNA gene amplicons from closely related species can be difficult to distinguish via melting curve analysis [23,38,39]. The newly developed qPCR assay is based on TaqMan technology and targets a single-copy gene which is highly conserved within the species but absent in other closely related species. The assay overcomes the limitations of previous qPCR assays based on 16S rRNA gene and SYBR chemistry.

Recently, a qPCR assay for the simultaneous detection of *P. larvae* and *M. plutonius* was developed, targeting the *tnp60* gene and *napA* pseudogene, respectively [28]. In the present study, the *tnp60* assay proved unsuitable for the detection and quantification of *P. larvae* because of the variable copy number of the *tnp60* gene, variations between multiple *tnp60* paralogs, and mismatches in the primer/probe binding sites (data not shown).

Here, dPCR was used for initial calibration of qPCR, allowing precise and absolute quantification of *P. larvae* spores in honey and hive debris. Compared with other techniques for the calibration of qPCR, such as plate counting, flow cytometry, and microscopy counting, dPCR represents a *P. larvae*-specific calibration method that circumvents the bias in spore germination. Moreover, it uses the same DNA template, chemistry, and PCR conditions as qPCR, maximizing the comparability of dPCR and qPCR. This was confirmed by the high agreement and strong correlation between qPCR and dPCR observed in this study. Because dPCR is more costly and time-consuming, has a lower sample throughput, and is characterized by a narrower dynamic range compared with qPCR [29], our results suggest that qPCR can effectively replace dPCR for routine detection/quantification of *P. larvae* in honey and hive debris samples. The newly developed qPCR assay is based on widely used TaqMan chemistry and can be extended to other bee-related samples after validation and calibration.

The constructed qPCR assay targeting the metalloproteinase gene showed high performance, with LOD values of 8 spores/g honey and 188 spores/mL hive debris, reaching the theoretical LOD of three targets per reaction [37] both for honey and hive debris. Martínez et al. [26] reported LOD of 2 spores/g of honey, whereas Rossi et al. [23] reported 10 spores/g of honey or hive debris. Both previous assays target the 16S rRNA gene, which is present in eight copies per *P. larvae* genome, and are thus expected to have higher sensitivity compared with the assay developed here. However, as mentioned above, assays based on the 16S rRNA gene and SYBR technology suffer from specificity issues and are less appropriate for reliable quantification. Previous assays employed different methods for qPCR calibration, PCR reagents, and conditions and are, therefore, not directly comparable with the metalloproteinase qPCR assay. In addition, they performed no qPCR replicates [26] or performed only technical replicates [23], thereby not accounting for the inherent biological variability and losses during sample preparation and DNA extraction. An important limitation of the assay developed by Rossi et al. [23], who optimized the assay developed by Martínez et al. [26], is its calibration using plate counting. The metalloproteinase qPCR assay was calibrated using dPCR and validated in accordance with qPCR publication guidelines [37], also considering the inherent variability of *P. larvae* spore counts in complex samples. The assay thus employed validation of the entire process from sample preparation to qPCR.

For honey, the average spore germination rate was 0.52% and showed a broad range (0.04–6.05%), reiterating the previously described discrepancy between molecular and cultivation-based spore counts, which can exceed 1 log unit [22]. The determined germination rate is markedly lower than the previously determined rate of 6%, which was based on the comparison of plate and microscopy counts [40]. Of note, germination rates are also highly dependent on the selection of cultivation/germination media [21,22]; thus, the differences between qPCR and plate counts observed in this study could be decreased if different growth media were to be used. Plate counting had a markedly lower sensitivity (75.8%) than qPCR, whereas none of the culture-positive samples was negative by qPCR, which is in agreement with previous findings [22]. Although qPCR does not enable the distinction between live and dead cells/spores and may lead to an overestimation of qPCR counts, previous work confirmed that two-thirds of culture-negative samples were positive after re-cultivation on modified growth media supplemented with germinant agents [22].

In this study, a significant correlation between the spore counts and the presence/absence of clinical symptoms was observed for both sample types. Spore counts did not differ significantly between the affected colonies with varying levels of disease severity. In the analysis of honey samples that were collected from the brood chamber of individual colonies, a threshold for distinguishing between clinically affected and asymptomatic apiaries at 0.5 probability could be set at ~1000 spores/g honey by qPCR. Spore counts in the hive debris samples collected from asymptomatic colonies varied greatly, making it difficult to establish such a threshold. Of note, all the affected colonies harbored >5 log spores/mL debris by qPCR. The asymptomatic colonies with >5 log spores/mL debris originated from apiaries with histories of AFB (most from apiaries with currently active AFB and fewer from currently AFB-negative apiaries). Gende et al. [14] showed a correlation between higher *P. larvae* spore counts in adult bees from clinically affected colonies compared with asymptomatic colonies and proposed a threshold of ~3000 spores per adult bee for the presence of clinical symptoms in the colony, as determined by plate counting. In this study, spore counts in the asymptomatic colonies also differed significantly with respect to the history of AFB, especially for the hive debris samples. Therefore, the establishment of a threshold (spores per sample unit) to distinguish between affected and unaffected colonies should be considered as a guide rather than a fixed value. Such a threshold is of great importance for future AFB surveillance by qPCR, because it identifies at-risk colonies or apiaries that should undergo clinical examination and isolation of *P. larvae* for final confirmation of AFB.

In Slovenia, there is a very high density of honeybee colonies, on average more than 10 per km^2 (data from the National Register of Apiaries for 2020, Ministry of Agriculture, Forestry, and Food). Together with the activities of beekeepers (e.g., migratory beekeeping, exchange of equipment, or unreported activities), this promotes the rapid spread of bee diseases. The newly developed qPCR assay provides a cost- and time-efficient means for a country-wide surveillance of AFB. It would enable early detection of at-risk colonies with a high discovery rate. Even if colonies with spore counts close to the threshold are not (yet) showing clinical symptoms, strict control measures should be implemented in such colonies/apiaries (e.g., shaking bees onto new frames, improved hygiene of the beekeeping equipment, strict control of honeybee transfer, and clinical examination of colonies in at-risk apiaries). Early detection and enumeration of *P. larvae* spores provides information on the actual AFB status of honeybee colonies, independent of clinical symptoms; even if they are not yet observed, measures can be activated to prevent the development and spread of this dangerous infectious disease. If the recognition of clinical symptoms is left only to the knowledge and skills of beekeepers who report suspected AFB to the authorities, important time could be wasted for preventing, or at least reducing, the spread of *P. larvae*. When clinical symptoms of AFB are already detected and the presence of *P. larvae* is confirmed, much more effort is required, and great economic losses are observed in the sanitation of the affected apiary and the management of the AFB zones.

In this study, the genotype of *P. larvae* isolates from honey samples was not considered; therefore, the effect of different ERIC types on the disease severity could not be assessed. ERIC types differ in their virulence and influence the disease status [4,31,41]. Thus, future studies on the association between spore counts and disease severity should include *P. larvae* isolates of different ERIC types.

5. Conclusions

In summary, we describe here a novel TaqMan-based qPCR assay for reliable quantification of *P. larvae* in bee-related samples (honey and hive debris) which was calibrated using dPCR. A significant correlation was found between increased qPCR counts and the presence of AFB clinical symptoms, both for honey and hive debris samples. The culture-based method, which is routinely used for AFB surveillance, was shown to be unreliable for detection/quantification of *P. larvae*, due to poor and inconsistent spore germination. The metalloproteinase qPCR assay allows a more sensitive, rapid, and specific detection/quantification of *P. larvae* and will lead to improved surveillance of AFB.

Supplementary Materials: The following are available online at https://www.mdpi.com/article/10.3390/insects12111034/s1: Figure S1. *Paenibacillus larvae* metalloproteinase primer/probe binding sites. Table S1. Association between AFB clinical symptoms and *Paenibacillus larvae* counts determined by qPCR and plate counting for honey and hive debris samples. Table S2. Scale for grading the severity of AFB clinical symptoms. Table S3. In silico qPCR inclusivity and exclusivity. Table S4. In vitro qPCR inclusivity and exclusivity. Table S5. Comparison of dPCR and qPCR for honey and hive debris samples.

Author Contributions: Conceptualization, D.K., B.P., U.Z., L.Ž. and M.P.O.; data curation, D.K. and B.P.; formal analysis, D.K., B.P. and T.K.; funding acquisition, D.K. and M.O.; investigation, D.K., B.P., U.Z., I.Z., M.G., L.Ž., J.A. and M.P.O.; methodology, D.K., B.P., U.Z., I.Z., M.G., J.A., M.O. and M.P.O.; project administration, D.K.; supervision, D.K.; validation, D.K., B.P. and U.Z.; visualization, D.K., B.P. and T.K.; writing—original draft, D.K. and B.P.; writing—review and editing, D.K., B.P., U.Z., I.Z., M.G., L.Ž., T.K., J.A., M.O. and M.P.O. All authors have read and agreed to the published version of the manuscript.

Funding: The authors acknowledge financial support from the Slovenian Research Agency ("Genotyping of the American foulbrood causative agent as a tool to determine the transmission routes of the disease"—project no. V4-1804; "Animal health, environment, and food safety"—research core funding no. P4-0092) and the Ministry of Agriculture, Forestry, and Food of the Republic of Slovenia (project no. V4-1804).

Institutional Review Board Statement: Not applicable.

Informed Consent Statement: Not applicable.

Data Availability Statement: Experimental dataset supporting the reported results is provided in Tables S1 and S5.

Acknowledgments: Maja Lepen and Olga Kosar (both Veterinary Faculty, University of Ljubljana) are acknowledged for managing the strain collection/database.

Conflicts of Interest: The authors declare no conflict of interest.

References

1. Klein, A.-M.; Vaissière, B.E.; Cane, J.H.; Steffan-Dewenter, I.; Cunningham, S.A.; Kremen, C.; Tscharntke, T. Importance of pollinators in changing landscapes for world crops. *Proc. R. Soc. B.* **2007**, *274*, 303–313. [CrossRef]
2. Neov, B.; Georgieva, A.; Shumkova, R.; Radoslavov, G.; Hristov, P. Biotic and abiotic factors associated with colonies mortalities of managed honey bee (*Apis mellifera*). *Diversity* **2019**, *11*, 237. [CrossRef]
3. Hristov, P.; Shumkova, R.; Palova, N.; Neov, B. Factors associated with honey bee colony losses: A mini-review. *Vet. Sci.* **2020**, *7*, 166. [CrossRef]
4. Genersch, E.; Forsgren, E.; Pentikäinen, J.; Ashiralieva, A.; Rauch, S.; Kilwinski, J.; Fries, I. Reclassification of *Paenibacillus larvae* subsp. *pulvifaciens* and *Paenibacillus larvae* subsp. *larvae* as *Paenibacillus larvae* without subspecies differentiation. *Int. J. System. Evol. Microbiol.* **2006**, *56*, 501–511. [CrossRef] [PubMed]
5. Genersch, E. American Foulbrood in honeybees and its causative agent, *Paenibacillus larvae*. *J. Invertebr. Pathol.* **2010**, *103*, 10–19. [CrossRef]
6. Spivak, M.; Gilliam, M. Hygienic behaviour of honey bees and its application for control of brood diseases and varroa. Part I. Hygienic behaviour and resistance to American foulbrood. *Bee World* **1998**, *79*, 124–134. [CrossRef]
7. Spivak, M.; Reuter, G.S. Resistance to American foulbrood disease by honey bee colonies Apis mellifera bred for hygienic behavior. *Apidologie* **2001**, *32*, 555–565. [CrossRef]

8. *PISRS (Pravno-Informacijski Sistem Republike Slovenije), Pravilnik o Ukrepih za ugotavljanje, Zatiranje, Obveščanje in Preprečevanje hude Gnilobe Čebelje Zalege*; Pravno-Informacijski Sistem Republike Slovenije: Ljubljana, Slovenia, 2020; p. 5. Available online: http://www.pisrs.si/Pis.web/pregledPredpisa?id=PRAV13788 (accessed on 1 July 2021).
9. Nordström, S.; Forsgren, E.; Fries, I. Comparative diagnosis of American foulbrood using samples of adult honey bees and honey. *J. Apic. Sci.* **2002**, *46*, 5–13. Available online: http://www.jas.org.pl:81/Comparative-diagnosis-of-American-foulbrood-using-samples-of-adult-honey-bees-and-honey,0,125.html (accessed on 30 July 2021).
10. Gillard, M.; Charriere, J.D.; Belloy, L. Distribution of *Paenibacillus larvae* spores inside honey bee colonies and its relevance for diagnosis. *J. Invertebr. Pathol.* **2008**, *99*, 92–95. [CrossRef] [PubMed]
11. Bassi, S.; Carpana, E.; Bergomi, P.; Galletti, G. Detection and quantification of *Paenibacillus larvae* spores in samples of bees, honey and hive debris as a tool for American foulbrood risk assessment. *Bull. Insectology* **2018**, *71*, 235–241. Available online: http://www.bulletinofinsectology.org/pdfarticles/vol71-2018-235-241bassi.pdf (accessed on 30 July 2021).
12. Lindström, A.; Fries, I. Sampling of adult bees for detection of American foulbrood (*Paenibacillus larvae* subsp. *larvae*) spores in honey bee (Apis mellifera) colonies. *J. Apicult. Res.* **2005**, *44*, 82–86. [CrossRef]
13. Lindström, A. Distribution of *Paenibacillus larvae* spores among adult honey bees (Apis mellifera) and the relationship with clinical symptoms of American foulbrood. *Microb. Ecol.* **2008**, *56*, 253–259. [CrossRef]
14. Gende, L.; Satta, A.; Ligios, V.; Ruiu, L.; Buffa, F.; Fernandez, N.; Churio, S.; Eguaras, M.; Fiori, M.; Floris, I. Searching for an American foulbrood early detection threshold by the determination of *Paenibacillus larvae* spore load in worker honey bees. *Bull. Insectology* **2011**, *64*, 229–233. Available online: http://www.bulletinofinsectology.org/pdfarticles/vol64-2011-229-233gende.pdf (accessed on 30 July 2021).
15. Forsgren, E.; Laugen, A.T. Prognostic value of using bee and hive debris samples for the detection of American foulbrood disease in honey bee colonies. *Apidologie* **2014**, *45*, 10–20. [CrossRef]
16. Erban, T.; Ledvinka, O.; Kamler, M.; Nesvorna, M.; Hortova, B.; Tyl, J.; Titera, D.; Markovic, M.; Hubert, J. Honeybee (*Apis mellifera*)-associated bacterial community affected by American foulbrood: Detection of *Paenibacillus larvae* via microbiome analysis. *Sci. Rep.* **2017**, *7*, 5084. [CrossRef]
17. Stephan, J.G.; de Miranda, J.R.; Forsgren, E. American foulbrood in a honeybee colony: Spore-symptom relationship and feedbacks between disease and colony development. *BMC Ecol.* **2020**, *20*, 15. [CrossRef]
18. Lindström, A.; Korpela, S.; Fries, I. The distribution of *Paenibacillus larvae* spores in adult bees and honey and larval mortality, following the addition of American foulbrood diseased brood or spore-contaminated honey in honey bee (*Apis mellifera*) colonies. *J. Invertebr. Pathol.* **2008**, *99*, 82–86. [CrossRef] [PubMed]
19. Forsgren, E.; Stevanovic, J.; Fries, I. Variability in germination and in temperature and storage resistance among *Paenibacillus larvae* genotypes. *Vet. Microbiol.* **2008**, *129*, 342–349. [CrossRef] [PubMed]
20. Goodwin, R.M.; McBrydie, H.M.; Borowik, O. Effect of honey bee extracts, and honey on the in vitro germination of Paenibacillus larvae spores. *J. Apicult. Res.* **2013**, *52*, 58–59. [CrossRef]
21. Beims, H.; Janke, M.; von der Ohe, W.; Steinert, M. Rapid identification and genotyping of the honeybee pathogen *Paenibacillus larvae* by combining culturing and multiplex quantitative PCR. *Open Vet. J.* **2020**, *10*, 53–58. [CrossRef] [PubMed]
22. Crudele, S.; Rocchiuti, L.; Ruberto, A.; Rossi, F. Quantitative PCR (qPCR) vs culture-dependent detection to assess honey contamination by *Paenibacillus larvae*. *J. Apicult. Res.* **2020**, *59*, 218–222. [CrossRef]
23. Rossi, F.; Amadoro, C.; Ruberto, A.; Ricchiuti, L. Evaluation of quantitative PCR (qPCR) *Paenibacillus larvae* targeted assays and definition of optimal conditions for its detection/quantification in honey and hive debris. *Insects* **2018**, *9*, 165. [CrossRef] [PubMed]
24. Han, S.-H.; Lee, D.-B.; Lee, D.-W.; Kim, E.-H.; Yoon, B.-S. Ultra-rapid real-time PCR for the detection of *Paenibacillus larvae*, the causative agent of American Foulbrood (AFB). *J. Invertebr. Pathol.* **2008**, *99*, 8–13. [CrossRef] [PubMed]
25. Chagas, S.S.; Vaucher, R.A.; Brandelli, A. Detection of *Paenibacillus larvae* by real-time PCR. *Acta Sci. Vet.* **2010**, *38*, 251–256. Available online: https://www.cabi.org/isc/FullTextPDF/2010/20103356040.pdf (accessed on 30 July 2021). [CrossRef]
26. Martínez, J.; Simon, V.; Gonzalez, B.; Conget, P. A real-time PCR-based strategy for the detection of *Paenibacillus larvae* vegetative cells and spores to improve the diagnosis and the screening of American foulbrood. *Lett. Appl. Microbiol.* **2010**, *50*, 603–610. [CrossRef] [PubMed]
27. Quintana, S.; Fernández, N.J.; Pagnuco, I.; Medici, S.; Eguaras, M.J.; Gende, L.B. Report of a real-time PCR assay for *Paenibacillus larvae* DNA detection from spores of scale samples. *Rev. Arg. Prod. Anim.* **2017**, *37*, 83–88. Available online: https://ri.conicet.gov.ar/bitstream/handle/11336/55325/CONICET_Digital_Nro.1e426611-2f4b-49f3-a681-e75bc46f6925_A.pdf?sequence=2&isAllowed=y (accessed on 30 July 2021).
28. Dainat, B.; Grossar, D.; Ecoffey, B.; Haldemann, C. Triplex real-time PCR method for the qualitative detection of European and American foulbrood in honeybee. *J. Microbiol. Methods* **2018**, *146*, 61–63. [CrossRef]
29. Basu, A.S. Digital assays Part I: Partitioning statistics and digital PCR. *SLAS Technol.* **2017**, *22*, 369–386. [CrossRef] [PubMed]
30. Bakonyi, T.; Derakhshifar, I.; Grabensteiner, E.; Nowotny, N. Development and evaluation of PCR assays for the detection of *Paenibacillus larvae* in honey samples: Comparison with isolation and biochemical characterization. *Appl. Environ. Microbiol.* **2003**, *69*, 1504–1510. [CrossRef] [PubMed]
31. Beims, H.; Bunk, B.; Erler, S.; Mohre, K.I.; Spröer, C.; Silke, C.S.; Pradella, S.; Günther, G.; Rohde, M.; von der Ohe, W.; et al. Discovery of *Paenibacillus larvae* ERIC V: Phenotypic and genomic comparison to genotypes ERIC I-IV reveal different inventories of

virulence factors which correlate with epidemiological prevalences of American Foulbrood. *Int. J. Med. Microbiol.* **2020**, *310*, e151394. [CrossRef]
32. Yoon, S.H.; Ha, S.M.; Kwon, S.; Lim, J.; Kim, Y.; Seo, H.; Chun, J. Introducing EzBioCloud: A taxonomically united database of 16S rRNA and whole genome assemblies. *Int. J. Syst. Evol. Microbiol.* **2017**, *67*, 1613–1617. [CrossRef]
33. Vaerman, J.L.; Saussoy, P.; Ingargiola, I. Evaluation of real-time PCR data. *J. Biol. Regul. Homeost. Agents* **2004**, *18*, 212–214. Available online: https://www.gene-quantification.de/vaerman-qpcr-data-analysis-2006.pdf (accessed on 6 August 2021).
34. Mehle, N.; Nikoli, P.; Gruden, K.; Ravnikar, M.; Dermastia, M. Real-time PCR for specific detection of three phytoplasmas from the apple proliferation group. In *Phytoplasma. Methods in Molecular Biology (Methods and Protocols)*; Dickinson, M., Hodgetts, J., Eds.; Humana Press: Totowa, NJ, USA, 2012; Volume 938, pp. 269–281. ISBN 978-1-62703-089-2. [CrossRef]
35. Berdal, K.G.; Holst-Jensen, A. Roundup Ready®soybean event-specific real-time quantitative PCR assay and estimation of the practical detection and quantification limits in GMO analyses. *Eur. Food Res. Technol.* **2001**, *213*, 432–438. [CrossRef]
36. R Core Team. The R Project for Statistical Computing. 2021. Available online: https://www.R-project.org/ (accessed on 18 October 2021).
37. Bustin, S.A.; Benes, V.; Garson, J.A.; Hellemans, J.; Huggett, J.; Kubista, M.; Mueller, R.; Nolan, T.; Pfaffl, M.W.; Shipley, G.L.; et al. The MIQE guidelines: Minimum information for publication of quantitative real-time PCR experiments. *Clin. Chem.* **2009**, *55*, 611–622. [CrossRef] [PubMed]
38. Cao, H.; Shockey, J.M. Comparison of TaqMan and SYBR Green qPCR methods for quantitative gene expression in tung tree tissues. *J. Agric. Food Chem.* **2012**, *60*, 12296–12303. [CrossRef] [PubMed]
39. Tajadini, M.; Panjehpour, M.; Javanmard, S.H. Comparison of SYBR Green and TaqMan methods in quantitative real-time polymerase chain reaction analysis of four adenosine receptor subtypes. *Adv. Biomed. Res.* **2014**, *3*, 85. [CrossRef] [PubMed]
40. Dingman, D.W.; Stahly, D.P. Medium promoting sporulation of *Bacillus larvae* and metabolism of medium components. *Appl. Environ. Microbiol.* **1983**, *46*, 860–869. [CrossRef]
41. Genersch, E.; Ashiralieva, A.; Fries, I. Strain- and genotype-specific differences in virulence of *Paenibacillus larvae* subsp. *larvae*, the causative agent of American foulbrood disease in honey bees. *Appl. Environ. Microbiol.* **2005**, *71*, 7551–7555. [CrossRef]

insects

Article

Effects of Disinfectants on Bacterium *Paenibacillus larvae* in Laboratory Conditions

Ivana Tlak Gajger [1,*], Zlatko Tomljanović [2], Franco Mutinelli [3], Anna Granato [3] and Josipa Vlainić [4,*]

1 Faculty of Veterinary Medicine, University of Zagreb, Heinzelova 55, 10000 Zagreb, Croatia
2 Ministry of Agriculture, Ulica Grada Vukovara 78, 10000 Zagreb, Croatia; zlatko.tomljanovic@mps.hr
3 Istituto Zooprofilattico Sperimentale delle Venezie, Viale dell'Università, 10, 35020 Legnaro, Italy; fmutinelli@izsvenezie.it (F.M.); agranato@izsvenezie.it (A.G.)
4 Institute Ruđer Bošković, Bijenička cesta 54, 10000 Zagreb, Croatia
* Correspondence: itlak@vef.unizg.hr (I.T.G.); josipa.vlainic@irb.hr (J.V.); Tel.: +385-91-2390-041 (I.T.G.)

Simple Summary: American foulbrood is a highly infectious disease that can harm the beekeeping sector if it becomes clinically visible. It is caused by the bacterium *Paenibacillus larvae*, and its spores are resistant to various disinfectants. It is important to ensure effective final disinfection following eradication measures at apiary in order to prevent the disease from reoccurring. A study was conducted to test ten commercially available disinfectants commonly used in beekeeping, as well as those with proven efficacy in the medicinal and veterinary sectors, on different strains of *P. larvae* bacterium. Early diagnosis methods and proper control measures can help minimize the disease's clinical signs and its incidence.

Abstract: American foulbrood is an infectious disease of the honeybee brood that causes multiple types of damage to beekeeping. The causative agent of the disease is the bacterium *Paenibacillus larvae*, which forms resistant infective spores and is viable for decades. After the eradication measures have been implemented, in cases of clinically visible disease, it is necessary to conduct effective final disinfections of equipment and tools. This study aimed to determine the effect of ten commercially available and commonly used disinfectants on certified strains of *P. larvae* under laboratory conditions, as well as to compare the obtained results among individual genotypes of *P. larvae*. Selected products were tested by determining the zone of inhibition using an agar diffusion test, a suspension test for viable bacteria, a surface disinfectant test, and a sporicidal effect in the suspension test. Incidin Oxy-Foam S and Sekusept Aktiv are both effective against all examined genotypes of *P. larvae*. Despadac and Despadac Secure have a bactericidal effect, but their sporocidal effect is not as satisfactory as that of Genox. Genoll does not exhibit a sporicidal effect, and Ecocide S at 1%, Bee protect H forte, and Bee protect F did not exhibit a satisfactory sporocidal effect. Additionally, EM® PROBIOTIC FOR BEES did not exhibit any bactericidal effect. The effective application of control measures and proper application of final disinfection can reduce the reoccurrence of visible clinical signs of disease, whereas methods of early diagnosis can significantly reduce the incidence of the disease.

Keywords: *Apis mellifera*; American foulbrood; *Paenibacillus larvae*; spores; disinfectants

Citation: Tlak Gajger, I.; Tomljanović, Z.; Mutinelli, F.; Granato, A.; Vlainić, J. Effects of Disinfectants on Bacterium *Paenibacillus larvae* in Laboratory Conditions. *Insects* **2024**, *15*, 268. https://doi.org/10.3390/insects15040268

Academic Editor: Brian Johnson

Received: 18 March 2024
Revised: 8 April 2024
Accepted: 11 April 2024
Published: 12 April 2024

1. Introduction

American foulbrood (AFB) is a severe infectious disease of honeybee colonies that threatens modern beekeeping [1]. The causative agent is the Gram-positive bacterium *Paenibacillus larvae* [2], which forms oval infectious spores. Remains of infected dead larvae contain billions of resilient spores, which can remain viable on combs, honeybee products, equipment, tools, and the apiary environment for decades. Vegetative forms of *P. larvae* are sensitive to scorching, drying, and disinfectants. According to Bakonyi et al., the infectious form of the bacterium presents as spores, and bent honeybee larva become susceptible to infection at the age when they actively take food [3]. Clinical recognition of the disease is possible based on the appearance of alterations in the honeybee brood. In the case of suspected AFB, cappings are wrinkled and retracted with dark spots and holes [4,5]. A lattice brood is noticeable, and the diseased larva turns into a shapeless, brown, and viscous ropy mass [6]. In the advanced stage, the rest of the dead larva fit along the lower wall of the honeycomb cell. A honeybee colony is officially declared infected if clinical examination has established changes typical of the disease and laboratory microscopic examination of the decayed larvae allows for identifying the *P. larvae* spores. Eradication measures involve stamping out and burning infected colonies and associated equipment, as well as the final disinfection of equipment and tools. Burning is the fastest, best, and most expensive way to combat AFB. Sometimes it is possible to save adult bees as an artificial swarm, housed in the new or disinfected hive on comb foundations [7]; however, relapses of clinically visible disease are possible due to poor implementation of final disinfection.

Bednář et al., described various forms of physical and chemical disinfection of beekeeping tools and equipment [8]. Also, they state that the success of disinfection measures in the apiary depends on the correct choice of disinfectants, the spectrum of microorganisms, the recommended concentration of working solutions, the method of application, and exposure to the disinfectant. It is important that the material is disinfected in regard to its possible damage and the possible effect on the environment. Dobbelaere et al. refute the common opinion of beekeepers that the method of burning wooden parts of beekeeping equipment and accessories with flames is a sufficient disinfection measure [9]. Therefore, comprehensive disinfection of the wooden parts of the hive is possible by combining different methods of heat application, such as immersion of wooden parts in microcrystalline wax (150 °C, 10 min) or using high concentrations of disinfectants [10,11]. However, high concentrations of disinfectants are not economically and environmentally friendly, so implementing preventive zoo-hygienic measures and regularly replacing at least 25–30% of dark, old honeycombs is a very important way to mechanically remove *P. larvae* spores and other pathogens [12–17]. Sporicidal effects on *P. larvae* were determined for glutaraldehyde and sodium hypochlorite, 0.5% aqueous sodium hypochlorite solution, 1.1% caustic soda solution, and gamma radiation [18–20]. Kiriamburi et al. examined the biocidal effects of two commercial disinfectants, Virkon® and Disinfection for beekeeping® [21]. Furthermore, extracts of various plants (flavonoids, alkaloids, terpenes, essential oils) exerted successful in vitro inactivation of *P. larvae* in several studies [22–26].

The study aimed to determine the effect of ten commercially available and commonly used disinfectants in beekeeping and veterinary medicine in general, on certified strains of *P. larvae*, in laboratory conditions, and to compare the obtained results among different ERIC (enterobacterial repetitive intergenic consensus classification) genotypes.

2. Materials and Methods

2.1. Selection and Cultivation of Microorganisms

The used strains of *P. larvae* were from the German Collection of Microorganisms and Cell Cultures (DSMZ, Braunschweig, Germany): genotypes (DSM 7030 (ERIC I), DSM 25430a (ERIC II), LMG 16252 (ERIC III) and LMG 16247 (ERIC IV).

P. larvae strains were cultured on a solid nutrient medium Columbia sheep blood agar and in a liquid nutrient medium Brain−Heart Infusion (BHI). To prepare the liquid nutrient medium, 37 g of BHI medium, 3 g of yeast extract, and 1 L of H_2O were used. The *P. larvae* were grown in the liquid nutrient medium on a shaker (New Brunswick Innova 4340 Incubator shaker, New Brunswick, NJ, USA) at a temperature of 37 °C and a speed of 200 rpm. The incubation period was 48 h.

2.2. Disinfectant Effect Test

Disinfectants were selected based on the recommendations of producers and beekeepers, as well as their availability on the market. The following disinfectants and food additives were used: Bee Protect products (Bee Protect H forte and Bee Protect F); (Honey

Bee Pro I Agro Simpa d.o.o., Sisak, Croatia), which contains sucrose, macronutrients and organic acids; Genox and Genoll with foam (Genox Aquagen, Zagreb, Croatia), which contain hypochlorite acid, sodium chloride and hypochlorite ion; Despadac® (Laboratories Calier, Barcelona, Spain) and Despadac Secure® (Laboratories Calier, Barcelona, Spain) with active ingredients didecil-dimethyl ammonia chrysanthemum, glutaraldehyde in different ratios; Ecocid® S (Krka d.d., Novo mesto, Slovenia) with active substances of potassium peroxymonosulfate, sodium dodecyl benzenesulfonate, and sulfamic acid; Sekusept® Aktiv (Ecolab, Zagreb, Croatia) and Incidin® Oxyfoam S (Ecolab, Zagreb, Croatia) with peracetic acid; and EM® probiotic for bees (EMRO, Okinawa, Japan), which is mixture of microbials. Selected products were tested by (1) determining the zone of inhibition in the agar diffusion test, (2) suspension tests for viable bacteria, (3) surface disinfectant tests, and (4) sporicidal effects in the suspension tests for genotypes of *P. larvae* (ERIC I to ERIC IV).

2.2.1. Agar Diffusion

An agar diffusion test was used to preliminarily determine which disinfectant meets the minimum criterion of reducing the number of viable bacteria. One bacterial colony was transferred from the solid nutrient medium plate to the BHI liquid nutrient medium optimized for the growth of *P. larvae* to obtain viable vegetative forms after 48 h of incubation with constant agitation. Then, bacterial cultures were diluted to 0.6 McFarland units determined by a nephelometer (Biosan Ltd., Riga, Latvia). The bacteria were diluted with a medium in a ratio of 1:9 and used in further experiments. In solid blood agar, two or more (depending on the expected diameter of the inhibition zone) wells were punctured, and individually tested disinfectants were added to them, i.e., phosphate–buffered saline (PBS), which in this case served as a negative control. To facilitate the diffusion of disinfectants into the agar, the plates were left for four hours at a temperature of 4 °C and subsequently moved to the incubator at a temperature of 37 °C. After incubation for 48 h, the diameters of the inhibition zones were measured. In further research, only substances (i.e., disinfectants) that showed efficiency in the preliminary agar diffusion test were used.

2.2.2. Sporocidal Effect of Disinfectants

Strains of *P. larvae* (ERIC I to IV) grown on solid blood agar at 30 °C were, according to the morphological peculiarities of typical colonies, suspended in 1 mL of PBS. Aliquots (0.1 mL) of bacteria were disposed of on solid blood agar plates and incubated at 30 °C for seven days to stimulate sporulation. The newly created colonies were collected with a sterile stick, suspended in PBS, and rinsed twice by centrifugation at 7500× *g* for 10 min. The resulting pellet was resuspended in PBS and kept at 4 °C until further use. The number of spores in the resulting suspension was determined spectrophotometrically (DEN-1, Biosan Ltd., Riga, Latvia) after heat treatment at 80 °C for 10 min (Thermomixer comfort, Eppendorf, Germany) to eliminate vegetative forms of bacteria. The number of spores was validated after plating of suspension and colony numbering. To determine the sporocidal effect, disinfectant was added at the concentration recommended in the manufacturer's instructions. Exposure times were set as 5, 15, 30, and 60 mins, except for Incidin Oxy Foam S, where a 1 min exposure was added (based on manufacturer declaration). After exposing the *P. larvae* spores to disinfectant, the spore suspension was filtered (Shott's bottle and Millipore vacuum pump used), and a filter (pore size of 0.45 μm) was placed in 2 mL of sterile PBS and vortexed (V-1, Biosan Ltd., Riga, Latvia) for two minutes to release spores. Then, 100 μL of suspension was seeded on Columbia sheep blood agar at 37 °C. After 48 h, colonies of *P. larvae* were counted, with Koch's method of counting bacterial colonies being used to determine the number of *P. larvae* spores that survived the disinfectant exposure, which serves to determine the number of living cells on the principle that one colony grown on a solid nutrient medium indicates one live spore. Seeding using the method of dilution of the sample enabled more accurate determination and counting of the actual number of colonies grown, as the number of colonies grown corresponds to the number of bacterial cells, i.e., spores in the sample (the number of colonies is indicated as the number of colony-forming units (*Colony Forming Units*, CFUs). We mathematically obtained the exact number of viable spores as the ratio of the number of bacterial colonies divided by the volume of the planted sample and divided by the reciprocal value of the dilution planted on the substrate. The ratio of the number of bacterial colonies grown in the control and treated group creates the logarithm (Log) reduction.

2.2.3. Determination of ATP Level

P. larvae were grown in the liquid nutrient medium BHI on the shaker to ensure ideal conditions for the growth and reproduction of the bacterial population. After the cultivation step for 48 h at 37 °C with agitation, the suspension of bacteria, prepared at a concentration of 0.6 McFarland units, was used in further research in dilution 1:10.

To determine the amount of ATP, we used the EnSURE (EnSURE Multi-Parameter Luminometer, Hygiena, Germany) instrument along with the Super Snap High Sensitivity ATP test (EnSURE, Hygiena, Germany), which detects low concentrations of ATP in the tested sample. The sampling aimed to examine the number of bacteria in the control sample before and after exposure of *P. larvae* to a disinfectant in the time sequence, depending on the time of exposure. The procedure is very simple and fast and involves immersing the test stick in the sample. According to the manufacturer's instructions, the sample is combined with the test solution, and then placed in a luminometer where the ATP level is spectrophotometrically determined.

2.2.4. Effects of the Disinfectant on the Contaminant Surface

To determine whether the selected disinfectant works on the surface, a clean dry hard surface free from organic pollution and microorganisms was experimentally contaminated with *P. larvae*. After the surface was dried, it was treated with the selected disinfectant according to the manufacturer's instructions regarding the concentration and length of exposure. At the end of the required time, the stick of the Super Snap High Sensitivity ATP test (Hygiena, Potsdam, Germany) was used to swab a surface of 10 cm^2. The level of ATP was determined using an EnSURE device (Hygiena, Potsdam, Germany).

In addition, the surface was sampled with a sterile test stick (10 cm^2), then placed in PBS and vortexed, while 100 µL of the suspension was seeded on solid blood agar and incubated for 72 h at 37 °C. After the end of incubation time, counting of bacterial colonies was used to determine the effect of a disinfectant on *P. larvae* on solid surfaces.

2.2.5. Statistical Data Processing

Results were presented as the mean ± standard error of at least three repetitions of each sample and used test. As the results in the control and treated groups for all four *P. larvae* genotypes (ERIC I–IV) were the same, they are presented as one for each test performed, e.g., results for sporocidal effect for all four genotypes are presented in one column. Statistical analysis was conducted using GraphPad Prism 9.0 (GraphPad Software, La Jolla, CA, USA). The existence of a statistically significant difference between the results was determined using a one-way analysis of variance (one-way ANOVA) with Tukey's post hoc test.

3. Results

A summary of the results is presented in Table 1. All results were compared with the control groups.

Despadac (1% solution) shows inhibitory activity for the growth of *P. larvae*, and the mean diameter of the inhibition zone was 19.25 mm, while Despadac Secure (10% solution) showed an inhibition zone of 14.75 mm. These results for Despadac Secure are statistically significant (ANOVA, F (8.32) = 68.30; $p < 0.0001$) in relation to the control group. The results show that disinfectants whose main active component is active oxygen (Ecocide S, Krka, and Sekusept Aktiv, Ecolab) have a statistically significant effect on the vegetative forms, creating a relatively wide band for an inhibition zone of bacterial growth inhibition

(ANOVA, F (2.17) = 35.15; $p < 0.0001$). Ecocide S caused a mean inhibition zone diameter of 41.63 ± 7.289 mm, while Sekusept Aktiv caused the formation of an inhibition zone with a width of 44.63 ± 12.19 mm. Incidin OxyFoam S caused the formation of a relatively wide band of the *P. larvae* growth inhibition zone, which was 47.52 ± 12.19 mm ($p < 0.0001$). Genox, at a concentration of 10%, showed a weaker disinfecting effect than an undiluted product; however, the producer recommends dilution of a product, even at a larger scale (up to 1% which in our study shows no efficacy,). The diameter of the inhibition zones caused by Genoll is significantly smaller (9.50 ± 1.04 mm). A single analysis of variance showed that Genox 100% and Genox 10% had a significant inhibitory effect on the growth of *P. larvae* ($p < 0.0001$) as well as Genoll ($p = 0.0088$). The application of Bee Protect F caused the formation of an inhibition zone of 24.75 ± 3.09 mm, and the application of the product Bee Protect H Forte inhibition zone of 36.75 ± 1.89 mm. Both described results compared to negative control values are statistically significant ($p < 0.0001$). The influence of effective microorganisms on the *P. larvae* showed that the visible zones of inhibition are only a few millimeters wide, so no additional analyses were made for the product EM® PROBIOTIC FOR BEES. These results are presented in Figure 1.

Table 1. Summary of results on disinfectant effects of ten commercially available products on bacterium *P. larvae* for all examined genotypes (ERIC I–ERIC IV).

Bacterium *Paenibacillus larvae*		**Method for Determination of Disinfectant Effects**									
		Suspension test for viable bacteria				Surface disinfectant test				Sporicidal suspension test	
Disinfectant product	Inhibition zone diameter (mm)	Determination of the amount of ATP (ATP units)								Logarithms of reducing spores (log 10)	
		5	15	30	60	5	15	30	60	30	60
		Exposure to disinfectant (min)									
Bee Protect H forte	36.75	-	-	-	172.00	-	-	-	-	-	<1
Bee Protect F	24.75	-	-	-	102.20	-	-	-	-	-	<1
Genox 100%	30.10	71.80	333.50	483.50	513.50	76.00	102.10	481.80	440.40	3	3
Genox 10%	23.20	64.20	90.40	151.20	175.60	61.10	120.40	154.30	170.20	1	1
Genoll 10%	9.50	72.33	93.20	101.60	107.50	68.00	94.20	103.40	120.30	-	-
Despadac	19.25	52.40	247.50	775.50	-	84.20	182.20	543.00	-	2	-
Despadac Secure	14.75	98.80	275.50	755.00	-	102.60	192.40	490.00	-	1	-
Ecocid S	41.63	80.20	346.50	539.80	838.50	50.80	340.00	542.80	840.20		6
Sekusept aktiv 2%	44.63	893.80	942.40	930.00	875.50	843.20	896.80	860.00	864.20	6	5
Sekusept aktiv 1%	-	75.20	102.00	297.50	880.00	52.40	68.20	360.40	875.60	3	1
Incidin Oxyfoam S	47.52	830.70	801.10	-	-	826.60	836.40	-	-	6	6
EM® probiotic for bees	-	-				-				-	

In the suspension test with the vegetative form of *P. larvae*, the effectiveness of the disinfectant during exposure following the standards for bactericidal action was determined. The results show that both Despadac products show statistically significant bactericidal actions on *P. larvae* (indirectly measured by the amount of ATP in suspension) (ANOVA, F (8.31) = 26.74; $p < 0.0001$). More precisely, after 15 min of exposure of the bacteria to Despadac disinfectant, the value of the relative units of ATP increased to 247.50 ± 29.55, reaching 775.50 ± 110.70 after 30 min of exposure. After exposure to Despadac Secure for 15 min, an increase in relative ATP units to 275.50 ± 22.16 was observed, reaching 755.0 ± 132.60 after 30 min. The results show that active oxygen-based products have a statistically significant bactericidal effect on *P. larvae* (ANOVA, F (12.44) = 46.18, $p < 0.0001$) depending on the time of exposure. In Sekusept Aktiv assets at 2% concentration, the effect was significant after 5 min of exposure (893.80 ± 68.66 RU, $p < 0.0001$) and in 1% concentration after 30 min (297.50 RU ± 28.10; $p < 0.0001$). The action of Ecocid S was observed after 15 min (346.50 ± 33.98) and after 30 min (539.80 ± 83.54) ($p < 0.0001$). After 60 min of exposure to vegetative forms of *P. larvae* at both applied concentrations of Sekusept Aktiv, as well as Ecocide S, a very similar number of relative units was determined, which indicates their significant bactericidal action (values of relative units as follows: 875.50 ± 84.95, 880.0 ± 97.21 and 838.50 ± 69.65; $p < 0.0001$). A statistically significant effect was caused by the disinfectant Sekusept Aktiv in 2% concentration after 5 min of exposure ($p < 0.0001$), while Sekusept Aktiv in 1% concentration and Ecocid S showed a statistically significant effect after 30 min ($p < 0.0001$). For Incidin OxyFoam S, the measured value after 1 min (exposure time referred by manufacturer) was 830.70 ± 61.39 ($p < 0.0001$), then 830.70 ± 83.70 after 5 min, while, after 15 min, a significant increase was observed to 801.10 ± 81.21 (ANOVA $F(3.20) = 18.05$, $p < 0.0001$). Genoll did not affect *P. larvae,* and the values of the relative units in the time frame of treatment were as follows: 72.33 ± 12.82; 93.20 ± 19.02; 101.60 ± 12.25 and 107.50 ± 15.47, respectively. Similarly, Genox's 10% was efficient at the longest period of exposure (191.30 ± 10.33, $p = 0.03$). On the other hand, undiluted Genox significantly affected bacteria after 15 min (333.50 ± 10.24; $p < 0.0001$), 30 min (483.50 ± 56.76; $p < 0.0001$) and after 60 min of exposure (513.50 ± 54.94; $p < 0.0001$). The results of the action of Bee Protect H Forte on the vegetative form of *P. larvae* showed that there was a significant increase in the amount of free ATP, which is an indicator of the decomposition of *P. larvae*. Consequently, there was a statistically significant decrease in the number of bacteria (ANOVA, $F(2.15) = 13.79$; $p = 0.0004$). Results are presented in Figure 2.

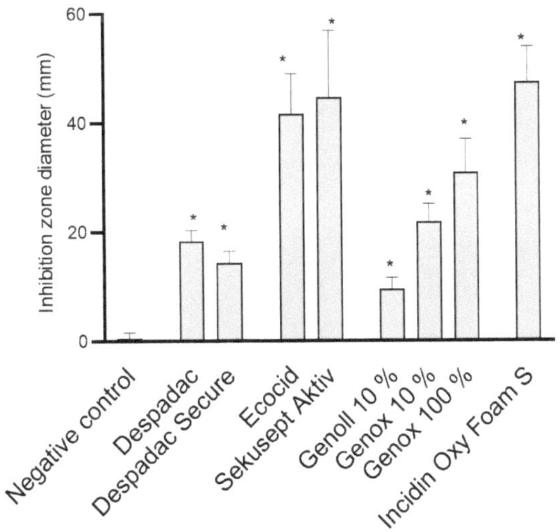

Figure 1. Summary of the diameter of the inhibition zone caused by examined disinfectant products on the growth of *P. larvae* bacteria. The results are shown as the mean ± standard error of at least three repetitions for each genotype. All data from the control group and for specific disinfectants are presented in one column, as the values for all genotypes (ERIC I–ERIC IV) were similar. For the sake of clarity, BeeProtect products and EM are not presented as they are not effective; * $p < 0.0001$.

A statistically significant difference between the experimental and control group was determined after surface treatment for 30 min with Despadac (543.0 ± 67.77) and Despadac Secure (490.00 ± 66.71) (ANOVA, $p < 0.0001$; for both groups) for all genotypes of *P. larvae.* A significant decrease in the number of *P. larvae* after surface treatment with Ecocid S was observed after 15 min (340.00 ± 32.90; $p = 0.0006$), and after 30 and 60 min of exposure, respectively, an even more significant increase in the number of relative ATP units was observed (542.80 ± 82.94; 840.20 ± 68.55). Sekusept Aktiv disinfectant (1%) significantly reduces the number of viable *P. larvae* after 60 min (875.50 ± 84.95; $p < 0.0001$). After one minute of surface treatment with Incidin OxyFoam S disinfectant, the number of *P. larvae* was significantly reduced (577.70 ± 57.72, $p < 0.0001$), and a further decrease was observed after 5 min (826.60 ± 68.11, $p < 0.0001$) and 15 min (836.40 ± 42.58, $p < 0.0001$), respectively. A significant effect on reducing the number of bacteria was observed by Genox only in 100% concentration after 30 min (481.80 ± 40.86; $p < 0.0001$) and 60 min of action (440.40 ± 46.95, $p < 0.0001$). After treating the surface with the Bee Protect line, even after 60 min of exposure, there was no increase in the value of the relative number of ATP units, nor a decrease in the number of viable bacteria. Results are presented in Figure 2.

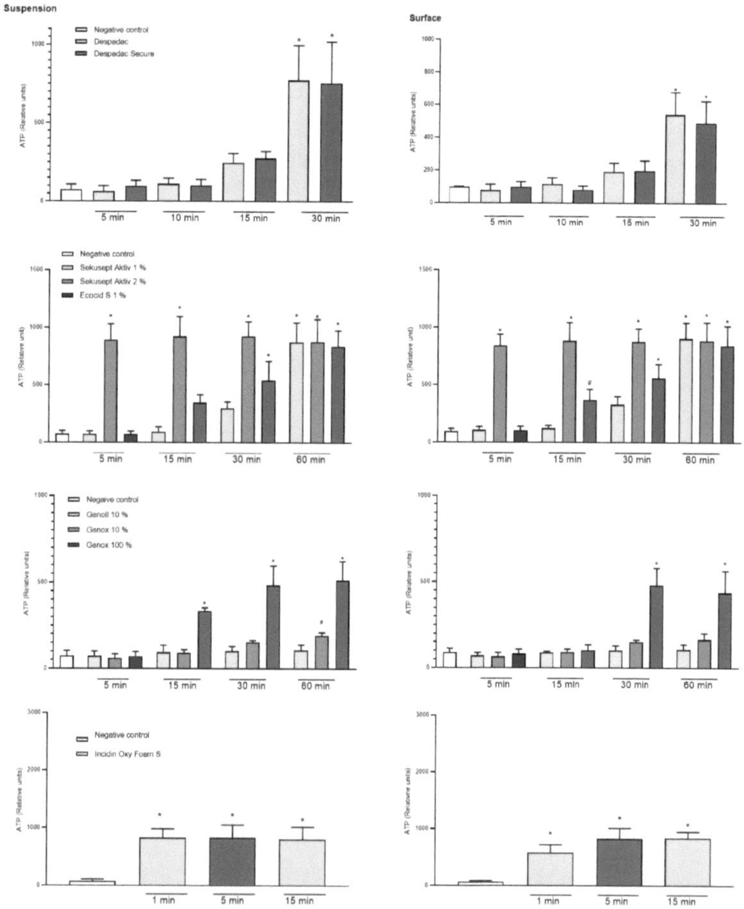

Figure 2. Summary of the relative units of ATP caused by disinfectant products Despadac, Sekusept Aktiv, Ecocid S, Genox and Genoll, and Incidin Oxy Foam S on the growth of *P. larvae* bacteria. Results for all genotypes treated with disinfectant are merged into one column. The results are shown as the mean ± standard error of at least three repetitions; * $p < 0.0001$; # $p < 0.001$.

The sporicidal effect of disinfectants from the Despadac group was not observed after 10 min exposure, while, after 30 min, it amounted to a one Log reduction for Despadac Secure, i.e., a two Log reduction for Despadac. After 60 min of action of both disinfectants, the number of spores was reduced by 99%, i.e., disinfectants caused a reduction in the number of spores by two logarithms. Disinfectants with oxygen as an active substance show a sporicidal effect after 30 min, namely a six Log reduction for Sekusept Aktiv (2%) and three Log reduction for Ecocid S (1%). After 60 min of action, both Ecocid S and Sekusept Aktiv saw a six and five Log reduction, respectively, while Sekusept Aktiv at 1% concentration reduces the number of spores only by one logarithm. After application of Incidin Oxy Foam S, an active substance hydrogen peroxide, the sporocidal effect after 30 min of exposure was six logarithms. The results showed that the disinfectant Genoll has no sporicidal effect. Genox at 10% caused a reduction in the number of spores of *P. larvae* by one logarithm, as well as at 100% concentration for three logarithms during exposure lasting 30 and 60 min, respectively. Bee Protect products do not show a sporicidal or negligible effect, as they also reduce the number of spores by less than one logarithm during exposure of 60 min. All four genotypes of *P. larvae* (ERIC I to ERIC IV) in the sporicidal suspension tests showed equal sensitivity to all the biocides tested.

We applied 1 million microorganisms onto a dry and smooth surface (square 10 cm^2) and, following disinfectant action, found a reduced number of viable bacteria (using CFU method) for all disinfectants applied (Figure 3). We found the reduction to 100672 CFU for undiluted Genox (100%) after 15 min and 9976 CFU after 30 min, which was similar to the reduction achieved after application of Despadac Secure. The decreases in Despadac reached 10025 CFU and 11 CFU, respectively. Sekusept Aktiv (1%) and Ecocid s (1%) reduced the number of viable bacteria after 15 min for 4 Log (cca 100 CFU remained viable) and after 30 min for 6 Log (cca 1 CFU remained). Incidin OxyFoam S at decreased the number of CFU: after 15 min to 90.70 and after 30 min to 1, corresponding to a five and six Log reduction.

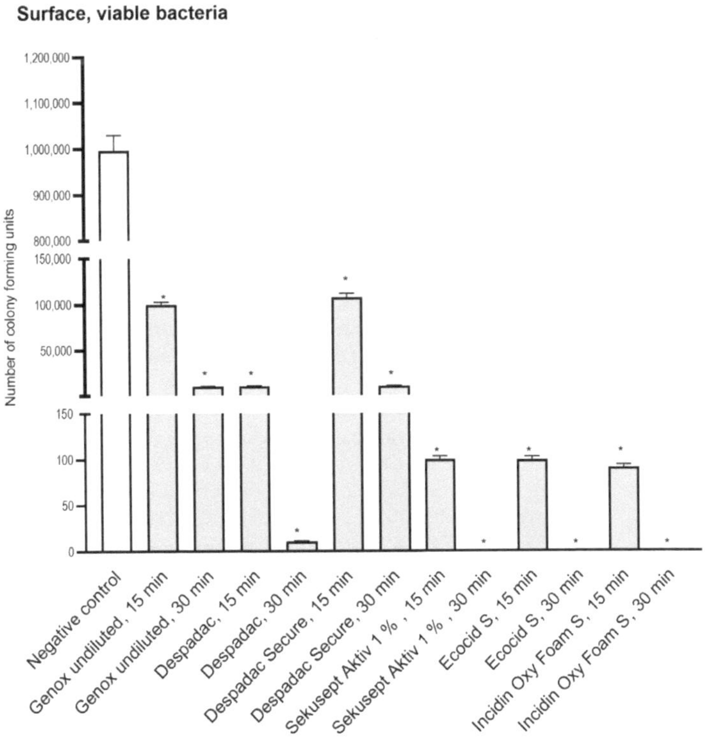

Figure 3. Results of a number of Colony Forming Units following application of disinfectant onto the surface. For the sake of clarity, results obtained following 5 and 60 min are omitted as they are comparable to those obtained after 15 and 30 min, which are presented. All genotypes produced were susceptible to disinfectants at the same level and thus are grouped into one column. The statistically significant decrease for CFU for all groups was * $p < 0.0001$ versus control group.

4. Discussion

Although AFB is a notifiable disease present all over the world, till now there have been no uniform measures nor guidelines that address the effectiveness of disinfectants against bacteria *P. larvae*; thus, we performed a series of experiments using several products with different active ingredients to find the most appropriate for use in apiaries, keeping in mind effectiveness, user-friendly features, and the overall cost.

Hypochlorous acid is an active ingredient in sodium hypochlorite that reacts quickly to various substances, such as proteins, DNA, lipids, thiols, and disulfides. It is believed that this reaction prevents the germination of *P. larvae* spores by affecting the inner membrane of exposed spores [26].

This study found that two disinfectant products, Genox and Genoll, had an inhibitory effect on vegetative forms of *P. larvae* in an agar diffusion test. The result for Genox depended on the concentration applied. In the suspension test, undiluted Genox had a significant inhibitory effect after 15 min of administration that was dependent on the duration of exposure. A similar result was observed on surfaces after 30 min that was not dependent on contact time. It is important to note that the effect of Genox weakened slightly after 60 min, as all chlorine-based biocides have a time-limited effect due to wear during exposure to environmental factors. Hypochlorite-based disinfectants are also consumed more in the presence of organic matter, and their effectiveness is affected by the presence of proteins, nucleic acids, or other organic substances in the environment [27].

The microbicidal effect of chlorine is achieved by damaging functional molecules, such as DNA or proteins [28], and depends on the oxidative potential. A study found that dry spores of bacteria are more sensitive to NaOCl than wet spores [18] and that there are no germinating spores after 40 min of exposure in dry spores. In contrast, in a humid environment, an effective concentration of only above 0.025% of active chlorine was found. However, in a study of sporicidal action on *P. larvae* microenvironments in wet conditions, Genox 10% reduces the number of germinating spores by one logarithm, or Genox 100% for three logarithms after 30 min of contact. This disinfectant does not have a desirable sporicidal profile due to the long time it takes to achieve a sporicidal effect, which is also very limited. To achieve a good sporicidal activity, it would be expected to reduce the number of spores by a minimum of five Log reduction in real contact time. Sporicidal activity of 30 min or more requires immersion of contaminated equipment in a disinfectant solution, making disinfection difficult, time consuming, and relatively expensive.

The antimicrobial properties of hydrogen peroxide are attributed to the creation of hydroxyl radicals. When present in sufficient concentrations, hydroxyl radicals can harm nucleic acids, proteins, and lipids [29]. Killing spores with hydrogen peroxide does not involve damaging their DNA, unlike vegetative forms of bacteria. In a study of commercially available disinfectants with different chemical compositions, we selected the product Incidin OxyFoam S, as it has a declared sporicidal action and a stabilized hydrogen peroxide content of 1.5 g per 100 g of liquid. The results showed that this disinfectant product, following the manufacturer's declaration, has antimicrobial effects on vegetative forms of *P. larvae* after just one minute of exposure in both suspension and surface tests. Additionally, the sporicidal action of Incidin OxyFoam S was determined to reduce by six Log after 30 min of contact.

A previously published study revealed that a 7% solution of hydrogen peroxide can inactivate spores of bacteria after six hours of exposure [30]. In another study, authors found that spores of *Bacillus* spp. exposed to 10% hydrogen peroxide for one hour were unable to germinate [31]. However, these studies were conducted using hydrogen peroxide that was not mixed with other substances or auxiliary agents, which is different from our research. No formulations of hydrogen peroxide that are stabilized and contain additional agents for oxidation and enhanced action have been patented and proven to be more effective. They act on the proteins from spore sheaths, create hydroxyl free radicals, and oxidize membrane lipids, enabling the disinfectant to act inside the cell, affecting vegetative forms of bacteria.

Hydrogen peroxide is highly effective in killing bacteria and spores in different environments, such as hospitals and industrial settings. It is used in the form of gas through dispersing and nebulization devices, either alone or in combination with silver. Although the mechanism of inactivation of bacteria and spores by hydrogen peroxide gas is complex and not fully understood, research has shown that the gas penetrates deeper into spore structures, leading to the oxidation of essential amino acids required for spore germination [32].

In the domestic market, only one product has been declared safe for bees and used in beekeeping. This study focuses on two available products—Despadac and Despadac Secure—both of which contain didecyl-dimethyl ammonium chloride (as quaternary ammonium salts of the 3rd generation; 1955) and glutaraldehyde as active components. Quaternary ammonium salts have been in use for many years in hand antiseptics, disinfectants, and preservatives in wood processing or for the preparation of eye drops. They have a wide range of possible chemical structures that can effectively reduce the number of bacteria or inhibit and suppress their reproduction. The findings of this study indicate that the Despadac product line exhibited bactericidal properties after the bacteria were exposed for 30 min to this disinfectant.

However, there was only a slight decrease in the number of germinating spores of *P. larvae*, ranging from one or two logarithms. Such a result is not entirely satisfactory for disinfection purposes, but it may suggest that Despadac products could be used for sanitation. Similarly, a previous study showed that 0.06% of third-generation quartile ammonium salt in the Carrier test did not have a disinfectant effect on the spores *Bacillus stearothermophilus* [33]. This suggests that dodecyl-dimethyl may not be effective against spores, as the spore sheath is composed of proteins, such as keratin and biguanides, on which quartile ammonium chloride does not act [18]. Quaternary ammonium salts are used as biocides, and they work by interacting with the cytoplasmic membrane of bacteria. This interaction results in a range of processes, such as adsorption on the cell wall, penetration into the cell, reaction with the cytoplasmic membrane (lipid or protein), and disorganization of the membrane. These processes lead to leakage of intracellular material, degradation of proteins and nucleic acids, and ultimately, cell wall lysis caused by autolytic enzymes [34]. It has been shown that the selection of formulations and methods of application of disinfection affects the effectiveness of quaternary ammonium salts, and relatively few studies have been conducted in which their effectiveness is evaluated in practical conditions. In addition, we should consider the possible determination of chemical residues (ecological dynamics) and the emergence of resistance to repeatedly applied disinfectants [35]. However, the research objectives did not meaningfully confirm the possible occurrence of resistance to the action of quaternary ammonium salts because they do not have fully comparable data, as quaternary ammonium salts exist today in seven generations and, in combination with other substances (commercial preparations), can have a changed effect.

According to Regulation (EC) No 396/2005 of the European Parliament and the Council [36] (ANON, 2005), didecyldimethylammonium-chloride has been authorized as an active substance in plant protection products for use only on ornamental plants. However, its authorization has been revoked, as its use may lead to the appearance of residues in food for humans and animals. As it has been observed that their use in plant protection products leads to the appearance of residues in food, minimum residue levels have been proposed. Considering that disinfectant products used in beekeeping must be safe because honeybee colonies produce food that must not contain residues of a chemical origin, we believe that Despadac products are not a suitable choice for disinfection in beekeeping. Moreover, we have not found any record of these products being registered as biocides in the Republic of Croatia, raising questions about their regular distribution that require further investigation. In addition, this active substance is not listed in the regulation (EU) No 528/2012 of the European Parliament and the Council [37] concerning making biocidal products available on the market.

The study indicates that disinfectants containing peracetic acid are the most effective against vegetative forms and spores of *P. larvae* bacteria. These disinfectants have been tested in independent microbiological laboratories and have shown bactericidal properties compared to rod bacteria, as well as sporicidal action in relatively short exposures. Peracetic acid is a potent biocide even at low concentrations ranging from 0.0001% to 0.2%.

Peracetic acid shows an advantage over other types of active substance disinfectants because it remains effective even in the presence of organic residues and is broken down into non-toxic and non-mutagenic substances—acetic acid and oxygen. Thus, it ensures a high degree of disinfection effect in a short contact period. We used these desirable properties of disinfectants, along with the stability of the prepared solution over a long period, to investigate the effect of two commercial products on the vegetative forms of *P. larvae* and their spores. Notably, researchers found that the action of peracetic acid as a disinfectant is not dependent on the temperature of the environment, although, at higher temperatures, the results are manifested by weaker antimicrobial activity [38]. The results from the same study showed that, during the entire experimental period of 24 days, Sekusept Aktiv solution had the same disinfection potential for at least four consecutive days. Studies have also shown that the effectiveness of peracetic acid is different depending on whether the microorganisms are in suspension or on the surface, which was not the case in our research. Kunigk et al., (2001) determined that the kinetics of bacterial cell destruction are twofold and may correspond to the contact time of peracetic acid from 2 to 25 min, and the other 25 to 35 min of contact [39]. Similar to ours is a study in which the effectiveness of peracetic acid was shown, which was constant in the first 30 min of contact of disinfectant with bacteria, but subsequently increased [40].

Sekusept Aktiv is a powder used in human medicine to disinfect instruments; however, in veterinary medicine, it is not yet widely used. Unlike this product, and the same active component, Ecocid S, produced by the Slovenian company Krka, has been present on the market for many years, and as such it is relatively often used as a disinfectant in veterinary medicine. Research shows that bactericidal concentrations of Ecocid S disinfectant cause the destruction of the cell wall and cytoplasmic membrane of mycobacteria, as synonymous with a microorganism with very high resistance to disinfectants and external factors. The results also showed that Ecocid S causes decay of the granular components of the cytoplasm with the formation of fine granular inclusions and vacuoles. These irreversible changes in the structural elements of mycobacteria led to the deterioration of bacterial cells [41].

The study of Kiriamburi et al., aimed to examine and compare the biocidal effect of two disinfectant products: "Disinfection for beekeeping" (DFB) (Swienty, Sønderborg, Denmark) and Virkon S (Lanxess, Berlin, Germany) on *P. larvae* spores [22]. Previous studies have shown that the 1% solution of Virkon (active ingredient potassium peroxymonosulfate) has a bactericidal effect on *Pseudomonas aeruginosa, Escherichia coli, Staphylococcus aureus, Enterococcus hirae* and *Mycobacterium smegmatis* in suspension tests and Carrier tests on *P. aeruginosa, E. coli, S. aureus* and *E. hirae*. However, this same concentration of Virkon did not show an inhibitory effect on spores and fungi, i.e., research has shown that the concentration and required time are not following the guidelines for sporicidal and fungicidal action [42]. In their paper, the same authors concluded that 1% Virkon is effective only against vegetative bacteria, yeasts, and viruses, and should therefore be considered a low-level disinfectant [42]. The improved formulation of Virkon S has been shown in studies to kill up to 80% of *P. larvae* spores [6]; however, in the Kiriamburi et al., study, the biocidal effect ranged from 88.6 to 96.8% after 30 min of treatment duration [22].

The suspension test showed that Bee protect H forte had a bactericidal effect, but the test on surfaces did not show the same effect. Additionally, the Bee protect products did not meet the set standards for an overall sporicidal effect in the suspension test. Although these products can increase the effect with the time of exposure of bacteria to disinfectant, they are not recommended for use in the final disinfection of equipment, accessories, and apiary after the sanitation of clinically visible AFB. However, they can be used as an aid in disinfection despite being declared as food supplements for bees.

It should be emphasized that, in disinfecting terms, reducing the number of spores of *P. larvae* by two logarithms (99%) does not represent a sufficient degree of disinfection due to the presence of a large number of spores. In addition, by reducing their number, it may be possible to reduce the extent of infection and/or the rate of development of the disease within one honeybee colony, but it certainly does not lead to a significant reduction in the incidence of the disease.

Effective microorganisms (EMRO, Kishaba, Japan) contain dozens of strains of microorganisms (bacteria, fungi, and mold) that are considered both beneficial for the physiological functioning of the organism and achieving balance in the environment and nature. Effective microorganisms are used in agriculture, forestry, animal husbandry, aquaculture, beekeeping, environmental protection and medicine [43–46]. The benefit of microorganisms in water purification have been documented for many years [43], as well as in the reduction in unpleasant odors on farms [47]. It has also been shown that this formulation of microorganisms applied in tumor cell culture leads to cell apoptosis [48]. In this study, the effect was investigated by using a dietary supplement for bees EM® PROBIOTIC FOR BEES. The achieved inhibitory effect was minimal, and therefore this product cannot be considered to be a disinfectant or biocide. However, the action of effective microorganisms in a living organism is multifaceted, as this product works to preserve normal microflora in the gastrointestinal system of mammals and insects based on the exclusion of pathogenic microorganisms and competitive antagonistic action [49] by changing metabolic pathways through enhancing the action of digestive enzymes while reducing the activity of bacterial enzymes and ammonia formation [50] and having positive effects on the immune system and gut microbiome of bees [51].

5. Conclusions

The products Incidin OxyFoam S and Sekusept Aktiv (when used at a concentration of 2%) have demonstrated a satisfactory sporicidal effect on all four genotypes (ERIC I to ERIC IV) of *P. larvae*. However, Despadac and Despadac Secure showed a bactericidal effect, but their sporicidal effect is not as satisfactory as that of Genox. On the other hand, the product Genoll with foam does not exhibit any sporicidal effect; additionally, the products Ecocide S at a concentration of 1%, as well as Bee protect H forte and Bee protect F, did not exhibit a satisfactory sporicidal effect on *P. larvae*. The food additive EM® PROBIOTIC FOR BEES did not exhibit a bactericidal effect. Therefore, we strongly recommend considering the effectiveness of a disinfectant before use and choosing an appropriate option to reduce the reoccurrence of a disease in apiary.

Author Contributions: Conceptualization, I.T.G. and J.V.; methodology, I.T.G., J.V. and A.G.; investigation, I.T.G., Z.T., J.V., A.G. and F.M.; writing—original draft preparation, I.T.G. and Z.T.; writing—review and editing, I.T.G., J.V., A.G. and F.M.; visualization, J.V.; funding acquisition, I.T.G. All authors have read and agreed to the published version of the manuscript.

Funding: This research received no external funding.

Data Availability Statement: Data were presented in the manuscript.

Acknowledgments: The authors would like to thank G. Husinec for her technical assistance.

Conflicts of Interest: The authors declare no conflicts of interest.

References

1. Forsgren, E.; Laugen, A.T. Prognostic value of using bee and hive debris samples for the detection of American foulbrood disease in honey bee colonies. *Apidologie* **2014**, *45*, 10–20. [CrossRef]
2. Genersch, E.; Forsgren, E.; Pentikäinen, J.; Ashiraileva, A.; Rauch, S.; Kilwinski, J.; Fries, I. Reclassification of *Paenibacillus larvae* subsp. *pulfifaciens* and *Paenibacillus larvae* subsp. *Paenibacillus larvae* without subspecies differentiation. *Int. J. Syst. Evol. Microbiol.* **2006**, *56*, 501–511. [CrossRef] [PubMed]
3. Bakonyi, T.; Irmgard, D.; Grabensteiner, E.; Nowotny, N. Development and evaluation of PCR assays for the detection of *Paenibacillus larvae* in honey samples: Comparison with isolation and biochemical characterization. *Appl. Environ. Microbiol.* **2003**, *69*, 1504–1510. [CrossRef] [PubMed]

4. Spivak, M.; Reuter, G.S. Resistance to American foulbrood disease by honey bee colonies *Apis mellifera* bred for hygienic behaviour. *Apidologie* **2021**, *32*, 555–565. [CrossRef]
5. De Graaf, D.C.; De Vos, P.; Heyndrickx, M.; Van Trappen, S.; Peiren, N.; Jacobs, F.J. Identification of *Paenibacillus larvae* to the subspecies level: An obstacle for AFB diagnosis. *J. Invertebr. Pathol.* **2006**, *91*, 115–123. [CrossRef] [PubMed]
6. Hansen, H.; Brødsgaard, C.J. American foulbrood: A review of its biology, diagnosis and control. *Bee World* **1999**, *80*, 5–23. [CrossRef]
7. Genersch, E. Honey bee pathology: Current threats to honey bees and beekeeping. *Appl. Microbiol. Biotechnol.* **2010**, *87*, 87–97. [CrossRef] [PubMed]
8. Bednář, M.; Dolánek, J.; Haklová, M.; Jaša, T.; Kamler, F.; Titěra, D.; Veselý, V. *Hygiene in the Apiary*; Titera, D., Ed.; Bee Research Institute Dol: Dol, Czech Republic, 2009; pp. 14–20.
9. Dobbelaere, W.; Graaf, D.C.; Peeters, J.E.; Jacobs, F.J. Development of a fast and reliable diagnostic method for American foulbrood disease (*Paenibacillus larvae* subsp. *larvae*) using a 16S rRNA gene based PCR. *Apidologie* **2021**, *32*, 363–370. [CrossRef]
10. Alippi, A.M.; Lopez, A.C.; Aguilar, O.M. A PCR-based method that permits specific detection of *Paenibacillus larvae* subsp. *larvae*, the cause of American Foulbrood of honey bees, at the subspecies level. *Lett. Appl. Microbiol.* **2004**, *39*, 25–33. [CrossRef]
11. Dobbelaere, W.; Graaf, D.C.; Reybroeck, E.; Desmedt, E.; Peters, J.E.; Jacobs, F.J. Disinfection of wooden structures contaminated with *Paenibacillus larvae* subsp. *larvae* spores. *J. Appl. Microbiol.* **2001**, *91*, 211–216. [CrossRef]
12. Budge, G.E.; Barret, B.; Jones, B.; Pietravalle, S.; Marris, G.; Chantawannakul, P.; Thwaites, R.; Hall, J.; Cuthbertson, A.G.S.; Brown, M.A. The occurrence of *Melissococcus plutonius* in healthy colonies of *Apis mellifera* and the efficacy of European foulbrood control measures. *J. Invertebr. Pathol.* **2010**, *105*, 164–170. [CrossRef] [PubMed]
13. Vidal-Naquet, N. *Honeybee Veterinary Medicine: Apis mellifera L.*; 5m Publishing: Sheffield, UK, 2015.
14. Tlak Gajger, I.; Mañes, A.M.; Formato, G.; Mortarino, M.; Toporcak, J. Veterinarians and beekeeping: What roles, expectations and future perspectives?—A review paper. *Vet. Arhiv.* **2021**, *91*, 437–443. [CrossRef]
15. Tlak Gajger, I.; Kosanović, M.; Bilandžić, N.; Sedak, M.; Čalopek, B. Variations in lead, cadmium, arsenic, and mercury concentrations during honeybee wax processing using casting technology. *Arh. Hig. Rada Toksikol.* **2016**, *67*, 223–228. [CrossRef] [PubMed]
16. Tlak Gajger, I.; Kosanović, M.; Oreščanin, V.; Kos, S.; Bilandžić, N. Mineral content in honeybee wax combs as a measurement of the impact of environmental factors. *Bull. Environ. Contam. Toxicol.* **2019**, *103*, 697–703. [CrossRef] [PubMed]
17. Kosanović, M.; Bilandžić, N.; Sedak, M.; Kos, S.; Tlak Gajger, I. Koncentracije arsena, kadmija i žive u pčelinjem vosku (*Apis mellifera*) tijekom njegove prerade iz saća u satne osnove. *Vet. Stanica* **2019**, *50*, 19–25.
18. Okayama, A.; Sakogava, T.; Nakajima, C.; Hayama, T. Sporicidal activities of disinfectants of *Paenibacillus larvae*. *J. Vet. Med. Sci.* **1997**, *59*, 953–954. [CrossRef] [PubMed]
19. APHA. *Hive Cleaning and Sterilization*; Animal and Plant Health Agency, National Bee Unit, National Agri–Food Innovation Campus: Sand Hutton, UK, 2014; pp. 1–15.
20. Tsourkas, P.K. *Paenibacillus larvae* bacteriophages: Obscure past, promising future. *Microb. Genom.* **2020**, *6*, e000329. [CrossRef] [PubMed]
21. González, M.J.; Marioli, J.M. Antibacterial activity of water extracts and essential oils of various aromatic plants against *Paenibacillus larvae*, the causative agent of American Foulbrood. *J. Invertebr. Pathol.* **2010**, *104*, 209–213. [CrossRef]
22. Kiriamburi, J.; Muturi, J.; Mugweru, J.; Forsgren, E.; Nilsson, A. Short communication: Efficacy of two commercial disinfectants on *Paenibacillus larvae* spores. *Front. Vet. Sci.* **2022**, *12*, 884425. [CrossRef]
23. Flesar, J.; Havlik, J.; Kloucek, P.; Rada, V.; Titera, D.; Bednar, M.; Stropnicky, M.; Kokoska, L. *In vitro* growth-inhibitory effect of plant-derived extracts and compounds against *Paenibacillus larvae* and their acute oral toxicity to adult honey bees. *Vet. Microbiol.* **2010**, *145*, 129–133. [CrossRef]
24. Santos, R.C.V.; Lopes, L.Q.S.; Alves, C.F.S.; Fausto, V.P.; Pizzutti, K.; Barboza, V.; Souza, M.E.; Raffin, R.P.; Gomes, P.; Takamatsu, D.; et al. Antimicrobial activity of tea tree oil nanopartcles against American and European foulbrood disease agents. *J. Asia-Pac. Entomol.* **2014**, *17*, 342–347. [CrossRef]
25. Vaucher, R.A.; Giongo, J.L.; Bolzan, L.P.; Côrrea, M.S.; Fausto, V.P.; Alves, C.F.S.; Lopes, L.Q.S.; Boligon, A.A.; Athayde, M.L.; Moreira, A.P.; et al. Antimicrobial activity of nanostructured Amazonian oils against *Paenibacillus* species and their toxicity on larvae and adult worker bees. *J. Asia-Pac. Entomol.* **2015**, *18*, 205–210. [CrossRef]
26. Young, S.B.; Setlow, P. Mechanisms of killing of *Bacillus subtilis* spores by hypochlorite and chlorine dioxide. *Appl. Environ. Microbiol.* **2003**, *95*, 54–67. [CrossRef] [PubMed]
27. Goda, H.; Yamaoka, H.; Nakayama-Imaohji, H.; Kawata, H.; Horiuchi, I.; Fujita, Y.; Nagao, T.; Tada, A.; Terada, A.; Kuwahara, T. Microbicidal effects of weakly acidified chlorous acid water against feline calicivirus and Clostridium difficile spores under protein-rich conditions. *PLoS ONE* **2017**, *12*, e0176718. [CrossRef] [PubMed]
28. Fukuzaki, S. Mechanisms of actions of sodium hypochlorite in cleaning and disinfection processes. *Biocontrol Sci.* **2006**, *11*, 147–157. [CrossRef] [PubMed]
29. Juven, B.J.; Pierson, M.D. Antibacterial effects of hydrogen peroxide and methods for its detection and quantitation. *J. Food Prot.* **1996**, *59*, 1233–1241. [CrossRef] [PubMed]
30. Sattar, S.A.; Springthorpe, V.S.; Rochon, M.A. A product based on accelerated and stabilized hydrogen peroxide: Evidence for broad-spectrum germicidal activity. *Can. J. Infect. Control* **1998**, *13*, 123–130.

31. Wardle, M.D.; Renninger, G.M. Bactericidal effect of hydrogen peroxide on spacecraft isolates. *Appl. Microbiol.* **1975**, *30*, 710–711. [CrossRef]
32. Finnegan, M.; Linley, E.; Denyer, S.P.; Mcdonnell, G.; Simons, C.; Maillard, J.Y. Mode of action of hydrogen peroxide and other oxidizing agents: Differences between liquid and gas forms. *J. Antimicrob. Chemother.* **2010**, *65*, 2108–2115. [CrossRef]
33. Best, M.; Springthorpe, V.S.; Sattar, S.A. Feasibility of a combined carrier test for disinfectants: Studies with a mixture of five types of microorganisms. *Am. J. Infect. Control* **1994**, *22*, 152–162. [CrossRef]
34. Mcdonnell, G. Biocides: Modes of action and mechanisms of resistance. In *Disinfection and Decontamination: Principles, Applications and Related Issues*; Manivannan, G., Ed.; CRC Press: Boca Raton, FL, USA, 2007; pp. 87–124.
35. Gerba, C.P. Quaternary ammonium biocides: Efficacy in application. *Appl. Environ. Microbiol.* **2015**, *81*, 464–469. [CrossRef] [PubMed]
36. The European Parliament; The Council of the European Union. Regulation (EC) No 396/2005 of the European Parliament and of the Council of 23 February 2005 on maximum residue levels of pesticides in or on food and feed of plant and animal origin and amending Council Directive 91/414/EEC. *Off. J. Eur. Union* **2005**, *70*, 1–16. Available online: http://data.europa.eu/eli/reg/2005/396/oj (accessed on 15 March 2024).
37. The European Parliament; The Council of the European Union. Regulation (EU) No 528/2012 of the European Parliament and of the Council of 22 May 2012 Concerning the Making Available on the Market and Use of Biocidal Products. *Off. J. Eur. Union* **2012**, *167*, 1–123.
38. Sudhaus, N.; Pina-Pérez, M.C.; Martínez, A.; Klein, G. Inactivation kinetics of spores of *Bacillus cereus* strains treated by a peracetic acid-based disinfectant at different concentrations and temperatures. *Foodborne Pathog. Dis.* **2012**, *9*, 442–452. [CrossRef] [PubMed]
39. Kunigk, L.; Almeida, M.C.B. Action of peracetic acid on *Escherichia coli* and *Staphylococcus aureus* in suspension or settled on stainless steel surfaces. *Braz. J. Microbiol.* **2001**, *32*, 38–41. [CrossRef]
40. Zhao, X.; Cheng, K.; Hao, J.; Liu, D. Preparation of peracetic acid from hydrogen peroxide, part II: Kinetics for spontaneous decomposition of peracetic acid in the liquid phase. *J. Mol. Catal. A Chem.* **2008**, *284*, 58–68. [CrossRef]
41. Paliy, A.P. Antibacterial effect of "Ecocide C" disinfectant against mycobacteria. *Ukr. J. Ecol.* **2018**, *8*, 141–147. [CrossRef]
42. Herńndez, A.; Martró, E.; Matas, L.; Martín, M.; Ausina, V. Assessment of *in-vitro* efficacy of 1% Virkon against bacteria, fungi, viruses and spores by means of AFNOR guidelines. *J. Hosp. Infect.* **2000**, *46*, 203–209.
43. Zakaria, Z.; Gairola, S.; Mohd Shariff, N. Effective microorganisms (EM) technology for water quality restoration and potential for sustainable water resources and management. In Proceedings of the 5th International Congress on Environmental Modelling and Software, Ottawa, ON, Canada, 5–8 July 2010; pp. 1–8.
44. Namsivayam, S.K.R.; Narendrakumar, G.; Kumar, J.A. Evaluation of effective microorganism (EM) for treatment of domestic sewage. *J. Exp. Sci.* **2011**, *2*, 30–32.
45. Yun-Hee, M.; Kwang-Bae, L.; Young-Jun, K.; Yoon-Mo, K. Current status of EM (Effective Microorganisms) utilization. *KSBBJ* **2011**, *26*, 365–373.
46. Tlak Gajger, I.; Vlainić, J.; Šoštarić, P.; Prešern, J.; Bubnič, J.; Smodiš Škerl, M.I. Effects on some therapeutical, biochemical, and immunological parameters of honey bee (*Apis mellifera*) exposed to probiotic treatments, in field and laboratory conditions. *Insects* **2020**, *11*, 638. [CrossRef] [PubMed]
47. Lee, J.H.; Joeng, J.O.; Park, S.H. A Study on the composting process of food waste by seeding the isolated effective microorganism. *J. Environ. Health Sci.* **2000**, *26*, 1–10.
48. Chui, C.H.; Hau, D.K.P.; Lau, F.Y.; Cheng, G.Y.M.; Wong, R.S.M.; Gambari, R.; Kok, S.H.L.; Lai, K.B.; Teo, I.T.N.; Leung, T.W.T.; et al. Apoptotic potential of the concentrated effective microorganism fermentation extract on human cancer cells. *Int. J. Mol. Med.* **2006**, *17*, 279–284. [CrossRef] [PubMed]
49. Kizerwetter-Świda, M.; Binek, M. Assessment of potentially probiotic properties of *Lactobacillus* strains isolated from chickens. *Pol. J. Vet. Sci.* **2016**, *19*, 15–20. [CrossRef] [PubMed]
50. Yoon, C.; Na, C.S.; Park, J.H.; Han, S.K.; Nam, Y.M.; Kwon, J.T. Effect of feeding multiple probiotics on performance and fecal noxious gas emission in broiler chicks. *Korean J. Poult. Sci.* **2004**, *3*, 229–235.
51. Tlak Gajger, I.; Nejedli, S.; Cvetnić, L. Influence of Probiotic Feed Supplement on *Nosema* spp. Infection Level and the Gut Microbiota of Adult Honeybees (*Apis mellifera* L.). *Microorganisms* **2023**, *11*, 610. [CrossRef]

insects

MDPI

Article

Effects of *Agaricus bisporus* Mushroom Extract on Honey Bees Infected with *Nosema ceranae*

Uros Glavinic [1,*], Milan Rajkovic [1], Jovana Vunduk [2,3], Branislav Vejnovic [1], Jevrosima Stevanovic [1], Ivanka Milenkovic [3] and Zoran Stanimirovic [1]

1 Faculty of Veterinary Medicine, University of Belgrade, Bulevar Oslobodjenja 18, 11000 Belgrade, Serbia; mrajkovic@vet.bg.ac.rs (M.R.); branislavv@vet.bg.ac.rs (B.V.); rocky@vet.bg.ac.rs (J.S.); zoran@vet.bg.ac.rs (Z.S.)

2 Faculty of Agriculture, Institute for Food Technology and Biochemistry, University of Belgrade, Nemanjina 6, 11080 Belgrade, Serbia; vunduk@agrif.bg.ac.rs

3 Ekofungi Ltd., Padinska Skela bb, 11000 Belgrade, Serbia; ekofungi@gmail.com

* Correspondence: uglavinic@vet.bg.ac.rs

Simple Summary: *Nosema ceranae* affects honey bee (*Apis mellifera* L.) causing nosemosis disease that often induces serious problems in apiculture. Antibiotic fumagillin is the only licenced treatment against nosemosis, but its effectiveness is questioned and its usage associated with risk of bee mortality and appearance of residues in bee products. In search for alternative treatment for the control of nosemosis, water crude extract of *Agaricus bisporus* was tested on bees in laboratory (cage) experiments. Bee survival and food consumption were monitored together with *Nosema* infection level and expression of five genes (abaecin, hymenoptaecin, defensin, apidaecin, and vitellogenin) were evaluated in bees sampled on days 7 and 15. Apart from the gene for defensin, the expression of all tested genes was up-regulated in bees supplemented with *A. bisporus* extract. Both anti-*Nosema* and immune protective effects of *A. bisporus* extract were observed when supplementation started at the moment of *N. ceranae* infection or preventively (before or simultaneously with the *Nosema* infection).

Citation: Glavinic, U.; Rajkovic, M.; Vunduk, J.; Vejnovic, B.; Stevanovic, J.; Milenkovic, I.; Stanimirovic, Z. Effects of *Agaricus bisporus* Mushroom Extract on Honey Bees Infected with *Nosema ceranae*. *Insects* **2021**, *12*, 915. https://doi.org/10.3390/insects12100915

Academic Editors: Ivana Tlak Gajger and Franco Mutinelli

Received: 30 August 2021
Accepted: 20 September 2021
Published: 7 October 2021

Abstract: *Agaricus bisporus* water crude extract was tested on honey bees for the first time. The first part of the cage experiment was set for selecting one concentration of the *A. bisporus* extract. Concentration of 200 μg/g was further tested in the second part of the experiment where bee survival and food consumption were monitored together with *Nosema* infection level and expression of five genes (abaecin, hymenoptaecin, defensin, apidaecin, and vitellogenin) that were evaluated in bees sampled on days 7 and 15. Survival rate of *Nosema*-infected bees was significantly greater in groups fed with *A. bisporus*-enriched syrup compared to those fed with a pure sucrose syrup. Besides, the anti-*Nosema* effect of *A. bisporus* extract was greatest when applied from the third day which coincides with the time of infection with *N. ceranae*. Daily food consumption did not differ between the groups indicating good acceptability and palatability of the extract. *A. bisporus* extract showed a stimulative effect on four out of five monitored genes. Both anti-*Nosema* and nutrigenomic effects of *A. bisporus* extract were observed when supplementation started at the moment of *N. ceranae* infection or preventively (before or simultaneously with the infection).

Keywords: honey bee; *Nosema ceranae*; mushroom extract; *Agaricus bisporus*; immune-related gene expression

1. Introduction

The honey bee (*Apis mellifera* L. (Hymenoptera: Apidae)) is a pollinator important for agricultural production [1] and providing essential macro- and micronutrients [2]. Considerable colony losses in the United States and Europe have significant negative economic and environmental consequences, and can be caused by many factors, most often pathogens and parasites, environmental pollution, nutritional stress, and inadequate beekeeping management [3–5].

One of the most common diseases of honey bees is nosemosis caused by microsporidia of the genus *Nosema* [3] with three species that may affect *A. mellifera: N. ceranae, N. apis,* and *N. neummani.* Currently, *N. ceranae* is one of the most widespread microsporidium causing infections in honey bees [6] with the different symptoms compared with those induced with *N. apis* [7]. *N. ceranae* is primarily a pathogen of honey bee midgut. However, this endoparasite or its DNA was also found in other tissues [8–10] and hemolymph [11]. Pathological changes in the epithelial cells of the midgut lead to digestive and metabolic disorders causing malnutrition in adult bees [12], diarrhea, decreased honey production, and increased mortality in winter bees [13]. There are reports that *N. ceranae* induces changes in the carbohydrate metabolism [14–17], which may increase nutritional and energetic stress [3,15–20]. Infection with *N. ceranae* compromises immunity [21] and induces oxidative stress in bees [5], especially in combination with pesticides [22,23]. It has also been shown that infection of bees with *Nosema* spp. can reduce the effectiveness of acaricides used for the control of honey bee mite *Varroa destructor* [24]. In addition to bee diseases, many problems in beekeeping are caused by chemical substances used in the control of bee pathogens due to the many side effects on adult worker bees (increased mortality, shortened lifespan, down-regulation of immune-related genes, induction of toxic stress and up-regulation of detoxification-related genes, decreased worker mobility, learning, memory and trophallaxis, increased *Nosema ceranae* infection rate), bee brood (increased egg, larval and pupal mortality, reduced brood survival and brood area, delayed larval development, delayed adult emergence), reproduction (reduced drone production, longevity and weight, decreased number and viability of drone sperm, increased queen mortality, failure in queen's development, reduced queen mating success, queen pupal weight, queen adult weight, queen cell acceptance), consequently leading to weakening of bee colonies and increase in susceptibility to diseases even if used at recommended doses (reviewed in [25]).

The control of nosemosis includes a wide range of beekeeping practices [3] and application of chemicals, such as fumagillin, an antibiotic derived from the fungus *Aspergillus fumigatus* [26,27]. Although fumagillin use is prohibited in most of European countries due to risk of its commercial formulations (Fumagilin-B, Fumidil-B,) to leave residues in bee products [28–30], to affect food safety [31,32], and to increase bee mortality [30], however, it is still licensed in the United States [33], Canada [34], and Argentina [35]. Moreover, its effectiveness has been questioned [28], imposing a constant effort of scientists to discover alternative substances, which would be effective against *Nosema* spp. Among many natural compounds tested [5,36–47], some of them showed promising anti-*Nosema* effects such as algae and fungus extracts [5,22,36–39], chitosan and peptidoglycan [40], naringenin, sulforaphane and carvacrol [41], probiotics [42,46], natural substances from Brassicaceae defatted seed meals [43], and natural-based commercial formulations [5,44,47,48]. Diet with the addition of plant-based, amino acid and vitamin supplements, led to reduction of *N. ceranae* spore number [5,48], immunostimulation, and reduction of oxidative stress in *N. ceranae*-infected bees [5], and consequently to the improved condition of bees. Interest in mushrooms has been increasing among researchers due to their nutritional and medicinal properties; mushrooms of the genus *Agaricus* contain numerous biologically-active substances (such as glucans, mannan, lentinan, schizophyllan, scleroglucan) with certain antitumor, antidiabetic, and immunostimulatory effects [49–56]. Stevanovic et al. [38] observed a favorable effect of *A. blazei* extract on the strength parameters of bee colonies in a field experiment, while Glavinic et al. [5] revealed the positive effect of *A. blazei* extract in laboratory experiment. The extract stimulated the expression of genes important for immunity, reducing oxidative stress caused by *N. ceranae* and consequently reducing *N. ceranae* infection. Beneficial effects on honey bees were observed for other mushroom species, too; extracts derived from the mycelia of *Fomes fomentarius, Ganoderma applanatum, G. Resinaceum,* and *Trametes versicolor* demonstrated strong anti-viral efficacy against two bee viruses (deformed wing virus—DWV) and Lake Sinai virus—LSV) in both laboratory and field trials [57].

Having in mind the described positive effects of *A. blazei* on honey bees and its potential in *Nosema* control, we decided to assess the effect of *A. bisporus* extract on bee survival, *Nosema* infection, and the expression of genes important for bee immunity in a laboratory experiment. *A. bisporus* is the most cultivated mushroom species among Western countries which makes this research and its application in beekeeping more cost-effective compared to the research and application of *A. blazei*.

2. Materials and Methods

2.1. *Agaricus bisporus* Extract Preparation

Water extract of the commercially-cultivated white button mushroom (*Agaricus bisporus*, strain A15) was prepared according to the procedure described by Klaus et al. [58]. In summary, dry powdered fruit bodies of mushrooms were extracted with distilled water at 121 °C, 1.2 bar, for 60 min. After being filtered, the liquid part was evaporated up to 1/3 of its volume and precipitated with 96% ethanol overnight. The material was then centrifuged, dried at 40 °C, powdered, and kept in the refrigerator (RK6333W, Gorenje, Velenje, Slovenia) before analysis.

2.2. Bees and the Design of Cage Experiment

During April 2020, frames containing areas with sealed brood were taken from three honey bee colonies at the experimental apiary of the Faculty of Veterinary Medicine, University of Belgrade. The frames were placed in net bags (to prevent the dissipation of hatched bees) and kept overnight in a preset incubator (Inkubatori, Pozarevac, Serbia) at the temperature 34 ± 1 °C and humidity 66 ± 1%. The next morning, the newly emerged worker bees were collected from the frames and transferred to experimental cages (60 bees per experimental cage: 10 bees were sampled for the RNA extraction, 20 for *Nosema* spore counting and the remaining 30 bees served for monitoring the survival) and each group comprised three cages (replicates). Cages were specifically designed by Glavinic et al. [21] for this experiment. Briefly, a plastic jar was punched to allow entry of air, placed with the lid down, and equipped with a plastic strainer inserted in the center of the lid allowing bees to take the syrup from the small petri dish placed below. The whole experiment was repeated, and the results were merged into a single dataset.

2.3. The Selection of an Appropriate Extract Concentration

To determine the optimal concentration of the extract, only the survival of bees was monitored in the first part of the experiment. Newly emerged bees were divided into four experimental groups. The control group (C group) received pure sucrose syrup (50% w/v). The remaining three groups (Table 1) received sucrose syrup (50% w/v) supplemented with *A. bisporus* extract at concentrations of 100 μg/g (Abs-100 group), 200 μg/g (Abs-200 group) and 400 μg/g (Abs-400 group). The experiment lasted 15 days. The dead bees were counted and removed from the cages on a daily basis.

Table 1. Experimental design.

GROUP [1]	Day of Starting the Treatment [2]	*N. ceranae* Infection Day [2]	Sampling Day [2]	
NI	–	–	7	15
I	–	3	7	15
Abs	1	–	7	15
I-Abs1	1	3	7	15
I-Abs3	3	3	7	15
I-Abs6	6	3	7	15

[1] Bees were non-infected (NI) or infected with *N. ceranae* (I) and treated with *A. bisporus* extract (Abs). [2] Days after bee emergence.

2.4. Effects of *A. bisporus* Extract on Nosema Infection

In the second part of the cage experiment, the effect of *A. bisporus* extract at a concentration of 200 μg/g was evaluated. Six groups were established and all were fed with 50% (*w*/*v*) sucrose syrup. There were two control groups: non-infected (NI) and infected with *N. ceranae* (I). The remaining four experimental groups received sucrose syrup enriched with *A. bisporus* extract either from the first day (in group Abs, those are non-infected and in *Nosema*-infected, group I-Abs1), or from the third or sixth day (groups I-Abs3 and I-Abs6, respectively, both *Nosema*-infected (Table 1)).

2.5. Inoculum Preparation, Experimental Infection, and Bee Sampling

Inoculum preparation and experimental infection of bees were performed according to the previously described methodology [5,21]. Briefly, the inoculum was prepared by crushing the abdomens of *N. ceranae*-infected bees in distilled water. Number of spores were determined according to Cantwell [59]. Freshly-prepared spore suspension with 99% viability, assessed with 4% trypan blue (Sigma–Aldrich, Steinheim, Germany), mixed with 50% sucrose (Centrohem, Stara Pazova, Serbia) solution was used to obtain 1×10^6 spores/mL final concentration. Bees in the infected control group (I) and treated groups (I-Abs1, I-Abs3, and I-Abs6) were infected with *N. ceranae* spores on day 3 [5]. From each cage, on days 7 and 15, bees were sampled for counting *Nosema* spores and gene expression analysis (Table 1). The remaining bees from the cage were used to determine the survival rate. Dead bees were removed from the cage and counted daily. During the experiment, syrup consumption was measured by weighing the food before and after the bees had been fed for 24 h [60]. According to these data, average consumption/bee/day was calculated.

2.6. Nosema Spore Counting

On the day of sampling (day 7 and day 15), 10 bees were taken from each cage. The abdomen of each bee was placed in a 1.5 mL tube and macerated in 1 mL of distilled water in Tissue Lyser II (QIAGEN, Hilden, Germany) for 1 min at 25 Hz. The number of *N. ceranae* spores was determined using a hemocytometer described in Cantwell [59] and OIE [61], and used also in our previous studies [5,21–23].

2.7. RNA Extraction and cDNA Synthesis

The total RNA was extracted with Quick-RNA MiniPrep Kit (Zymo Research, Irvine, CA, USA). A single bee was placed in a 1.5 mL tube with 500 μL of Genomic Lysis Buffer. Homogenization was performed in a TissueLyser II (QIAGEN, Germany) for 1 min at 25 Hz using a 3 mm tungsten carbide bead (Qiagen, Hilden, Germany). Other extraction steps were performed according to the manufacturer's instructions. In-column DNase treatment (treatment with DNase I Reaction Mix) was applied for all samples during the extraction process with the aim to remove contaminating DNA. cDNA was immediately generated from extracted RNA using the RevertAid™ First Strand cDNA Synthesis Kit (Thermo Fisher Scientific, Vilnius, Lithuania).

2.8. Real-Time Quantitative PCR

SYBR green method was used for real-time PCR (qPCR) amplification in a 20 μL reaction mixture with "FastGene® IC Green 2 x qPCR Universal Mix" (Nippon Genetics, Düren, Germany) according to instructions of the manufacturer and modifications described by Glavinic et al. [5]. A specific primer pair was used for each gene (Table 2). The qPCR reactions were performed in a Rotor-Gene Q 5 plex (Qiagen, Valencia, CA, USA). Gene expression level of abaecin, hymenoptaecin, defensin, apidaecin, and vitellogenin was determined using the $2^{-\Delta\Delta CT}$ method described in Galvinic et al. [5] while β-actin was used as an internal control for normalization of each gene expression.

Table 2. Primer used for qPCR analysis.

Primer	Sequence 5′–3′	Reference
Beta actin-F	TTGTATGCCAACACTGTCCTTT	[62]
Beta actin-R	TGGCGCGATGATCTTAATTT	
Abaecin-F	CAGCATTCGCATACGTACCA	[63]
Abaecin-R	GACCAGGAAACGTTGGAAAC	
Hymenopt-F	CTCTTCTGTGCCGTTGCATA	[63]
Hymenopt-R	GCGTCTCCTGTCATTCCATT	
Defensin-F	TGCGCTGCTAACTGTCTCAG	[63]
Defensin-R	AATGGCACTTAACCGAAACG	
ApidNT-F	TTTTGCCTTAGCAATTCTTGTTG	[62]
ApidNT-R	GTAGGTCGAGTAGGCGGATCT	
VgMC-F	AGTTCCGACCGACGACGA	[62]
VgMC-R	TTCCCTCCCACGGAGTCC	

3. Statistical Methods

The survival of bees was monitored by the number of dead bees per day in each experimental group. The data on the survival distribution obtained in the Kaplan–Meier survival estimator were compared in the log-rank test. The results for *N. ceranae* spores and gene expression were tested for normality by using Shapiro–Wilk's test. Due to the normal distribution of data (Shapiro–Wilk's test, $p > 0.05$), groups were compared in two-way ANOVA with repeated measures in one factor, followed by Tukey's test within and Sidak's test between groups over time. The levels of significance below 0.05 ($p < 0.05$) were considered significant. The analyses were done using GraphPad Prism 7.0 (GraphPad Software, San Diego, CA, USA).

4. Results

4.1. Bee Survival and Food Consumption

In the first part of the experiment, which was conducted to determine the optimal concentration of the extract for further investigation, no significant differences in bee survival ($p > 0.05$) were detected between all groups (Figure 1). The lowest number of dead bees was in the group treated with 200 µg/g of *A. bisporus* extract (Abs-200). Consequently, this concentration was used in the second part of the experiment.

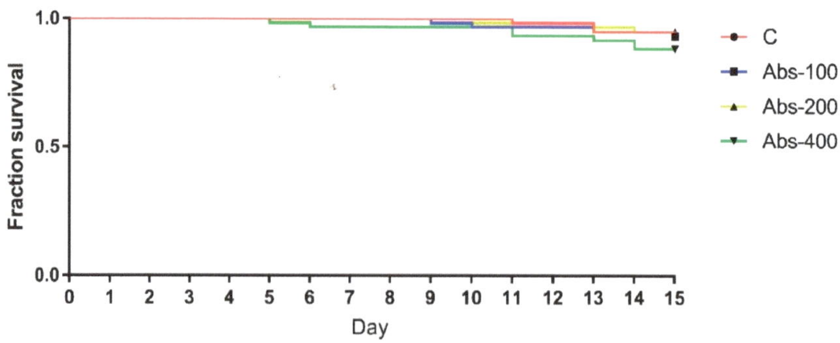

Figure 1. Survival of bees treated with *A. bisporus* extract: 100 µg/g (Abs-100), 200 µg/g (Abs-200), 400 µg/g (Abs-400), and control (C) group.

The number of dead bees in the second part of the experiment (Figure 2) was significantly higher in the infected control group (I) compared to NI ($p = 0.009$), Abs ($p = 0.003$), I-Abs1 ($p = 0.036$), I-Abs3 ($p = 0.018$), and not significantly different ($p = 0.185$) in comparison with group I-Abs6 (log-rank test). The survival of bees in the non-infected control group (NI) did not differ statistically compared to the groups of Abs ($p = 0.714$), I-Abs1 ($p = 0.552$), I-Abs3 ($p = 0.760$), and I-Abs6 ($p = 0.173$).

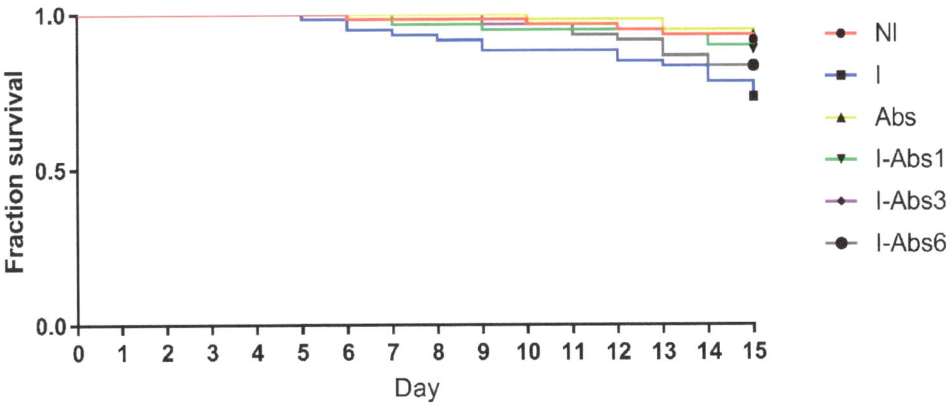

Figure 2. Survival of bees infected with *N. ceranae* (I), non-infected but treated with *A. bisporus* extract (Abs), bees infected with *N. ceranae* and treated with *A. bisporus* extract from day 1 (group I-Abs1), day 3 (I-Abs3), and day 6 (I-Abs6), as well as bees from the control non-infected (NI) group.

There were no differences in daily food consumption between the control group and all treatment groups. Mean consumption across all groups was 121.63 mg/bee/day (Figure S1).

4.2. Quantification of *N. ceranae* Spores

In samples of bees from the control non-infected group (NI) and non-infected but treated with *A. bisporus* extract (Abs), no spores of *N. ceranae* were detected at any time of sampling. By comparing the number of spores in groups infected with *N. ceranae* on the 7th and 15th day of sampling, Shidak's multiple comparison test showed a significantly higher ($p < 0.001$) number of spores on day 15 compared to day 7. In bee samples from day 7 of the experiment, the Tukey test revealed no significant differences in the number of spores among all groups ($p > 0.05$). In contrast, on day 15 (Figure 3), the number of spores was significantly higher in group I compared to all other groups ($p < 0.001$). The lowest number of spores was detected in group I-Abs3, significantly lower than in groups I and I-Abs6 ($p < 0.001$) and I-Abs1 ($p < 0.05$).

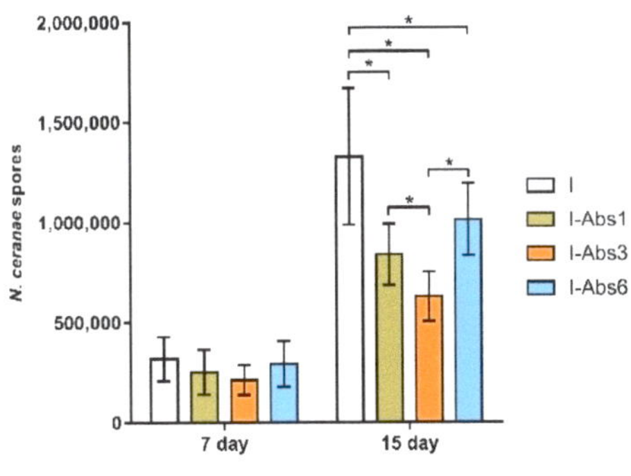

Figure 3. Number of *N. ceranae* spores in infected group (I) and groups of bees infected with *N. ceranae* and treated with *A. bisporus* extract from day 1 (I-Abs1), day 3 (I-Abs3), and day 6 (I-Abs6). * $p < 0.05$.

4.3. Gene Expression Analyses

On day 7, the highest expression levels for the majority of monitored genes were in groups Abs and I-Abs6, in contrast to the infected control (I) group where the lowest gene expression was detected (Figure 4). However, the most important and most significant changes in expression levels were obtained on day 15 (Figures 4 and 5). According to the Tukey test, abaecin expression levels (Figure 5A) were significantly lower in group I compared to all other groups ($p < 0.001$), but also in group I-Abs6 compared to I-Abs1 ($p < 0.05$). Gene expression for hymenoptaecin (Figure 5B), on day 15 of the experiment, was significantly lower in group I ($p < 0.001$) compared to all other groups. However, pairwise comparisons of all treated groups revealed that the expression of the hymenoptaecin gene was significantly higher in the I-Abs1 group compared to I-Abs3 ($p < 0.001$) and I-Abs6 ($p < 0.01$). The expression of apidaecin (Figure 5C) and vitellogenin (Figure 5D) was also found to be lower in the control infected group (I) compared to all other groups ($p < 0.01$). The expression of the gene for apidaecin did not differ among the groups treated with *A. bisporus* extract, while the level for vitellogenin was higher ($p < 0.01$) in the group I-Abs3 compared to the groups I-Abs1 and Abs. Regarding gene expression for defensin, although the lowest level was in the group I (Figure 5E), the Tukey test showed no significant differences between the experimental groups.

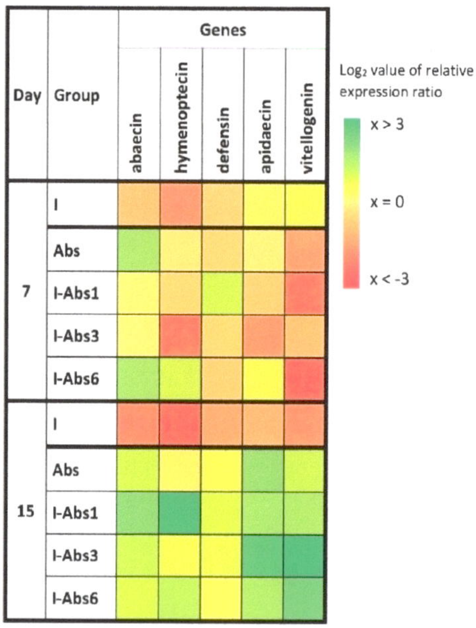

Figure 4. Heatmap of immune−related genes (means of Log2 of relative expression ratios for abaecin, hymenoptaecin, defensin, apidaecin and vitellogenin) at different time points in experimental groups. *N. ceranae* infected control (I), group treated with extract (Abs), and groups infected and treated with *A. bisporus* extract from day 1 (I-Abs1), day 3 (I-Abs3), and day 6 (I-Abs6). Range log2 value of relative expression ratio is given in the legend on the right indicating up- or down-regulation of the expression. Group names are indicated in Table 1.

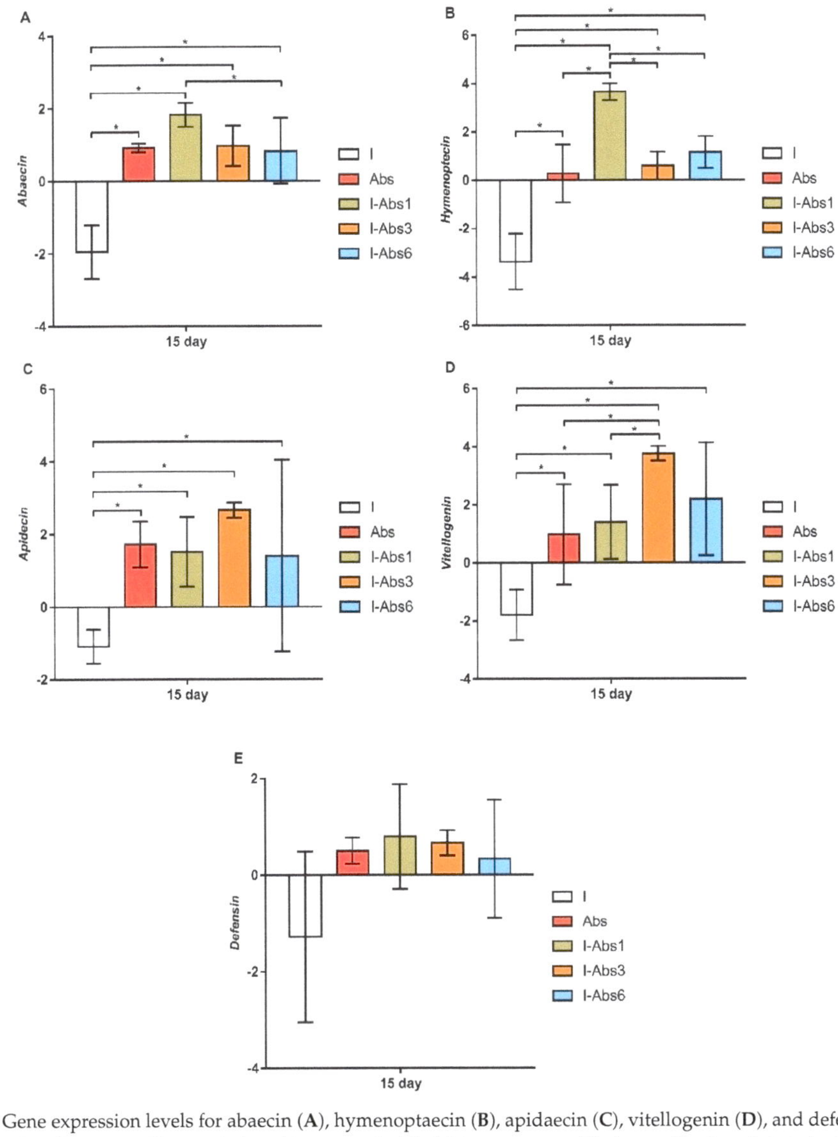

Figure 5. Gene expression levels for abaecin (**A**), hymenoptaecin (**B**), apidaecin (**C**), vitellogenin (**D**), and defensin (**E**) in bees infected with *N. cerane* (I), non−infected but treated with *A. bisporus* extract (Abs), as well as bees infected with *N. cerane* and treated with *A. bisporus* extract from day 1 (I-Abs1), day 3 (I-Abs3) and day 6 (I-Abs6) of the experiment. * $p < 0.05$.

5. Discussion

Nosemosis is a common and widespread disease that negatively affects adult bees [37]. Infection with *N. ceranae* is especially dangerous due to its implied role in colony losses [3]. There is a relatively small number of safe preparations used as alternatives to fumagillin in nosemosis control. Thus, there is a constant effort of scientists to discover new natural substances with beneficial effects in the control of this disease [5,36–47].

This is the first study in which *A. bisporus* extract was tested on honey bees. In the first part of the experiment, none of the tested concentrations of *A. bisporus* extract (100, 200, and 400 μg/g) increased bee mortality compared to control. This has been expected, bearing in mind already reported beneficial effects of other *Agaricus* spp. extracts on bees [5,38] as well as of other natural-based dietary supplements [21,48,64–67]. Despite the absence of significant differences among groups, the lowest number of dead bees was in the group fed with 200 μg/g of *A. bisporus* extract. This result, along with the economic aspect (optimal ratio between the extract amount and the benefit), led us to choose this concentration for further testing. In this experiment, a significantly lower survival rate in the group of infected bees (I) compared to the non-infected group (Figure 2) suggests undoubted fatal effects of *N. ceranae* infection, which is consistent with previous studies [5,21,68,69]. However, feeding bees with the addition of *A. bisporus* extract starting from day 1 and 3 at a concentration of 200 μg/g significantly increased the survival rate of infected bees compared to infected bees (group I) fed without the addition of mushroom extract (Figure 2). This result is in accordance with our previous findings where an extract of related fungus *A. blazei* induced better bee survival [5]. In a field experiment, Stevanović et al. [38] revealed the beneficial impact of *A. blazei* extract on the strength of honey bee colonies. However, when the extract was applied to infected bees starting from day 6, it did not contribute significantly to mortality decrease compared to group I (Figure 2). This finding indicates a better protective-preventive effect (application before or at the moment of infection) compared to the effect obtained by post-infection application. Having in mind that the main action of medicinal mushroom metabolites is immunomodulation [70], this result was expected. In addition, the extract used in this study was rich in polysaccharides (more than 68%) which are the most reported immunomodulatory compounds from mushrooms [71,72]. The absence of significant differences in bee mortality between the control non-infected group (NI) and all groups that received *A. bisporus* extract in the diet (Abs, I-Abs1, I-Abs3, and I-Abs6) is an additional confirmation of the extract's positive effects on bee survival regardless of the application period, similarly as previously reported for *A. blazei* extract [5], plant extracts (*Aristotelia chilensis, Ugni molinae, Gevuina avellan*), and propolis [73].

The level of nosema infection was monitored by counting spores on days 7 and 15 of the experiment. The presence of spores in infected groups, and its absence in non-infected groups (NI and Abs) suggest that there was no cross-contamination between the experimental groups, and the design of the cage [21] is adequate for such research. This confirms that the experiment has been performed accurately and in line with COLOSS BEEBOOK recommendations [74]. A significantly higher number of spores was detected on day 15 compared to day 7 in all infected groups. Analysis of the number of spores in the samples collected on day 7 did not reveal significant difference among infected groups. This finding was expected since the infection was still in the initial stage [75]. In contrast, on day 15, a significantly higher number of spores was in group I (Figure 3) compared to all other groups that received the extract. Comparing groups infected with the *Nosema* and treated with *A. bisporus* extract (I-Abs1, I-Abs3, and I-Abs6), the lowest spore number was detected in the group I-Abs3 (Figure 3). Thus, we can conclude that the extract of white button mushroom has a noticeable anti-*Nosema* effect. Moreover, this effect is the highest when applied to start from the third day which coincides with the time of infection with *N. ceranae*. Such effects have already been demonstrated for *A. blazei* [5], algae extract also rich in polysaccharides [37], and other supplements [21,73]. Hayman et al. [76] explained microsporidia reduction by inhibition of adhesion to the target cells. Vunduk et al. [77] also showed that mushroom extract exhibits antiadhesion and biofilm formation effect against foodborne bacteria. The extract's more beneficial effect in groups IAbs-1 and I-Abs3 could be explained by the absence of infected epithelial cells with *N. ceranae* [75], which gives a better opportunity for more efficient action (adhesion inhibition) of the extract.

Infection with *N. ceranae* suppressed most of the immune-related genes examined in this study (Figures 4 and 5), which is in line with previous reports of the immunosuppressive effect of this endoparasite [5,21,78–80]. The expression level of all genes, except defensin, was significantly lower in group I compared to all other experimental groups (Figures 4 and 5). This result is not surprising considering the amount of *N. ceranae* spores detected at the end of the experiment (highest spore load in group I—Figure 3) but also the bee mortality (highest number of dead bees in group I—Figure 2). The gene expression levels in all the groups supplemented with *A. bisporus* extract (Abs, I-Abs1, I-Abs3, and I-Abs6) indicate its stimulatory effect on the expression of all monitored immune genes except defensin (Figures 4 and 5). This finding is consistent with the previous study demonstrating the stimulating effect of *A. blazei* on honey bee immunity [5,39]. Higher levels of immune-related gene expression in infected and treated groups compared to the infected control group (I) indicate the immunoprotective and immunomodulating effect of *A. bisporus* extract. Moreover, higher gene expression levels (Figures 4 and 5) in infected and treated groups, and lower spore loads in those groups (Figure 3) revealed an inverse correlation between spore loads and levels of immune-related gene expression. Gene expression levels on day 15 compared to day 7 were significantly higher for all tested genes in group I-Abs3 (except the one for defensin); significantly higher for all tested genes in group I-Abs1 (except those for defensin and apidaecin); significantly higher only for vitellogenin gene in I-Abs6 group (Figure 4), and significantly lower for all tested genes in the infected control group (I). Greater stimulation of gene expression in group I-Abs1 and I-Abs3 compared to I-Abs6 indicates better efficacy of the extract before and at the moment of infection with *N. ceranae*. A better protective effect is expected as the extract is rich in polysaccharides [5,36–38], while the anti-*Nosema* effect is more typical for antibiotics such as fumagillin [5,29,30,33–35]. Moreover, we used crude water extract which, besides polysaccharides (almost 70%), contains proteins (5.31%) and phenolic compounds (2.7%) as well [71,81]. These metabolites express different modes of action. Proteins are often bound to polysaccharides and cumulative immunomodulatory effect can be expected. Phenolics, as fungal secondary metabolites, mainly act as antioxidants, and can be important in the early stage of infection development [82].

6. Conclusions

This is the first study of the effects of *A. bisporus* extract on honey bees that showed the potential for improving the survival and the immunity of bees infected with *N. ceranae*. The effect on immunity has been demonstrated through increased expression of immune-related genes in both non-infected bees and bees infected with *N. ceranae*. Besides, *A. bisporus* extract was effective in reducing the number of *Nosema* spores. The best results were observed when the extract was applied at the moment of infection or preventively (before or simultaneously with the *Nosema* infection). Finally, the tested extract has a strong stimulatory effect on gene expression in both *N. ceranae*-infected and uninfected bees. In the absence of adequate and safe natural-based therapy for nosemosis, we can conclude that *A. bisporus* extract has a great potential for use in the control of this bee disease and needs to be further investigated in the field experiments.

Supplementary Materials: The following are available online at https://www.mdpi.com/article/10.3390/insects12100915/s1, Figure S1: Daily food consumption per bee in the non-infected group (NI), infected group (I) and the groups infected with *N. ceranae* and treated with *A. bisporus* extract from the day 1 (I-Abs1), day 3 (I-Abs3) and day 6 (I-Abs6).

Author Contributions: Conceptualization: Z.S., U.G., and I.M.; Design of experiment and methodology: U.G., and Z.S.; Laboratory analysis: U.G., J.V., M.R., and B.V.; Data curation: U.G., M.R., and B.V.; Writing, review and editing: U.G., M.R., Z.S., J.S., J.V., B.V., and I.M. All authors have read and agreed to the published version of the manuscript.

Funding: The study was supported by the Ministry of Education, Science and Technological Development of the Republic of Serbia (contract no. 451-03-9/2021-14/200143 for the project led by Zoran Stanimirovic).

Institutional Review Board Statement: Ethical review and approval were waived for this study, due to the study of invertebrates only (honey bees are exempted from ethical review).

Data Availability Statement: The data presented in this study are available on request from the corresponding author. The data are not publicly available due to the excessive data size.

Conflicts of Interest: The authors declare no conflict of interest.

References

1. Burnham, A.J. Scientific Advances in Controlling Nosema ceranae (*Microsporidia*) Infections in Honey Bees (*Apis mellifera*). *Front. Vet. Sci.* **2019**, *6*, 79. [CrossRef]
2. Reeves, A.M.; O'Neal, S.T.; Fell, R.D.; Brewster, C.C.; Anderson, T.D. In-Hive Acaricides Alter Biochemical and Morphological Indicators of Honey Bee Nutrition, Immunity, and Development. *J. Insect Sci.* **2018**, *18*. [CrossRef]
3. Stanimirović, Z.; Glavinić, U.; Ristanić, M.; Aleksić, N.; Jovanović, N.M.; Vejnović, B.; Stevanović, J. Looking for the causes of and solutions to the issue of honey bee colony losses. *Acta Vet.* **2019**, *69*, 1–31. [CrossRef]
4. Barroso-Arévalo, S.; Vicente-Rubiano, M.; Puerta, F.; Molero, F.; Sánchez-Vizcaíno, J.M. Immune related genes as markers for monitoring health status of honey bee colonies. *BMC Vet. Res.* **2019**, *15*, 72. [CrossRef]
5. Glavinic, U.; Stevanovic, J.; Ristanic, M.; Rajkovic, M.; Davitkov, D.; Lakic, N.; Stanimirovic, Z. Potential of Fumagillin and *Agaricus blazei* Mushroom Extract to Reduce *Nosema ceranae* in Honey Bees. *Insects* **2021**, *12*, 282. [CrossRef] [PubMed]
6. Araneda, X.; Cumian, M.; Morales, D. Distribution, epidemiological characteristics and control methods of the pathogen Nosema ceranae Fries in honey bees *Apis mellifera* L. (Hymenoptera, Apidae). *Arch. Med. Vet.* **2015**, *47*, 129–138. [CrossRef]
7. Martín-Hernandez, R.; Bartolomé, C.; Chejanovsky, N.; Le Conte, Y.; Dalmon, A.; Dussaubat, C.; García-Palencia, P.; Meana, A.; Pinto, M.A.; Soroker, V.; et al. Nosema ceranae in *Apis mellifera*: A 12 years postdetection perspective. *Environ. Microbiol.* **2018**, *20*, 1302–1329. [CrossRef] [PubMed]
8. Chen, Y.P.; Evans, J.; Murphy, C.; Gutell, R.; Zuker, M.; Gundensen-Rindal, D.; Pettis, J.S. Morphological, Molecular, and Phylogenetic Characterization ofNosema ceranae, a Microsporidian Parasite Isolated from the European Honey Bee, *Apis mellifera*. *J. Eukaryot. Microbiol.* **2009**, *56*, 142–147. [CrossRef] [PubMed]
9. Gisder, S.; Hedtke, K.; Möckel, N.; Frielitz, M.C.; Linde, A.; Genersch, E. Five-year cohort study of Nosema spp. in Germany: Does climate shape virulence and assertiveness of Nosema ceranae? *Appl. Environ. Microb.* **2010**, *76*, 3032–3038. [CrossRef]
10. Copley, T.; Jabaji, S. Honeybee glands as possible infection reservoirs of Nosema ceranae and Nosema apis in naturally infected forager bees. *J. Appl. Microbiol.* **2011**, *112*, 15–24. [CrossRef]
11. Glavinic, U.; Stevanovic, J.; Gajic, B.; Simeunovic, P.; Đuric, S.; Vejnovic, B.; Stanimirovic, Z. Nosema ceranae DNA in honey bee haemolymph and honey bee mite Varroa destructor. *Acta Vet.-Beograd* **2014**, *64*, 349–357.
12. Gajger, I.; Kozaric, Z.; Berta, D.; Nejedli, S.; Petrinec, Z. Effect of the Herbal preparation Nozevict on the mid-gut structure of honeybees (Apis melífera) infected with *Nosema* sp. Spores. *Vet. Med.* **2011**, *56*, 344–351. [CrossRef]
13. Traver, B.E.; Fell, R.D. Nosema ceranae in drone honey bees (*Apis mellifera*). *J. Invertebr. Pathol.* **2011**, *107*, 234–236. [CrossRef]
14. Dussaubat, C.; Brunet, J.-L.; Higes, M.; Colbourne, J.; Lopez, J.; Choi, J.H.; Hernández, R.M.; Botías, C.; Cousin, M.; McDonnell, C.; et al. Gut Pathology and Responses to the Microsporidium Nosema ceranae in the Honey Bee *Apis mellifera*. *PLoS ONE* **2012**, *7*, e37017. [CrossRef] [PubMed]
15. Mayack, C.; Naug, D. Energetic stress in the honeybee *Apis mellifera* from Nosema ceranae infection. *J. Invertebr. Pathol.* **2009**, *100*, 185–188. [CrossRef]
16. Aliferis, K.A.; Copley, T.; Jabaji, S. Gas chromatography–mass spectrometry metabolite profiling of worker honey bee (*Apis mellifera* L.) hemolymph for the study of Nosema ceranae infection. *J. Insect Physiol.* **2012**, *58*, 1349–1359. [CrossRef]
17. Vidau, C.; Panek, J.; Texier, C.; Biron, D.G.; Belzunces, L.P.; Le Gall, M.; Broussard, C.; Delbac, F.; El Alaoui, H. Differential proteomic analysis of midguts from Nosema ceranae-infected honeybees reveals manipulation of key host functions. *J. Invertebr. Pathol.* **2014**, *121*, 89–96. [CrossRef]
18. Mayack, C.; Naug, D. Parasitic infection leads to decline in hemolymph sugar levels in honeybee foragers. *J. Insect Physiol.* **2010**, *56*, 1572–1575. [CrossRef]
19. Hernández, R.M.; Botías, C.; Barrios, L.; Martínez-Salvador, A.; Meana, A.; Mayack, C.; Higes, M. Comparison of the energetic stress associated with experimental Nosema ceranae and Nosema apis infection of honeybees (*Apis mellifera*). *Parasitol. Res.* **2011**, *109*, 605–612. [CrossRef]
20. Aufauvre, J.; Misme-Aucouturier, B.; Viguès, B.; Texier, C.; Delbac, F.; Blot, N. Transcriptome Analyses of the Honeybee Response to Nosema ceranae and Insecticides. *PLoS ONE* **2014**, *9*, e91686. [CrossRef]
21. Glavinic, U.; Stankovic, B.; Draskovic, V.; Stevanovic, J.; Petrovic, T.; Lakic, N.; Stanimirovic, Z. Dietary amino acid and vitamin complex protects honey bee from immunosuppression caused by Nosema ceranae. *PLoS ONE* **2017**, *12*, e0187726. [CrossRef]
22. Glavinic, U.; Tesovnik, T.; Stevanovic, J.; Zorc, M.; Cizelj, I.; Stanimirovic, Z.; Narat, M. Response of adult honey bees treated in larval stage with prochloraz to infection with Nosema ceranae. *PeerJ* **2019**, *7*, e6325. [CrossRef]
23. Tesovnik, T.; Zorc, M.; Ristanić, M.; Glavinić, U.; Stevanović, J.; Narat, M.; Stanimirović, Z. Exposure of honey bee larvae to thiamethoxam and its interaction with Nosema ceranae infection in adult honey bees. *Environ. Pollut.* **2020**, *256*, 113443. [CrossRef] [PubMed]
24. Botías, C.; Barrios, L.; Nanetti, A.; Meana, A.; Martín-Hernández, R.; Garrido-Bailón, E.; Higes, M. *Nosema* spp. parasitization decreases the effectiveness of acaricide strips (Apivar®) in treating varroosis of honey bee (*Apis mellifera* iberiensis) colonies. *Environ. Microbiol. Rep.* **2011**, *4*, 57–65. [CrossRef]
25. Tihelka, E. Effects of synthetic and organic acaricides on honey bee health: A review. *Slov. Vet.-Res.* **2018**, *55*. [CrossRef]

26. Katznelson, H.; Jamieson, C.A. Control of Nosema Disease of Honeybees with Fumagillin. *Science* **1952**, *115*, 70–71. [CrossRef] [PubMed]
27. Bailey, L. Effect of Fumagillin upon Nosema apis (Zander). *Nature* **1953**, *171*, 212–213. [CrossRef] [PubMed]
28. Huang, W.-F.; Solter, L.F.; Yau, P.; Imai, B.S. Nosema ceranae Escapes Fumagillin Control in Honey Bees. *PLOS Pathog.* **2013**, *9*, e1003185. [CrossRef] [PubMed]
29. Heever, J.P.V.D.; Thompson, T.; Otto, S.; Curtis, J.M.; Ibrahim, A.; Pernal, S.F. Evaluation of Fumagilin-B® and other potential alternative chemotherapies against Nosema ceranae-infected honeybees (*Apis mellifera*) in cage trial assays. *Apidologie* **2015**, *47*, 617–630. [CrossRef]
30. Heever, J.P.V.D.; Thompson, T.; Otto, S.; Curtis, J.M.; Ibrahim, A.; Pernal, S.F. The effect of dicyclohexylamine and fumagillin on Nosema ceranae-infected honey bee (*Apis mellifera*) mortality in cage trial assays. *Apidologie* **2015**, *47*, 663–670. [CrossRef]
31. Stanimirovic, Z.; Stevanovic, J.; Bajic, V.; Radovic, I. Evaluation of genotoxic effects of fumagillin (dicyclohexylamine) by citogenetic tests in vivo. *Mut. Res.-Genet. Toxicol. Environ. Mutagenes.* **2007**, *628*, 1–10. [CrossRef] [PubMed]
32. Stevanovic, J.; Stanimirović, Z.; Radakovic, M.; Stojić, V. In vitro evaluation of the clastogenicity of fumagillin. *Environ. Mol. Mutagenes.* **2008**, *49*, 594–601. [CrossRef] [PubMed]
33. Higes, M.; Nozal, M.J.; Alvaro, A.; Barrios, L.; Meana, A.; Hernández, R.M.; Bernal, J.L. The stability and effectiveness of fumagillin in controlling Nosema ceranae (Microsporidia) infection in honey bees (*Apis mellifera*) under laboratory and field conditions. *Apidologie* **2011**, *42*, 364–377. [CrossRef]
34. McCallum, R.; Olmstead, S.; Shaw, J.; Glasgow, K. Evaluating efficacy of Fumagilin-B® against Nosemosis and tracking seasonal srends of Nosema spp. in Nova Scotia Honey bee colonies. *J. Apic. Sci.* **2020**, *64*, 277–286.
35. Sarlo, E.G.; Medici, S.K.; Porrini, M.P.; Garrido, P.M.; Floris, I.; Eguaras, M.J. Comparison between different fumagillin dosage and evaluation method in the apiary control of Nosemosis type C. *Redia* **2011**, *94*, 39–44.
36. Charistos, L.; Parashos, N.; Hatjina, F. Long term effects of a food supplement HiveAlive™ on honey bee colony strength and Nosema ceranae spore counts. *J. Apic. Res.* **2015**, *54*, 420–426. [CrossRef]
37. Roussel, M.; Villay, A.; Delbac, F.; Michaud, P.; Laroche, C.; Roriz, D.; El Alaoui, H.; Diogon, M. Antimicrosporidian activity of sulphated polysaccharides from algae and their potential to control honeybee nosemosis. *Carbohydr. Polym.* **2015**, *133*, 213–220. [CrossRef]
38. Stevanovic, J.; Stanimirovic, Z.; Simeunovic, P.; Lakic, N.; Radovic, I.; Sokovic, M.; Van Griensven, L.J. The effect of Agaricus brasiliensis extract supplementation on honey bee colonies. *Acad. Bras. Ciências* **2018**, *90*, 219–229. [CrossRef]
39. Glavinic, U. The effects of various antimicrobials and supplements on the expression of immune-related genes, oxidative stress and survival of honey bee *Apis mellifera* infected with microsporidium Nosema ceranae. Ph.D. Thesis, Faculty of Veterinary Medicine, University of Belgrade, Belgrade, Serbia, 2019.
40. Valizadeh., P.; Guzman-Novoa, E.; Goodwin, P.H. Effect of immune inducers on Nosema ceranae multiplication and their impact on Honey Bee (*Apis mellifera* L.) survivorship and behaviors. *Insects* **2020**, *11*, 572. [CrossRef]
41. Borges, D.; Guzman-Novoa, E.; Goodwin, P.H. Control of the microsporidian parasite Nosema ceranae in honey bees (*Apis mellifera*) using nutraceutical and immuno-stimulatory compounds. *PLoS ONE* **2020**, *15*, e0227484. [CrossRef]
42. Borges., D.; Guzman-Novoa., E.; Goodwin, P.H. Effects of prebiotics and probiotics on honey bees (*Apis mellifera*) infected with the microsporidian parasite Nosema ceranae. *Microorganisms* **2021**, *9*, 481. [CrossRef] [PubMed]
43. Nanetti, A.; Ugolini, L.; Cilia, G.; Pagnotta, E.; Malaguti, L.; Cardaio, I.; Matteo, R.; Lazzeri, L. Seed meals from Brassica nigra and Eruca sativa control artificial Nosema ceranae infections in *Apis mellifera*. *Microorganisms* **2021**, *9*, 949. [CrossRef]
44. Cilia, G.; Garrido, C.; Bonetto, M.; Tesoriero, D.; Nanetti, A. Effect of Api-Bioxal® and ApiHerb® treatments against Nosema ceranae infection in *Apis mellifera* investigated by two qPCR methods. *Vet. Sci.* **2020**, *7*, 125. [CrossRef] [PubMed]
45. Cilia, G.; Fratini, F.; Tafi, E.; Turchi, B.; Mancini, S.; Sagona, S.; Nanetti, A.; Cerri, D.; Felicioli, A. Microbial profile of the ven-triculum of honey bee (*Apis mellifera* ligustica Spinola, 1806) fed with veterinary drugs, dietary supplements and non-protein amino acids. *Vet. Sci.* **2020**, *7*, 76. [CrossRef] [PubMed]
46. El Khoury, S.; Rousseau, A.; Lecoeur, A.; Cheaib, B.; Bouslama, S.; Mercier, P.-L.; Demey, V.; Castex, M.; Giovenazzo, P.; Derome, N. Deleterious Interaction Between Honeybees (*Apis mellifera*) and its Microsporidian Intracellular Parasite Nosema ceranae Was Mitigated by Administrating Either Endogenous or Allochthonous Gut Microbiota Strains. *Front. Ecol. Evol.* **2018**, *6*. [CrossRef]
47. Michalczyk, M.; Sokół, R. Estimation of the influence of selected products on co-infection with N. apis/N. ceranae in *Apis mellifera* using real-time PCR. *Invertebr. Reprod. Dev.* **2018**, *62*, 92–97. [CrossRef]
48. Jovanovic, N.M.; Glavinic, U.; Delic, B.; Vejnovic, B.; Aleksic, N.; Mladjan, V.; Stanimirovic, Z. Plant-based supplement containing B-complex vitamins can improve bee health and increase colony performance. *Prev. Vet.-Med.* **2021**, *190*, 105322. [CrossRef]
49. Ishii, P.L.; Prado, C.K.; de Oliveira Mauro, M.; Carreira, C.M.; Mantovani, M.S.; Ribeiro, L.R.; Dichi, J.B.; Oliveira, R.J. Eval-uation of Agaricus blazei in vivo for antigenotoxic, anticarcinogenic, phagocytic and immunomodulatory activities. *Regul. Toxicol. Pharm.* **2011**, *59*, 412–422. [CrossRef] [PubMed]
50. Soković, M.; Ćirić, A.; Glamočlija, J.; Nikolić, M.; Van Griensven, L.J.L.D. Agaricus Blazei Hot Water Extract Shows Anti Quorum Sensing Activity in the Nosocomial Human Pathogen Pseudomonas Aeruginosa. *Molecules* **2014**, *19*, 4189–4199. [CrossRef]
51. Stojković, D.; Reis, F.S.; Glamočlija, J.; Ćirić, A.; Barros, L.; Van Griensven, L.J.; Ferreira, C.F.R.I.; Soković, M. Cultivated strains of Agaricus bisporus and A. brasiliensis: Chemical characterization and evaluation of antioxidant and antimicrobial properties for the final healthy product–natural preservatives in yoghurt. *Food Funct.* **2014**, *5*, 1602–1612. [CrossRef]

52. Stojkovic, D.; Smiljkovic, M.; Ciric, A.; Glamočlija, J.; Van Griensven, L.; Ferreira, I.; Sokovic, M. An insight into antidiabetic properties of six medicinal and edible mushrooms: Inhibition of α-amylase and α-glucosidase linked to type-2 diabetes. *S. Afr. J. Bot.* **2019**, *120*, 100–103. [CrossRef]
53. da Silva de Souza, A.C.; Correa, V.G.; Goncalves, G.A.; Soares, A.A.; Bracht, A.; Peralta, R.M. Agaricus blazei bioactive compounds and their effects on human health: Benefits and controversies. *Curr. Pharm. Des.* **2017**, *23*, 2807–2834. [CrossRef] [PubMed]
54. Gargano, M.L.; van Griensven, L.J.; Isikhuemhen, O.S.; Lindequist, U.; Venturella, G.; Wasser, S.P.; Zervakis, G.I. Medicinal mushrooms: Valuable biological resources of high exploitation potential. *Plant Biosyst.* **2017**, *151*, 548–565. [CrossRef]
55. Gallego, P.; Rojas, Á.; Falcón, G.; Carbonero, P.; García-Lozano, M.R.; Gil, A.; Grande, L.; Cremades, O.; Romero-Gómez, M.; Bautista, D.J.; et al. Water-soluble extracts from edible mushrooms (*Agaricus bisporus*) as inhibitors of hepatitis C viral replication. *Food Funct.* **2019**, *10*, 3758–3767. [CrossRef]
56. Vunduk, J.; Kozarski, M.; Djekic, I.; Tomašević, I.; Klaus, A. Effect of modified atmosphere packaging on selected functional characteristics of Agaricus bisporus. *Eur. Food Res. Technol.* **2021**, *247*, 829–838. [CrossRef]
57. Stamets, P.E.; Naeger, N.L.; Evans, J.D.; Han, J.O.; Hopkins, B.K.; Lopez, D.; Moershel, H.M.; Nally, R.; Sumerlin, D.; Taylor, A.W.; et al. Extracts of polypore mushroom mycelia reduce viruses in honey bees. *Sci. Rep.* **2018**, *8*, 13936. [CrossRef] [PubMed]
58. Klaus, A.; Kozarski, M.; Niksic, M.; Jakovljevic, D.; Todorovic, N.; Van Griensven, L.J.L.D. Antioxidative activities and chemical char-acterization of polysaccharides extracted from the basidiomycete Schizophyllum commune. *LWT-Food Sci. Technol.* **2011**, *44*, 1–7. [CrossRef]
59. Cantwell, G.E. Standard methods for counting Nosema spores. *Am. Bee J.* **1970**, *110*, 222–223.
60. Williamson, S.M.; Wright, G.A. Exposure to multiple cholinergic pesticides impairs olfactory learning and memory in honey-bees. *J. Exp. Biol.* **2013**, *216*, 1799–1807.
61. OIE–Office International Des Epizooties. Manual of Diagnostic Tests and Vaccines for Terrestrial Animals. Chapter 2.2.4. Nosemosis of Honey Bees. 2018. Available online: https://www.oie.int/fileadmin/Home/eng/Health_standards/tahm/3.02.04_NOSEMOSIS_FINAL.pdf (accessed on 30 August 2021).
62. Simone, M.; Evans, J.D.; Spivak, M. Resin collection and social immunity in honey bees. *Evolution* **2009**, *63*, 3016–3022. [CrossRef]
63. Evans, J.D.; Aronstein, K.; Chen, Y.P.; Hetru, C.; Imler, J.L.; Jiang, H.; Kanost, M.; Thompson, G.J.; Zou, Z.; Hultmark, D. Immune pathways and defence mechanisms in honey bees *Apis mellifera*. *Insect Mol. Biol.* **2006**, *15*, 645–656. [CrossRef]
64. Jehlík, T.; Kodrík, D.; Krištůfek, V.; Koubová, J.; Sábová, M.; Danihlík, J.; Tomčala, A.; Frydrychová, R.Č. Effects of *Chlorella* sp. on biological characteristics of the honey bee *Apis mellifera*. *Apidologie* **2019**, *50*, 564–577. [CrossRef]
65. Ricigliano, V.A.; Simone-Finstrom, M. Nutritional and prebiotic efficacy of the microalga Arthrospira platensis (spirulina) in honey bees. *Apidologie* **2020**, *51*, 898–910. [CrossRef]
66. Ricigliano, V.A.; Ihle, K.E.; Williams, S.T. Nutrigenetic comparison of two Varroa-resistant honey bee stocks fed pollen and spirulina microalgae. *Apidologie* **2021**, *52*, 873–886. [CrossRef]
67. Shumkova, R.; Balkanska, R.; Hristov, P. The Herbal Supplements NOZEMAT HERB® and NOZEMAT HERB PLUS®: An Alternative Therapy for N. ceranae Infection and Its Effects on Honey Bee Strength and Production Traits. *Pathogens* **2021**, *10*, 234. [CrossRef]
68. Williams, G.R.; Shutler, D.; Burgher-MacLellan, K.L.; Rogers, R.E. Infra-population and-community dynamics of the parasites Nosema apis and Nosema ceranae, and consequences for honey bee (*Apis mellifera*) hosts. *PLoS ONE* **2014**, *9*, e99465. [CrossRef]
69. Huang, W.F.; Solter, L.; Aronstein, K.; Huang, Z. Infectivity and virulence of Nosema ceranae and Nosema apis in commercially available North American honey bees. *J. Invertebr. Pathol.* **2015**, *124*, 107–113. [CrossRef]
70. Lull, C.; Wichers, H.J.; Savelkoul, H.F. Antiinflammatory and immunomodulating properties of fungal metabolites. *Mediat. Inflamm.* **2005**, *2*, 63–80. [CrossRef]
71. Vunduk, J.; Djekic, I.; Petrović, P.; Tomašević, I.; Kozarski, M.; Despotović, S.; Nikšić, M.; Klaus, A. Challenging the difference between white and brown Agaricus bisporus mushrooms: Science behind consumers choice. *Br. Food J.* **2018**, *120*, 1381–1394. [CrossRef]
72. Zhao, S.; Gao, Q.; Rong, C.; Wang, S.; Zhao, Z.; Liu, Y.; Xu, J. Immunomodulatory effects of edible and medicinal mushrooms and their bioactive immunoregulatory products. *J. Fungi* **2020**, *6*, 269. [CrossRef]
73. Arismendi, N.; Vargas, M.; López, M.D.; Barría, Y.; Zapata, N. Promising antimicrobial activity against the honey bee parasite Nosema ceranae by methanolic extracts from Chilean native plants and propolis. *J. Apic. Res.* **2018**, *57*, 522–535. [CrossRef]
74. Fries, I.; Chauzat, M.P.; Chen, Y.P.; Doublet, V.; Genersch, E.; Gisder, S.; Higes, M.; McMahon, P.D.; Martín-Hernández, R.; Natsopoulou, M.; et al. Standard methods for Nosema research. *J. Apicult. Res.* **2013**, *52*, 1–28. [CrossRef]
75. Higes, M.; Garcia-Palencia, P.; Martín-Hernández, R.; Meana, A. Experimental infection of *Apis mellifera* honeybees with Nosema ceranae (Microsporidia). *J. Invertebr. Pathol.* **2007**, *94*, 211–217. [CrossRef]
76. Hayman, J.R.; Southern, T.R.; Nash, T.E. Role of sulfated glycans in adherence of the microsporidian Encephalitozoon intestinalis to host cells in vitro. *Infect. Immun.* **2005**, *73*, 841–848. [CrossRef] [PubMed]
77. Vunduk, J.; Wan-Mohtar, W.A.A.Q.I.; Mohamad, S.A.; Halim, N.H.A.; Dzomir, A.Z.M.; Žižak, Ž.; Klaus, A. Polysaccharides of Pleurotus flabellatus strain Mynuk produced by submerged fermentation as a promising novel tool against adhesion and biofilm formation of foodborne pathogens. *LWT-Food Sci. Technol.* **2019**, *112*, 108221. [CrossRef]

78. Antunez, K.; Martín-Hernández, R.; Prieto, L.; Meana, A.; Zunino, P.; Higes, M. Immune-suppression in the honey bee (*Apis mellifera*) following infection by Nosema ceranae (Microsporidia). *Environ. Microbiol.* **2009**, *11*, 2284–2290. [CrossRef]
79. Chaimanee, V.; Chantawannakul, P.; Chen, Y.; Evans, J.D.; Pettis, J.S. Differential expression of immune genes of adult honey bee (*Apis mellifera*) after inoculated by Nosema ceranae. *J. Insect. Physiol.* **2012**, *58*, 1090–1095. [CrossRef] [PubMed]
80. Badaoui, B.; Fougeroux, A.; Petit, F.; Anselmo, A.; Gorni, C.; Cucurachi, M.; Cersini, A.; Granato, A.; Cardeti, G.; Formato, G.; et al. RNA-sequence analysis of gene expression from honeybees (*Apis mellifera*) infected with Nosema ceranae. *PLoS ONE* **2017**, *12*, e0173438. [CrossRef] [PubMed]
81. Djekic, I.; Vunduk, J.; Tomašević, I.; Kozarski, M.; Petrovic, P.; Niksic, M.; Pudja, P.; Klaus, A. Total quality index of Agaricus bisporus mushrooms packed in modified atmosphere. *J. Sci. Food Agric.* **2017**, *97*, 3013–3021. [CrossRef] [PubMed]
82. Kozarski, M.; Klaus, A.; Jakovljevic, D.; Todorovic, N.; Vunduk, J.; Petrović, P.; Niksic, M.; Vrvic, M.M.; Van Griensven, L. Antioxidants of edible mushrooms. *Molecules* **2015**, *20*, 19489–19525. [CrossRef]

insects

MDPI

Article

Effects of Thiamethoxam-Dressed Oilseed Rape Seeds and *Nosema ceranae* on Colonies of *Apis mellifera iberiensis*, L. under Field Conditions of Central Spain. Is Hormesis Playing a Role?

Elena Alonso-Prados [1,*], Amelia Virginia González-Porto [2], Carlos García-Villarubia [3], José Antonio López-Pérez [3], Silvia Valverde [4], José Bernal [4], Raquel Martín-Hernández [5,6] and Mariano Higes [6]

1 Unidad de Productos Fitosanitarios, Instituto Nacional de Investigación y Tecnología Agraria y Alimentaria (INIA, CSIC), 28040 Madrid, Spain
2 Laboratorio de Mieles y Productos de las Colmenas, Centro de Investigación Apícola y Agroambiental (CIAPA), Instituto Regional de Investigación y Desarrollo Agroalimentario y Forestal (IRIAF), Consejería de Agricultura de la Junta de Comunidades de Castilla-La Mancha, 19180 Marchamalo, Spain; avgonzalezp@jccm.es
3 Experimentación e Investigación Hortícola Centro de Investigación Apícola y Agroambiental, Centro de Investigación Apícola y Agroambiental (CIAPA), Instituto Regional de Investigación y Desarrollo Agroalimentario y Forestal (IRIAF), Consejería de Agricultura de la Junta de Comunidades de Castilla-La Mancha, 19180 Marchamalo, Spain; cgarciav@jccm.es (C.G.-V.); jalopezp@jccm.es (J.A.L.-P.)
4 Instituto Universitario CINQUIMA, Analytical Chemistry Group, Universidad de Valladolid, 47011 Valladolid, Spain; silviaval2308@gmail.com (S.V.); jose.bernal@uva.es (J.B.)
5 Instituto de Recursos Humanos para la Ciencia y la Tecnología (INCRECYT–ESF/EC-FSE), Fundación Parque Científico y Tecnológico de Castilla–La Mancha, 02006 Albacete, Spain; rmhernandez@jccm.es
6 Laboratorio de Patología Apícola, Centro de Investigación Apícola y Agroambiental (CIAPA), Instituto Regional de Investigación y Desarrollo Agroalimentario y Forestal (IRIAF), Consejería de Agricultura de la Junta de Comunidades de Castilla-La Mancha, 19180 Marchamalo, Spain; mhiges@jccm.es
* Correspondence: aprados@inia.csic.es

Citation: Alonso-Prados, E.; González-Porto, A.V.; García-Villarubia, C.; López-Pérez, J.A.; Valverde, S.; Bernal, J.; Martín-Hernández, R.; Higes, M. Effects of Thiamethoxam-Dressed Oilseed Rape Seeds and *Nosema ceranae* on Colonies of *Apis mellifera iberiensis*, L. under Field Conditions of Central Spain. Is Hormesis Playing a Role? *Insects* **2022**, *13*, 371. https://doi.org/10.3390/insects13040371

Academic Editors: Ivana Tlak Gajger and Franco Mutinelli

Received: 17 March 2022
Accepted: 6 April 2022
Published: 9 April 2022

Simple Summary: The collapse of the honey bee colonies is a complex phenomenon in which different factors may participate in an interrelated manner (e.g., pathogen interactions, exposure to chemicals, beekeeping practices, climatology, etc.). In light of the current debate regarding the interpretation of field and monitoring studies in prospective risk assessments, here we studied how exposure to thiamethoxam affects honey bee colonies in Central Spain when applied as a seed treatment to winter oilseed rape, according to the good agricultural practice in place prior to the EU restrictions. Under the experimental conditions, exposure to thiamethoxam, alone or in combination with other stressors, did not generate and maintain sufficient chronic stress as to provoke honey bee colony collapse. The stress derived from exposure to thiamethoxam and honey bee pathogens was compensated by adjustments in the colony's dynamics, and by an increase in the worker bee population, a behavior known as hormesis. An analysis of the factors underlying this phenomenon should be incorporated into the prospective risk assessment of plant protection products in order to improve the future interpretation of field studies and management practices.

Abstract: To study the influence of thiamethoxam exposure on colony strength and pathogen prevalence, an apiary (5 colonies) was placed in front of a plot sown with winter oilseed rape (wOSR), just before the flowering phase. Before sowing, the seeds were treated with an equivalent application of 18 g thiamethoxam/ha. For comparison, a second apiary (5 colonies) was located in front of a separate 750 m plot sown with untreated wOSR. Dead foragers at the entrance of hives were assessed every 2–3 days throughout the exposure period, while the colony strength (number of combs covered with adult honey bees and brood) and pathogens were monitored each month until the following spring. Foraging on the wOSR crop was confirmed by melissopalynology determination of the corbicular pollen collected periodically, while the chemical analysis showed that exposure to thiamethoxam was mainly through nectar. There was an increase in the accumulation of dead bees in the apiary exposed to thiamethoxam relating with the control, which was coped with an increment of bee brood surface and adult bee population. However, we did not find statistically significant differences between apiaries ($\alpha = 0.05$) in terms of the evolution of pathogens. We discuss these results under hormesis perspective.

Keywords: honey bees; Spain; thiamethoxam; *Nosema ceranae*; hormesis; *Brassica napus*; oilseed rape; seed treatment; EPPO; viruses

1. Introduction

In the past decade, significant losses of honey bee colonies have been reported, mainly in Europe and North America [1–11]. The consensus is that several factors are involved in this phenomenon: nutritional stress, pathogens, hive management practices, exposure to multiple xenobiotic residues (pesticides and veterinary drugs), invasive species, seasonal weather changes and genetic variability [7,12–19]. However, exposure to chemicals and the most prevalent pathogens are without doubt the main drivers in the phenomenon of honey bee colony loss [20].

Among the most relevant chemicals, much attention has been paid to neonicotinoids [21] in recent years due to their widespread use [22–25]. These pesticides selectively act on the insect's central nervous system (CNS) as they are agonists of post-synaptic acetylcholine receptors [26–29]. Neonicotinoids have a low molecular weight, they are moderately-to-highly water soluble and they have a low octanol:water partitioning coefficient. These physico-chemical properties favor the efficient translocation of these chemicals through the plant xylem, such that they can systemically control piercing-sucking insects [24]. However, these characteristics also mean they have a high capacity for environmental transport [30–32]. Depending on the toxophore, these compounds are classified as N-nitroguanidines (imidacloprid, thiamethoxam, clothianidin and dinotefuran), nitromethylenes (nitenpyram), or N-cyano-amidines (acetamiprid and thiacloprid) neonicotinoids, with the former representing the class that has produced most concern for bees [33]. Many sub-lethal effects of N-nitroguanidine neonicotinoids have been identified in bees, influencing on orientation and olfactory learning, flight, queen performance, honey bee colony physiology and the bee's immune system [34–47]. In particular, thiamethoxam (TMX) causes damage in the midgut, brain cells and Malpighian tubules of bees [48–50], affecting several biological process such as oxidative phosphorylation and tyrosine metabolism [51], as well as altering the gut microflora [52].

On the other hand, *Nosema ceranae* and *Varroa destructor* have been intensively studied in relation to the phenomenon of honey bee colony losses. *N. ceranae* is found worldwide [14,53,54], it has a high pathogenic capacity [55,56] and it is transmitted horizontally in the colony, through food, water, trophallaxis and cleaning duties (reviewed in [54]). It is an obligate intracellular parasite that depends on the host's energy for reproduction, and it alters the physiology and behavior of its host to ensure conditions are optimal for it to complete its biological cycle (reviewed in [54]). As such, *Apis mellifera* individuals infected by *N. ceranae* experience: (1) a strong carbohydrate demand; (2) altered lipid and amino acid metabolism; (3) an altered vitellogenin/juvenile hormone equilibrium; (4) a reduction in hexamerins and major royal jelly proteins; (5) an upregulation of the octampamide pathway; (6) a suppressed immune response; (7) impaired apoptosis of ventricular epithelial cells [54,57,58], and (8) changes in midgut microbiota [52]. As a consequence of all these alterations, *N. ceranae* will perturb colony homeostasis if the infection is maintained over time [59]. In this way, infected bees generate higher Ethyl Olate (EO) titers and undertake more flight activity than non-infected bees [60]. The constant reduction in bee number due to the tissue damage, energetic stress, and altered flight behavior due to infection accelerates age polyethism in young bees to cover the energy demands of the colony [61,62]. However, precocious foragers (FBs) are less effective, so higher number of FBs will be needed to fulfil the colony's demands, reducing the time each bee can dedicate to hive tasks. Thus, if the buffering capacity of the colony is exceeded then there is a strong risk of collapse [54].

For its part, *V. destructor*, one of the most important pathogens of honey bees [63–65], not only damages the bee by feeding on its hemolymph and fat body [66,67] but also, it acts as a vector for a number of viruses [68]. As a result, this pathogen impairs the responses of the immune system [69] and affects the nervous system [70,71].

Honey bees exposed to neonicotinoids may be more susceptible to *N. ceranae* and *V. destructor* infections and that of other viruses [72]. Indeed, *N. ceranae*-infected honey bees suffer increased mortality under laboratory conditions when exposed to sub-lethal concentrations of imidacloprid [73,74], thiacloprid [75,76] or TMX [77]. Similarly, the damage caused by *N. ceranae* in midgut cells [78] may make infected honey bees more susceptible to neonicotinoid exposure. The negative synergistic effects of *Varroa* and neonicotinoid insecticides have been related to immune suppression, which impairs antiviral immune barriers [47] and memory retention [79].

However, as all these effects are rarely observed under real world conditions [80,81], they remain somewhat controversial.

Here, we present the results of a field study undertaken in accordance with good beekeeping practice (prior to European restrictions) in Central Spain that set out to determine the evolution of honey bee colonies and the prevalence of pathogens following exposure to TMX through the pollen and nectar of seed-treated oilseed rape. The stress provoked by both TMX and pathogens was compensated by the dynamics of the colonies, mainly through an increase in the bee population. We discuss the possible reasons for these responses from the hormesis point of view and in the context of the prospective risk assessment of plant protection protocols.

2. Materials and Methods

2.1. Experimental Design

The experiments were designed in accordance with the EPPO 170 standards [82]. Colonies of *Apis mellifera iberiensis* from the same apiary located at the 'Centro de Investigación Apícola y Agroambiental" (CIAPA, Marchamalo, Guadalajara, Spain) and with naturally mated sister queens of the same age, were used to monitor the presence of queens, and to analyze and control the sanitary status of the colonies in order to ensure the colony's strength at the beginning of the study (10 brood chamber combs covered by bees). The experimental colonies were randomly divided into two groups of five colonies each, and they were placed in front of two plots of 2 ha (750 m apart), that were sown with winter oilseed rape (wOSR, Ginfizz variety), just before their flowering phase (Figure 1).

Sowing took place on 29 October 2014 at a density of 75 seeds m^{-2}. In one plot, the seeds were treated with Cruiser 350 FS© before sowing (TMX 35% p/v FS, 1200 mL/100 kg seed), while both fields were treated with Butisan S© (metazachlor 50% p/v SC) at 1.5 L/ha on 31 November, receiving no further phytosanitary treatments during the study. The fields were not treated with plant protection products for at least 2 years and before sowing, the soil of the plots was sampled and characterized as described in Section 2.2.

The exposure phase lasted from 8 April to 11 May 2015, at the beginning of which an upper box was placed in each hive, which contained empty combs ready to be used by the honey bees. The stored pollen (beebread) and honey in the brood chambers was removed to stimulate foraging activity during the exposure period, and to ensure that the pollen and nectar came from the experimental plots [83]. After the exposure phase, the colonies were placed 35 km away from the area of exposure in order to monitor them until the following spring (Figure 1).

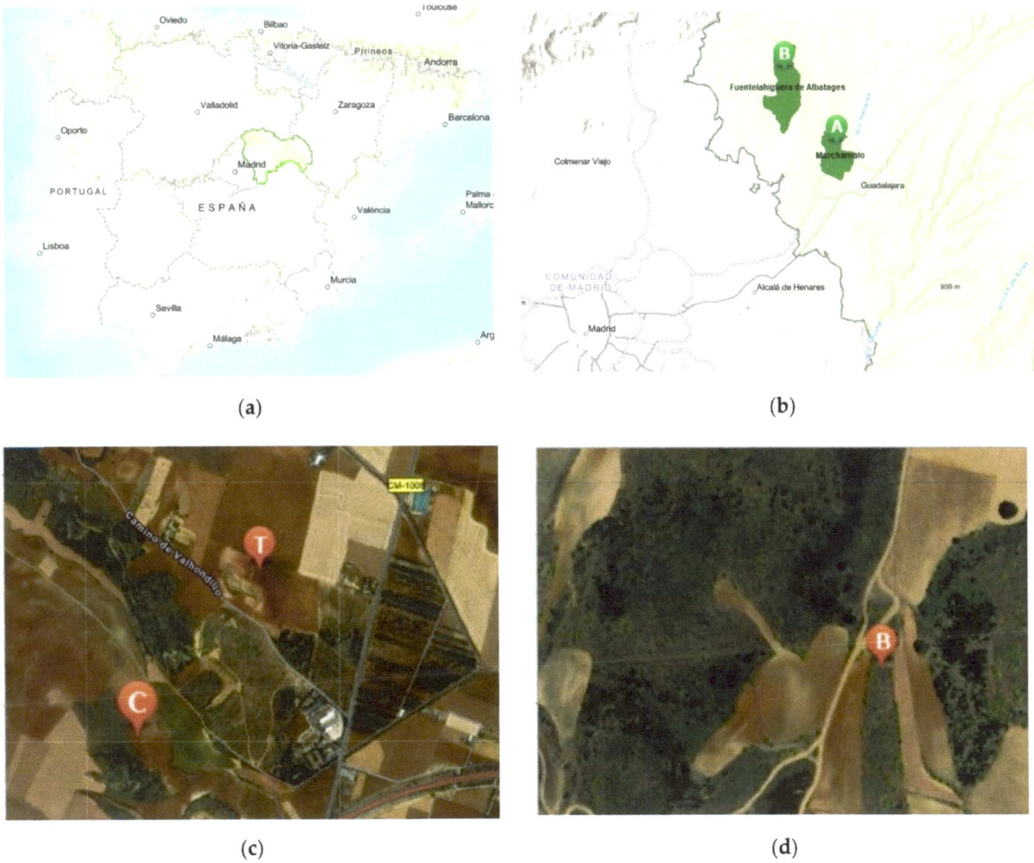

Figure 1. Site of the experimental study: (**a**) Map of Spain with the Province of Guadalajara highlighted in green; (**b**) Location of the apiaries during (A: Marchamalo) and after (B: Fuentelahiguera de Albatages) the exposure phase; (**c**) Detailed location of the apiaries during the exposure period in Marchamalo (C: control, T: treatment); (**d**) Detailed location of the apiaries after the exposure period in Fuentelahiguera de Albatages (B).

To avoid winter collapse due to nosemosis C, the colonies were treated exceptionally with Fumidil B on 17 September 2015 under controlled conditions. The treatment consisted of 4 doses of 30 mg fumagillin per colony dissolved in 250 cc of sugar syrup (50% sugar/distilled water), administered in plastic bags placed over the brood at one-week intervals to ensure full consumption. This posology was effective without leaving residues in the honey [84]. *V. destructor* was controlled at the beginning of September with Apitraz© (amitraz a.m.), employed at the recommended dose and in accordance with current Spanish legislation [85].

2.2. Soil Characterization and Climatological Data

The soil surface layer (0–20 cm) was characterized following ISRIC protocols [86] and using three samples gathered randomly in each plot. The anion analysis (F^-, Cl^-, NO_2^-, NO_3^-, PO_4^{3-}, SO_4^{2-}) was carried out at room temperature using a Metrohm Compact IC model 761 ion chromatograph (Metrohm, Herisau, Switzerland). Na^+ and K^+ were measured using a flame photometry SHERWOOD 410 apparatus, and Ca^{2+} and Mg^{2+} using atomic absorption spectrophotometry (ANALYTIKJENA NOVAA 300). Electrical conductivity (EC), pH, and humidity were measured with an EC meter CRISON micro CM2000, a pH meter CRISON GLP 21, and a thermogravimetric balance KERN DBS, respectively. Bouyoucos' densitometry method was followed to characterize the soil texture [87], and the organic matter and carbonate content was determined with the Walkley–Black and acid neutralization methods [86], respectively.

A Walter–Lieth diagram [88] was developed with historical weather data obtained from the meteorological station at Marchamalo from the regional Irrigation Advisory Service (SIAR-CREA) network to compare it with data during the period of study. These diagrams illustrate the rainfall and temperature changes throughout the year in standardized charts which provide brief summaries of the average climatic variables and their time course. The diagrams were drawn up with the diagnostic tool of the Worldwide Bioclimatic Classification System, 1996–2021 [89].

2.3. Sampling Schedule

The number of dead worker bees was counted in box dead-bee traps [90] at the entrance of hives every 2–3 days throughout the exposure period. Colony strength was checked by determining the number of combs covered by brood and bees once every two months [59]. The honey production in each colony was evaluated in the harvesting season as the difference in the weight of the combs before and after honey extraction [91].

Corbicular pollen loads were collected using standard traps, activated for 24 h at the beginning (8 April), in the middle (15 April), and at the end of blooming (29 April). For melissopalynological and chemical analyses, at the end of the exposure period, beebread and honey samples were collected individually from one frame of the brood chambers and the upper boxes, respectively. Beebread was extracted aseptically, removing the wax from the combs as described previously [92–94], and storing the samples at −80 °C in the dark until they were analyzed. The honey was sampled using a sterile spatula and stored at −20 °C.

To identify viruses and quantify the *Nosema* spp. parasitic load in the colonies, both non-foragers, (IBs, bees sampled from no-brood combs, $n > 60$), and foragers (FBs, bees arriving at the colony after closing the hive entrance for 30 min, $n > 60$) were collected separately from each colony (Table 1) and frozen (−20 °C).

Table 1. Field schedule.

Sampling Date	Forager Death	Colony Strength	Pathogens	Bee Pollen	Beebread	Honey
30 March 2015	X	X	X			
8 April 2015	X			X		
10 April 2015	X					
13 April 2015	X	X		X		
15 April 2015	X					
17 April 2015	X					
20 April 2015	X					
22 April 2015	X					
24 April 2015	X					
27 April 2015	X					
29 April 2015	X			X		
4 May 2015	X					
6 May 2015	X					
8 May 2015	X					
11 May 2015	X	X	X		X	X
15 July 2015		X	X			
15 September 2015		X	X			
17 September 2015						
15 November 2015		X	X			
15 March 2016		X	X			

2.4. Palynological and Melissopalynological Assessment

To confirm the type of foraging flora, corbicular pollen, beebread and honey samples were analyzed according to previously described methods [95,96]. Corbicular pollen samples were weighed and stored at -20 ± 2 °C until further analysis. Pollen loads were separated based on color and texture to identify and determine the contribution of each botanical type into the samples. The pollen grains were isolated from each sample and cleaned using the Erdtam method [97]. One aliquot from each of the ten pollen loads of the same "color-texture" class was placed onto a slide in glycerin jelly mounting medium and examined microscopically in order to ensure the homogeneity of identification. The proportion of each pollen type was calculated by dividing the weight obtained for each color fraction by the overall weight of the sample analyzed [95].

Pollen was extracted from the beebread by diluting 0.5 g in 10 mL of 0.5% acidulated water (96% sulfuric acid) and centrifuging at 2500 r.p.m. for 15 min. The pellet was washed with double-distilled water and centrifuged twice, and the sediment was placed onto a slide in glycerin jelly mounting medium and examined microscopically to identify the pollen. The frequency of Brassica type pollen grains (*Brassica* t. pollen) was expressed as a percentage of the total pollen grains.

The honey samples were treated chemically with 0.5% acidified water (sulfuric acid 96%), and a qualitative and quantitative analysis was performed on the sediment recovered from 5 g aliquots [98]. Between 300 and 1200 pollen grains were counted in each sample, and the pollen grains were identified and classified on the basis of the identification keys [99,100]. All the reference collection pollen slides used were available at the honey laboratory at the CIAPA.

2.5. Chemical Analysis

Chemical analyses were carried out as described elsewhere [101–103]. FLUKA-PESTANAL analytical standards of clothianidin (CLO; purity 99.9%), TMX (purity 99.6%), and TMX-d3 (isotope-labeled standard; purity $\geq$ 98.0%) were purchased from Sigma-Aldrich Laborchemikalien (Seelze, Germany). LC grade methanol (MeOH), and acetonitrile (ACN) were supplied by Lab Scan Ltd. (Dublin, Ireland). Formic acid (FA), ammonium formate, and ethyl acetate (EA) were obtained from Sigma-Aldrich Chemie Gbmh (Steinheim, Germany); sodium chloride, trisodium citrate dihydrate, and trisodium citrate sesquihydrate were supplied by Panreac (Barcelona, Spain). Meanwhile, primary secondary amine (PSA) and octadecylsilane (C_{18}) were provided by Supelco (Bellefonte, PA, USA). Finally, ultrapure water was obtained using Milipore Mili-RO plus and Mili-Q systems (Bedford, MA, USA). All chemicals used were of analytical grade.

Briefly, to quantify the levels of TMX and its metabolite clothianidin (CLO) in the seeds [101], insecticides were extracted with a mixture of acetonitrile (ACN) and sodium chloride (60:40, v/v). After centrifugation (5810 R refrigerated bench-top Eppendorf centrifuge; Hamburg, Germany), the supernatant was collected and concentrated at 60 °C (R-210/215 rotary evaporator from Buchi, Flawil, Switzerland). The dry extract was reconstituted with 1 mL of a mixture of 0.1% (v/v) FA in ACN and 0.1% (v/v) FA in water (25:75, v/v). The extract was filtered and injected onto a liquid chromatography coupled to a diode array detector (DAD; Agilent Technologies 1200 series; Palo Alto, Santa Clara, CA, USA) column (Kinetex C_{18}, 150 × 4.6 mm, 2.6 µm, 100 Å; Phenomenex, Torrance, CA, USA) system. The limits of quantifications (LOQs) were set at 0.11 and 0.15 g/kg for TMX and CLO, respectively [101].

In relation to the analysis in bee pollen, it was employed as a sample treatment modified QuEChERS (quick, easy, cheap, effective, rugged, and safe) method. A representative amount of sample (1.0 g) was mixed with 2 mL of water and 6 mL of ACN, and the resulting mixture was shaken. Then, magnesium sulfate (1.0 g), sodium chloride (0.5 g), and tri-sodium citrate dihydrate (0.8 g) was added to the mixture, and after centrifugation (10,000 rpm, 10 °C, 5 min), 2 mL of the supernatant was transferred to an Eppendorf tube. After that, magnesium sulfate (150 mg), PSA (25 mg), and C_{18} (25 mg) were added to the tube, which was again centrifuged. Supernatant (1 mL) was collected and evaporated to dryness (60 °C). A further reconstitution of the dry extract was performed with 1 mL of a MeOH and water (80:20, *v*/*v*) mixture, and the resulting extract was filtered and analyzed by using an ultra-high performance liquid chromatography (UHPLC; ACQUITY, Waters, Milford, MA, USA) coupled to quadrupole time-of-flight mass spectrometry (qTOF; maXis impact, Bruker Daltonik GmbH, Bremen, Germany) system equipped with an electrospray interface (ESI), which was operated in positive mode. The obtained LOQs were 2.1 ng/g for TMX and 3.9 ng/g for CLO [102].

Finally, two different sample treatments were employed when determining TMX and CLO depending on the botanical origin of honey. Insecticides were extracted from light color honeys (multifloral and rosemary) by using a modified QuEChERS protocol. Briefly, 5.0 g of honey was mixed with 10 mL of water and 10 mL of an ACN and EA (70:30, *v*/*v*) mixture, and after shaking, different amounts of magnesium sulfate (2.0 g), sodium acetate (1.0 g), trisodium citrate dihydrate (1.5 g), and trisodium citrate sesquihydrate were added. Then, the mixture was shaken in an ultrasound device (J.P. Selecta S.A., Barcelona, Spain), centrifuged (10,000 rpm, 10 °C, 5 min); the supernatant was collected and evaporated to dryness (60 °C), while the dry extract was reconstituted with 1 mL of a MeOH and water (80:20, *v*/*v*) mixture, being the resulting extract filtered and analyzed by an ACQUITY UHPLC (Waters) coupled to a Xevo TQ-S (triple quadrupole, QqQ) mass spectrometer (Waters) equipped with an ESI interface (positive mode). The obtained LOQs were 0.06 ng/g for TMX and 0.20 ng/g for CLO [102]. On the other hand, dark honeys (heather) required a solid-phase extraction (SPE) procedure. A representative amount of dark honey (5.0 g) was diluted in 10 mL of ammonium formate (10 mmol/L) in water. The resulting mixture was loaded onto a polymeric SPE cartridge (Strata® X; Phenomenex), which was previously conditioned with 5 mL of MeOH and 5 mL of water. Then, after 5 min of drying time, the insecticides were eluted from the cartridge with 4 mL of an ACN and EA (80:20, *v*/*v*) mixture. The extract was evaporated to dryness, and the dry residue reconstituted with 1 mL of a MeOH and water (80:20, *v*/*v*) mixture, filtered and analyzed by using the same UHPLC-QqQ system that was employed for analyzing light honeys. The LOQs were 0.06 ng/g for TMX and 0.20 ng/g for CLO [103].

2.6. Identification of Pathogens

2.6.1. Detection of Varroa Mites

At each sampling time, a total of 60 FBs and 60 IBs were examined individually per colony, using sterile tweezers to detect malformations and collect parasitic mites to be identified macroscopically. A honey bee colony was considered to be infested with *V. destructor* when at least 1 *Varroa* mite was found in the sample. The rate of infestation of the bee colony was estimated by assessing the number of *Varroa* mites relative to the number of adult bees in each sample, and it was expressed as the number of *Varroa* mites/100 bees/sample.

2.6.2. Pathogen Screening

DNA and RNA were extracted at each sampling time (see Table 1) as described previously [104]. In summary, the abdomen was removed from 30 FB and 30 IB bees in each colony, placing them individually in a well of a 96-well plate containing glass beads (2 mm: Sigma®, St. Louis, MO, USA). The tissue was macerated in AL buffer (Qiagen® 19075, Hilden, Germany) and incubated with proteinase K under the same conditions as described in [104]. DNA and RNA purification were performed simultaneously in a Biosprint 96 workstation (Quiagen®, Hilden, Germany) according to the BS96 DNA Tissue extraction protocol. Total nucleic acids (RNA and DNA) were eluted in 100 µL of elution buffer.

An aliquot of 75 µL was used to detect *Nosema* spp. according to an earlier procedure [105]. To identify *N. ceranae* and *N. apis*, an internal control (IPC) based on mitochondrial cytochrome oxidase subunit I (COI) was used as a target in the triplex PCR assay [105].

The percentage of FBs or IBs parasitized by Nosema was estimated as 100 × number of positives/30.

The remaining 25 μL aliquots of the total nucleic acids eluted from each FB and IB sample from each colony and sampling time were pooled immediately and subjected to DNase I digestion to remove DNA (Quiagen® kit 79254), as described previously [104]. The total RNA recovered was used immediately to generate first strand cDNA with the iScriptTM cDNA Synthesis Kit (Biorad®, Hercules, CA, USA). The resultant cDNA was used to analyze the viruses present. Black Queen Cell Virus (BQCV) and Deformed Wings Virus (DWV) were detected by Real Time-PCR as detailed in [104], and using the primers and probes described by [106]. The Acute Bee Paralysis Virus-Kashmir Bee Virus-Israeli Acute Paralysis Virus (AKI) complex was detected by following the procedure described in [107]. In all cases, negative and positive controls were run in parallel for each step (sample processing, nucleic acid extraction, reverse transcription, and RT-PCR and PCR analysis).

2.7. Statistical Analysis

The statistical analysis was made according to the recommendations in [108,109]. Colony strength or pathogen parameters were compared between the two groups of colonies at the end of the exposure period (n(control) = n(TMX) = 5) with a one tailed Mann–Whitney U test (exact approach, alpha = 0.05). The Mann–Whitney U test compares the distribution of the ranks of the two populations, and to interpret this as a comparison of medians, the distributions of the two populations are assumed to be the same shape [110]. Moreover, we reported the effect size for non-parametric tests following the approach given elsewhere [111]. Mann–Whitney U tests and effect sizes were generated using the Real Statistics Resource Pack software [112] and transformed to the Cohen's d coefficient as described previously [113]. Finally, we used G-Power 3.1.9.6 [114] to estimate the power (1-β) of the statistical tests.

To explore the evolution of the percentage of *Nosema* spp. infection in each honey bee colony, we used a Friedman test. This test is the non-parametric alternative to one-way ANOVA with repeated measures [115] and it was implemented with Statgraphics Centurion 18.

3. Results

3.1. Soil Characterization and Climatic Conditions

The 0–20 cm soil horizons of the control and treatment plots were classified as slightly acid clay and neutral clay loam, respectively, in both cases with low organic matter content and soil conductivity. The chemical characterization of the soil showed the CEC is dominated by Ca^{2+} (Table S1). The soils are derived from Miocene-tertiary aged fluvio-alluvial sediments from Henares River Basin [116], and following the classification criteria of the World Reference Base (WRB) for Soil Resources [117], they could tentatively be classified as calcic luvisols.

Based on 12 years of climatological data (period 2010–2021) from the nearest station to the experimental plots (Figure S1a), and on the bioclimatic indices estimated with the diagnostic tool of the Worldwide Bioclimatic Classification System [89], the climate in the experimental plots falls into the Upper Mesomediterranean Low Dry Blioclimatic belt, with a freezing period extending from December to March. During the period of exposure, the mean temperature was 13.0 ± 2.0 °C in April and 17.8 ± 2.5 °C in May 2015. The relative humidity ranged from 38.8 to 92.9 (mean = 61.58 ± 13.33) and the total rainfall was 29.3 mm, concentrated in 9 days and with 52.56% of the precipitation falling on 26 April 2015.

When compared to historic average climate data (Figure S1a), the Walter–Lieth diagrams show that autumn 2014 (Figure S1b) was wetter than the mean values for the period between 2010 and 2021. However, 2015 was characterized as being dryer and hotter than average, and with a wider range of temperatures (Figure S1c). By contrast, spring 2016 was more temperate and significantly wetter, especially in April (Figure S1d).

3.2. Palynological and Melissopalynological Assessments

The *Brassica* t. pollen content varied in the corbicular pollen sampled in April (Figure 2a). In addition to *Brassica* t. pollen, *Prunus* t. and *Cistus* t. pollens were also predominant in the samples at the beginning of the month. These taxa are typical of regression succession steps of the chorological series *Bupleuro rigidi-Querceto rotundifoliae sigmetum* in the Castellano-Maestrazgo-Manchega province [118]. In the middle of April, *Brassica* t. pollen was replaced by pollen from the surroundings, mainly *Cistus* t pollen, which was in turn replaced by *Brassica* t. pollen at the end of the month. At the end of the exposure period, the *Brassica* t. pollen content in the beebread was above 50% in all the colonies except two (C-01 and C-08), and the *Brassica* t. content of the honey in the cells of the frame introduced at the beginning of the exposure period ranged from 50.3% to 94.2% (Figure 2b).

(a) (b)

Figure 2. Melissopalynological analysis: (**a**) The % (*w*/*w*) *Brassica* t. loads in corbicular pollen at the beginning (8 April 2015), middle (13 April 2015), and end (29 April 2015) of winter oilseed rape (wOSR) flowering; (**b**) The frequency (%) of *Brassica* t. pollen in beebread and honey : C-01 to C-05 = Control hives; C-06 to C-10 = hives exposed to TMX.

3.3. Chemical Analysis

The chemical analyses revealed the level of TMX residues in treated seeds was 4660 mg/kg seed, equivalent to an application of 18 g TMX/ha. The TMX residues in bee pollen (corbicular pollen) were below the LOQ in all samples, while in the honey samples, it varied between below the limit of detection (<LOD) and 144 µg/kg (Table S13 in [103]).

3.4. Field Monitoring of the Honey Bee Colonies

During the exposure period, the accumulated honey bee death was higher in the treatment apiary (TMX) as the combs became covered with worker honey bees and brood (Figure 2, Table 2). Subsequently, both the worker honey bee population and the brood balanced out in both apiaries (Figure 3).

The mean number of dead bees was 127.6 $\pm$ 44.1 and 214.4 $\pm$ 116.5 in the control and treatment (TMX) apiaries, respectively. Likewise, the average number of combs covered with adult honey bees and honey bee brood was 12.2 $\pm$ 1.095 and 9 $\pm$ 4.96 in the controls, respectively, and 15.2 $\pm$ 2.05 and 11 $\pm$ 1.82 in the treatment apiaries. The same trend was observed in terms of production over the period of exposure, reaching 24.75 $\pm$ 15.05 (control) and 31.98 $\pm$ 12.21 kg (TMX) at the end of the exposure phase, respectively (Table 2).

Table 2. Descriptive statistics of the variables of strength in control and treatment (TMX) apiaries at the end of the exposure period.

Variable	Descriptive Statistics	Control (n = 5)		TMX (n = 5)
Accumulated deaths	Min	80		92
	Max	192		394
	Median	129		214
	Mean	127.6		214.4
	Standard deviation (SD)	44.106		116.468
	Variation coefficient (%)	34.565		54.323
	Mean 95% confidence interval	[72.836; 182.364]		[69.785; 359.015]
	Power (1-β)		0.379	
	Cohen's d		0.863	
Number of combs with worker bees	Min	11		11
	Max	14		16
	Median	12		16
	Mean	12.2		15
	Standard deviation (SD)	1.095		2.236
	Mean 95% confidence interval	[10.84; 13.56]		[12.22; 17.77]
	Variation coefficient (%)	8.979		14.907
	Power (1-β)		0.541	
	Cohen's d		1.245	
Number of combs with brood	Min	2		9
	Max	13		13
	Median	10.5		11.0
	Mean	9.0		11.0
	Standard deviation (SD)	4.966		1.826
	Variation coefficient (%)	55.184		16.598
	Mean 95% confidence interval	[1.097; 16.903]		[8.095; 13.905]
	Power (1-β)		0.103	
	Cohen's d		0.310	
Honey production (kg)	Min	10.24		13.02
	Max	45.22		45.58
	Median	16.204		32.034
	Mean	24.755		31.977
	Standard deviation (SD)	15.075		12.21
	Variation coefficient (%)	60.894		38.184
	Mean 95% confidence interval	[6.038; 43.473]		[16.816; 47.137]
	Power (1-β)		0.165	
	Cohen's d		0.475	

Due to the high dispersion of the data within the apiaries (Table 2), at the end of the exposure period, a one-tailed Mann–Whitney U test did not find significant differences in the dead bees accumulated (Mann–Whitney U = 6; p-value = 0.111), the number of combs with bee brood (U = 6.5; p-value = 0.343) or in the honey produced (U = 9; p-value = 0.274). As such, the statistical power varied between 0.183 and 0.397 (Table 2). Regarding the number of combs covered with bees, the test did reveal a significant difference (U = 4.5; p-value = 0.048; 1-β = 0.72).

3.5. Identification of Pathogens

While *N. apis* was not detected in the colonies at any point in the study, the percentage of infection of *N. ceranae* increased in both non-forager (IB) and forager bee (FB) populations until fumagillin treatment was applied (Figure 4). The percentage of infection was significantly different between the distinct honey bee populations (Friedman Test statistic = 71.4394; p-value = 1.132 10–14) but not between apiaries. The maximum percentage infection was observed at the end of summer, with a mean of 62.6 ± 10.6% of FBs infected and a mean of 17.6 ± 5.06% of IBs.

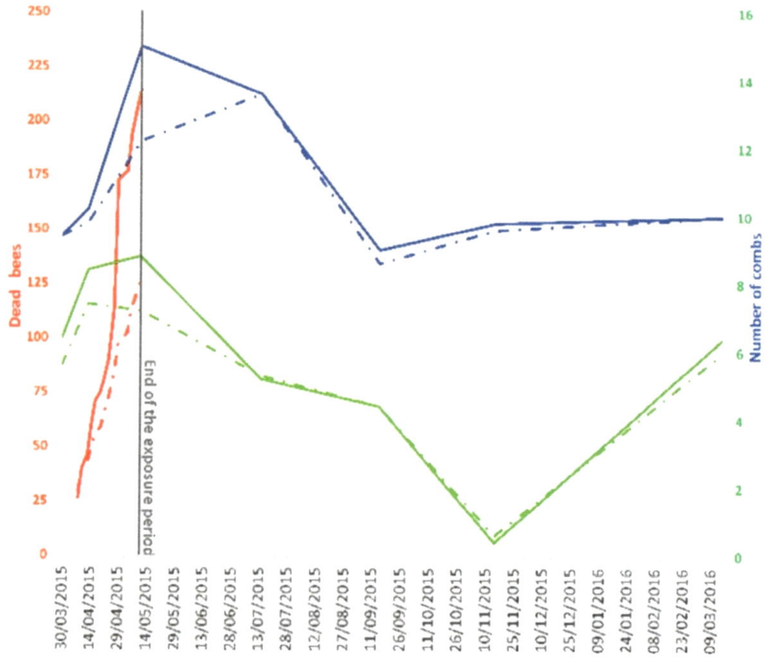

Figure 3. Colony strength parameters throughout the study in control and treatment (TMX) apiaries: mean accumulated dead bees at the entrance of the colonies (— •— control; — TMX) and mean number of combs covered with bees (—• — control;—TMX) or brood (—• — control;— TMX).

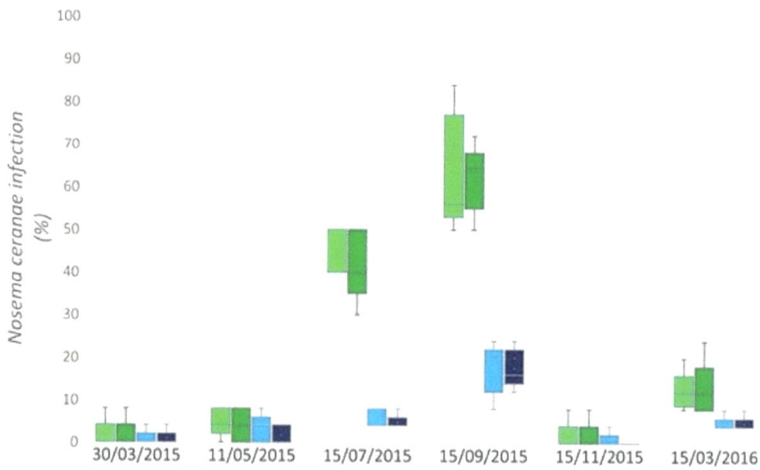

Figure 4. Box and Whisker plots of *N. ceranae* infection in non-forager honey bees (IB; control; TMX) and forager honey bees (FB; control; TMX).

Varroa mite was not detected in any honey bee sampled and DWV was found at a low frequency until the end of the summer, yet it was detected in all the colonies just before wintering (Table 3). By contrast, BQCV was detected throughout the experiment (Table 3), whereas the AKI complex was not identified in any sample.

Table 3. Detections of Black queen cell virus (BQCV) and Deformed wing virus (DWV) during the study in non-forager (IB) and forager bees (FB).

Hive	30 March 2015		11 May 2015		15 July 2015		15 September 2015		17 November 2015		15 March 2016	
	IB	FB	IB	FB	IB	FB	IB	FB	IB	FB	IB	FB
C 01	BQCV	-	BQCV	BQCV	BQCV	BQCV	-	-	DWV	DWV	DWV	DWV
C 02	BQCV	-	BQCV	-	BQCV	BQCV	BQCV	-	BQCV DWV	DWV	-	-
C 03	BQCV	BQCV-DWV	BQCV	BQCV	BQCV	BQCV	-	BQCV	BQCV-DWV	DWV	-	-
C 04	BQCV	BQCV-DWV	BQCV	DWV	BQCV	BQCV	BQCV	BQCV	BQCV-DWV	DWV	BQCV	-
C 05	DWV	BQCV	BQCV-DWV	-	DWV	BQCV	BQCV	-	BQCV-DWV	DWV	BQCV	-
C 06	BQCV	BQCV	BQCV	-	BQCV	BQCV	-	-	BQCV-DWV	DWV	BQCV	-
C 07	BQCV	-	BQCV	-	BQCV	BQCV	-	BQCV	BQCV-DWV	DWV	BQCV	-
C 08	BQCV	-	BQCV	BQCV	BQCV	BQCV	-	-	BQCV-DWV	DWV	-	DWV
C 09	-	BQCV	-	-	BQCV	BQCV	BQCV	-	BQCV-DWV	DWV	-	-
C 10	-	BQCV-DWV	-	BQCV	BQCV	BQCV	BQCV	-	BQCV-DWV	DWV	-	-

4. Discussion

The foraging activity of honey bees in the experimental fields was guaranteed during the period of exposure due to the melliferous potential of the oilseed rape crop. Its attractiveness to honey bees is derived from the features of its flowers, their fragrance and nutritive nectar, and the high flower density of its plants, opening successively for 3–4 weeks [119]. The amount of *Brassica* t. in the bee pollen varied throughout the month of April, decreasing in the middle of April when the bulk flowering of other species took place, such as the plants of the genus *Cistus* [120,121], a producer of high quality pollen [122,123]. The chemical analysis confirmed that the honey bees were exposed to TMX through nectar rather than pollen [103]. The levels of residues in wOSR plants may have been influenced by the time between the sowing (Autumn 2014) and the exposure time (Spring 2015), and possibly due to soil leaching. Given their physico-chemical properties [24], TMX and its metabolite chlothianidin have high potential for lixiviation. This fact, together with the precipitation that fell in the days following drilling and the low organic matter content of the soils, may have favored the low retention of the residues of these chemicals in soils.

Several detections above LOQ were found in the honey samples of the control apiary [103], however, this does not invalidate the results. Thus, an increment on the mean accumulated deaths in the apiary located in front of the treated plot was observed when compared with the control at the end of the exposure period. Taking into account that exposure during foraging is not a discrete process but distributional [124], the mean colony exposure remained low even though effects were seen in a number of bees. The loss of forager bees can accelerate behavioral changes aimed at increasing food gathering in order to maintain the energy balance within the colony [125]. These changes bring about an increase in bee brood to compensate the interactions within the colony [126], referred to as a hormetic-like effect [127–130]. Hormesis is common in arthropods [131,132] and it is a dose-dependent phenomenon where exposure to high levels of a stressor are inhibitory, whereas low doses are stimulatory [133]. Such effects have been considered to be an example of evolutionary fitness [134] whereby organisms undergo metabolic adaptations in response to the interactions between the diverse stressors they are exposed to. Indeed, it has been postulated that such responses may be driven by antioxidants and oxidative stress [135]. Chronic stress on the colony has an important influence on colony dynamics [136], and if it surpasses a threshold, there is a risk of a destructive response, in this case, derived from the less efficient foraging of young bees [137]. This imbalance in the colony can drive the queen towards the limits of her egg-laying capacity [138], which may be accelerated by the interaction of several stressors. This behavior is similar to the Allee effect, which is related to the extinction of populations [139–141].

In terms of neonicotinoids, a significant increase in capped bee brood has been documented recently in small nucleus colonies exposed to low concentrations of chlothianidin over a period of 7 weeks [126]. It was reported that colonies needed to produce 1.57-fold more larvae to maintain a stable population when exposed to a concentration of 1 μg clothianidin/L syrup, and that this effort increases as does exposure. Alternatively, effects that compensate nosemosis C [59] were evident in colonies during phase 2 of the disease, but if this does not control the stress provoked by the parasite in phase 3, it can eventually lead to colony collapse. Under the conditions of our field study, *N. ceranae* infection developed similarly over time in the FB and IB populations in both apiaries, with a relative infection below 5% at the end of the exposure period. The infection in the colonies had evolved to phase 2 (replacement) of nosemosis C by the end of the summer season [59]. Similar behavior has been described in colonies affected by varroosis, which might fail to survive because of their effort to redress the disturbed homeostasis [81]. Nevertheless, the pressure exerted by *V. destructor* was sufficiently well-controlled in the present study through the mandatory autumn treatment. In fact, according to the Spanish Ministry of Agriculture's monitoring program, the prevalence of *V. destructor* was <1% in more than 80% of the colonies sampled in Castilla la Mancha in autumn 2014 and spring 2015 [142].

Immune defenses are costly to the individual, entering into a trade-off with other life-history traits such as reproduction, growth, and self-maintenance [134,143]. Suppression of the immune system may be favored by the interaction of several stressors, such as *N. ceranae* and TMX [77,144], and by high environmental temperatures [145], which could explain the prevalence of the BQCV virus in the colonies throughout the study. Significant increases in DWV loads have been documented in honey bees exposed to clothianidin, the metabolite of TMX, which has been related to inhibition of dorsal-1a gene transcription [36]. However, this increment was not observed here, with an increase in DWV only apparent in the autumn. These results are consistent with previous data [146] and they may be related to potential routes of transmission other than the main one through *Varroa* mite [147,148]. Moreover, the high environmental temperature in the spring and summer of 2015, together with a possible increase in thermogenesis due to exposure to TMX, might influence the inhibition of DWV reproduction, in part due to the host synthesis of heat shock proteins [149–153] and possibly by exceeding the optimal growth temperature range of this virus [154]. Nevertheless, further research is needed to understand the influence of high temperatures on the complex relationships between different traits of honey bee colonies and their pathogens.

Hormetic effects may not be statistically significant at the field level, in part because of their low magnitude and in part because of inadequate replication in experiments [155]. However, the difference between statistical significance and the biological relevance of an effect are not necessarily linked. Significance is a statistical measure reflecting whether an observed effect is likely to have occurred by chance alone [156]. EPPO standards (170) mean that statistical analysis may not be feasible in field studies due to the inherent variability of the end-points assessed and the limitations on replication [82]. Based on a Cohen's d of 0.310, the lowest of our experimental results, it would be necessary to analyze 99 colonies/treatment to increase the statistical power up to 80%. Considering the possibility of establishing 8 hives/plot, this means a total of 12–13 plots per treatment. With this design, and according to the analysis carried out recently by the European Food Safety Authority [157], a significant effect between 7–8% could be detected. Few field studies in the literature follow such an exhaustive study design and they have generally failed to find deleterious effects at a colony level as a result of the use of neonicotinoids under good agricultural practices [158–163]. However, an increase in colony size is not always found as a result of neonicotinoid use, as observed here, suggesting that environmental conditions could play a key role in colony behavior.

5. Conclusions

Hormesis is common in arthropods and should be integrated into the dose response model as a continuum. However, the cost of the compensatory behavior adopted by the honey bee colonies over the long term is difficult to assess from an ecological point of view due to the complex relationships between the different traits affected. Further research is needed to understand the mechanisms underlining these compensation events, and the influence of environmental conditions on them, in order to define more appropriate management programs for this key pollinator.

Supplementary Materials: The following supporting information can be downloaded at: https://www.mdpi.com/article/10.3390/insects13040371/s1, Table S1: Characteristics of the soil at the experimental plots (mean ± sd, $n = 3$), Figure S1: Walter–Lieth diagrams for the period 2010–2021 (a), 2014 (b), 2015 (c), and 2016 (d), data taken from the Marchamalo meteorological station (40°40′ N 3°12′ W).

Author Contributions: Conceptualization, M.H., E.A.-P. and R.M.-H.; data curation, M.H. and E.A.-P.; formal analysis: E.A.-P.; funding acquisition, M.H. and J.B.; investigation, E.A.-P., A.V.G.-P., C.G.-V., J.A.L.-P., S.V., J.B., R.M.-H. and M.H.; methodology, E.A.-P., A.V.G.-P., C.G.-V., J.A.L.-P., S.V., J.B., R.M.-H. and M.H.; project administration, M.H. and J.B. resources, E.A.-P., A.V.G.-P., C.G.-V., J.A.L.-P., S.V., J.B., R.M.-H. and M.H; supervision, E.A.-P. and M.H.; validation, M.H., E.A.-P., R.M.-H., A.V.G.-P. and J.B. writing—original draft preparation, E.A.-P., A.V.G.-P., C.G.-V., J.A.L.-P., S.V., J.B., R.M.-H. and M.H.; writing—review and editing, E.A.-P., A.V.G.-P., C.G.-V., J.A.L.-P., S.V., J.B., R.M.-H. and M.H. All authors have read and agreed to the published version of the manuscript.

Funding: This research was funded by INIA-FEDER funds (RTA2013-00042-C10-03 and RTA2013-00042-C10-06).

Institutional Review Board Statement: The EU Directive 2010/63/EU on the protection of animals used for scientific purposes laid down the ethical framework for the use of animals in scientific experiments. The scope of this directive also includes specific invertebrate species (e.g., cephalopods) but not insects. Thus, according to European Legislation, no specific permits were required for the studies described.

Informed Consent Statement: Not applicable.

Data Availability Statement: The data presented in this study are available on reasonable request from the corresponding author.

Acknowledgments: The authors wish to thank V. Albendea, T. Corrales, M. Gajero, and C. Uceta at the honey bee pathology laboratory for their technical support, and L. Barrios at Departamento de Estadística, CTI, CSIC for the revision and recommendations.

Conflicts of Interest: The authors declare no conflict of interest. The funders had no role in the design of the study; in the collection, analyses or interpretation of data; in the writing of the manuscript, or in the decision to publish the results.

References

1. Brodschneider, R.; Gray, A.; Adjlane, N.; Ballis, A.; Brusbardis, V.; Charrière, J.-D.; Chlebo, R.; Coffey, M.F.; Dahle, B.; de Graaf, D.C.; et al. Multi-country loss rates of honey bee colonies during winter 2016/2017 from the COLOSS survey. *J. Apic. Res.* **2018**, *57*, 452–457. [CrossRef]
2. Brodschneider, R.; Gray, A.; van der Zee, R.; Adjlane, N.; Brusbardis, V.; Charrière, J.-D.; Chlebo, R.; Coffey, M.F.; Crailsheim, K.; Dahle, B.; et al. Preliminary analysis of loss rates of honey bee colonies during winter 2015/16 from the COLOSS survey. *J. Apic. Res.* **2016**, *55*, 375–378. [CrossRef]
3. Gray, A.; Adjlane, N.; Arab, A.; Ballis, A.; Brusbardis, V.; Charrière, J.-D.; Chlebo, R.; Coffey, M.F.; Cornelissen, B.; da Costa, C.A.; et al. Honey bee colony winter loss rates for 35 countries participating in the COLOSS survey for winter 2018–2019, and the effects of a new queen on the risk of colony winter loss. *J. Apic. Res.* **2020**, *59*, 744–751. [CrossRef]
4. Gray, A.; Brodschneider, R.; Adjlane, N.; Ballis, A.; Brusbardis, V.; Charrière, J.-D.; Chlebo, R.; Coffey, M.F.; Cornelissen, B.; Amaro da Costa, C.; et al. Loss rates of honey bee colonies during winter 2017/18 in 36 countries participating in the COLOSS survey, including effects of forage sources. *J. Apic. Res.* **2019**, *58*, 479–485. [CrossRef]
5. Lee, K.V.; Steinhauer, N.; Rennich, K.; Wilson, M.E.; Tarpy, D.R.; Caron, D.M.; Rose, R.; Delaplane, K.S.; Baylis, K.; Lengerich, E.J.; et al. A national survey of managed honey bee 2013–2014 annual colony losses in the USA. *Apidologie* **2015**, *46*, 292–305. [CrossRef]
6. Neumann, P.; Carreck, N.L. Honey bee colony losses. *J. Apic. Res.* **2010**, *49*, 1–6. [CrossRef]
7. Van der Zee, R.; Gray, A.; Pisa, L.; de Rijk, T. An Observational Study of Honey Bee Colony Winter Losses and Their Association with Varroa destructor, Neonicotinoids and Other Risk Factors. *PLoS ONE* **2015**, *10*, e0131611. [CrossRef]

8. Van der Zee, R.; Pisa, L.; Andonov, S.; Brodschneider, R.; Charrière, J.-D.; Chlebo, R.; Coffey, M.F.; Crailsheim, K.; Dahle, B.; Gajda, A.; et al. Managed honey bee colony losses in Canada, China, Europe, Israel and Turkey, for the winters of 2008–2009 and 2009–2010. *J. Apic. Res.* **2012**, *51*, 100–114. [CrossRef]
9. VanEngelsdorp, D.; Caron, D.; Hayes, J.; Underwood, R.; Henson, M.; Rennich, K.; Spleen, A.; Andree, M.; Snyder, R.; Lee, K.; et al. A national survey of managed honey bee 2010–11 winter colony losses in the USA: Results from the Bee Informed Partnership. *J. Apic. Res.* **2012**, *51*, 115–124. [CrossRef]
10. VanEngelsdorp, D.; Hayes, J., Jr.; Underwood, R.M.; Pettis, J. A Survey of Honey Bee Colony Losses in the U.S., Fall 2007 to Spring 2008. *PLoS ONE* **2009**, *3*, e4071. [CrossRef]
11. VanEngelsdorp, D.; Hayes, J.; Underwood, R.M.; Pettis, J.S. A survey of honey bee colony losses in the United States, Fall 2008 to Spring 2009. *J. Apic. Res.* **2010**, *49*, 7–14. [CrossRef]
12. Carreck, N.L. A beekeeper's perspective on the neonicotinoid ban. *Pest Manag. Sci.* **2017**, *73*, 1295–1298. [CrossRef] [PubMed]
13. Clermont, A.; Eickermann, M.; Kraus, F.; Georges, C.; Hoffmann, L.; Beyer, M. A survey on some factors potentially affecting losses of managed honey bee colonies in Luxembourg over the winters 2010/2011 and 2011/2012. *J. Apic. Res.* **2014**, *53*, 43–56. [CrossRef]
14. Goulson, D.; Nicholls, E.; Botías, C.; Rotheray, E.L. Bee declines driven by combined stress from parasites, pesticides, and lack of flowers. *Science* **2015**, *347*, 1255957. [CrossRef] [PubMed]
15. Hristov, P.; Shumkova, R.; Palova, N.; Neov, B. Factors Associated with Honey Bee Colony Losses: A Mini-Review. *Vet. Sci.* **2020**, *7*, 166. [CrossRef]
16. Johnson, R.M. Honey Bee Toxicology. *Annu. Rev. Entomol.* **2015**, *60*, 415–434. [CrossRef]
17. Potts, S.G.; Biesmeijer, J.C.; Kremen, C.; Neumann, P.; Schweiger, O.; Kunin, W.E. Global pollinator declines: Trends, impacts and drivers. *Trends Ecol. Evol.* **2010**, *25*, 345–353. [CrossRef]
18. Switanek, M.; Crailsheim, K.; Truhetz, H.; Brodschneider, R. Modelling seasonal effects of temperature and precipitation on honey bee winter mortality in a temperate climate. *Sci. Total Environ.* **2017**, *579*, 1581–1587. [CrossRef]
19. VanEngelsdorp, D.; Meixner, M.D. A historical review of managed honey bee populations in Europe and the United States and the factors that may affect them. *J. Invertebr. Pathol.* **2010**, *103*, S80–S95. [CrossRef]
20. Neov, B.; Georgieva, A.; Shumkova, R.; Radoslavov, G.; Hristov, P. Biotic and Abiotic Factors Associated with Colonies Mortalities of Managed Honey Bee (*Apis mellifera*). *Diversity* **2019**, *11*, 237. [CrossRef]
21. Yamamoto, I.a.C.; John, E. *Nicotinoid Insecticides and the Nicotinic Acetylcholine Receptor*; Springer: Tokyo, Japan, 1999.
22. Elbert, A.; Haas, M.; Springer, B.; Thielert, W.; Nauen, R. Applied aspects of neonicotinoid uses in crop protection. *Pest Manag. Sci.* **2008**, *64*, 1099–1105. [CrossRef]
23. Jeschke, P. Status and outlook for acaricide and insecticide discovery. *Pest Manag. Sci.* **2021**, *77*, 64–76. [CrossRef] [PubMed]
24. Jeschke, P.; Nauen, R. Neonicotinoids—From zero to hero in insecticide chemistry. *Pest Manag. Sci.* **2008**, *64*, 1084–1098. [CrossRef] [PubMed]
25. Jeschke, P.; Nauen, R.; Beck, M.E. Nicotinic Acetylcholine Receptor Agonists: A Milestone for Modern Crop Protection. *Angew. Chem. Int. Ed.* **2013**, *52*, 9464–9485. [CrossRef] [PubMed]
26. Matsuda, K.; Buckingham, S.D.; Kleier, D.; Rauh, J.J.; Grauso, M.; Sattelle, D.B. Neonicotinoids: Insecticides acting on insect nicotinic acetylcholine receptors. *Trends Pharmacol. Sci.* **2001**, *22*, 573–580. [CrossRef]
27. Matsuda, K.; Kanaoka, S.; Akamatsu, M.; Sattelle, D.B. Diverse Actions and Target-Site Selectivity of Neonicotinoids: Structural Insights. *Mol. Pharmacol.* **2009**, *76*, 1–10. [CrossRef]
28. Tomizawa, M.; Casida, J.E. Neonicotinoid Insecticide Toxicology: Mechanisms of Selective Action. *Annu. Rev. Pharmacol. Toxicol.* **2005**, *45*, 247–268. [CrossRef]
29. Tomizawa, M.; Lee, D.L.; Casida, J.E. Neonicotinoid Insecticides: Molecular Features Conferring Selectivity for Insect versus Mammalian Nicotinic Receptors. *J. Agric. Food Chem.* **2000**, *48*, 6016–6024. [CrossRef]
30. Bonmatin, J.M.; Giorio, C.; Girolami, V.; Goulson, D.; Kreutzweiser, D.P.; Krupke, C.; Liess, M.; Long, E.; Marzaro, M.; Mitchell, E.A.D.; et al. Environmental fate and exposure; Neonicotinoids and fipronil. *Environ. Sci. Pollut. Res.* **2015**, *22*, 35–67. [CrossRef]
31. Borsuah, J.F.; Messer, T.L.; Snow, D.D.; Comfort, S.D.; Mittelstet, A.R. Literature Review: Global Neonicotinoid Insecticide Occurrence in Aquatic Environments. *Water* **2020**, *12*, 3388. [CrossRef]
32. Wood, T.J.; Goulson, D. The environmental risks of neonicotinoid pesticides: A review of the evidence post 2013. *Environ. Sci. Pollut. Res.* **2017**, *24*, 17285–17325. [CrossRef] [PubMed]
33. Van der Sluijs, J.P.; Simon-Delso, N.; Goulson, D.; Maxim, L.; Bonmatin, J.-M.; Belzunces, L.P. Neonicotinoids, bee disorders and the sustainability of pollinator services. *Curr. Opin. Environ. Sustain.* **2013**, *5*, 293–305. [CrossRef]
34. Blacquière, T.; Smagghe, G.; van Gestel, C.A.M.; Mommaerts, V. Neonicotinoids in bees: A review on concentrations, side-effects and risk assessment. *Ecotoxicology* **2012**, *21*, 973–992. [CrossRef] [PubMed]
35. Decourtye, A.; Devillers, J. Ecotoxicity of neonicotinoid insecticides to bees. *Adv. Exp. Med. Biol.* **2010**, *683*, 85–95. [CrossRef] [PubMed]
36. Di Prisco, G.; Cavaliere, V.; Annoscia, D.; Varricchio, P.; Caprio, E.; Nazzi, F.; Gargiulo, G.; Pennacchio, F. Neonicotinoid clothianidin adversely affects insect immunity and promotes replication of a viral pathogen in honey bees. *Proc. Natl. Acad. Sci. USA* **2013**, *110*, 18466–18471. [CrossRef] [PubMed]

37. Goulson, D. Review: An overview of the environmental risks posed by neonicotinoid insecticides. *J. Appl. Ecol.* **2013**, *50*, 977–987. [CrossRef]
38. Hatjina, F.; Papaefthimiou, C.; Charistos, L.; Dogaroglu, T.; Bouga, M.; Emmanouil, C.; Arnold, G. Sublethal doses of imidacloprid decreased size of hypopharyngeal glands and respiratory rhythm of honeybees in vivo. *Apidologie* **2013**, *44*, 467–480. [CrossRef]
39. Henry, M.; Béguin, M.; Requier, F.; Rollin, O.; Odoux, J.-F.; Aupinel, P.; Aptel, J.; Tchamitchian, S.; Decourtye, A. A Common Pesticide Decreases Foraging Success and Survival in Honey Bees. *Science* **2012**, *336*, 348–350. [CrossRef]
40. Lundin, O.; Rundlöf, M.; Smith, H.G.; Fries, I.; Bommarco, R. Neonicotinoid Insecticides and Their Impacts on Bees: A Systematic Review of Research Approaches and Identification of Knowledge Gaps. *PLoS ONE* **2015**, *10*, e0136928. [CrossRef]
41. Schneider, C.W.; Tautz, J.; Grünewald, B.; Fuchs, S. RFID Tracking of Sublethal Effects of Two Neonicotinoid Insecticides on the Foraging Behavior of *Apis mellifera*. *PLoS ONE* **2012**, *7*, e30023. [CrossRef]
42. Tackenberg, M.C.; Giannoni-Guzmán, M.A.; Sanchez-Perez, E.; Doll, C.A.; Agosto-Rivera, J.L.; Broadie, K.; Moore, D.; McMahon, D.G. Neonicotinoids disrupt circadian rhythms and sleep in honey bees. *Sci. Rep.* **2020**, *10*, 17929. [CrossRef] [PubMed]
43. Tavares, D.A.; Dussaubat, C.; Kretzschmar, A.; Carvalho, S.M.; Silva-Zacarin, E.C.M.; Malaspina, O.; Bérail, G.; Brunet, J.-L.; Belzunces, L.P. Exposure of larvae to thiamethoxam affects the survival and physiology of the honey bee at post-embryonic stages. *Environ. Pollut.* **2017**, *229*, 386–393. [CrossRef] [PubMed]
44. Tavares, D.A.; Roat, T.C.; Carvalho, S.M.; Silva-Zacarin, E.C.M.; Malaspina, O. In vitro effects of thiamethoxam on larvae of Africanized honey bee *Apis mellifera* (Hymenoptera: Apidae). *Chemosphere* **2015**, *135*, 370–378. [CrossRef] [PubMed]
45. Tosi, S.; Nieh, J.C. A common neonicotinoid pesticide, thiamethoxam, alters honey bee activity, motor functions, and movement to light. *Sci. Rep.* **2017**, *7*, 15132. [CrossRef]
46. Whitehorn, P.R.; O'Connor, S.; Wackers, F.L.; Goulson, D. Neonicotinoid Pesticide Reduces Bumble Bee Colony Growth and Queen Production. *Science* **2012**, *336*, 351–352. [CrossRef]
47. Annoscia, D.; Di Prisco, G.; Becchimanzi, A.; Caprio, E.; Frizzera, D.; Linguadoca, A.; Nazzi, F.; Pennacchio, F. Neonicotinoid Clothianidin reduces honey bee immune response and contributes to Varroa mite proliferation. *Nat. Commun.* **2020**, *11*, 5887. [CrossRef]
48. Catae, A.F.; Roat, T.C.; De Oliveira, R.A.; Nocelli, R.C.; Malaspina, O. Cytotoxic effects of thiamethoxam in the midgut and malpighian tubules of Africanized *Apis mellifera* (Hymenoptera: Apidae). *Microsc. Res. Tech.* **2014**, *77*, 274–281. [CrossRef]
49. Gregorc, A.; Silva-Zacarin, E.C.M.; Carvalho, S.M.; Kramberger, D.; Teixeira, E.W.; Malaspina, O. Effects of *Nosema ceranae* and thiametoxam in *Apis mellifera*: A comparative study in Africanized and Carniolan honey bees. *Chemosphere* **2016**, *147*, 328–336. [CrossRef]
50. Oliveira, R.A.; Roat, T.C.; Carvalho, S.M.; Malaspina, O. Side-effects of thiamethoxam on the brain andmidgut of the africanized honeybee *Apis mellifera* (Hymenopptera: Apidae). *Environ. Toxicol.* **2014**, *29*, 1122–1133. [CrossRef]
51. Shi, T.-F.; Wang, Y.-F.; Liu, F.; Qi, L.; Yu, L.-S. Sublethal Effects of the Neonicotinoid Insecticide Thiamethoxam on the Transcriptome of the Honey Bees (Hymenoptera: Apidae). *J. Econ. Entomol.* **2017**, *110*, 2283–2289. [CrossRef]
52. Paris, L.; Peghaire, E.; Moné, A.; Diogon, M.; Debroas, D.; Delbac, F.; El Alaoui, H. Honeybee gut microbiota dysbiosis in pesticide/parasite co-exposures is mainly induced by *Nosema ceranae*. *J. Invertebr. Pathol.* **2020**, *172*, 107348. [CrossRef] [PubMed]
53. Higes, M.; Martín-Hernández, R.; Meana, A. *Nosema ceranae* in Europe: An emergent type C nosemosis. *Apidologie* **2010**, *41*, 375–392. [CrossRef]
54. Martín-Hernández, R.; Bartolomé, C.; Chejanovsky, N.; Le Conte, Y.; Dalmon, A.; Dussaubat, C.; García-Palencia, P.; Meana, A.; Pinto, M.A.; Soroker, V.; et al. *Nosema ceranae* in *Apis mellifera*: A 12 years postdetection perspective. *Environ. Microbiol.* **2018**, *20*, 1302–1329. [CrossRef] [PubMed]
55. García-Palencia, P.; Martín-Hernández, R.; González-Porto, A.-V.; Marin, P.; Meana, A.; Higes, M. Natural infection by *Nosema ceranae* causes similar lesions as in experimentally infected caged-worker honey bees (*Apis mellifera*). *J. Apic. Res.* **2010**, *49*, 278–283. [CrossRef]
56. Higes, M.; García-Palencia, P.; Martín-Hernández, R.; Meana, A. Experimental infection of *Apis mellifera* honeybees with *Nosema ceranae* (Microsporidia). *J. Invertebr. Pathol.* **2007**, *94*, 211–217. [CrossRef]
57. Goblirsch, M. *Nosema ceranae* disease of the honey bee (*Apis mellifera*). *Apidologie* **2018**, *49*, 131–150. [CrossRef]
58. Paris, L.; El Alaoui, H.; Delbac, F.; Diogon, M. Effects of the gut parasite *Nosema ceranae* on honey bee physiology and behavior. *Curr. Opin. Insect Sci.* **2018**, *26*, 149–154. [CrossRef]
59. Higes, M.; Martín-Hernández, R.; Botías, C.; Bailón, E.G.; González-Porto, A.V.; Barrios, L.; del Nozal, M.J.; Bernal, J.L.; Jiménez, J.J.; Palencia, P.G.; et al. How natural infection by *Nosema ceranae* causes honeybee colony collapse. *Environ. Microbiol.* **2008**, *10*, 2659–2669. [CrossRef]
60. Dussaubat, C.; Maisonnasse, A.; Alaux, C.; Tchamitchan, S.; Brunet, J.-L.; Plettner, E.; Belzunces, L.P.; Le Conte, Y. *Nosema* spp. Infection Alters Pheromone Production in Honey Bees (*Apis mellifera*). *J. Chem. Ecol.* **2010**, *36*, 522–525. [CrossRef]
61. Goblirsch, M.; Huang, Z.Y.; Spivak, M. Physiological and Behavioral Changes in Honey Bees (*Apis mellifera*) Induced by *Nosema ceranae* Infection. *PLoS ONE* **2013**, *8*, e58165. [CrossRef]
62. Ares, A.M.; Nozal, M.J.; Bernal, J.L.; Martín-Hernández, R.; Higes, M.; Bernal, J. Liquid chromatography coupled to ion trap-tandem mass spectrometry to evaluate juvenile hormone III levels in bee hemolymph from *Nosema* spp. infected colonies. *J. Chromatogr. B* **2012**, *899*, 146–153. [CrossRef] [PubMed]

63. Vilarem, C.; Piou, V.; Vogelweith, F.; Vétillard, A. *Varroa destructor* from the Laboratory to the Field: Control, Biocontrol and IPM Perspectives—A Review. *Insects* **2021**, *12*, 800. [CrossRef] [PubMed]
64. Traynor, K.S.; Mondet, F.; de Miranda, J.R.; Techer, M.; Kowallik, V.; Oddie, M.A.Y.; Chantawannakul, P.; McAfee, A. *Varroa destructor*: A Complex Parasite, Crippling Honey Bees Worldwide. *Trends Parasitol.* **2020**, *36*, 592–606. [CrossRef] [PubMed]
65. Rosenkranz, P.; Aumeier, P.; Ziegelmann, B. Biology and control of *Varroa destructor*. *J. Invertebr. Pathol.* **2010**, *103*, S96–S119. [CrossRef]
66. Ramsey, S.D.; Ochoa, R.; Bauchan, G.; Gulbronson, C.; Mowery, J.D.; Cohen, A.; Lim, D.; Joklik, J.; Cicero, J.M.; Ellis, J.D.; et al. *Varroa destructor* feeds primarily on honey bee fat body tissue and not hemolymph. *Proc. Natl. Acad. Sci. USA* **2019**, *116*, 1792–1801. [CrossRef]
67. Weinberg, K.P.; Madel, G. The influence of the mite *Varroa jacobsoni* Oud. on the protein concentration and the haemolymph volume of the brood of worker bees and drones of the honey bee *Apis mellifera* L. *Apidologie* **1985**, *16*, 421–436. [CrossRef]
68. Sammataro, D.; Gerson, U.; Needham, G. Parasitic Mites of Honey Bees: Life History, Implications, and Impact. *Annu. Rev. Entomol.* **2000**, *45*, 519–548. [CrossRef]
69. Yang, X.; Cox-Foster, D.L. Impact of an ectoparasite on the immunity and pathology of an invertebrate: Evidence for host immunosuppression and viral amplification. *Proc. Natl. Acad. Sci. USA* **2005**, *102*, 7470–7475. [CrossRef]
70. Kralj, J.; Brockmann, A.; Fuchs, S.; Tautz, J. The parasitic mite *Varroa destructor* affects non-associative learning in honey bee foragers, *Apis mellifera* L. *J. Comp. Physiol. A* **2007**, *193*, 363–370. [CrossRef]
71. Kralj, J.; Fuchs, S. Parasitic *Varroa destructor* mites influence flight duration and homing ability of infested *Apis mellifera* foragers. *Apidologie* **2006**, *37*, 577–587. [CrossRef]
72. Coulon, M.; Dalmon, A.; Di Prisco, G.; Prado, A.; Arban, F.; Dubois, E.; Ribière-Chabert, M.; Alaux, C.; Thiéry, R.; Le Conte, Y. Interactions Between Thiamethoxam and Deformed Wing Virus Can Drastically Impair Flight Behavior of Honey Bees. *Front. Microbiol.* **2020**, *11*, 722. [CrossRef] [PubMed]
73. Alaux, C.; Brunet, J.-L.; Dussaubat, C.; Mondet, F.; Tchamitchan, S.; Cousin, M.; Brillard, J.; Baldy, A.; Belzunces, L.P.; Le Conte, Y. Interactions between *Nosema* microspores and a neonicotinoid weaken honeybees (*Apis mellifera*). *Environ. Microbiol.* **2010**, *12*, 774–782. [CrossRef] [PubMed]
74. Pettis, J.S.; vanEngelsdorp, D.; Johnson, J.; Dively, G. Pesticide exposure in honey bees results in increased levels of the gut pathogen *Nosema*. *Naturwissenschaften* **2012**, *99*, 153–158. [CrossRef] [PubMed]
75. Doublet, V.; Labarussias, M.; de Miranda, J.R.; Moritz, R.F.A.; Paxton, R.J. Bees under stress: Sublethal doses of a neonicotinoid pesticide and pathogens interact to elevate honey bee mortality across the life cycle. *Environ. Microbiol.* **2015**, *17*, 969–983. [CrossRef]
76. Vidau, C.; Diogon, M.; Aufauvre, J.; Fontbonne, R.; Viguès, B.; Brunet, J.-L.; Texier, C.; Biron, D.G.; Blot, N.; El Alaoui, H.; et al. Exposure to Sublethal Doses of Fipronil and Thiacloprid Highly Increases Mortality of Honeybees Previously Infected by *Nosema ceranae*. *PLoS ONE* **2011**, *6*, e21550. [CrossRef]
77. Tesovnik, T.; Zorc, M.; Ristanić, M.; Glavinić, U.; Stevanović, J.; Narat, M.; Stanimirović, Z. Exposure of honey bee larvae to thiamethoxam and its interaction with *Nosema ceranae* infection in adult honey bees. *Environ. Pollut.* **2020**, *256*, 113443. [CrossRef]
78. Dussaubat, C.; Brunet, J.-L.; Higes, M.; Colbourne, J.K.; Lopez, J.; Choi, J.-H.; Martín-Hernández, R.; Botías, C.; Cousin, M.; McDonnell, C.; et al. Gut pathology and responses to the microsporidium *Nosema ceranae* in the honey bee *Apis mellifera*. *PLoS ONE* **2012**, *7*, e37017. [CrossRef]
79. Morfin, N.; Goodwin, P.H.; Guzman-Novoa, E. The combined effects of *Varroa destructor* parasitism and exposure to neonicotinoids affects honey bee (*Apis mellifera* L.) Memory and Gene Expression. *Biology* **2020**, *9*, 237. [CrossRef]
80. Odemer, R. *Effects of Chronic Pesticide and Pathogen Exposure on Honey Bee (Apis mellifera L.) Health at the Colony Level*; Kommunikations, Informations und Medienzentrum der Universität Hohenheim: Stuttgart, Germany, 2018.
81. Van Dooremalen, C.; Cornelissen, B.; Poleij-Hok-Ahin, C.; Blacquière, T. Single and interactive effects of *Varroa destructor*, *Nosema* spp., and imidacloprid on honey bee colonies (*Apis mellifera*). *Ecosphere* **2018**, *9*, e02378. [CrossRef]
82. European and Mediterranean Plant Protection Organization. PP 1/170 (4): Side-effects on honeybees. *EPPO Bull.* **2010**, *40*, 313–319. [CrossRef]
83. Schulz, D.J.; Huang, Z.-Y.; Robinson, G.E. Effects of colony food shortage on behavioral development in honey bees. *Behav. Ecol. Sociobiol.* **1998**, *42*, 295–303. [CrossRef]
84. Higes, M.; Nozal, M.J.; Alvaro, A.; Barrios, L.; Meana, A.; Martín-Hernández, R.; Bernal, J.L.; Bernal, J. The stability and effectiveness of fumagillin in controlling *Nosema ceranae* (Microsporidia) infection in honey bees (*Apis mellifera*) under laboratory and field conditions. *Apidologie* **2011**, *42*, 364–377. [CrossRef]
85. Ministerio de Agricultura, Pesca y Alimentación. *Real Decreto 608/2006, de 19 de Mayo, por el que se Establece y Regula un Programa Nacional de Lucha y Control de las Enfermedades de las Abejas de la Miel*; Ministerio de Agricultura: Madrid, Spain, 2006.
86. International Soil Reference and Information Centre. *Procedures for Soil Analysis*, 6th ed.; van Reeuwijk, L., Ed.; ISRIC: Wageningen, The Netherlands, 2002.
87. Bouyoucos, G.J. A Recalibration of the Hydrometer Method for Making Mechanical Analysis of Soils1. *Agron. J.* **1951**, *43*, 434–438. [CrossRef]
88. Walter, H.; Lieth, H. *Klimadiagramm—Weltatlas*; VEB Gustav Fisher: Jena, Germany, 1967.

89. Rivas-Martinez, S. Worldwide Bioclimatic Classification System, 1996–2021. Available online: http://www.globalbioclimatics.org/default.htm (accessed on 7 July 2021).
90. Pérez, J.L.; Higes, M.; Suárez, M.; Llorente, J.; Meana, A. Easy ways to determine honey bee mortality using dead-bee traps. *J. Apic. Res.* **2001**, *40*, 25–28. [CrossRef]
91. Botías, C.; Martín-Hernández, R.; Días, J.; García-Palencia, P.; Matabuena, M.; Juarranz, Á.; Barrios, L.; Meana, A.; Nanetti, A.; Higes, M. The effect of induced queen replacement on *Nosema* spp. infection in honey bee (*Apis mellifera iberiensis*) colonies. *Environ. Microbiol.* **2012**, *14*, 845–859. [CrossRef] [PubMed]
92. Bernal, J.; Garrido-Bailón, E.; Del Nozal, M.J.; González-Porto, A.V.; Martín-Hernández, R.; Diego, J.C.; Jiménez, J.J.; Bernal, J.L.; Higes, M. Overview of pesticide residues in stored pollen and their potential effect on bee colony (*Apis mellifera*) losses in Spain. *J. Econ. Entomol.* **2010**, *103*, 1964–1971. [CrossRef] [PubMed]
93. Cepero, A.; Ravoet, J.; Gómez-Moracho, T.; Bernal, J.L.; Del Nozal, M.J.; Bartolomé, C.; Maside, X.; Meana, A.; González-Porto, A.V.; de Graaf, D.C.; et al. Holistic screening of collapsing honey bee colonies in Spain: A case study. *BMC Res. Notes* **2014**, *7*, 649. [CrossRef]
94. Higes, M.; Martín-Hernández, R.; Garrido-Bailón, E.; González-Porto, A.V.; García-Palencia, P.; Meana, A.; Del Nozal, M.J.; Mayo, R.; Bernal, J.L. Honeybee colony collapse due to *Nosema ceranae* in professional apiaries. *Environ. Microbiol. Rep.* **2009**, *1*, 110–113. [CrossRef]
95. Hidalgo, M.I.B.; Pacheco, M.L.; Pacheco, J. Origen floral de las cargas de polen recogidas por *Apis mellifera* L. en Álora (Málaga, España). *Acta Bot. Malacit.* **1990**, *15*, 33–44. [CrossRef]
96. Louveaux, J.; Maurizio, A.; Vorwohl, G. Methods of Melissopalynology. *Bee World* **1978**, *59*, 139–157. [CrossRef]
97. Erdtman, G. Morphology—Taxonomy—Ecology. An Introduction to the Study of Pollen Grains and Spores. In *Handbook of Palynology*; Munksgaard: Copenhagen, Denmark, 1969.
98. Ohe, W.V.D.; Oddo, L.P.; Piana, M.L.; Morlot, M.; Martin, P. Harmonized methods of melissopalynology. *Apidologie* **2004**, *35*, S18–S25.
99. Faegri, K.; Iversen, J. *Textbook of Pollen Analysis*, 4th ed.; John Wiley & Sons: Chichester, UK, 1989.
100. Valdés, B.; Fernández, I.D.M.J. *Atlas Polínico de Andalucía Occidental*; Instituto de Desarrollo Regional de la Universidad de Sevilla y Excma, Diputación de Cádiz: Sevilla, Spain, 1987.
101. Valverde, S.; Ares, A.M.; Bernal, J.L.; Nozal, M.J.; Bernal, J. Effect of the Storage Conditions (Light and Temperature) on the Detection of Thiamethoxam and Clothianidin Content in Rapeseeds by LC-DAD. *Food Anal. Methods* **2018**, *11*, 161–169. [CrossRef]
102. Valverde, S.; Bernal, J.L.; Martín, M.T.; Nozal, M.J.; Bernal, J. Fast determination of neonicotinoid insecticides in bee pollen using QuEChERS and ultra-high performance liquid chromatography coupled to quadrupole time-of-flight mass spectrometry. *Electrophoresis* **2016**, *37*, 2470–2477. [CrossRef] [PubMed]
103. Valverde, S.; Ibáñez, M.; Bernal, J.L.; Nozal, M.J.; Hernández, F.; Bernal, J. Development and validation of ultra high performance-liquid chromatography–tandem mass spectrometry based methods for the determination of neonicotinoid insecticides in honey. *Food Chem.* **2018**, *266*, 215–222. [CrossRef]
104. Jara, L.; Ruiz, C.; Martín-Hernández, R.; Muñoz, I.; Higes, M.; Serrano, J.; De la Rúa, P. The Effect of Migratory Beekeeping on the Infestation Rate of Parasites in Honey Bee (*Apis mellifera*) Colonies and on Their Genetic Variability. *Microorganisms* **2021**, *9*, 22. [CrossRef]
105. Martín-Hernández, R.; Botías, C.; Bailón, E.G.; Martínez-Salvador, A.; Prieto, L.; Meana, A.; Higes, M. Microsporidia infecting *Apis mellifera*: Coexistence or competition. Is *Nosema ceranae* replacing *Nosema apis*? *Environ. Microbiol.* **2012**, *14*, 2127–2138. [CrossRef]
106. Chantawannakul, P.; Ward, L.; Boonham, N.; Brown, M. A scientific note on the detection of honeybee viruses using real-time PCR (TaqMan) in *Varroa* mites collected from a Thai honeybee (*Apis mellifera*) apiary. *J. Invertebr. Pathol.* **2006**, *91*, 69–73. [CrossRef]
107. Francis, R.; Kryger, P. Single Assay Detection of Acute Bee Paralysis Virus, Kashmir Bee Virus and Israeli Acute Paralysis Virus. *J. Apic. Sci.* **2012**, *56*, 137–146. [CrossRef]
108. Medrzycki, P.; Giffard, H.; Aupinel, P.; Belzunces, L.P.; Chauzat, M.-P.; Claßen, C.; Colin, M.E.; Dupont, T.; Girolami, V.; Johnson, R.; et al. Standard methods for toxicology research in *Apis mellifera*. *J. Apic. Res.* **2013**, *52*, 1–60. [CrossRef]
109. Pirk, C.W.W.; de Miranda, J.R.; Kramer, M.; Murray, T.E.; Nazzi, F.; Shutler, D.; van der Steen, J.J.M.; van Dooremalen, C. Statistical guidelines for *Apis mellifera* research. *J. Apic. Res.* **2013**, *52*, 1–24. [CrossRef]
110. Divine, G.W.; Norton, H.J.; Barón, A.E.; Juarez-Colunga, E. The Wilcoxon–Mann–Whitney Procedure Fails as a Test of Medians. *Am. Stat.* **2018**, *72*, 278–286. [CrossRef]
111. Fritz, C.O.; Morris, P.E.; Richler, J.J. Effect size estimates: Current use, calculations, and interpretation. *J. Exp. Psychol. Gen.* **2012**, *141*, 2–18. [CrossRef] [PubMed]
112. Zaiontz, C. Real Statistics Using Excel. Available online: www.real-statistics.com (accessed on 15 March 2021).
113. Lenhard, W.; Lenhard, A. Calculation of Effect Sizes. Available online: https://www.psychometrica.de/effect_size.html (accessed on 6 June 2021).
114. Faul, F.; Erdfelder, E.; Lang, A.-G.; Buchner, A. G*Power 3: A flexible statistical power analysis program for the social, behavioral, and biomedical sciences. *Behav. Res. Methods* **2007**, *39*, 175–191. [CrossRef] [PubMed]
115. Friedman, M. The Use of Ranks to Avoid the Assumption of Normality Implicit in the Analysis of Variance. *J. Am. Stat. Assoc.* **1937**, *32*, 675–701. [CrossRef]

116. Instituto Geológico y Minero de España. *IGME_Geode_50 (MapServer): Continuous Digital Geological Map of Spain, Scale 1:50.000*; Instituto Geológico y Minero de España: Madrid, Spain, 2014.
117. World Reference Base, I.W.G. *World Reference Base for Soil Resources 2014, Update 2015 International Soil Classification System for Naming Soils and Creating Legends for Soil Maps*; FAO: Rome, Italy, 2015; Volume 106.
118. Rivas-Martinez, S. *Memoria del Mapa de Series de Vegetación de España*; Serie Técnica; Ministerio de Agricultura, Pesca y Alimentación; I.C.O.N.A: Madrid, Spain, 1987.
119. Nedić, N.; Mačukanović-Jocić, M.; Rančić, D.; Rørslett, B.; Šoštarić, I.; Stevanović, Z.D.; Mladenović, M. Melliferous potential of *Brassica napus* L. subsp. napus (Cruciferae). *Arthropod-Plant Interact.* **2013**, *7*, 323–333. [CrossRef]
120. Talavera, S.; Gibbs, P.E.; Herrera, J. Reproductive biology of *Cistus ladanifer* (Cistaceae). *Plant Syst. Evol.* **1993**, *186*, 123–134. [CrossRef]
121. Ortiz, P.L. The Cistaceae as food resources for honey bees in SW Spain. *J. Apic. Res.* **1994**, *33*, 136–144. [CrossRef]
122. Atsalakis, E.; Chinou, I.; Makropoulou, M.; Karabournioti, S.; Graikou, K. Evaluation of Phenolic Compounds in *Cistus creticus* Bee Pollen from Greece. Antioxidant and Antimicrobial Properties. *Nat. Prod. Commun.* **2017**, *12*, 1934578X1701201141. [CrossRef]
123. Papaefthimiou, D.; Papanikolaou, A.; Falara, V.; Givanoudi, S.; Kostas, S.; Kanellis, A.K. Genus Cistus: A model for exploring labdane-type diterpenes' biosynthesis and a natural source of high value products with biological, aromatic, and pharmacological properties. *Front. Chem.* **2014**, *2*, 35. [CrossRef]
124. Sponsler, D.B.; Johnson, R.M. Mechanistic modeling of pesticide exposure: The missing keystone of honey bee toxicology. *Environ. Toxicol. Chem.* **2017**, *36*, 871–881. [CrossRef]
125. Huang, Z.-Y.; Robinson, G.E. Regulation of honey bee division of labor by colony age demography. *Behav. Ecol. Sociobiol.* **1996**, *39*, 147–158. [CrossRef]
126. Schott, M.; Sandmann, M.; Cresswell, J.E.; Becher, M.A.; Eichner, G.; Brandt, D.T.; Halitschke, R.; Krueger, S.; Morlock, G.; Düring, R.-A.; et al. Honeybee colonies compensate for pesticide-induced effects on royal jelly composition and brood survival with increased brood production. *Sci. Rep.* **2021**, *11*, 62. [CrossRef]
127. Calabrese, E.J.; Baldwin, L.A. The Hormetic Dose-Response Model Is More Common than the Threshold Model in Toxicology. *Toxicol. Sci.* **2003**, *71*, 246–250. [CrossRef]
128. Luckey, T.D. Insecticide Hormoligosis. *J. Econ. Entomol.* **1968**, *61*, 7–12. [CrossRef] [PubMed]
129. Southam, C.M.; Ehrlich, J. Effects of Extract of western red-cedar heartwood on certain wood-decaying fungi in culture. *Phytopathology* **1943**, *33*, 517–524.
130. Kendig, E.L.; Le, H.H.; Belcher, S.M. Defining hormesis: Evaluation of a complex concentration response phenomenon. *Int. J. Toxicol.* **2010**, *29*, 235–246. [CrossRef] [PubMed]
131. Cohen, E. Pesticide-mediated homeostatic modulation in arthropods. *Pestic. Biochem. Physiol.* **2006**, *85*, 21–27. [CrossRef]
132. Guedes, R.N.C.; Cutler, G.C. Insecticide-induced hormesis and arthropod pest management. *Pest Manag. Sci.* **2014**, *70*, 690–697. [CrossRef]
133. Calabrese, E.J.; Blain, R. The occurrence of hormetic dose responses in the toxicological literature, the hormesis database: An overview. *Toxicol. Appl. Pharm.* **2005**, *202*, 289–301. [CrossRef]
134. Costantini, D.; Metcalfe, N.B.; Monaghan, P. Ecological processes in a hormetic framework. *Ecol. Lett.* **2010**, *13*, 1435–1447. [CrossRef]
135. Berry, R.; López-Martínez, G. A dose of experimental hormesis: When mild stress protects and improves animal performance. *Comp. Biochem. Physiol. Part A Mol. Integr. Physiol.* **2020**, *242*, 110658. [CrossRef] [PubMed]
136. Bryden, J.; Gill, R.J.; Mitton, R.A.A.; Raine, N.E.; Jansen, V.A.A. Chronic sublethal stress causes bee colony failure. *Ecol. Lett.* **2013**, *16*, 1463–1469. [CrossRef] [PubMed]
137. Perry, C.J.; Søvik, E.; Myerscough, M.R.; Barron, A.B. Rapid behavioral maturation accelerates failure of stressed honey bee colonies. *Proc. Natl. Acad. Sci. USA* **2015**, *112*, 3427–3432. [CrossRef] [PubMed]
138. Higes, M.; Meana, A.; Bartolomé, C.; Botías, C.; Martín-Hernández, R. *Nosema ceranae* (Microsporidia), a controversial 21st century honey bee pathogen. *Environ. Microbiol. Rep.* **2013**, *5*, 17–29. [CrossRef]
139. Saitanis, C.J.; Agathokleous, E. Stress response and population dynamics: Is Allee effect hormesis? *Sci. Total Environ.* **2019**, *682*, 623–628. [CrossRef]
140. Dennis, B.; Kemp, W.P. How Hives Collapse: Allee Effects, Ecological Resilience, and the Honey Bee. *PLoS ONE* **2016**, *11*, e0150055. [CrossRef]
141. Courchamp, F.; Clutton-Brock, T.; Grenfell, B. Inverse density dependence and the Allee effect. *Trends Ecol. Evol.* **1999**, *14*, 405–410. [CrossRef]
142. Ministerio de Agricultura, Pesca y Alimentación. Informes de Resultados del Programa de Vigilancia Sobre la Pérdida de Colonias de Abejas. Available online: https://www.mapa.gob.es/es/ganaderia/temas/sanidad-animal-higiene-ganadera/sanidad-animal/enfermedades/otras-enfermedades-abejas/otras-enf-abejas.aspx (accessed on 11 November 2021).
143. Sadd, B.M.; Schmid-Hempel, P. Perspective: Principles of ecological immunology. *Evol. Appl.* **2009**, *2*, 113–121. [CrossRef]
144. Liu, Z.; Li, S.; Li, H. Interactions between sublethal doses of thiamethoxam and *Nosema ceranae* in the honey bee, *Apis mellifera*. *J. Apic. Res.* **2021**, *60*, 717–725. [CrossRef]
145. McKinstry, M.; Chung, C.; Truong, H.; Johnston, B.A.; Snow, J.W. The heat shock response and humoral immune response are mutually antagonistic in honey bees. *Sci. Rep.* **2017**, *7*, 8850. [CrossRef]

146. Straub, L.; Williams, G.R.; Vidondo, B.; Khongphinitbunjong, K.; Retschnig, G.; Schneeberger, A.; Chantawannakul, P.; Dietemann, V.; Neumann, P. Neonicotinoids and ectoparasitic mites synergistically impact honeybees. *Sci. Rep.* **2019**, *9*, 8159. [CrossRef] [PubMed]
147. Locke, B.; Semberg, E.; Forsgren, E.; de Miranda, J.R. Persistence of subclinical deformed wing virus infections in honeybees following *Varroa* mite removal and a bee population turnover. *PLoS ONE* **2017**, *12*, e0180910. [CrossRef] [PubMed]
148. Möckel, N.; Gisder, S.; Genersch, E. Horizontal transmission of deformed wing virus: Pathological consequences in adult bees (*Apis mellifera*) depend on the transmission route. *J. Gen. Virol.* **2011**, *92*, 370–377. [CrossRef] [PubMed]
149. Potts, R.; Clarke, R.M.; Oldfield, S.E.; Wood, L.K.; de Ibarra, N.H.; Cresswell, J.E. The effect of dietary neonicotinoid pesticides on non-flight thermogenesis in worker bumble bees (*Bombus terrestris*). *J. Insect Physiol.* **2018**, *104*, 33–39. [CrossRef] [PubMed]
150. Tosi, S.; Démares, F.J.; Nicolson, S.W.; Medrzycki, P.; Pirk, C.W.W.; Human, H. Effects of a neonicotinoid pesticide on thermoregulation of African honey bees (*Apis mellifera scutellata*). *J. Insect Physiol.* **2016**, *93*, 56–63. [CrossRef] [PubMed]
151. Bordier, C.; Dechatre, H.; Suchail, S.; Peruzzi, M.; Soubeyrand, S.; Pioz, M.; Pélissier, M.; Crauser, D.; Conte, Y.L.; Alaux, C. Colony adaptive response to simulated heat waves and consequences at the individual level in honeybees (*Apis mellifera*). *Sci. Rep.* **2017**, *7*, 3760. [CrossRef] [PubMed]
152. Dalmon, A.; Peruzzi, M.; Le Conte, Y.; Alaux, C.; Pioz, M. Temperature-driven changes in viral loads in the honey bee *Apis mellifera*. *J. Invertebr. Pathol.* **2019**, *160*, 87–94. [CrossRef]
153. McMenamin, A.J.; Daughenbaugh, K.F.; Flenniken, M.L. The Heat Shock Response in the Western Honey Bee (*Apis mellifera*) is Antiviral. *Viruses* **2020**, *12*, 245. [CrossRef]
154. Ryan, S.J.; Carlson, C.J.; Mordecai, E.A.; Johnson, L.R. Global expansion and redistribution of Aedes-borne virus transmission risk with climate change. *PLoS Negl. Trop. Dis.* **2019**, *13*, e0007213. [CrossRef]
155. Cutler, G.C.; Rix, R.R. Can poisons stimulate bees? Appreciating the potential of hormesis in bee–pesticide research. *Pest Manag. Sci.* **2015**, *71*, 1368–1370. [CrossRef]
156. EFSA Scientific Committee. Statistical Significance and Biological Relevance. *EFSA J.* **2011**, *9*, 2372. [CrossRef]
157. Authority, E.F.S.; Ippolito, A.; Focks, A.; Rundlöf, M.; Arce, A.; Marchesi, M.; Neri, F.M.; Rortais, A.; Szentes, C.; Auteri, D. Analysis of background variability of honey bee colony size. *EFSA Supporting Publ.* **2021**, *18*, 6518E. [CrossRef]
158. Cutler, G.C.; Scott-Dupree, C.D.; Sultan, M.; McFarlane, A.D.; Brewer, L. A large-scale field study examining effects of exposure to clothianidin seed-treated canola on honey bee colony health, development, and overwintering success. *PeerJ* **2014**, *2*, e652. [CrossRef] [PubMed]
159. Flores, J.M.; Gámiz, V.; Gil-Lebrero, S.; Rodríguez, I.; Navas, F.J.; García-Valcárcel, A.I.; Cutillas, V.; Fernández-Alba, A.R.; Hernando, M.D. A three-year large scale study on the risk of honey bee colony exposure to blooming sunflowers grown from seeds treated with thiamethoxam and clothianidin neonicotinoids. *Chemosphere* **2021**, *262*, 127735. [CrossRef] [PubMed]
160. Hernando, M.D.; Gámiz, V.; Gil-Lebrero, S.; Rodríguez, I.; García-Valcárcel, A.I.; Cutillas, V.; Fernández-Alba, A.R.; Flores, J.M. Viability of honeybee colonies exposed to sunflowers grown from seeds treated with the neonicotinoids thiamethoxam and clothianidin. *Chemosphere* **2018**, *202*, 609–617. [CrossRef]
161. Rolke, D.; Fuchs, S.; Grünewald, B.; Gao, Z.; Blenau, W. Large-scale monitoring of effects of clothianidin-dressed oilseed rape seeds on pollinating insects in Northern Germany: Effects on honey bees (*Apis mellifera*). *Ecotoxicology* **2016**, *25*, 1648–1665. [CrossRef]
162. Woodcock, B.A.; Bullock, J.M.; Shore, R.F.; Heard, M.S.; Pereira, M.G.; Redhead, J.; Ridding, L.; Dean, H.; Sleep, D.; Henrys, P.; et al. Country-specific effects of neonicotinoid pesticides on honey bees and wild bees. *Science* **2017**, *356*, 1393–1395. [CrossRef]
163. Osterman, J.; Wintermantel, D.; Locke, B.; Jonsson, O.; Semberg, E.; Onorati, P.; Forsgren, E.; Rosenkranz, P.; Rahbek-Pedersen, T.; Bommarco, R.; et al. Clothianidin seed-treatment has no detectable negative impact on honeybee colonies and their pathogens. *Nat. Commun.* **2019**, *10*, 692. [CrossRef]

insects

MDPI

Article

Screening of Honey Bee Pathogens in the Czech Republic and Their Prevalence in Various Habitats

Petr Mráz [1,*], Marian Hýbl [1], Marek Kopecký [1], Andrea Bohatá [1], Irena Hoštičková [1], Jan Šipoš [2], Kateřina Vočadlová [1] and Vladislav Čurn [1]

[1] Faculty of Agriculture, University of South Bohemia in Ceske Budejovice, Studentska 1668, 370 05 Ceske Budejovice, Czech Republic; mario.eko@seznam.cz (M.H.); mkopecky@zf.jcu.cz (M.K.); bohata@zf.jcu.cz (A.B.); jelini00@zf.jcu.cz (I.H.); katerina.vocadlova@gmail.com (K.V.); curn@zf.jcu.cz (V.Č.)

[2] Faculty of Agronomy, Mendel University in Brno, Zemedelska 1, 613 00 Brno, Czech Republic; jan.sipos@mendelu.cz

* Correspondence: mrazpe01@zf.jcu.cz

Simple Summary: Worldwide, mass losses of honey bee colonies are being observed more frequently in recent times. Except for the overuse of pesticides, one of the main reasons for high honey bee colony collapse is diseases. For this reason, nationwide screening of common pathogens involving viruses, bacterial, fungal, and protozoa pathogens was performed in three different types of habitat including agroecosystems, towns, and national parks. The most frequent eukaryotic pathogens were Trypanosomatids and *N. ceranae* and in the case of viruses DWV-A and ABPV. In addition, the association between the occurrence of particular pathogens and winter colony losses was found. Although the differences in mortality between individual habitats were not significant, results of this study suggest a significant correlation between DWV-B and DWC-C occurrence and mortality of bee colonies, despite their relatively low occurrence.

Citation: Mráz, P.; Hýbl, M.; Kopecký, M.; Bohatá, A.; Hoštičková, I.; Šipoš, J.; Vočadlová, K.; Čurn, V. Screening of Honey Bee Pathogens in the Czech Republic and Their Prevalence in Various Habitats. *Insects* **2021**, *12*, 1051. https://doi.org/10.3390/insects12121051

Academic Editors: Ivana Tlak Gajger, Franco Mutinelli and Cristina Botías

Received: 1 November 2021
Accepted: 23 November 2021
Published: 24 November 2021

Abstract: Western honey bee (*Apis mellifera*) is one of the most important pollinators in the world. Thus, a recent honey bee health decline and frequent honey bee mass losses have drawn attention and concern. Honey bee fitness is primarily reduced by pathogens, parasites, and viral load, exposure to pesticides and their residues, and inadequate nutrition from both the quality and amount of food resources. This study evaluated the prevalence of the most common honey bee pathogens and viruses in different habitats across the Czech Republic. The agroecosystems, urban ecosystems, and national park were chosen for sampling from 250 colonies in 50 apiaries. Surprisingly, the most prevalent honey bee pathogens belong to the family Trypanosomatidae including *Lotmaria passim* and *Crithidia mellificae*. As expected, the most prevalent viruses were DWV, followed by ABPV. Additionally, the occurrence of DWV-B and DWV-C were correlated with honey bee colony mortality. From the habitat point of view, most pathogens occurred in the town habitat, less in the agroecosystem and least in the national park. The opposite trend was observed in the occurrence of viruses. However, the prevalence of viruses was not affected by habitat.

Keywords: *Apis mellifera*; deformed wing virus; screening; trypanosomatids

1. Introduction

The western honey bee (*Apis mellifera*) is one of the most important pollinators of many agricultural crops and wild plants worldwide. Overall, annual economic evaluation of the pollination service was quantified in 2005 to 153 billion euros, representing a yield of about 10% of global agriculture production [1]. Considering the ecological and economical importance of pollination, the widespread honey bee colony losses are a worrying phenomenon [2]. Researchers have found many factors that are a potential cause of honey bee collapse including viral [3], fungal [4], and bacterial diseases [5] together with the use of pesticides [6]. Other factors leading to the collapse of honey bee colonies are parasites, chemical treatments (amitraz, tau-fluvalinate, coumaphos, antibiotics), nutritional stress (pollen monodiete), and others [7,8]. Some stressors act synergistically such as *Nosema apis* and some pesticides [9]. The collapse of honey bee colonies is thus probably caused by combinations of multiple factors. Therefore, it is necessary to look at and deal with the health of honey bee colonies comprehensively [10].

Recently, however, viral diseases have largely contributed to bee colony losses. The most common and most dangerous virus is a deformed wing virus (DWV). This single-stranded RNA virus is a member of *Iflaviridae* [11] and creates highly genetically heterogeneous forms known as quasispecies, which can exist as several master variants [12]. One of them is type A (DWV-A), which has been attributed to the global decline in honey bees [13–15]. Another variant is type B (DWV-B), known as Varroa destructor virus-1 (VDV-1) [16,17], since it was isolated from the Varroa mite for the first time [18]. The third master variant is type C (DWV-C). However, its impact on honey bees is still unclear [12]. Other common honey bee viruses are slow bee paralysis virus (SBPV), acute bee paralysis virus (ABPV), chronic bee paralysis virus (CBPV), black queen cell virus (BQCV), sacbrood virus (SBV), Lake Sinai virus (LSV), and Macula-like virus (MLV) [19]. Their increasing distribution is mainly due to the ubiquity of the *Varroa destructor* mite, which serves as a vector and transmits viruses [10], both directly on honey bees and indirectly on other insect pollinators [20].

Another dangerous pathogen is the bacteria *Paenibacillus larvae* causing the disease called American foulbrood (AFB). Several genotypes (ERIC I-V) of this bacteria are known, and each has its specific properties such as virulence or distribution area [21]. American foulbrood is one of the most infectious honeybee diseases spread worldwide [22]. In some countries (USA, Canada, Argentina), it is allowed to use antibiotics against AFB. However, antibiotic treatment can only mitigate the symptoms but not eliminate the disease. Moreover, the antibiotics leave residues in the honey and their use in beekeeping is prohibited in many countries [23]. Given that the spores of this bacterium are very resilient and remain viable for more than 35 years, the only effective provision against the spread of *P. larvae* is to burn the infected hives together with combustible beekeeping equipment. It is essential to monitor infected habitats and their surroundings for a long time [5].

The bacterium *Melissococcus plutonius*, the causal agent for European foulbrood, has a similar infection course and method of control. It often appears together with other bacteria, so-called secondary invaders. This pathogen causes great problems, especially in the UK and Switzerland [24,25]. However, *M. plutonius* has been recorded in the Czech Republic in 2015 after a long time [26].

Important parasites are also pathogenic fungi *Nosema apis* and *Nosema ceranae,* which cause disease of the digestive tract of adult honey bees. At present, this disease is considered one of the main causes of the collapse of honey bee colonies during the winter period [10,27].

So far, less attention has been drawn to fungal diseases such as chalkbrood disease caused by entomopathogenic fungus *Ascosphaera apis* [28]. It causes mummification of bee larvae in the hive, resulting in weakening the colony and increasing susceptibility to other pathogens. Under suitable environmental conditions, the reproductive potential of the pathogen increases [29]. In some cases, it can even cause the death of bee colonies [30]. In addition, its worldwide distribution and its frequent occurrence make it an economically significant disease on a global scale [31,32].

Recently, of concern is also an infection by parasitic protozoa *Crithidia mellificae* and *Lotmaria passim* belonging to the order Trypanosomatida [33], which were previously considered relatively harmless [34]. However, it turns out that they can cause significant losses of honey bee colonies, especially with co-infection with *Nosema ceranae* [35–37]. Castelli et al. [38] also reported an association between the infected colonies and higher level of *V. destructor* infestation. Furthermore, honey bees have a highly conserved and specialized intestinal microbiome [39] that might be disrupted by trypanosomatids [40]. *L. passim* species has only recently been described [33] and now represents the dominant trypanosomatids species [37], which has already been detected in the Czech Republic [40].

All of the above-mentioned pathogens contribute to the deaths of honey bee colonies. In particular, they have a significant negative effect on the bees' winter generation, which, due to stronger immunity and longevity, ensures the survival of honey bee colonies during winter. However, since the winter generation of bees is weakened, the length of their lives is significantly reduced, which might subsequently lead to honey bee colony losses [41].

To inhibit pathogens within the congenital and social immunity and for the proper development of honey bee brood, the quality of honey bee nutrition represented by pollen is crucial. In particular, its diverse composition with a broader range of biologically active substances significantly contributes to strengthening the bee detoxification capacity [42], immunity, and resistance to overcome some diseases [43] or viral infections [44]. In contrast, the low diversity of food resources can cause malnutrition and, together with the cocktail of pesticides applicable on the fields, can shorten the life of the winter generation of bees. This can disrupt the immune response of bees, which are then more susceptible to pathogens, parasites, and other stressors. This situation occurs more often in intensively cultivated agricultural areas where a significant change in the landscape has been made, leading to a reduction in biodiversity [45]. Very specific are urban areas, which have recently become increasingly popular for beekeeping. These are mainly characterized by a built-up area and high human disturbances. Nevertheless, urban areas also contain parks, gardens, and other seminatural areas, which provide honey bees with continual nectar and pollen flow [46]. Protected areas are represented by a less anthropogenically influenced landscape characterized by a high diversity of vegetation providing rich food resources and a low level of chemical contamination [47].

This study aims to evaluate the prevalence of the main honey bee pathogens in the Czech Republic, depending on different types of habitats representing various anthropogenic burdens as well as to determine the possible impact of individual pathogens and their co-infection on the honey bee colony losses during the winter period.

2. Materials and Methods

2.1. Sampling

Samplings were carried out from selected apiaries placed in different landscapes across the Czech Republic in the fall of 2019. Agroecosystems, urban ecosystems, and national parks were chosen concerning different urban burdens to sample biological material from 250 hives in 50 apiaries (22 apiaries in agroecosystems, 22 apiaries in urban ecosystems, and six apiaries in the national park). From each apiary, five beehives were randomly chosen. Approximately 50 honey bees were collected from the brood frame of each beehive and immediately frozen on dry ice. The samples were stored at −80 °C until processing. All brood frames from the tested colonies were checked for symptoms of bacterial bee brood diseases. The colony losses were assessed in spring 2020 (the percentage of collapsed colonies of the whole apiaries).

2.2. Characterization of Different Types of Habitat

The town habitat in the Czech Republic involves especially built-up area of towns with houses and factories and is affected by increased industrial contamination and high levels of traffic. Therefore, it represents the highest urban burdens. This habitat also includes town parks and gardens. The agroecosystems are characterized by large areas of fields with agricultural crops, especially monocultures, a high rate of landscape fragmentation and agrochemical contamination. In addition, low diversity of bee food sources as well as short-term availability of food due to intensive agricultural management is typical. National parks, as the most potential honey bee-friendly environment with minimal human disturbance is characterized by flowery meadows, pastures, and forests. Habitat is characterized by an absence of industry, a low degree of landscape fragmentation, and a rich diversity of flowers, which are a good source of food for bees. Agricultural management is possible only through an ecological approach without the use of pesticides.

2.3. Sample Preparation and Nucleic Acid Purification

Samples for RNA (detection of DWV-A, DWV-B, DWV-C, BQCV, CBPV, ABPV, SBV, LSV, MLV) and DNA (detection of *Nosema apis, Nosema ceranae, Paenibacillus larvae, Melissococcus plutonius, Ascosphaera apis, Crithidia mellificae, Lotmaria passim*) purification were collected as a bulk of approximately 250 bees from five hives in each location, frozen in dry ice, and stored at −80 °C. After homogenization in liquid nitrogen, aliquotes for separate RNA and DNA purification were made.

According to the manufacturer's instructions, total RNA was extracted using the TRI Reagent (MRC, Montgomery, OH, USA). Contaminating DNA was removed using the DNA-free TM Kit (Ambion, supplied by ThermoFisher Scientific, Loughborough, UK). BioSpec Nano (Shimadzu, Nakagyo-ku, Kyoto, Japan) was used to quantify RNA (OD260) and to assess sufficient quality (OD260/280 ratio and OD260/230 ratio). cDNA templates were prepared using a Standard Reverse Transcription Protocol (Promega, Madison, WI, USA) and OligodT primer and stored at −20 °C until use.

DNA was extracted using a modified CTAB method. Homogenized tissue was resuspended in CTAB buffer (2% CTAB, 100 mM Tris pH 8.0, 20 mM EDTA pH 7.8, 1.4 M NaCl) with 1% β-mercaptoethanol and incubated at 65 °C for 10 min. The solution was extracted with 500 µL chloroform:isoamylalcohol (24:1) and precipitated in 250 µL of 2-propanol at −20 °C for 30 min. After washing with 1 mL of 70% ethanol, the pellet was resuspended in 150 µL of TE buffer (10 mM Tris pH 8.0, 1 mM EDTA pH 7.8) and stored in 4 °C until use.

2.4. PCR Conditions

The RT-PCR (detection of DWV-A, DWV-B, DWV-C, BQCV, CBPV, ABPV, SBV, LSV, MLV) was performed on the QuantStudio™ 6 Flex Real-Time PCR System (Applied Biosystems, supplied by ThermoFisher scientific, Loughborough, UK) using Power SYBR® Green PCR Master Mix (Applied Biosystems, supplied by ThermoFisher Scientific, Loughborough, UK) in a 96-well reaction plate using parameters recommended by the manufacturer (2 min at 50 °C, 10 min at 95 °C, and 40 cycles of 15 s 95 °C, 1 min of 60 °C, 15 s at 95 °C, 1 min at 60 °C, and 15 s at 95 °C). The no-template controls were included. Positive samples were considered a true positive using a Ct cutoff of 36 cycles. The specificity of amplification was determined by dissociation curve analyses and sequencing of randomly selected positive samples. The sequence of the primer, orientation, annealing temperature, and references are shown in Table 1.

The PCR (detection of *Nosema apis, Nosema ceranae, Paenibacillus larvae, Melissococcus plutonius, Ascosphaera apis, Crithidia mellificae, Lotmaria passim*) was performed on the Eppendorf Mastercycler PRO system (Eppendorf, Hamburg, DE) in 25 µL volume containing 1× PPP Master Mix (Top-Bio, Vestec, Czech Republic), 10 pmol each forward and backward primer, and 2 µL of DNA template using the following cycling conditions: denaturation at 95 °C for 5 min, 40 cycles of 30 s 95 °C, 45 s of TA, 1 min at 72 °C; and a final extension at 72 °C for 10 min. PCR products were visualized by 1.5% agarose gel electrophoresis and stained with ethidium bromide solution (Merck Life Science, Darmstadt, Germany). The specificity of amplification was determined by sequencing randomly selected positive samples. The sequence of the primer, orientation, annealing temperature, and references are shown in Table 1.

Table 1. Primers for PCR analysis.

Gene	Sequences 5′-3′	TA [°C]	Reference
Nosema apis	F: GGGGGCATGTCTTTGACGTACTATGTA R: GGGGGGCGTTTAAAATGTGAAACAACTATG	62	[48]
Nosema ceranae	F: CGGCGACGATGTGATATGAAAATATTAA R: CCCGGTCATTCTCAAACAAAAAACCG	62	[48]
Paenibacillus larvae	F: GCTCTGTTGCCAAGGAAGAA R: AGGCGGAATGCTTACTGTGT	55	[49]
Melissococcus plutonius	F: GAAGAGGAGTTAAAAGGCGC R: TTATCTCTAAGGCGTTCAAAGG	55	[50]
Ascosphaera apis	F: TGTGTCTGTGCGGCTAGGTG R: GCTAGCCAGGGGGGAACTAA	60	[51]
Crithidia mellificae	F: AGTTTGAGCTGTTGGATTTGTT R: AACCTATTACAGGCACAGTTGC	56	[52]
Lotmaria passim	F: TGACTTGAATTAGCAAGCATGGGATAACA R: CCTTTAGGCTACCGTTTCGGCTTTTGTTGGT	60	[53]
DWV-A	F: CGTCGGCCTATCAAAG R: CTTTTCTAATTCAACTTCACC	60	[54]
DWV-B	F: GCCCTGTTCAAGAACATG R: CTTTTCTAATTCAACTTCACC	60	[54]
DWV-C	F: TACTAGTGCTGGTTTTCCTTT R: ATAAGTTGCGTGGTTGAC	60	[54]
BQCV	F: GGACGAAAGGAAGCCTAAAC R: ACTAGGAAGAGACTTGCACC	48	[48]
CBPV	F: AACCTGCCTCAACACAGGCAAC R: ACATCTCTTCTTCGGTGTCAGCC	60	[55]
ABPV	F: TGAGAACACCTGTAATGTGG R: ACCAGAGGGTTGACTGTGTG	48	[56]
SBV	F: GGATGAAAGGAAATTACCAG R: CCACTAGGTGATCCACACT	48	[56]
LSV	F: CKTGCGGNCCTCATTTCTTCATGTC R: CATGAATCCAAKGTCAAAGGTRTCGT	60	[57]
MLV	F: ATCCCTTTTCAGTTCGCT R: AGAAGAGACTTCAAGGAC	60	[58]

2.5. Statistical Analysis

To evaluate whether pathogen occurrence and species richness differ among honey bee colonies and habitat types, we used separate generalized liner mixed-effects models (GLMM) [59]. In the case when species richness was used as dependent variable, GLMM with a Gaussian error distribution was used. When the pathogen occurrence or honey bee mortality rate was used as the dependent variable, binomial error distribution with logit link function was used. In each model, we specify habitat types and pathogen species as fixed factors and the owner of the honey bee colony was used as a factor with a random intercept effect. To compare the means within a particular fixed factor, the Tukey multiple comparison test with Bonferroni adjustment of p-values was used. Data were analyzed in the R program (R Development Core Team 2020).

To visualize and test the association between the mortality rate of honey bees and species composition of pathogens, partial canonical correspondence analysis (pCCA) was used with the habitat type as the covariable. We used this type of covariable to eliminate the possible confounding effect of habitat type on the mortality of honey bees regardless of the pathogen species composition. The significance of the canonical axis was tested with a restricted Monte Carlo permutation test for the time series with 2000 permutations. All ordination analyses were conducted by the statistical software CANOCO, v. 5 [60].

3. Results

The proportion of eukaryotic pathogen occurrence significantly differs between town habitat and national park, whereas the lowest rate of pathogen occurrence has been observed in the national park and the highest in the towns. A moderate rate of pathogen burden has been observed in agroecosystems. However, this habitat did not differ significantly between urban areas or national parks (Figure 1a, Table 2). The species richness of eukaryotic honey bee pathogens did not significantly differ between the tested habitats (Figure 1b, Table 2).

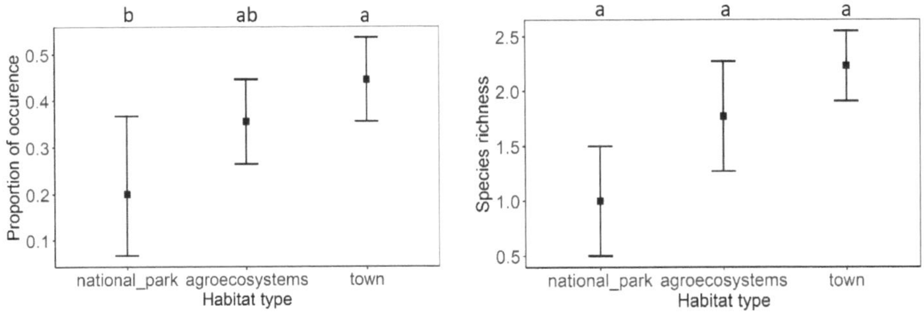

(a) (b)

Figure 1. (**a**) The proportion of eukaryotic pathogen occurrence in different types of habitats and (**b**) the proportion of eukaryotic pathogen richness in different types of habitats. Black squares represent means and the error bars represent 95% confidence intervals. Significant differences (<0.05) are indicated by different letters.

Table 2. The results of the analysis of deviance (likelihood-ratio test) testing the partial effect of habitat type and pathogen species identity on the species richness and occurrence of pathogens in the honey bee colonies. Likelihood-ratio analysis testing of whether the Akaike information criterion (AIC) of the full model significantly increased after a particular explanatory variable was excluded from the model.

	Df.	AIC	LRT	Pr (Chi)
	Dependent variable: species occurrence			
Full model		243.68		
Eukaryote	4	332.25	96.570	<0.0001
Habitat	2	246.76	7.081	0.02899
Full model		398.26		
Virus	9	494.96	114.695	<0.0001
Habitat	2	396.69	2.423	0.2977
	Dependent variable: number of eukaryotic species			
Full model		156.78		
Habitat	2	157.23	4.453	0.107
	Dependent variable: number of virus types			
Full model		181.88		
Habitat	2	180.52	2.642	0.267

In all types of habitat, the same species of eukaryotic pathogens dominated. In all cases, the most dominant species were *L. passim* and *N. ceranae*, followed by *C. mellificae*, and the lowest occurrence rate had *M. plutonius* and *P. larvae*. No clinical symptoms of bacterial brood diseases were observed. In contrast, *A. apis* and *N. apis* were not detected at all (Figure 2).

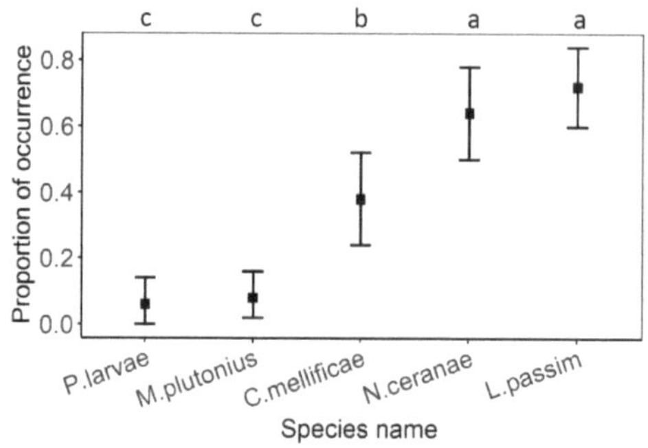

Figure 2. The comparison of proportion of eukaryotic pathogen occurrence regardless of habitat. Black squares represent means and the error bars represent 95% confidence intervals. Significant differences (<0.05) are indicated by different letters.

In the case of individual habitats, all five tested pathogens were detected in a town habitat. The most prevalent pathogens were *L. passim* and *N. ceranae*, followed by *C. mellificae*. Bacteria *P. larvae* and *M. plutonius* only had a low prevalence. The most dominated species in the agroecosystems were *N. ceranae*, *L. passim*, and *C. mellificae*. *M. plutonius* occurred significantly less and *P. larvae* were not detected at all. In the case of national parks, only *L. passim* and *N. ceranae* were detected (Figure 3).

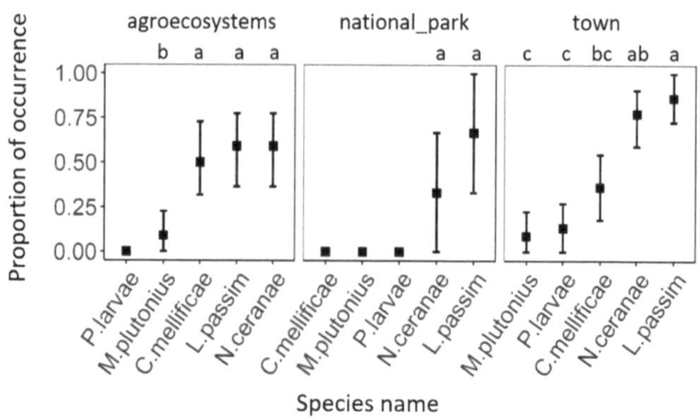

Figure 3. The comparison of proportion of eukaryotic pathogen occurrence within each habitat type. Black squares represent means and the error bars represent 95% confidence intervals. Significant differences (<0.05) are indicated by different letters.

Viral pathogen occurrence and species richness did not significantly differ between individual habitats (Figure 4 and Table 2). Generally, the most abundant viruses were DWV-A and ABPV, followed by DWV-B and LSV. Less frequent viruses were MLV, SBV, CBPV, DWV-C, and BQCV (Figure 5). A similar pattern was observed in all types of habitats. Only DWV-A dominated in the agroecosystems (Figure 6).

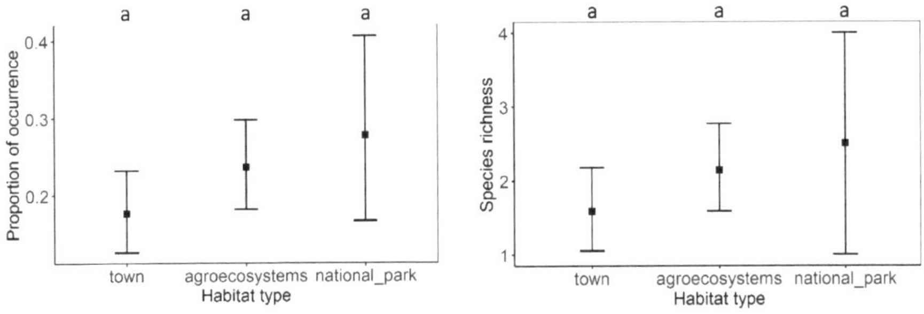

(a) (b)

Figure 4. (**a**) The proportion of viral pathogens occurrence in different types of habitats and (**b**) comparison of species richness of viral pathogens between different types of habitats. Black squares represent means and the error bars represent 95% confidence intervals. Significant differences (<0.05) are indicated by different letters.

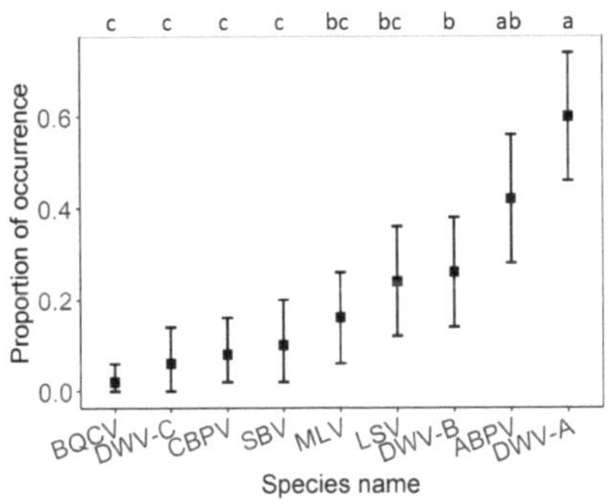

Figure 5. The comparison of proportion of viral pathogen occurrence regardless of habitat. Black squares represent means and the error bars represent 95% confidence intervals. Significant differences (<0.05) are indicated by different letters.

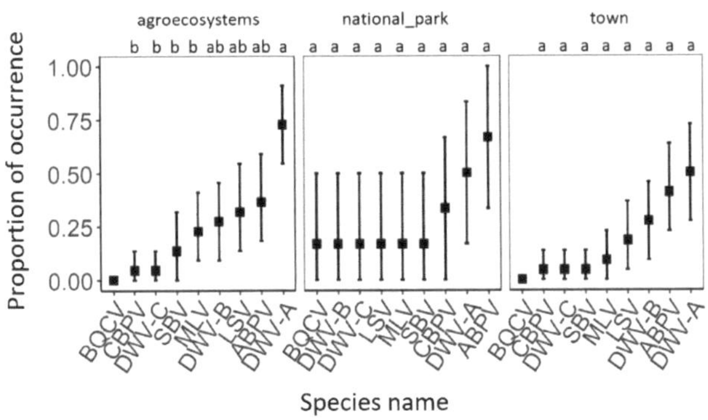

Figure 6. The comparison of proportion of viral pathogen occurrence within each habitat type. Black squares represent means and the error bars represent 95% confidence intervals. Significant differences (<0.05) are indicated by different letters.

Differences winter mortality rates in honey bee colonies between habitats were not statistically significant (Figure 7) due to a small number of samples from national parks and high confidence interval from the data. However, the average winter mortality in town (24.51%) and agroecosystem (21.50%) habitats were twice as high as in national parks (11.11%).

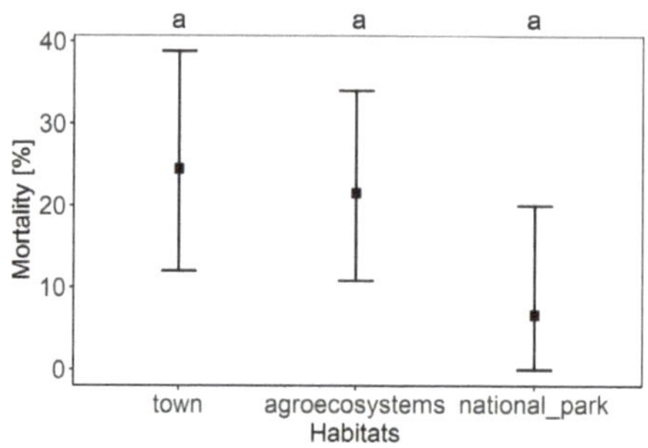

Figure 7. The comparison of honey bee winter mortality rate according to type of habitat. Black squares represent means and the error bars represent 95% confidence intervals.

Based on the results of pCCA species, structures of all pathogens (i.e., species composition and their abundances) were significantly associated with honey bee mortality (pseudo-F = 1.8, p = 0.053, test of all canonical axes, R^2 = 3.73%). In the separate pCCA analyses evaluating association only between viruses and honey bee mortality, we found that the assemblage composed only with viruses (pseudo-F = 2.2, p = 0.037, test of all canonical axes, R^2 = 5.28%) had a closer relationship to mortality than the assemblage composed only with eukaryotes (pseudo-F = 0.3, p = 0.881, test of all canonical axes, R^2 = 0.80%). The pCCA diagram revealed that the closest association with honey bee mortality was shown by DWV-C and DWV-B viruses (Figure 8).

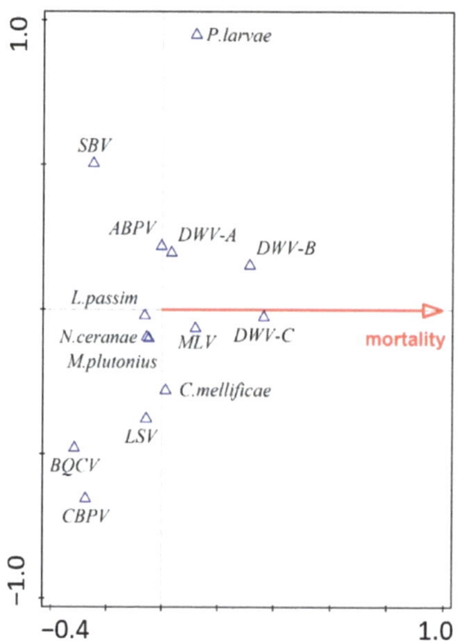

Figure 8. Partial canonical correspondence analysis biplot with the habitat type as a covariable showing the strength of the association of individual pathogens with the mortality rate.

4. Discussion

In the study, the prevalence of several honey bee pathogens was detected including viruses, fungal, protozoa, and bacterial pathogens on different types of habitats. The most frequently detected pathogens belonged to the family Trypanosomatida, in particular, *Lotmaria passim* (72%) and *Crithidia mellificae* (38%). Both protozoa significantly shorten the life of bees and are therefore thought to cause significant bee colony losses [61]. Of even more concern is that trypanosomatids affect the composition of the symbiotic bacterial taxa of bees [40]. However, little is known about the full extent of the harmfulness and mechanism of pathogenesis of these two pathogens [38,62]. Other studies have shown an even higher risk of trypanosomatids when co-infected with *N. ceranae* [35–37]. In addition, it led to a reduction in immune gene expression [37]. The high incidence of trypanosomatids is similar in other European countries [36,62].

The prevalent pathogen is also *Nosema ceranae* (64%), often associated with colony losses, especially in Mediterranean areas [4,63,64]. However, its occurrence has also been recorded in the temperate zone to a lesser extent [10] and with less impact [65,66]. In this study, *N. ceranae* has not been significantly associated with colony losses (Figure 8). This pathogen occurred independently of the habitat type observed. On the other hand, *Nosema apis* was not detected at all. The decline in *N. apis* and the spread of *N. ceranae* is a well-known and long-lasting trend taking place globally [67–71]. However, the complete absence of *N. apis* in the nationwide screening is a novelty. We attribute this to the displacement of the more aggressive *N. ceranae* due to its higher virulence [68,69]. *Ascosphaera apis* was also not been detected. It is an opportunistic pathogen that occurs in the colony, especially in stressful situations such as thermal discomfort [29]. Higher prevalence was recorded in humid areas, and, for example, in China [72] and northern Thailand [30], the fungal pathogen causes great damage.

Bacterial diseases occurred only to a lesser extent and only in urban areas (*P. larvae* and *M. plutonius*) and agroecosystems (*M. plutonius*). They did not occur in the national parks at all. *P. larvae* commonly occurs across the whole Czech Republic, especially in Moravia, and the dominant genotype is ERIC II (80.4%) over ERIC I (19.4%) [73]. The outbreak of European foulbrood caused by *M. plutonius* was observed in 2015 after 40 years in the Czech Republic. Since then, the occurrence persists, but with a very low prevalence [74]. In contrast, in some countries such as England [75], France [76], and Switzerland [77], bacterial disease very often occurs. These two bacterial diseases are very infectious and can cause great economic losses. Therefore, the government often monitors its prevalence, and in many cases, there is an effort to eliminate them through strict rules.

In the case of viral diseases, at least one of the tested honey bee viruses were detected in 74% of cases, while two or more viruses were present in one-third of the tested apiaries. The most prevalent honey bee virus was the deformed wing virus (DWV). There are multiple variants of DWV that include type A [11], type B (Varroa destructor virus-1 (VDV-1) [14,18], and type C [12]. These variants have a different impact on honey bee colonies, and their virulence is not clear. Whereas some studies claim DWV-A has higher virulence [16,78,79], other studies claim DWV-B has the same or even higher virulence [17,80–82]. Since the variant DWV-B can replicate in Varroa mites, the viral load is usually higher in honey bee tissues than in other DWV variants [78,83]. DWV-C is associated with DWV-A and has been indicated as a contributing factor in overwintering losses of honey bee colonies [78,79]. Our study reports DWV-A as the most frequent variant (60%) in the Czech Republic (Figure 5). Surprisingly, similar results where variant DWV-A dominated have been reported from the USA [79,83], whereas variant B dominated in Europe [78,80,84]. However, despite their low prevalence, only DWV-B (26%) and C (6%) variants were significantly associated with the overwintering losses (Figure 8). Other authors have also concluded that these variants are associated with winter colony losses [17,85].

The second most prevalent virus was ABPV, which was detected in half of the tested colonies. This virus has commonly been detected in Germany [10], the USA [3], Switzerland [86], and Belgium [87] and its co-infection with DWV is attributed to overwintering losses [10]. The LSV (24%) virus is also a major concern, especially in the USA [88]. However, its prevalence is also high in Europe [36]. One of the recently identified honey bee viruses is MLV (16%), which is associated with the mite *V. destructor* [89]. However, its virulence and impact on honey bees are still unclear [90]. Its high prevalence has been observed in France [89], Belgium [36], and Syria [91]. The occurrence of SBV (10%), CBPV (8%), and BQVC (2%) was only minor, especially in urban areas and agroecosystems. The presences of these viruses were not significantly related to the decline of honey bee colonies in the Czech Republic.

The lowest occurrence of eukaryotic pathogens was detected in the national parks, higher occurrence in the agroecosystems, and the highest occurrence in town habitats (Figure 1). This probably corresponds with a high density of bee colonies in the landscape [92] because the number of bee colonies per km^2 in the Czech Republic is one of the highest in the world (>8 honey bee colonies/km^2) [93]. According to these results, Taric [94] also found a higher parasitic burden in commercially kept colonies than traditionally kept colonies, which are mostly situated in natural areas. The richness of individual pathogens was in the same trend, where only two eukaryotic pathogens were present in the national parks. At the same time, four of them occurred in the agroecosystems and five in the towns.

The opposite trend was observed for viruses. All nine tested viruses were present in the national parks, while in agroecosystems and towns, there were eight species. However, these differences were not statistically significant. The study shows that the occurrence of honey bee pathogens, and especially viruses, did not differ between the tested habitats. In addition, the viruses also spread quickly among other species of wild pollinators, which can cause problems with species composition and affect trophic bonds and ecosystem stability [20,84,95].

Differences in the mortality between habitats were not statistically significant. The results were not significant probably due to the low number of samples from the national parks. One of the reasons for colony mortality in national parks is probably due to the high prevalence of viruses as in other habitats (DWV-B and DWV-C), which were associated with colony mortality. The next issue is the trading of bee queens or whole colonies and the migratory management of colonies [96]. This is connected with colony density, which is usually lower in natural parks. This might be another reason for lower honey bee eukaryotic pathogen occurrence in natural parks. At localities with a high bee density, bee colonies cannot avoid sharing food resources, which represent hotspots of infections [97].

5. Conclusions

The most prevalent eukaryotic pathogens in the population of *A. mellifera* in the Czech Republic were *L. passim* and *N. ceranae*, followed by *C. mellificae*. This trend was valid in all types of monitored habitats. In contrast, *P. larvae* and *M. plutonius* were detected only sporadically. *N. apis* and *A. apis* were not detected at all.

The most prevalent viruses were DWV-A and ABPV in all types of tested habitats. On the other hand, BCQV, SBV, and DWV-C were the least prevalent, except in national parks, where the occurrence of all the monitored viruses was relatively uniform.

Of all the monitored eukaryotic and viral pathogens, only DWV-C and DWV-B were significantly associated with colony mortality.

Author Contributions: Conceptualization, P.M. and M.H.; Data curation, M.K.; K.V., I.H. and J.Š.; Formal analysis, A.B., V.Č. and J.Š.; Investigation, P.M., M.K. and I.H.; Methodology, P.M. and K.V.; Supervision, A.B. and V.Č.; Writing—original draft, P.M. and M.H. All authors have read and agreed to the published version of the manuscript.

Funding: This research was funded by Ministry of Agriculture, grant number QK1910356, and University of South Bohemia, grant number GAJU027/2019/Z.

Institutional Review Board Statement: Not applicable.

Informed Consent Statement: Not applicable.

Data Availability Statement: The study did not report any data.

Conflicts of Interest: The authors declare no conflict of interest.

References

1. Gallai, N.; Salles, J.-M.; Settele, J.; Vaissière, B.E. Economic Valuation of the Vulnerability of World Agriculture Confronted with Pollinator Decline. *Ecol. Econ.* **2009**, *68*, 810–821. [CrossRef]
2. Stokstad, E. The Case of the Empty Hives. *Science* **2007**, *316*, 970–972. [CrossRef]
3. Cox-Foster, D.L.; Conlan, S.; Holmes, E.C.; Palacios, G.; Evans, J.D.; Moran, N.A.; Quan, P.-L.; Briese, T.; Hornig, M.; Geiser, D.M. A Metagenomic Survey of Microbes in Honey Bee Colony Collapse Disorder. *Science* **2007**, *318*, 283–287. [CrossRef]
4. Higes, M.; Martín-Hernández, R.; Garrido-Bailón, E.; González-Porto, A.V.; García-Palencia, P.; Meana, A.; Del Nozal, M.J.; Mayo, R.; Bernal, J.L. Honeybee Colony Collapse Due to *Nosema ceranae* in Professional Apiaries. *Environ. Microbiol. Rep.* **2009**, *1*, 110–113. [CrossRef]
5. Alonso-Salces, R.M.; Cugnata, N.M.; Guaspari, E.; Pellegrini, M.C.; Aubone, I.; De Piano, F.G.; Antunez, K.; Fuselli, S.R. Natural Strategies for the Control of *Paenibacillus larvae*, the Causative Agent of American Foulbrood in Honey Bees: A Review. *Apidologie* **2017**, *48*, 387–400. [CrossRef]
6. Frazier, M.; Mullin, C.; Frazier, J.; Ashcraft, S. What Have Pesticides Got to Do with It? *Am. Bee J.* **2008**, *148*, 521–524.
7. Mutinelli, F.; Granato, A. La Sindrome Del Collasso Della Colonia (Colony Collapse Disorder) Negli USA. *Aggiorn. Sulla Situaz. Attuale. Apoidea* **2007**, *4*, 175–185.
8. Sharpe, R.J.; Heyden, L.C. Honey Bee Colony Collapse Disorder Is Possibly Caused by a Dietary Pyrethrum Deficiency. *Biosci. Hypotheses* **2009**, *2*, 439–440. [CrossRef]
9. Alaux, C.; Ducloz, F.; Crauser, D.; Le Conte, Y. Diet Effects on Honeybee Immunocompetence. *Biol. Lett.* **2010**, *6*, 562–565. [CrossRef]
10. Genersch, E.; Von Der Ohe, W.; Kaatz, H.; Schroeder, A.; Otten, C.; Büchler, R.; Berg, S.; Ritter, W.; Mühlen, W.; Gisder, S. The German Bee Monitoring Project: A Long Term Study to Understand Periodically High Winter Losses of Honey Bee Colonies. *Apidologie* **2010**, *41*, 332–352. [CrossRef]
11. Lanzi, G.; De Miranda, J.R.; Boniotti, M.B.; Cameron, C.E.; Lavazza, A.; Capucci, L.; Camazine, S.M.; Rossi, C. Molecular and Biological Characterization of Deformed Wing Virus of Honeybees (*Apis mellifera* L.). *J. Virol.* **2006**, *80*, 4998–5009. [CrossRef] [PubMed]
12. Mordecai, G.J.; Wilfert, L.; Martin, S.J.; Jones, I.M.; Schroeder, D.C. Diversity in a Honey Bee Pathogen: First Report of a Third Master Variant of the Deformed Wing Virus Quasispecies. *ISME J.* **2016**, *10*, 1264–1273. [CrossRef] [PubMed]
13. Dainat, B.; Evans, J.D.; Chen, Y.P.; Gauthier, L.; Neumann, P. Dead or Alive: Deformed Wing Virus and *Varroa destructor* Reduce the Life Span of Winter Honeybees. *Appl. Environ. Microbiol.* **2012**, *78*, 981–987. [CrossRef]
14. Martin, S.J.; Highfield, A.C.; Brettell, L.; Villalobos, E.M.; Budge, G.E.; Powell, M.; Nikaido, S.; Schroeder, D.C. Global Honey Bee Viral Landscape Altered by a Parasitic Mite. *Science* **2012**, *336*, 1304–1306. [CrossRef] [PubMed]
15. Schroeder, D.C.; Martin, S.J. Deformed Wing Virus: The Main Suspect in Unexplained Honeybee Deaths Worldwide. *Virulence* **2012**, *3*, 589–591. [CrossRef]
16. Mordecai, G.J.; Brettell, L.E.; Martin, S.J.; Dixon, D.; Jones, I.M.; Schroeder, D.C. Superinfection Exclusion and the Long-Term Survival of Honey Bees in Varroa-Infested Colonies. *ISME J.* **2016**, *10*, 1182–1191. [CrossRef]
17. Natsopoulou, M.E.; McMahon, D.P.; Doublet, V.; Frey, E.; Rosenkranz, P.; Paxton, R.J. The Virulent, Emerging Genotype B of Deformed Wing Virus Is Closely Linked to Overwinter Honeybee Worker Loss. *Sci. Rep.* **2017**, *7*, 5242. [CrossRef]
18. Ongus, J.R.; Peters, D.; Bonmatin, J.-M.; Bengsch, E.; Vlak, J.M.; van Oers, M.M. Complete Sequence of a Picorna-like Virus of the Genus Iflavirus Replicating in the Mite *Varroa destructor*. *J. Gen. Virol.* **2004**, *85*, 3747–3755. [CrossRef]
19. McMahon, D.P.; Fürst, M.A.; Caspar, J.; Theodorou, P.; Brown, M.J.; Paxton, R.J. A Sting in the Spit: Widespread Cross-infection of Multiple RNA Viruses across Wild and Managed Bees. *J. Anim. Ecol.* **2015**, *84*, 615–624. [CrossRef]
20. Tehel, A.; Brown, M.J.; Paxton, R.J. Impact of Managed Honey Bee Viruses on Wild Bees. *Curr. Opin. Virol.* **2016**, *19*, 16–22. [CrossRef]
21. Beims, H.; Bunk, B.; Erler, S.; Mohr, K.I.; Spröer, C.; Pradella, S.; Günther, G.; Rohde, M.; von der Ohe, W.; Steinert, M. Discovery of *Paenibacillus larvae* ERIC V: Phenotypic and Genomic Comparison to Genotypes ERIC I-IV Reveal Different Inventories of Virulence Factors Which Correlate with Epidemiological Prevalences of American Foulbrood. *Int. J. Med Microbiol.* **2020**, *310*, 151394. [CrossRef]
22. Genersch, E. American Foulbrood in Honeybees and Its Causative Agent, *Paenibacillus larvae*. *J. Invertebr. Pathol.* **2010**, *103*, S10–S19. [CrossRef] [PubMed]
23. Reybroeck, W. Residues of Antibiotics and Chemotherapeutics in Honey. *J. Apic. Res.* **2018**, *57*, 97–112. [CrossRef]
24. Forsgren, E. European Foulbrood in Honey Bees. *J. Invertebr. Pathol.* **2010**, *103*, S5–S9. [CrossRef] [PubMed]
25. Lewkowski, O.; Erler, S. Virulence of *Melissococcus plutonius* and Secondary Invaders Associated with European Foulbrood Disease of the Honey Bee. *MicrobiologyOpen* **2019**, *8*, e00649. [CrossRef] [PubMed]
26. Erban, T.; Ledvinka, O.; Kamler, M.; Hortova, B.; Nesvorna, M.; Tyl, J.; Titera, D.; Markovic, M.; Hubert, J. Bacterial Community Associated with Worker Honeybees (*Apis mellifera*) Affected by European Foulbrood. *PeerJ* **2017**, *5*, e3816. [CrossRef] [PubMed]

27. Martín-Hernández, R.; Bartolomé, C.; Chejanovsky, N.; Le Conte, Y.; Dalmon, A.; Dussaubat, C.; García-Palencia, P.; Meana, A.; Pinto, M.A.; Soroker, V. *Nosema ceranae* in *Apis mellifera*: A 12 Years Postdetection Perspective. *Environ. Microbiol.* **2018**, *20*, 1302–1329. [CrossRef]
28. Spiltoir, C.F. Life Cycle of *Ascosphaera apis* (Pericystis Apis). *Am. J. Bot.* **1955**, 501–508. [CrossRef]
29. Mráz, P.; Hýbl, M.; Kopecký, M.; Bohatá, A.; Konopická, J.; Hoštičková, I.; Konvalina, P.; Šipoš, J.; Rost, M.; Čurn, V. The Effect of Artificial Media and Temperature on the Growth and Development of the Honey Bee Brood Pathogen *Ascosphaera apis*. *Biology* **2021**, *10*, 431. [CrossRef]
30. Chantawannakul, P.; Puchanichanthranon, T.; Wongsiri, S. Inhibitory Effects of Some Medicinal Plant Extracts on the Growth of *Ascosphaera apis*. In Proceedings of the III WOCMAP Congress on Medicinal and Aromatic Plants-Volume 4: Targeted Screening of Medicinal and Aromatic Plants, Economics 678, Chiang Mai, Thailand, 3–7 February 2003; pp. 183–189.
31. Ansari, M.J.; Al-Ghamdi, A.; Usmani, S.; Khan, K.A.; Alqarni, A.S.; Kaur, M.; Al-Waili, N. In Vitro Evaluation of the Effects of Some Plant Essential Oils on *Ascosphaera apis*, the Causative Agent of Chalkbrood Disease. *Saudi J. Biol. Sci.* **2017**, *24*, 1001–1006. [CrossRef]
32. Aronstein, K.A.; Murray, K.D. Chalkbrood Disease in Honey Bees. *J. Invertebr. Pathol.* **2010**, *103*, S20–S29. [CrossRef]
33. Schwarz, R.S.; Bauchan, G.R.; Murphy, C.A.; Ravoet, J.; de Graaf, D.C.; Evans, J.D. Characterization of Two Species of Trypanosomatidae from the Honey Bee *Apis mellifera*: *Crithidia mellificae* Langridge and McGhee, and *Lotmaria passim* n. Gen., n. Sp. *J. Eukaryot. Microbiol.* **2015**, *62*, 567–583. [CrossRef]
34. Langridge, D.; McGhee, R.B. *Crithidia mellificae* n. Sp. an Acidophilic Trypanosomatid of the Honey Bee *Apis mellifera*. *J. Protozool.* **1967**, *14*, 485–487. [CrossRef]
35. Vejnovic, B.; Stevanovic, J.; Schwarz, R.S.; Aleksic, N.; Mirilovic, M.; Jovanovic, N.M.; Stanimirovic, Z. Quantitative PCR Assessment of *Lotmaria passim* in *Apis mellifera* Colonies Co-Infected Naturally with *Nosema ceranae*. *J. Invertebr. Pathol.* **2018**, *151*, 76–81. [CrossRef]
36. Ravoet, J.; Maharramov, J.; Meeus, I.; De Smet, L.; Wenseleers, T.; Smagghe, G.; De Graaf, D.C. Comprehensive Bee Pathogen Screening in Belgium Reveals *Crithidia mellificae* as a New Contributory Factor to Winter Mortality. *PLoS ONE* **2013**, *8*, e72443.
37. Arismendi, N.; Caro, S.; Castro, M.P.; Vargas, M.; Riveros, G.; Venegas, T. Impact of Mixed Infections of Gut Parasites *Lotmaria passim* and *Nosema ceranae* on the Lifespan and Immune-Related Biomarkers in *Apis mellifera*. *Insects* **2020**, *11*, 420. [CrossRef] [PubMed]
38. Castelli, L.; Branchiccela, B.; Invernizzi, C.; Tomasco, I.; Basualdo, M.; Rodriguez, M.; Zunino, P.; Antúnez, K. Detection of *Lotmaria passim* in Africanized and European Honey Bees from Uruguay, Argentina and Chile. *J. Invertebr. Pathol.* **2019**, *160*, 95–97. [CrossRef]
39. Raymann, K.; Moran, N.A. The Role of the Gut Microbiome in Health and Disease of Adult Honey Bee Workers. *Curr. Opin. Insect Sci.* **2018**, *26*, 97–104. [CrossRef]
40. Hubert, J.; Bicianova, M.; Ledvinka, O.; Kamler, M.; Lester, P.J.; Nesvorna, M.; Kopecky, J.; Erban, T. Changes in the Bacteriome of Honey Bees Associated with the Parasite *Varroa destructor*, and Pathogens *Nosema* and *Lotmaria passim*. *Microb. Ecol.* **2017**, *73*, 685–698. [CrossRef]
41. Hristov, P.; Shumkova, R.; Palova, N.; Neov, B. Factors Associated with Honey Bee Colony Losses: A Mini-Review. *Vet. Sci.* **2020**, *7*, 166. [CrossRef] [PubMed]
42. Hýbl, M.; Mráz, P.; Šipoš, J.; Hoštičková, I.; Bohatá, A.; Čurn, V.; Kopec, T. Polyphenols as Food Supplement Improved Food Consumption and Longevity of Honey Bees (*Apis mellifera*) Intoxicated by Pesticide Thiacloprid. *Insects* **2021**, *12*, 572. [CrossRef]
43. Brodschneider, R.; Crailsheim, K. Nutrition and Health in Honey Bees. *Apidologie* **2010**, *41*, 278–294. [CrossRef]
44. Rutter, L.; Carrillo-Tripp, J.; Bonning, B.C.; Cook, D.; Toth, A.L.; Dolezal, A.G. Transcriptomic Responses to Diet Quality and Viral Infection in *Apis mellifera*. *BMC Genom.* **2019**, *20*, 412. [CrossRef] [PubMed]
45. Naug, D. Nutritional Stress Due to Habitat Loss May Explain Recent Honeybee Colony Collapses. *Biol. Conserv.* **2009**, *142*, 2369–2372. [CrossRef]
46. Garbuzov, M.; Schürch, R.; Ratnieks, F.L. Eating Locally: Dance Decoding Demonstrates That Urban Honey Bees in Brighton, UK, Forage Mainly in the Surrounding Urban Area. *Urban Ecosyst.* **2015**, *18*, 411–418. [CrossRef]
47. Hybl, M.; Mraz, P.; Sipos, J. Diversity of Bees (Apoidea) and Their Pesticide Contamination in Two Different Types of Agricultural Management. In Proceedings of the MendelNet, Brno, Czech Republic, 11–12 November 2020; Volume 27, pp. 216–221.
48. Ryba, S.; Titera, D.; Schodelbauerova-Traxmandlova, I.; Kindlmann, P. Prevalence of Honeybee Viruses in the Czech Republic and Coinfections with Other Honeybee Disease. *Biologia* **2012**, *67*, 590–595. [CrossRef]
49. Bakonyi, T.; Derakhshifar, I.; Grabensteiner, E.; Nowotny, N. Development and Evaluation of PCR Assays for the Detection of *Paenibacillus larvae* in Honey Samples: Comparison with Isolation and Biochemical Characterization. *Appl. Environ. Microbiol.* **2003**, *69*, 1504–1510. [CrossRef]
50. Govan, V.; Allsopp, M.; Davison, S. A PCR Detection Method for Rapid Identification of *Paenibacillus larvae*. *Appl. Environ. Microbiol.* **1999**, *65*, 2243–2245. [CrossRef]
51. Garrido-Bailón, E.; Higes, M.; Martínez-Salvador, A.; Antúnez, K.; Botías, C.; Meana, A.; Prieto, L.; Martín-Hernández, R. The Prevalence of the Honeybee Brood Pathogens A *Scosphaera apis*, P *Aenibacillus larvae* and M *Elissococcus plutonius* in S Panish Apiaries Determined with a New Multiplex PCR Assay. *Microb. Biotechnol.* **2013**, *6*, 731–739. [CrossRef]

52. Stevanovic, J.; Schwarz, R.S.; Vejnovic, B.; Evans, J.D.; Irwin, R.E.; Glavinic, U.; Stanimirovic, Z. Species-Specific Diagnostics of *Apis mellifera* Trypanosomatids: A Nine-Year Survey (2007–2015) for Trypanosomatids and Microsporidians in Serbian Honey Bees. *J. Invertebr. Pathol.* **2016**, *139*, 6–11. [CrossRef]
53. Arismendi, N.; Bruna, A.; Zapata, N.; Vargas, M. PCR-Specific Detection of Recently Described *Lotmaria passim* (Trypanosomatidae) in Chilean Apiaries. *J. Invertebr. Pathol.* **2016**, *134*, 1–5. [CrossRef]
54. Riveros, G.; Arismendi, N.; Zapata, N.; Evans, D.; Pérez, I.; Aldea, P.; Vargas, M. Occurrence, Prevalence and Viral Load of Deformed Wing Virus Variants in *Apis mellifera* Colonies in Chile. *J. Apic. Res.* **2020**, *59*, 63–68. [CrossRef]
55. Sguazza, G.H.; Reynaldi, F.J.; Galosi, C.M.; Pecoraro, M.R. Simultaneous Detection of Bee Viruses by Multiplex PCR. *J. Virol. Methods* **2013**, *194*, 102–106. [CrossRef] [PubMed]
56. Tentcheva, D.; Gauthier, L.; Zappulla, N.; Dainat, B.; Cousserans, F.; Colin, M.E.; Bergoin, M. Prevalence and Seasonal Variations of Six Bee Viruses in *Apis mellifera* L. and *Varroa destructor* Mite Populations in France. *Appl. Environ. Microbiol.* **2004**, *70*, 7185–7191. [CrossRef]
57. Iwanowicz, D.D.; Wu-Smart, J.Y.; Olgun, T.; Smart, A.H.; Otto, C.R.; Lopez, D.; Evans, J.D.; Cornman, R. An Updated Genetic Marker for Detection of Lake Sinai Virus and Metagenetic Applications. *PeerJ* **2020**, *8*, e9424. [CrossRef] [PubMed]
58. Gauthier, L.; Ravallec, M.; Tournaire, M.; Cousserans, F.; Bergoin, M.; Dainat, B.; de Miranda, J.R. Viruses Associated with Ovarian Degeneration in *Apis mellifera* L. Queens. *PLoS ONE* **2011**, *6*, e16217. [CrossRef]
59. Bates, D.; Mächler, M.; Bolker, B.; Walker, S. Fitting Linear Mixed-Effects Models Using Lme4. *arXiv* **2014**, arXiv:1406.5823.
60. ter Braak, C.J.; Smilauer, P. Canoco Reference Manual and User's Guide: Software for Ordination, Version 5.0. 2012. Available online: https://research.wur.nl/en/publications/canoco-reference-manual-and-users-guide-software-for-ordination-v (accessed on 22 September 2021).
61. Gómez-Moracho, T.; Buendía-Abad, M.; Benito, M.; García-Palencia, P.; Barrios, L.; Bartolomé, C.; Maside, X.; Meana, A.; Jiménez-Antón, M.D.; Olías-Molero, A.I. Experimental Evidence of Harmful Effects of *Crithidia mellificae* and *Lotmaria passim* on Honey Bees. *Int. J. Parasitol.* **2020**, *50*, 1117–1124. [CrossRef]
62. Michalczyk, M.; Bancerz-Kisiel, A.; Sokół, R. *Lotmaria passim* as Third Parasite Gastrointestinal Tract of Honey Bees Living in Tree Trunk. *J. Apic. Sci.* **2020**, *64*, 143–151. [CrossRef]
63. Cepero, A.; Ravoet, J.; Gómez-Moracho, T.; Bernal, J.L.; Del Nozal, M.J.; Bartolomé, C.; Maside, X.; Meana, A.; González-Porto, A.V.; de Graaf, D.C. Holistic Screening of Collapsing Honey Bee Colonies in Spain: A Case Study. *BMC Res. Notes* **2014**, *7*, 649. [CrossRef]
64. Higes, M.; Martín-Hernández, R.; Botías, C.; Bailón, E.G.; González-Porto, A.V.; Barrios, L.; Del Nozal, M.J.; Bernal, J.L.; Jiménez, J.J.; Palencia, P.G. How Natural Infection by *Nosema ceranae* Causes Honeybee Colony Collapse. *Environ. Microbiol.* **2008**, *10*, 2659–2669. [CrossRef] [PubMed]
65. Gisder, S.; Hedtke, K.; Möckel, N.; Frielitz, M.-C.; Linde, A.; Genersch, E. Five-Year Cohort Study of *Nosema* Spp. in Germany: Does Climate Shape Virulence and Assertiveness of *Nosema ceranae*? *Appl. Environ. Microbiol.* **2010**, *76*, 3032–3038. [CrossRef] [PubMed]
66. Stevanovic, J.; Stanimirovic, Z.; Genersch, E.; Kovacevic, S.R.; Ljubenkovic, J.; Radakovic, M.; Aleksic, N. Dominance of *Nosema ceranae* in Honey Bees in the Balkan Countries in the Absence of Symptoms of Colony Collapse Disorder. *Apidologie* **2011**, *42*, 49–58. [CrossRef]
67. Higes, M.; Meana, A.; Bartolomé, C.; Botías, C.; Martín-Hernández, R. *Nosema ceranae* (Microsporidia), a Controversial 21st Century Honey Bee Pathogen. *Environ. Microbiol. Rep.* **2013**, *5*, 17–29. [CrossRef] [PubMed]
68. Paxton, R.J.; Klee, J.; Korpela, S.; Fries, I. *Nosema ceranae* Has Infected *Apis mellifera* in Europe since at Least 1998 and May Be More Virulent than *Nosema apis*. *Apidologie* **2007**, *38*, 558–565. [CrossRef]
69. Gisder, S.; Schüler, V.; Horchler, L.L.; Groth, D.; Genersch, E. Long-Term Temporal Trends of *Nosema* Spp. Infection Prevalence in Northeast Germany: Continuous Spread of *Nosema ceranae*, an Emerging Pathogen of Honey Bees (*Apis mellifera*), but No General Replacement of *Nosema apis*. *Front. Cell. Infect. Microbiol.* **2017**, *7*, 301. [CrossRef]
70. Shumkova, R.; Georgieva, A.; Radoslavov, G.; Sirakova, D.; Dzhebir, G.; Neov, B.; Bouga, M.; Hristov, P. The First Report of the Prevalence of *Nosema ceranae* in Bulgaria. *PeerJ* **2018**, *6*, e4252. [CrossRef]
71. Ostroverkhova, N.V.; Konusova, O.L.; Kucher, A.N.; Kireeva, T.N.; Rosseykina, S.A. Prevalence of the Microsporidian *Nosema* Spp. in Honey Bee Populations (*Apis mellifera*) in Some Ecological Regions of North Asia. *Vet. Sci.* **2020**, *7*, 111. [CrossRef]
72. Li, Z.; You, X.; Wang, L.; Yan, Z.; Zhou, Z. Spore Morphology and Ultrastructure of an *Ascosphaera apis* Strain from the Honeybees (*Apis mellifera*) in Southwest China. *Mycologia* **2018**, *110*, 325–338. [CrossRef]
73. Biová, J.; Bzdil, J.; Dostálková, S.; Petrivalsky, M.; Brus, J.; Carra, E.; Danihlík, J. American Foulbrood in the Czech Republic: ERIC II Genotype of *Paenibacillus larvae* Is Prevalent. *Front. Vet. Sci.* **2021**, *8*, 698976. [CrossRef]
74. Erban, T.; Ledvinka, O.; Kamler, M.; Hortova, B.; Nesvorna, M.; Tyl, J.; Titera, D.; Markovic, M.; Hubert, J. European Foulbrood in Czechia after 40 Years: Application of next-Generation Sequencing to Analyze *Melissococcus plutonius* Transmission and Influence on the Bacteriome of *Apis mellifera*. *PeerJ Prepr.* **2017**, *4*, e2618v1.
75. Budge, G.E.; Barrett, B.; Jones, B.; Pietravalle, S.; Marris, G.; Chantawannakul, P.; Thwaites, R.; Hall, J.; Cuthbertson, A.G.; Brown, M.A. The Occurrence of *Melissococcus plutonius* in Healthy Colonies of *Apis mellifera* and the Efficacy of European Foulbrood Control Measures. *J. Invertebr. Pathol.* **2010**, *105*, 164–170. [CrossRef] [PubMed]
76. Roy, C.; Franco, S. Investigation of an Atypical Case of European Foulbrood in France. *Vet. Rec. Case Rep.* **2021**, *9*, e45. [CrossRef]

77. Grangier, V.; Belloy, L.; Charrière, J.-D.; Doherr, M.G.; Fritsche, A.; Waldvogel, A.S. Real-Time PCR as a Decision Aid in the Control of European Foulbrood. *J. Apic. Res.* **2015**, *54*, 366–372. [CrossRef]
78. Kevill, J.L.; de Souza, F.S.; Sharples, C.; Oliver, R.; Schroeder, D.C.; Martin, S.J. DWV-A Lethal to Honey Bees (*Apis mellifera*): A Colony Level Survey of DWV Variants (A, B, and C) in England, Wales, and 32 States across the US. *Viruses* **2019**, *11*, 426. [CrossRef]
79. Olgun, T.; Everhart, S.E.; Anderson, T.; Wu-Smart, J. Comparative Analysis of Viruses in Four Bee Species Collected from Agricultural, Urban, and Natural Landscapes. *PLoS ONE* **2020**, *15*, e0234431.
80. McMahon, D.P.; Natsopoulou, M.E.; Doublet, V.; Fürst, M.; Weging, S.; Brown, M.J.; Gogol-Döring, A.; Paxton, R.J. Elevated Virulence of an Emerging Viral Genotype as a Driver of Honeybee Loss. *Proc. R. Soc. B Biol. Sci.* **2016**, *283*, 20160811. [CrossRef]
81. Tehel, A.; Vu, Q.; Bigot, D.; Gogol-Döring, A.; Koch, P.; Jenkins, C.; Doublet, V.; Theodorou, P.; Paxton, R. The Two Prevalent Genotypes of an Emerging Infectious Disease, Deformed Wing Virus, Cause Equally Low Pupal Mortality and Equally High Wing Deformities in Host Honey Bees. *Viruses* **2019**, *11*, 114. [CrossRef]
82. Gisder, S.; Möckel, N.; Eisenhardt, D.; Genersch, E. In Vivo Evolution of Viral Virulence: Switching of Deformed Wing Virus between Hosts Results in Virulence Changes and Sequence Shifts. *Environ. Microbiol.* **2018**, *20*, 4612–4628. [CrossRef]
83. Ryabov, E.V.; Childers, A.K.; Chen, Y.; Madella, S.; Nessa, A.; Evans, J.D. Recent Spread of *Varroa destructor* Virus-1, a Honey Bee Pathogen, in the United States. *Sci. Rep.* **2017**, *7*, 17447. [CrossRef]
84. Manley, R.; Temperton, B.; Doyle, T.; Gates, D.; Hedges, S.; Boots, M.; Wilfert, L. Knock-on Community Impacts of a Novel Vector: Spillover of Emerging DWV-B from Varroa-infested Honeybees to Wild Bumblebees. *Ecol. Lett.* **2019**, *22*, 1306–1315. [CrossRef]
85. Kevill, J.L.; Highfield, A.; Mordecai, G.J.; Martin, S.J.; Schroeder, D.C. ABC Assay: Method Development and Application to Quantify the Role of Three DWV Master Variants in Overwinter Colony Losses of European Honey Bees. *Viruses* **2017**, *9*, 314. [CrossRef]
86. Berthoud, H.; Imdorf, A.; Haueter, M.; Radloff, S.; Neumann, P. Virus Infections and Winter Losses of Honey Bee Colonies (*Apis mellifera*). *J. Apic. Res.* **2010**, *49*, 60–65. [CrossRef]
87. Nguyen, B.K.; Ribière, M.; vanEngelsdorp, D.; Snoeck, C.; Saegerman, C.; Kalkstein, A.L.; Schurr, F.; Brostaux, Y.; Faucon, J.-P.; Haubruge, E. Effects of Honey Bee Virus Prevalence, *Varroa destructor* Load and Queen Condition on Honey Bee Colony Survival over the Winter in Belgium. *J. Apic. Res.* **2011**, *50*, 195–202. [CrossRef]
88. Faurot-Daniels, C.; Glenny, W.; Daughenbaugh, K.F.; McMenamin, A.J.; Burkle, L.A.; Flenniken, M.L. Longitudinal Monitoring of Honey Bee Colonies Reveals Dynamic Nature of Virus Abundance and Indicates a Negative Impact of Lake Sinai Virus 2 on Colony Health. *PLoS ONE* **2020**, *15*, e0237544. [CrossRef]
89. De Miranda, J.R.; Cornman, R.S.; Evans, J.D.; Semberg, E.; Haddad, N.; Neumann, P.; Gauthier, L. Genome Characterization, Prevalence and Distribution of a Macula-like Virus from *Apis mellifera* and *Varroa destructor*. *Viruses* **2015**, *7*, 3586–3602. [CrossRef]
90. McMenamin, A.J.; Flenniken, M.L. Recently Identified Bee Viruses and Their Impact on Bee Pollinators. *Curr. Opin. Insect Sci.* **2018**, *26*, 120–129. [CrossRef]
91. Abou Kubaa, R.; Molinatto, G.; Khaled, B.; Daher-Hjaij, N.; Heinoun, K.; Saponari, M. First Detection of Black Queen Cell Virus, *Varroa destructor* Macula-like Virus, *Apis mellifera* Filamentous Virus and *Nosema ceranae* in Syrian Honey Bees *Apis mellifera* Syriaca. *Bull. Insectol.* **2018**, *71*, 217–224.
92. von Büren, R.S.; Oehen, B.; Kuhn, N.J.; Erler, S. High-Resolution Maps of Swiss Apiaries and Their Applicability to Study Spatial Distribution of Bacterial Honey Bee Brood Diseases. *PeerJ* **2019**, *7*, e6393. [CrossRef]
93. Brodschneider, R.; Brus, J.; Danihlík, J. Comparison of Apiculture and Winter Mortality of Honey Bee Colonies (*Apis mellifera*) in Austria and Czechia. *Agric. Ecosyst. Environ.* **2019**, *274*, 24–32. [CrossRef]
94. Taric, E.; Glavinic, U.; Vejnovic, B.; Stanojkovic, A.; Aleksic, N.; Dimitrijevic, V.; Stanimirovic, Z. Oxidative Stress, Endoparasite Prevalence and Social Immunity in Bee Colonies Kept Traditionally vs. Those Kept for Commercial Purposes. *Insects* **2020**, *11*, 266. [CrossRef] [PubMed]
95. Tehel, A.; Streicher, T.; Tragust, S.; Paxton, R.J. Experimental Infection of Bumblebees with Honeybee-Associated Viruses: No Direct Fitness Costs but Potential Future Threats to Novel Wild Bee Hosts. *R. Soc. Open Sci.* **2020**, *7*, 200480. [CrossRef] [PubMed]
96. Simone-Finstrom, M.; Li-Byarlay, H.; Huang, M.H.; Strand, M.K.; Rueppell, O.; Tarpy, D.R. Migratory Management and Environmental Conditions Affect Lifespan and Oxidative Stress in Honey Bees. *Sci. Rep.* **2016**, *6*, 32023. [CrossRef] [PubMed]
97. Graystock, P.; Ng, W.H.; Parks, K.; Tripodi, A.D.; Muñiz, P.A.; Fersch, A.A.; Myers, C.R.; McFrederick, Q.S.; McArt, S.H. Dominant Bee Species and Floral Abundance Drive Parasite Temporal Dynamics in Plant-Pollinator Communities. *Nat. Ecol. Evol.* **2020**, *4*, 1358–1367. [CrossRef]

insects

MDPI

Article

Reliability of Morphological and PCR Methods for the Official Diagnosis of *Aethina tumida* (Coleoptera: Nitidulidae): A European Inter-Laboratory Comparison

Stéphanie Franco [1,*], Nicolas Cougoule [1], Amandine Tison [1], Aurélie Del Cont [1], Cristina Gastaldi [1], ILC Consortium [2,†] and Véronique Duquesne [1,*]

[1] Honey Bee Pathology Unit, European Union Reference Laboratory for Bee Health, French Agency for Food, Environmental and Occupational Health and Safety (ANSES), 06902 Sophia Antipolis, France; nicolas.cougoule@anses.fr (N.C.); amandine.tison@anses.fr (A.T.); aurelie.delcont@anses.fr (A.D.C.); cristina.gastaldi@anses.fr (C.G.)

[2] Inter Laboratory Comparison Consortium, European Union Reference Laboratory for Bee Health, 06902 Sophia Antipolis, France; eurl.bee@anses.fr

* Correspondence: stephanie.franco@anses.fr (S.F.); veronique.duquesne@anses.fr (V.D.); Tel.: +33-4-92-94-37-00 (S.F. & V.D.)

† The full author names and affiliations are listed in the Appendix D (Table A4).

Simple Summary: *Aethina tumida*, also called the Small Hive Beetle, is an insect that multiplies primarily in honeybee hives, causing honey losses and weakening colonies. It is native to sub-Saharan Africa and was introduced into different countries and continents over the last 20 years, posing a threat to beekeeping internationally. In case of introduction into a new area, officially approved laboratories (certified by government services) carry out analyses to confirm the outbreak. The reliability of the results is essential in the implementation of management measures. Therefore, a study was organised at the European level to compare the results between official laboratories for two types of methods, used routinely for the identification of *A. tumida*: morphological examination (form and structure) and DNA testing (genetics). The 22 participants analysed in a blinded way a panel of 12 samples (positive and negative samples). The results were very satisfactory, with the exception of one participant who encountered several anomalies for negative samples and especially for DNA tests, probably related to his inexperience with the method. This study proved the ability of laboratories and analytical methods to identify *A. tumida*, which is a key element in monitoring and managing this risk.

Citation: Franco, S.; Cougoule, N.; Tison, A.; Del Cont, A.; Gastaldi, C.; ILC Consortium; Duquesne, V. Reliability of Morphological and PCR Methods for the Official Diagnosis of *Aethina tumida* (Coleoptera: Nitidulidae): A European Inter-Laboratory Comparison. *Insects* **2022**, *13*, 33. https://doi.org/10.3390/insects13010033

Academic Editor: David Ward Roubik

Received: 15 November 2021
Accepted: 22 December 2021
Published: 28 December 2021

Abstract: The Small Hive Beetle (*Aethina tumida* Murray, 1867) is an invasive scavenger of honeybees. Originally endemic in sub-Saharan Africa, it is regulated internationally in order to preserve the areas still free from this species. To ensure the reliability of official diagnoses in case of introduction, an inter-laboratory comparison was organised on the identification of *A. tumida* by morphology and real-time PCR. Twenty-two National Reference Laboratories in Europe participated in the study and analysed 12 samples with adult coleopterans and insect larvae. The performance of the laboratories was evaluated in terms of sensitivity and specificity. Sensitivity was satisfactory for all the participants and both types of methods, thus fully meeting the diagnostic challenge of confirming all truly positive cases as positive. Two participants encountered specificity problems. For one, the anomaly was minor whereas, for the other, the issues concerned a larger number of results, especially real-time PCR, which probably were related to inexperience with this technique. The comparison demonstrated the reliability of official diagnosis, including the entire analytical process of *A. tumida* identification: from the first step of the analysis to the expression of opinions. The performed diagnostic tools, in parallel with field surveillance, are essential to managing *A. tumida* introduction.

Keywords: diagnosis; inter-laboratory comparison; morphology; real-time PCR; honey bee; *Aethina tumida*; Small Hive Beetle

1. Introduction

The Small Hive Beetle (SHB), *Aethina tumida* Murray, 1867 (Coleoptera: Nitidulidae), is an invasive scavenger of honeybee colonies native to sub-Saharan Africa [1]. Part of its biological cycle takes place in *Apis mellifera* colonies and part in the soil. Adult beetles, attracted by the smell of the hive, enter it to reproduce. They lay masses of eggs in wood crevices or inside combs that hatch into larvae. The predatory larvae grow by feeding on bee brood, pollen and honey. Their faeces cause fermentation processes in the hive, making the honey unfit for human consumption. Once they have grown to a sufficient size, after a few weeks, the larvae leave the hive to begin their pupation in the soil. The development of larvae in the hive may cause significant damage for beekeeping, which can, in the most severe cases, result in loss of the entire bee colony and harvest losses [2]. Cases of infestation of bumblebee colonies (*Bombus spp.*) and of solitary bee nests were also reported, but data are still lacking on the impact of *A. tumida* on these species [3,4]. Outside host nests, adult SHBs can feed on alternative sources, such as fruits or foraging on flowers [4,5].

The SHB disperses naturally by flight over unknown distances (possibly more than 10 km) [6]. However, over the last twenty years, it was introduced to different continents outside its natural range [6,7]. Migratory beekeeping, globalisation and international trade play a major role in its spread over long distances [8]. Trade-in wax could facilitate SHB invasion [9]. Therefore, in order to protect *A. tumida*-free areas, infestation by the SHB is regulated at the international level and it is listed as a notifiable disease to the World Organisation for Animal Health (OIE) [10].

In the European Union (EU), the first case of introduction was reported in Portugal in 2004, following the import of a bee queen from Texas, United States [11]. Early detection and effective control measures enabled eradication [12]. Ten years later, in September 2014, the presence of SHB was confirmed in the region of Calabria, in southern Italy [13,14]. Control measures were set up immediately, but visits to the apiaries in the same areas demonstrated that *A. tumida* had spread in this region [15], where it can now be considered as established. So far, containment measures have prevented the spread of SHB to the rest of the EU, which remains free of this parasite. In order to limit the spread of SHB and further introductions into the EU, infestation by *A. tumida* is subject to compulsory surveillance and notification according to the "Animal Health Law". Moreover, regulations apply to intra-EU trade and to imports from non-EU countries [16,17] to prevent SHB introduction.

Detection of *A. tumida* in apiaries relies mainly on inspection of honeybee colonies and installation of traps [18]. These two approaches enable the visual detection of adult coleopterans and larvae. Molecular analysis of hive debris is also an alternative way to detect the presence of SHB DNA [19], although the method needs to be validated in field conditions to better evaluate its sensitivity [3]. When suspicious specimens are detected, a differential diagnosis should be established with other beetles and insects living in the hive environment. Particularly, a distinction should be made with other beetles belonging to the Nitidulidae family, such as *Cychramus luteus* Fabricius, 1787 [20] and *Carpophilus lugubris* Murray, 1864 [21], which do not have a detrimental effect on bees but seek refuge in the hive and feed on pollen or debris. The larva of *A. tumida* can also be mistaken for larvae of the lesser wax moth, *Achroia grisella* Fabricius, 1794 (Lepidoptera: Pyralidae), as well as for the honeycomb moth, *Galleria mellonella* Linnaeus, 1758 (Lepidoptera: Pyralidae). These Lepidoptera are frequently found in colonies and on beekeeping equipment. They can cause damage to the combs of weak colonies and on frames that are not properly stored [22].

In case of suspicion, laboratory diagnosis is crucial to reliably and rapidly identify *A. tumida* and confirm the outbreak. This diagnosis is supported by the quality of the methods and by the competence of the laboratories officially approved to carry out analyses (i.e., certified by government services). Two types of methods are routinely used and recommended in the Manual of Diagnostic Tests and Vaccines for Terrestrial Animals of the OIE for the identification of *A. tumida* [7]: (i) morphological examination, which provides a result in a short amount of time through a low-cost technique and which is therefore particularly indicated for first-line diagnosis, and (ii) PCR, which is more sensitive and specific, and which is generally used as a second step.

In response to diagnostic issues and to ensure the quality of the analytical results obtained within the EU, the European Union Reference Laboratory (EURL) for Bee Health, located in the laboratory of the French Agency for Food, Environmental and Occupational Health and Safety (ANSES) in Sophia-Antipolis (France), organised an inter-laboratory comparison using morphology and molecular identification of *A. tumida* in October 2020. This was the first trial organised using these methods. All the EU National Reference Laboratories (NRLs) were invited to participate. Non-EU countries that were interested in the comparison were also included. The evaluation was supported by a blind analysis of a panel of 12 insect samples. The objective was to evaluate the sensitivity, specificity and accuracy of the results obtained by the laboratories with the analytical methods they routinely use for an official diagnosis.

2. Materials and Methods

2.1. Participating Laboratories

Twenty-two NRLs for Bee Health took part in the inter-laboratory comparison, 21 from EU member states and one from another European country (Table A4). Sixteen participants used morphological and PCR identification methods, whereas six used morphological identification only, depending on their diagnostic possibilities. Importantly, the latter have not yet implemented real-time PCR for the identification of *A. tumida*, and refer to another official laboratory, competent for this method, if confirmation is required.

2.2. Reference Methods for Panel Sample Characterisation

Two reference methods were used by the EURL to characterise and to check the homogeneity and the stability of the samples for comparison: (i) the morphological method (EURL procedure, also published in the OIE Manual), and (ii) the PCR method (EURL procedure, also published in the OIE Manual) [7,23,24]. The EURL is accredited by the French Accreditation Committee (COFRAC) for these two methods, in compliance with the international standard NF EN ISO/IEC 17025 on "General requirements for the competence of testing and calibration laboratories" [25]. Moreover, sequencing of the COI gene was also performed to determine or confirm the species of the panel specimens.

The morphological method consists of the visual examination of individuals (adults and/or larvae) with recording of specific morphological characteristics and, if necessary, comparison of the sample to be identified with a reference sample or detailed photographs. The identification relies on the assessment of eight criteria for the adult and three criteria for the larva (Table 1). A stereomicroscope (and/or magnifier) (minimum 40 $\times$ magnification) is used during the analysis. For the adult, if all the criteria are present, the final result of the examination is "positive". For the larva, as the number of criteria is limited, the presence of all the criteria should be considered a "suspicion" and must always be confirmed by PCR. If at least one criterion is absent, the result of the analysis is "negative". Finally, in some cases (e.g., damaged specimen) the result can be inconclusive because it is not possible to assess the presence of certain morphological identification criteria. In this situation, PCR identification must systematically be performed.

Table 1. Morphological identification of *Aethina tumida* Murray–Criteria for adult and larva diagnosis [7,23].

Criteria	Adult Form
1	Body divided into three parts: head, thorax and abdomen.
2	Three pairs of legs.
3	Presence of elytra [1].
4	Elytra not covering the entire abdomen: some abdominal segments are apparent in dorsal view.
5	Overall uniform body colour (no spots), light brown to black [2].
6	Antenna tips with compact, almost rounded club ends. The three terminal articles of the antennae, corresponding to the "clubs" [3], are narrowed between them, and their length is almost equal to their width.
7	Sharp latero-posterior tips of the pronotum [4].
8	Dimensions. Length: 4 to 7 mm (±1 mm). Width: 3 mm (±1 mm).
	Larval form
1	Three pairs of legs, one on each of the anterior segments, corresponding to the larva thorax.
2	All of the abdominal segments are bare and have no false legs (also called pseudopods) on their ventral part.
3	From the mesothorax [5], presence on each segment, of two dorsal tubercles on either side of the midline. These tubercles are finished with a short fine seta. They look like "spines".

[1] Elytra are sclerotised (meaning thickened) forewings covering the hind wings at rest in beetles and some other insects. [2] Depending on the maturity of the specimen, the colour of adult *A. tumida* varies from light brown/red brown after emergence to dark brown to black when fully mature. [3] In some beetle families, such as the Nitidulidae, the terminal articles of the antennae are larger and club-shaped. [4] The pronotum is the dorsal part of the first segment of the thorax. The first segment of the thorax, called the «prothorax», carries the first pair of legs on the ventral side. [5] The mesothorax corresponds to the second thoracic segment of the larva. It carries the second pair of legs. The prothorax corresponds to the first thoracic segment; it does not have a tubercle, its dorsal part (tergum) is sclerotised.

Real-time PCR was used to confirm specimen identification, using species-specific primers designed by Ward et al. [19] for the COI region of *A. tumida*. Briefly, specimens (adult or larva) are ground manually in an appropriate volume of phosphate buffer (1 mL for one adult specimen, 200 µL for one larva) after a rinse with the same buffer (as specimens are preserved in 70% ethanol). Eighty µL of the crushed suspension is used to extract DNA using a QIAamp DNA Mini kit (Qiagen, Hilden, Germany). The extracted DNA is eluted in 200 µL of elution buffer, according to the manufacturer's recommendations and stored at −20 °C ± 5 °C until further analysis. The realtime PCR is performed in a 25 µL reaction system containing 2x SsoAdvanced Universal Probes Supermix (Bio-Rad, Marnes-La-Coquette, France) 12.5 µL, forward (5′-TCTAAATACTACTTTCTTCGACCCATCR-3′) and reverse (5′-TCCTGGTAGAATTAAAATATAAACTTCTGG-3′) primers (100 µM) 0.4 µL, TaqMan® probe (5′-ATCCAATCCTATACCAACACTTATTTGATTCTTCGGAC-3′) (50 µM) 0.05 µL, TaqMan® Exogenous Internal Positive Control Mixture (Applied Biosystems, Waltham, MA, USA) 2.5 µL, 50× Internal Positive Control (IPC) DNA (Applied Biosystems, Waltham, MA, USA) 0.05 µL, nuclease-free H_2O 4.1 µL and 5 µL of DNA template. The PCR runs consist of an initial step of 3 min at 95 °C, followed by 40 successive cycles of 10 s at 95 °C and 30 s at 60 °C. The reaction is run on a CFX96 RealTime PCR system (Bio-Rad, Marnes-La-Coquette, France). DNA detection is expressed in Ct values. Positive and negative controls are used in each DNA extraction and PCR session and in every run, the non-template controls and the positive controls should have the expected results. The success of the amplification and the absence of inhibition is verified by the result of an exogenous IPC. The specific limit of positivity was characterised during the validation process. Thus, it was defined that a result is positive when the Ct value (cycle threshold) is under 35.

Sequencing of the mitochondrial cytochrome oxidase subunit I (COI) gene was performed on specimens to determine or confirm the species. A fragment of 710 bp was amplified with universal primers LCO1499 (5′-GGTCAACAAATCATAAAGATATTGG-3′)/HCO2198 (5′-TAAACTTCAGGGTGACCAAAAAATCA-3′) [26]. The amplification products were sequenced by the Sanger method, using the two primers previously mentioned. From the consensus sequences, a query of the NCBI databases using a Basic Local Alignment Search Tool (blast.ncbi.nlm.nih.gov) was performed.

2.3. Selection of Inter-Comparison Samples

The panel included two types of samples: (i) positive samples, i.e., adult and larval specimens of *A. tumida* Murray species (Coleoptera, Nitidulidae family), and (ii) negative samples, i.e., adult and larva specimens of insect species other than *A. tumida*.

The adult *A. tumida* specimens (A-POS) were obtained experimentally in the confined laboratory of Fera Science, Ltd. in 2019 (York, UK), whereas the *A. tumida* larvae (L-POS) were collected in Maryland (Beltsville, MD, USA) in 2013.

The negative species were selected based on different criteria of interest:

- The fact that the coleopteran species belong to the Nitidulidae family (i.e., same family as *A. tumida*), and therefore have similar morphological and molecular characteristics, and/or,
- The fact that the species were likely to be found in the honeybee hive environment, and/or,
- The fact that the species presented morphological features close to *A. tumida*, and,
- The availability of specimens in sufficient numbers to constitute the panels and to carry out homogeneity and stability tests.

Thus, four different adult species were selected (Table 2): *Tenebrio molitor* (A-NEG1), *Alphitobius diaperinus* (A-NEG2), *Epuraea luteola* (A-NEG3), and *Cryptolaemus montrouzieri* (A-NEG4). The negative larvae belonged to three different species: *Tenebrio molitor* (L-NEG1), *Galleria mellonella* (L-NEG2), and *Carpophilus dimidiatus* (L-NEG3).

Table 2. Nature and origin of the different specimens included in the sample panel for comparison. (Se: Sensitivity; Sp: Specificity; Ac: Accuracy).

Sample Code	Species			Origin	Criteria Evaluated
		ADULTS			
A-POS	*Aethina tumida* (Murray, 1867)		Coleoptera Nitidulidae Nitidulinae	Specimens obtained experimentally in the confined laboratory of Fera Science, Ltd. in 2019	Se, Ac
A-NEG1	*Tenebrio molitor* (Linnaeus, 1758) "Mealworm Beetle"		Coleoptera Tenebrionidae Tenebrioninae	Specimens obtained in 2018 by the company MICRONUTRIS providing insects for human consumption	Sp, Ac
A-NEG2	*Alphitobius diaperinus* (Panzer, 1797)		Coleoptera Tenebrionidae Tenebrioninae	Specimens collected in 2018 on the frame lid of honeybee hives located near a rabbit farm in Vendée (France)	Sp, Ac

Table 2. *Cont.*

Sample Code	Species			Origin	Criteria Evaluated
A-NEG3	*Epuraea luteola* (Erichson, 1843)		Coleoptera Nitidulidae Epuraeinae	Specimens collected in 2018 on *Citrus sinensis* in Corsica (France) and provided by the Plant Health Laboratory (ANSES, Montpellier, France)	Sp, Ac
A-NEG4	*Cryptolaemus montrouzieri* (Mulsant, 1853) "Ladybug"		Coleoptera Coccinellidae Coccinellidae	Specimens obtained in 2019 by the company BIOLINE AGROSCIENCES providing insects for biological control in the plant field	Sp, Ac
		LARVAE			
L-POS	*Aethina tumida* (Murray, 1867)		Coleoptera Nitidulidae Nitidulinae	Specimens collected in 2013 in Maryland (United States) and provided by J.S. Pettis	Se, Ac
L-NEG1	*Tenebrio molitor* (Linnaeus, 1758) "Mealworm beetle"		Coleoptera Tenebrionidae Tenebrioninae	Specimens obtained in 2018 by the company MICRONUTRIS providing insects for human consumption	Sp, Ac
L-NEG2	*Galleria mellonella*(Linnaeus, 1758) "Wax moth"		Lepidoptera Pyralidae Galleriinae	Specimens obtained experimentally in 2019 by the EURL, from hive frames coming from the EURL apiary and naturally infested by wax moth	Sp, Ac
L-NEG3	*Carpophilus dimidiatus* (Fabricius, 1792) "Sap Beetle"		Coleoptera Nitidulidae Carpophilinae	Specimens collected in 2018 on crops (*Prunus dulcis*) in Côte-d'Or (France), provided by the Plant Health Laboratory (ANSES, Montpellier, France)	Sp, Ac

Each batch of samples was characterised at the EURL by the two independent accredited methods described in Section 2.2 (morphology and PCR), and sequencing of the COI gene was performed for the negative specimens to determine the species.

In total, the panel distributed to the participants contained 12 samples for comparison: two positive and four negative adult coleopterans, two positive and three negative insect larvae, and a "lure", which was either positive or negative. The "lure" sample was not evaluated and was included in the panel to limit the risk of collusion.

Each sample in the panel consisted of one specimen (one adult or one larva) which was packaged in a 2 mL micro-tube, filled with non-denatured ethanol 70%. During the packaging process, the integrity of each specimen was visually controlled. When detected, damaged specimens (e.g., specimens with broken legs, incomplete abdomen segments or broken antennae) were excluded from the assay. The samples were stored at room temperature until shipment.

2.4. Homogeneity and Stability Tests

The homogeneity study was performed for all the sample batches packed in micro-tubes from December 2019 to January 2020. Considering the characteristics of the batches (origin of the specimens particularly) and the number of specimens prepared, the number of samples to be taken was defined specifically for each batch.

For the A-NEG2 and the L-POS batches, which were collected in the field and with a risk of non-homogeneity, the protocol strictly followed the recommendations of the international standard NF ISO 13528 on "Statistical methods for use in proficiency testing by interlaboratory comparisons" [27], which specifies that 10 samples are to be randomly taken per batch and analysed in duplicate. For the other batches, the risk of non-homogeneity was considered low, taking into consideration the following aspects:

- The specimens from batches A-POS, A-NEG1, L-NEG1, A-NEG4 were produced artificially in a controlled environment;
- The specimens from batches L-NEG3 and A-NEG3 were provided and previously analysed by specialised entomologists;
- The EURL produced the *G. mellonella* larvae (L-NEG2) artificially from combs coming from the ANSES apiary in Sophia-Antipolis (territory officially free from *A. tumida* at sampling date in 2019);
- The results of the homogeneity tests of a preliminary study carried out in 2018 were satisfactory for *A. tumida* specimens (A-POS) and *T. molitor* specimens (A-NEG1 and L-NEG1), of the same origin;
- The specimens were visually controlled when the tubes were prepared.

For these samples, the sampling strategy was designed according to the International Laboratory Accreditation Cooperation (ILAC) recommendation [28] and based on the number of samples prepared (Table A1). This protocol made it possible to reduce the sample size and to address the lack of availability of some specimens. The tubes were analysed, first by morphological examination, and then by molecular identification. The samples were analysed in duplicate by PCR (i.e., by analysing two collections per sample). All the results met the expected values of 100% positive results for positive samples, and 100% negative results for negative samples, which validated the homogeneity study.

Results from analyses carried out on samples coming from the EURL entomological collection and from preliminary data showed that the samples were stable in 70% ethanol at room temperature for several years. In order to confirm the stability of the samples during the assay, tests were also performed at the deadline for reporting the results. In compliance with standard NF ISO 13528, three samples from each batch were randomly taken and analysed by the two reference methods of the comparison (morphology and PCR, with a duplicate analysis for the latter). The results were consistent with the expected values, which confirmed the stability of the samples.

2.5. Process for the Interlaboratory Comparison

The trial was organised in compliance with the quality requirements described in international standard NF EN ISO/IEC 17043 on "General requirements for proficiency testing" [29]. The samples were packed and shipped between the EURL and NRLs in compliance with the International Air Transport Association (IATA) regulation for shipping and handling dangerous goods.

Each participating laboratory was anonymously coded with a 1- or 2-digit random number to ensure confidentiality of the results. Each sample included in the blind-test panel was coded with the attribution of a random number between 1 and 12. Participating laboratories received inter-comparison samples with a laboratory code on each tube. After receiving the package, the laboratories were required to store the samples at room temperature until analysis and to send back their results within 24 days after the shipment date of the panel samples.

They were asked to report the qualitative results corresponding to one or both of the methods used to analyse the samples. For morphology, three types of values could be rendered: "positive" (or "suspicion" for larvae), "negative" and "inconclusive" (i.e., impossibility to confirm the presence or absence of certain characteristics). For PCR, there were also three modalities: "positive", "negative" and "inhibited" (i.e., PCR reaction inhibited).

In order to conclude on the results of different analyses performed (morphology and PCR), guidelines were given for expressing opinions. They were based on the decision rules described in Table 3. In particular, these opinions allow for consideration of cases where the results are "inconclusive" in morphology or/and "inhibited" in PCR. These opinions are important for health authorities who must then decide what control measures to implement based on the analytical results.

Table 3. Decision rules for expressing opinions, taking into consideration morphology and PCR results.

Morphology Result	PCR Result			
	Analysis Not Performed	*Positive*	*Negative*	*Inhibited*
ADULT				
Positive	(1)	(1)	(3)	(1)
Negative	(2)	(3)	(2)	(2)
Inconclusive	(4)	(1)	(2)	(4)
LARVA				
Suspicion	(3)	(1)	(2)	(4)
Negative	(2)	(3)	(2)	(2)
Inconclusive	(4)	(1)	(2)	(4)
Opinion				
(1) Positive identification of the Small Hive Beetle, *A. tumida*.				
(2) Negative identification of the Small Hive Beetle, *A. tumida*.				
(3) Suspected identification of the Small Hive Beetle, *A. tumida*. Further analysis is required to ascertain the identification.				
(4) Inconclusive result of *A. tumida* identification.				

For each analytical method, three criteria of performance were assessed by adding the scores obtained for the different samples:

1. Sensitivity, i.e., the ability of the laboratory to give a positive result for a positive sample [30];
2. Specificity, i.e., the ability of the laboratory to give a negative result for a negative sample [30];
3. Accuracy, i.e., the closeness of agreement between the obtained results and the assigned values, definition adapted from international standard NF EN ISO 16140 [31].

The analytical conclusion was not evaluated but was analysed as complementary information.

2.6. Results Evaluation

First, the coherence of the inter-laboratory comparison was checked by analysing the raw results of all the participants. In a second step, the EURL assessed the individual performance of the laboratories and the overall performance of the NRL network.

The assigned values were designated beforehand in compliance with international standard NF ISO 13528. They were based on expert judgement and on the knowledge of the origin and the preparation of the specimens. Moreover, the status of the samples was controlled according to the two independent reference methods (morphology and PCR).

For each analytical method, the results of the participants were evaluated in compliance with international standard NF ISO 13528, by scoring each single result based on the assigned value Then, the three criteria for performance (sensitivity, specificity and accuracy) were assessed by summing, for each method and participant, the scores obtained for the different samples (Table A2). For each performance criterion, a total score of "0" was considered "satisfactory", whereas a total score of "1" was considered a "warning signal" and a total score "$\geq$2" was considered an "action signal", requiring the implementation of corrective actions.

The overall performance of the network was assessed taking into account all the results of the laboratories. The rate of sensitivity was calculated, corresponding to the percentage of true positive results obtained by all participating laboratories out of the total number of positive samples distributed and evaluated (i.e., with a positive assigned value). The rate of specificity was also evaluated corresponding to the percentage of true negative results obtained by all participating laboratories out of the total number of negative samples distributed and evaluated (i.e., with a negative assigned value). Finally, the rate of accuracy corresponded to the percentage of accurate results (true positives and true negatives) obtained by all participating laboratories out of the total number of samples distributed and evaluated. We calculated these rates for the two different techniques and for the analytical conclusion, applying the formulas given by international standard NF EN ISO 22117 on "Microbiology of the food chain-Specific requirements and guidance for proficiency testing by interlaboratory comparison" [30].

2.7. Information on Analytical Methods Employed by the Participants

The participants were asked about the analytical methods they used for the trial. They had to specify the reference of their methods (i.e., the official method disseminated by the EURL that is also published in the OIE Manual or another one, which could for example be published in the literature). If they used the EURL reference method for morphology, detailed results of the examination of each criterion (Table 1) had to be rendered. For PCR, laboratories were requested to report technical information on their extraction method, amplification kit and thermocycler used. In addition, they were asked to provide the Ct value obtained for each sample.

3. Results

3.1. Individual Laboratory Performance for the Morphological Identification of *A. tumida*

The results showed good consistency for all samples in the panel (Figure A1). They complied overall with the expected values. We observed a few disparities for three negative samples (A-NEG4, L-NEG2 and L-NEG3), but they concerned a limited number of laboratories.

Sensitivity was satisfactory for all the participants (score of 0) (Table A3). However, two laboratories encountered problems regarding specificity. Laboratory No. 11 gave a positive result for a negative larva sample (L-NEG2), leading to a score of 2, considered an "action signal". Laboratory No. 6 gave an inconclusive result for the negative larva sample L-NEG3, leading to a score of 1, considering the fact that one morphological criterion was mis-assessed. This score was considered a "warning signal".

Laboratories No. 9 and No. 16 also gave inconclusive results for samples A-NEG4 and L-NEG3, respectively as they were not able to assess the presence of a morphological criterion. However, the results were considered acceptable, given the explanation reported by the participants and the relevant results of the evaluation for the other criteria assessed. Thus, they were given a score of 0 for these samples.

3.2. *Individual Laboratory Performance for PCR Identification of A. tumida*

The PCR results also showed good consistency (Figure A2). With the exception of one laboratory, they complied with the expected values.

Sensitivity was satisfactory for all the participants (score of 0) (Table A3). One laboratory experienced problems regarding specificity. Laboratory No. 11 gave positive results for two negative adult samples (A-NEG1 and A-NEG4) and two negative larva samples (L-NEG1 and L-NEG3), leading to a total score of 8, considered an "action signal".

3.3. *Individual Laboratory Performance for Expressing Opinions*

The instructions specified that a conclusion had to be given for each sample, based on the results of the analyses carried out (Table 2).

The analytical conclusions were overall satisfactory for the positive and negative samples (Figure 1). However, it should be noted that three participants (Nos. 7, 8 and 20) that only used morphological identification concluded, for the positive larvae, "positive identification of SHB, *Aethina tumida*", whereas the expected answer was more precisely: "suspected identification of SHB, *Aethina tumida*. Further analysis is required to ascertain the identification". In this case, molecular identification is necessary to confirm the morphology result. For negative specimens, three laboratories (Nos. 6, 11 and 16) could not conclude with certainty on some samples because of the anomalies observed with their analytical results.

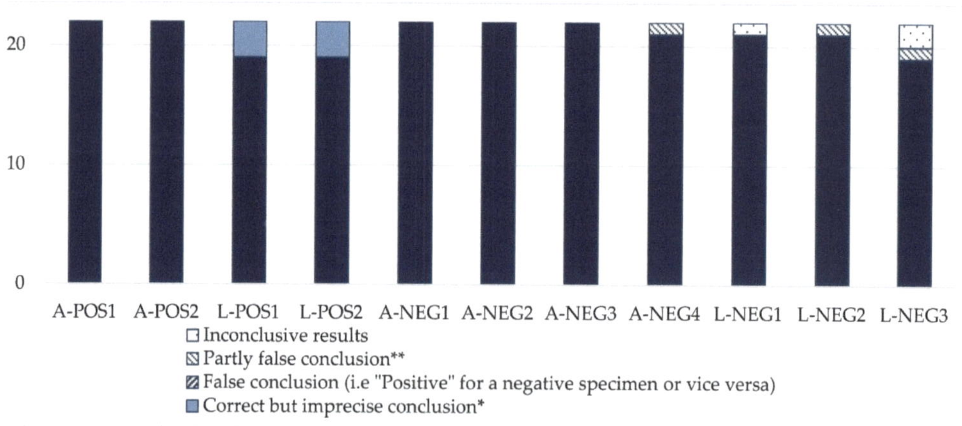

Figure 1. Analytical conclusions. * For larvae: "Positive identification of the Small Hive Beetle, *A. tumida*", instead of "Suspected identification of the Small Hive Beetle, *A. tumida*. Further analysis is required to ascertain the identification", as precisely expected in this case, where identification was based on morphology only. ** "Suspected, further analysis required in confirmation", whereas the specimen was negative.

3.4. *Overall Performance of the Participants*

The overall performance of the participants for morphological analysis reached a rate of 100% for sensitivity, 98.7% for specificity, and an overall accuracy rate of 99.2% (Table 4). The PCR results were also very satisfactory, with 100% sensitivity, 97.4% specificity and 98.3% accuracy. For the analytical conclusion, the rates ranged from 93.2% to 96.1%, due to the minor anomalies observed for the positive larvae, and to the problems regarding specificity identified for some negative specimens.

Table 4. Overall performance of the official laboratory network.

Predicted Result	Morphology		PCR		Analytical Conclusion	
	Compliant Results	Non-Compliant Results	Compliant Results	Non-Compliant Results	Compliant Results	Non-Compliant Results
Positive	88	0	88	0	88	82
Negative	152	2	150	4	148	6
Sensitivity	100% (88/88)		100% (88/88)		93.2% (82/88)	
Specificity	98.7% (152/154)		97.4% (150/154)		96.1% (148/154)	
Accuracy	99.2% ((88 + 152)/(88 + 154))		98.3% (88 + 150)/(88 + 154)		95.0% (88 + 148)/(88 + 154)	

3.5. Performance of the Methods Used for the Official Diagnosis of A. tumida

The 22 participants used the EURL protocol for morphological identification, which is also published in the OIE Manual [7,23].

Out of 16 participants who performed the PCR method to confirm the morphology results, 14 participants used the EURL protocol based on the publication by Ward, et al. [19], which is also referenced in the OIE Manual [7]. However, none of them used the same experimental conditions as those of the EURL in terms of the combination of extraction kit, amplification kit and thermocycler used. The Ct values obtained for the adult SHB specimens ranged from 21.3 to 30.8, and for the larva SHB specimens from 16.4 to 30.5. As mentioned above, one participant (No. 11) reported four positive results on negative specimens (Ct values ranging from 30.9 to 34.9). The other two participants (No. 10 and No. 18) used the PCR method described by Silacci, et al. [32]. The Ct values obtained for the adult SHB specimens ranged from 22 to 39.8, and for the larva SHB specimens from 16.4 to 30.5. The two participants who used this method had different results on the negative specimens. For one of them, no amplification was observed, while for the other Ct values ranging from 34.8 to 40.2 were reported. However, the analytical conclusion of this participant was in accordance with the expected identification. The Ct differences observed in the participants may be due to more or less efficient rinsing of the specimens and to the DNA extraction method. The evaluation of the participants was based on the complete process.

4. Discussion

In general, the surveillance of exotic diseases, also called epidemio-vigilance, aims to detect diseases as early as possible [33,34]. Thus, in a country (or a zone) free from *A. tumida*, the objective is usually to detect any outbreak at an early stage in order to increase the chances of eradication [18]. In this context, it is essential that the performance of official laboratories and tests are appropriate for this intended surveillance and/or control objectives [35]. Sensitivity and specificity are the two main criteria for the evaluation of this performance [33]. For an exotic pathogen and a high-risk disease, the need to have a high level of sensitivity is even more essential because the challenge is to confirm all truly positive cases as positive, and thus avoid missing the detection of a possible introduction [34]. The results of the inter-laboratory comparison showed that sensitivity was satisfactory (overall performance of 100%) for all 22 participants and for both types of methods, morphology and PCR), therefore fully meeting this diagnostic challenge.

Specificity is also an important criterion to eliminate the possibility of false-positive results and thus, to avoid false health alerts and unnecessary destruction of honeybee colonies [34]. The overall performance of the network was 98.9% for morphology and 97.4% for PCR. Two laboratories encountered problems with specificity. Laboratory No. 6 obtained a "warning signal" for their morphological identification (specificity score of 1) due to a misjudgement of criterion 3 on sample L-NEG3, corresponding to a larva of *C.*

dimidiatus. The "spines" are indeed absent in this species. Following the inter-laboratory comparison, the laboratory explained that they damaged the specimen when they took it out of the tube. The specimen was in fact stuck to the wall of the tube, and during removal, it was crushed and cut into two pieces. Therefore, the participant could not properly evaluate the morphological criteria. Analysing the cause showed that this anomaly has a very limited impact and can be evaluated as minor. However, this highlights that specimen handling is a critical point in the analysis. To detach the specimens from the tube, ethanol (70%) can be poured into the tube to loosen the specimen from the wall, possibly by closing the tube and shaking it gently. In addition, in case of mishandling, PCR remains the tool of choice to confirm the result of the morphological analysis. Laboratory No. 11 experienced specificity issues resulting in an "action signal" in both microscopy and PCR (specificity scores of 2 and 8, respectively). The larva sample L-NEG 2 was found to be positive after morphological identification, whereas the expected value was negative (and the PCR result was also negative). The specimen was a wax moth larva with "false legs" on its posterior segments (criterion 2 absent) and had no dorsal "spines" (criterion 3 absent). Moreover, the PCR results were positive for the four negative samples A-NEG1, A-NEG4, L-NEG1 and L-NEG3, whereas their morphological results were negative. The participant indicated that they were inexperienced and that they performed the PCR method for the first time. Furthermore, the participant mentioned that the method used did follow the guidelines described by the EURL; however, the combination of extraction method and reagents used for PCR had not yet been tested against the reference materials. The evaluation of sensitivity had therefore not been defined under the conditions described by this participant. These results demonstrate that the validation and adoption stages of the methods are crucial to ensure the reliability of the analytical results. These steps include participation in training and the use of reference materials [36]. In the framework of the EU network of NRLs, the EURL for Bee Health has organised two training courses to support laboratories in the adoption of the reference methods for *A. tumida* identification: one in 2014 on the morphological identification of *A. tumida*, and a more general one in 2016 on PCR diagnosis of honeybee diseases. Following this training, reference specimens and PCR-positive controls were distributed to laboratories. The inter-laboratory comparison organised in 2020 proved that the adoption process was an overall success for the participants, with the exception of one laboratory for which support (e.g., additional training) will be given.

The results of the comparison confirmed the reproducibility of the official methods for the morphological and molecular identification of *A. tumida*. All the participants strictly used the EURL reference method (OIE method) for morphological identification [7,23]. This method is suitable and reliable for routine official diagnosis in reference laboratories for bee diseases, but not necessarily for specialists when it comes to morphological identification of Coleoptera, which is a field that in itself requires specific skills and equipment. The precise morphological identification of *Aethina* species relies on the use of identification keys and on the expertise of entomologists specialising in Coleoptera, and in the family Nitidulidae itself [37]. Thus, the method may have limited performance for accurate identification of *Aethina* spp. in a research setting, but is highly suitable for diagnostic purposes and meets the operational requirements of international reference laboratories [35]. Diagnosis through PCR technology can be used to confirm the precise identification of the species in case of doubt. However, when a first case is detected in a territory, it would be advisable to implement PCR systematically to avoid any uncertainty or doubt.

The 87% of the laboratories that applied the PCR method corresponding to the EURL reference method (i.e., based on Ward, et al. [19] and also referenced by the OIE) used different types of experimental conditions. Eight different DNA extraction and eleven different PCR mix methods were used by the participating laboratories. This led to the use of 22 different protocols. The diversity of the experimental conditions used in this trial did not, however, affect the PCR result (93% satisfactory results), thus demonstrating the robustness of this method.

The comparison also included a part concerning the expression of opinions. The outcomes were satisfactory and correlated well with the analytical results. This aspect is crucial because it is on this basis that health authorities make decisions on the surveillance or control measures to be applied. However, it should be mentioned that while laboratory diagnostic performance is an important component of *A. tumida* surveillance, it is also essential to combine it with good sampling and thus with high sensitivity of field detection methods [38]. The OIE Manual describes several approaches based on the visual inspection of colonies and the use of traps [7], but they still need to be standardised. In addition, there is a lack of knowledge on their performance level in low-infested colonies. The sensitivity of these field methods may also vary depending on seasons, climate, environment conditions and colony strength [6,39]. Concerning the molecular approach, a method for the detection and the identification of *A. tumida* DNA in hive debris is described [19]. However, again, validation data are lacking to characterise its performance in the field in the context of low infestation rates [3].

Lastly, the assay demonstrated the feasibility of an inter-laboratory comparison organised at the international level on the official diagnosis of *A. tumida*, which is to our knowledge the first in this field. The comparison aimed to include the whole analytical process: the first step of morphological examination of the specimen, the PCR step in confirmation, and finally the expression of opinions. The samples were prepared and stored in ethanol (70%) at room temperature, as expected in routine analysis. The objective was to match the reality of analyses as closely as possible. Thus, the comparison allowed for inconclusive results and therefore a specific scoring system was established to evaluate participant performance, taking into account that specimens could be damaged or that the PCR could be inhibited. In addition, this original scoring system, based on the recommendations of international standard NF ISO 13528 [27], made it possible to prioritise the performance results and translate them into action or warning signals, similar to the calculation of the z-score, which is applied for quantitative trials.

The availability of relevant specimens is a difficulty in organising this type of trial. Specimens of *A. tumida* can quite easily be collected in infested areas or obtained experimentally. The cost to recover these specimens remains significant, however. The most difficult part is to get an appropriate amount and variety of negative specimens to include in the test. It would have been interesting to include more specimens of Nitidulidae, and particularly, some more morphologically close to *A. tumida*, or even other species of beetles frequently found in hives. The challenge is to have a sufficient number of them to constitute all the panels and to perform the homogeneity and stability tests. This practical constraint makes it necessary to adopt the recommendations of NF ISO 13528 for testing the homogeneity of the samples for comparison. Like in the field of plant health comparisons, a relevant risk analysis has to be performed to reduce the number of samples to be analysed according to this standard [40]. This was carried out specifically for each batch of samples, taking into consideration appropriate data available. Finally, it should be mentioned that one of the critical points for the organisation of this type of comparison, including morphological analysis, is the preparation of the samples. Careful examination of the specimens is clearly necessary to eliminate, as far as possible, those that are damaged and for which evaluation of all the criteria may not be possible.

5. Conclusions

In this study, we compared the performances of laboratories in identification methods of *Aethina tumida* across 22 European countries. This inter-laboratory comparison is, to our knowledge, the first organised at the international level on the official diagnosis of *A. tumida*.

It demonstrated the reliability of the reference methods, including the whole analytical process (morphology, PCR confirmation and opinions). These diagnostic tools are essential in the surveillance and management of *A. tumida* introductions in countries where its presence has not yet been shown and where early detection is crucial.

Avenues for improvement were identified, especially the inclusion of Coleoptera species more similar to *A. tumida* among the samples to be blind tested. However, in view of the difficulties encountered, in particular in constituting the sample panel, the study proved the feasibility of organising an international inter-laboratory comparison in this field.

Author Contributions: Conceptualization: S.F. and V.D.; Methodology: S.F. and V.D.; Investigation: all authors, especially, N.C., A.T. and A.D.C. for preparation of the samples, C.G. and ILC Consortium for data acquisition; Formal analysis: S.F. and V.D.; Writing—original draft preparation: S.F. and V.D.; Writing—review and editing: ILC Consortium, especially, B.D., H.G., A.G., H.K., S.M., M.O.S. and E.S. (Table A4). All authors have read and agreed to the published version of the manuscript.

Funding: This research was supported by the Directorate-General for Health and Food Safety (European Commission).

Institutional Review Board Statement: Not applicable.

Acknowledgments: The authors would like to thank: Raphaëlle Mouttet, Jeffery S. Pettis and Samuel Boucher, who provided negative coleopteran samples, which were particularly valuable for the study, and Christine Torres and Stéphane Groncin for administrative and IT support, respectively.

Conflicts of Interest: The authors declare that they have no conflicts of interest. The funders had no role in the design of the study; in the collection, analyses, or interpretation of data; in the writing of the manuscript, or in the decision to publish the results.

Appendix A. Information on Homogeneity Tests

Table A1. Sampling strategy for homogeneity tests and results.

Code	Species	Number of Samples Prepared	Number of Samples Tested	Results
A-POS	*Aethina tumida*	115	10	100% Pos.
A-NEG1	*Tenebrio molitor*	60	7	100% Neg.
A-NEG2	*Alphitobius diaperinus*	60	10	100% Neg.
A-NEG3	*Epuraea luteola*	50	5	100% Neg.
A-NEG4	*Cryptolaemus montrouzieri*	60	7	100% Neg.
L-POS	*Aethina tumida*	115	10	100% Pos.
L-NEG1	*Tenebrio molitor*	75	7	100% Neg.
L-NEG2	*Galleria mellonella*	60	7	100% Neg.
L-NEG3	*Carpophilus dimidiatus*	60	7	100% Neg.

Appendix B. Scoring Individual Performance

Table A2. Methodology of the score evaluation.

Step 1–Scoring of Each Single Results, According to the Following Process	
Score of "0"	Results that matched the assigned values exactly: - "Positive" result of the participant for a positive sample, - "Negative" result for a negative sample, - Or "inconclusive" result instead of positive or negative as assigned values, AND with justification (i.e., explanations and if possible, photos) regarding the state of the damaged sample.

Table A2. *Cont.*

Step 1–Scoring of Each Single Results, According to the Following Process	
Score of "1"	Results that did not match the assigned values exactly: - For morphological analysis, "inconclusive" result instead of positive or negative (as assigned values) AND without any justification regarding the state of the sample, - For PCR: "Inhibited" result instead of positive or negative (as assigned values).
Score of "2"	Results that did not match the assigned values: - "Positive" result of the participant instead of negative, - Or "Negative" result instead of positive.
Step 2–Assessment of the three criteria for individual performance	
Sensitivity	Sum of the scores obtained for all the positive samples of the panel.
Specificity	Sum of the scores obtained for all the negative samples of the panel.
Accuracy	Sum of the scores obtained for all the samples of the panel (positive and negative).
Step 3–Final performance evaluation	
For each performance criterion: - A total score of "0" is "Satisfactory", - A total score of "1" is a "warning signal", - A total score "≥ 2" is an "action signal", requiring the implementation of corrective actions.	

Appendix C. Detailed Results of the Participants

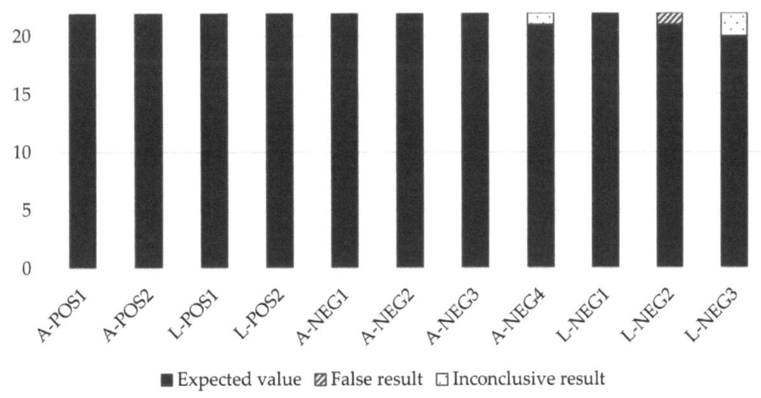

Figure A1. Consistency of the morphology results.

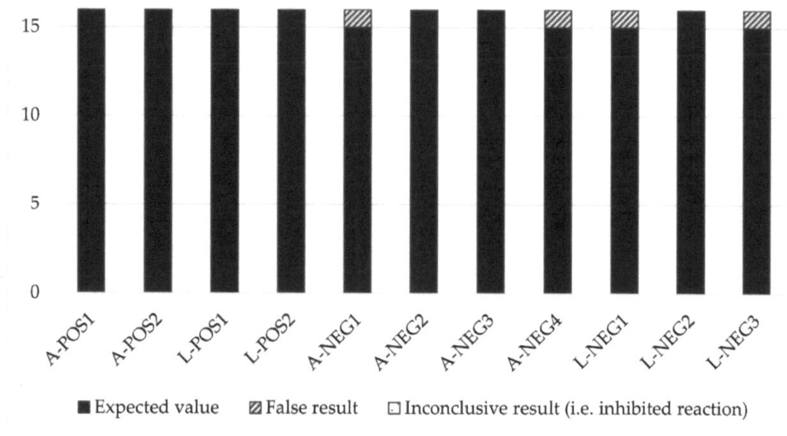

Figure A2. Consistency of the PCR results.

Table A3. Individual performance scores for morphology and PCR.

	Morphology			PCR		
Lab Code	**Acc [1]**	**Se [2]**	**Sp [3]**	**Acc [1]**	**Se [2]**	**Sp [3]**
1	0	0	0	0	0	0
2	0	0	0	0	0	0
3	0	0	0	0	0	0
4	0	0	0	0	0	0
5	0	0	0	0	0	0
6	**1**	0	**1**	0	0	0
7	0	0	0	NP	NP	NP
8	0	0	0	NP	NP	NP
9	0	0	0	0	0	0
10	0	0	0	0	0	0
11	**2**	0	**2**	**8**	0	**8**
12	0	0	0	NP	NP	NP
13	0	0	0	0	0	0
14	0	0	0	0	0	0
15	0	0	0	0	0	0
16	0	0	0	0	0	0
17	0	0	0	NP	NP	NP
18	0	0	0	0	0	0
19	0	0	0	0	0	0
20	0	0	0	NP	NP	NP
21	0	0	0	0	0	0
22	0	0	0	NP	NP	NP

[1] Accuracy; [2] Sensitivity; [3] Specificity. NP = Not performed. A total score of **0** correspond to satisfactory/acceptable results, a score of **1** to a warning signal, and a score ≥ **2** to an action signal. Unsatisfactory or partially satisfactory scores (i.e., "warning signal" or "action signal") are highlighted in bold.

Appendix D. Members of the ILC Consortium

Table A4. Complete membership of the ILC Consortium (names listed in alphabetical order).

Name	Institute	Email
Rosen Aleksandrov	National Diagnostic and Research Veterinary Medical Institute, Aquaculture, Fish and Bee Diseases, NRL for Bee Diseases, 1040 Sofia, Bulgaria	rosen_angelov1989@abv.bg
Gabriela Chioveanu	Institute for Diagnosis and Animal Health, NRL for Animal Health, 50557 Bucharest, Romania	gabriela.chioveanu@idah.ro
Mary Coffey	Department of Agriculture, Food and the Marine, Plant Health Laboratories, W23 X3PH Celbridge, Ireland	MaryF.Coffey@agriculture.gov.ie
Benjamin Dainat	Agroscope Liebefeld, Swiss Bee Research Centre, 3003 Bern, Switzerland	benjamin.dainat@agroscope.admin.ch
María Pilar Fernández Somalo	Laboratorio Central de Veterinaria, Ministerio de Agricultura, Pesca y Alimentación (Gobierno de España), D.G. Sanidad de la Producción Primaria, S.D.G. Sanidad e Higiene Animal y Trazabilidad, 28110 Algete (Madrid), Spain	mfsomalo@mapa.es
Miriam Filipova	Department of Molecular Biology and Epizootiology, Veterinary and Food Institute in Dolny Kubin, State Veterinary and Food Institute, NRL for Honeybee Health, 026 01 Dolny Kubin, Slovakia	miriam.filipova@svpu.sk
Heather Graham	Wageningen Bioveterinary Research, Wageningen University and Research, NRL for Bee Diseases, 8221 RA Lelystad, The Netherlands	heather.graham@wur.nl
Anna Granato	Istituto Zooprofilattico Sperimentale delle Venezie, NRL for Honey Bee Health, 35020 Legnaro (PD), Italy	agranato@izsvenezie.it
Sirpa Heinikainen	Kuopio laboratory, Veterinary Bacteriology and Pathology Unit, Laboratory and Research Division, Finnish Food Authority, 70210 Kuopio, Finland	sirpa.heinikainen@ruokavirasto.fi
Hemma Köglberger	Department for Apiculture and Bee protection, Ages-Austrian Agency for Health and Food Safety, 1220 Vienna, Austria	hemma.koeglberger@ages.at
Nóra Krejczinger	National Food Chain Safety Office, Food Chain Safety Laboratory Directorate, NRL of Parasitology, Fish and Bee Diseases, 1095 Budapest, Hungary	krnori@gmail.com
Merle Kuus	Veterinary and Food Laboratory, 51006 Tartu, Estonia	merle.kuus@vetlab.ee
Severine Matthijs	Sciensano, Enzootic, Vector-borne and Bee Diseases, 1180 Brussels, Belgium	severine.matthijs@sciensano.be

Table A4. *Cont.*

Name	Institute	Email
Konstantinos Oureilidis	Veterinary Laboratory of Kavala, Directorate of Veterinary Center of Thessaloniki, Ministry of Rural Development and Food, 64012 Kavala, Greece	ktekav@yahoo.gr
Zanda Ozolina	Parasitology group, Laboratory of Microbiology and Pathology, Institute of Food Safety, Animal Health and Environment "BIOR", LV-1076, Riga, Latvia	zanda.ozolina@bior.lv
Metka Pislak Ocepek	National Veterinary Institute, Veterinary Faculty, University of Ljubljana, Laboratory for Bee Health Care, Gerbičeva 60, 1000 Ljubljana, Slovenia	metka.pislakocepek@vf.uni-lj.si
Marc Oliver Schäfer	Friedrich Loeffler Institut-Germany, NRL for Bee Diseases, 17493 Greifswald-Insel Riems, Germany	marc.schaefer@fli.de
Emilia Semberg	Swedish University of Agricultural Sciences, Ecology, NRL for Honey Bee Health, 75007 Uppsala, Sweden	emilia.semberg@slu.se
Ivana Tlak Gajger	Laboratory for Honeybee Diseases APISlab, Department for Biology and Pathology of Fish and Bees, Faculty of Veterinary Medicine, University of Zagreb, 10 000 Zagreb, Croatia	ivana.tlak@vef.hr
Maria José Valério	Patologia Apícola, Patologia, Instituto Nacional de Investigacao Agraria e Veterinaria, I. P., 2780-157 Oeiras, Portugal	mjose.valerio@iniav.pt
Sarka Vyroubalova	Department of Pathological Morphology, State Veterinary Institute Olomouc, NRL for Honeybee Health, 77900 Olomouc, Czech Republic	svyroubalova@svuol.cz

References

1. Lundie, A.E. The small hive beetle *Aethina tumida*. *Sci. Bull.* **1940**, *220*, 30.
2. Cuthbertson, A.G.S.; Wakefield, M.E.; Powell, M.E.; Marris, G.; Anderson, H.; Budge, G.E.; Mathers, J.J.; Blackburn, L.F.; Brown, M.A. The small hive beetle *Aethina tumida*: A review of its biology and control measures. *Curr. Zool.* **2013**, *59*, 644–653. [CrossRef]
3. European Food Safety Authority (EFSA-Panel on Animal Health and Animal Welfare). Survival, spread and establishment of the small hive beetle (*Aethina tumida*). *EFSA J.* **2015**, *13*, 4328. [CrossRef]
4. Gonthier, J.; Papach, A.; Straub, L.; Campbell, J.; Williams, G.; Neumann, P. Bees and flowers: How to feed an invasive beetle species. *Ecol. Evol.* **2019**, *9*, 6422–6432. [CrossRef]
5. Buchholz, S.; Schäfer, M.O.; Spiewok, S.; Pettis, J.S.; Duncan, M.; Ritter, W.; Spooner-Hart, R.; Neumann, P. Alternative food sources of *Aethina tumida* (Coleoptera: Nitidulidae). *J. Apic. Res.* **2008**, *47*, 202–209. [CrossRef]
6. Neumann, P.; Pettis, J.S.; Schäfer, M.O. Quo vadis *Aethina tumida*? Biology and control of small hive beetles. *Apidologie* **2016**, *47*, 427–466. [CrossRef]
7. World Organisation for Animal Health (OIE). Infestation with *Aethina tumida* (Small Hive Beetle). In *Manual of Diagnostic Tests and Vaccines for Terrestrial Animals*, 9th ed.; OIE: Paris, France, 2019; pp. 750–764.
8. Mutinelli, F. The spread of pathogens through trade in honey bees and their products (including queen bees and semen): Overview and recent developments. *OIE Rev. Sci. Tech.* **2011**, *30*, 257–271. [CrossRef] [PubMed]
9. Idrissou, F.O.; Huang, Q.; Yañez, O.; Neumann, P. International beeswax trade facilitates small hive beetle invasions. *Sci. Rep.* **2019**, *9*, 10665. [CrossRef] [PubMed]
10. World Organisation for Animal Health (OIE). Infestation with *Aethina tumida* (Small Hive Beetle). In *Terrestrial Animal Health Code*; OIE: Paris, France, 2019; Volume 2, pp. 1–6.

11. Murilhas, A. *Aethina tumida* arrives in Portugal. Will it be eradicated? *EurBee Newsl.* **2005**, *2*, 7–9.
12. Valerio da Silva, M.J. The first report of *Aethina tumida* in the European Union, Portugal, 2004. *Bee World* **2014**, *91*, 90–91. [CrossRef]
13. Mutinelli, F.; Montarsi, F.; Federico, G.; Granato, A.; Ponti, A.M.; Grandinetti, G.; Ferré, N.; Franco, S.; Duquesne, V.; Riviere, M.P.; et al. Detection of *Aethina tumida* Murray (Coleoptera: Nitidulidae.) in Italy: Outbreaks and early reaction measures. *J. Apic. Res.* **2014**, *53*, 569–575. [CrossRef]
14. Palmeri, V.; Scirtò, G.; Malacrinò, A.; Laudani, F.; Campolo, O. A scientific note on a new pest for European honeybees: First report of small hive beetle *Aethina tumida* (Coleoptera: Nitidulidae) in Italy. *Apidologie* **2015**, *46*, 527–529. [CrossRef]
15. Istituto Zooprofilattico Sperimentale delle Venezie (IZSV). Aethina tumida in Italy: Updates. Available online: https://www.izsvenezie.it/aethina-tumida-in-italia/ (accessed on 17 June 2021).
16. European Commission. Commission implementing regulation (EU) 2018/1882 of 3 December 2018 on the application of certain disease prevention and control rules to categories of listed diseases and establishing a list of species and groups of species posing a considerable risk for the spread of those listed diseases. *Off. J. Eur. Union* **2018**, *61*, 21–29.
17. European Commission. Regulation (EU) 2016/429 of the European Parliament and of the Council of 9 March 2016 on transmissible animal diseases and amending and repealing certain acts in the area of animal health ("Animal Health Law"). *Off. J. Eur. Union* **2016**, *59*, 1–208.
18. European Reference Laboratory for Bee Health (EURL); Chauzat, M.-P.; Laurent, M.; Brown, M.; Kryger, P.; Mutinelli, F.; Roelandt, S.; Roels, S.; van der Stede, Y.; Schäfer, M.O.; et al. Guidelines for the Surveillance of the Small Hive Beetle (Aethina tumida) Infestation. Updated Version (April 2016). Available online: https://sitesv2.anses.fr/en/system/files/Guidelines_SHB_surveillance_EURL_V2.pdf (accessed on 19 April 2021).
19. Ward, L.; Brown, M.; Neumann, P.; Wilkins, S.; Pettis, J.; Boonham, N. A DNA method for screening hive debris for the presence of small hive beetle (*Aethina tumida*). *Apidologie* **2007**, *38*, 272–280. [CrossRef]
20. Neumann, P.; Ritter, W. A scientific note on the association of *Cychramus luteus* (Coleoptera: Nitidulidae) with honeybee (*Apis mellifera*) colonies. *Apidologie* **2004**, *35*, 665–666. [CrossRef]
21. Marini, F.; Mutinelli, F.; Montarsi, F.; Cline, A.R.; Gatti, E.; Audisio, P. First report in Italy of the dusky sap beetle, *Carpophilus lugubris*, a new potential pest for Europe. *J. Pest Sci.* **2013**, *86*, 157–160. [CrossRef]
22. Ellis, J.D.; Graham, J.R.; Mortensen, A. Standard methods for wax moth research. *J. Apic. Res.* **2013**, *52*, 1–17. [CrossRef]
23. European Reference Laboratory for Bee Health (EURL). Analytical Method for Animal Health, ANSES/SOP/ANA-I1.MOA.1500, Version 04: "Morphological Identification of the Small Hive Beetle (OIE Method)". Available online: https://sitesv2.anses.fr/en/system/files/Protocol_Instructions_SHB_morphological_Id_2.pdf (accessed on 15 June 2021).
24. European Reference Laboratory for Bee Health (EURL). Analytical Method for Animal Health, ANSES/SOP/ANA-I1.MOA.3700, Version 02: "Identification of the Small Hive Beetle Aethina tumida, Using Real Time PCR (In-House Method)". Available online: https://sitesv2.anses.fr/en/system/files/Protocol_Instructions_SHB_real-time_PCR_2.pdf (accessed on 15 June 2021).
25. International Organization for Standardization (ISO-ISO/CASCO Technical Committee on Conformity Assessment). ISO/IEC 17025: 2017 International Standard on General Requirements for the Competence of Testing and Calibration Laboratories. ISO, Switzerland, 2017. Available online: https://www.iso.org/standard/66912.html (accessed on 28 December 2021).
26. Folmer, O.; Black, M.; Hoeh, W.; Lutz, R.; Vrijenhoek, R. DNA primers for amplification of mitochondrial cytochrome c oxidase subunit I from diverse metazoan invertebrates. *Mol. Mar. Biol. Biotechnol.* **1994**, *3*, 294–299. [PubMed]
27. International Organization for Standardization (ISO-ISO/TC69/SC6 Technical Committee on Measurement Methods and Results). ISO 13528:2015 International Standard on Statistical Methods for Use in Proficiency Testing by Interlaboratory Comparison. ISO, Switzerland, 2015. Available online: https://www.iso.org/standard/56125.html (accessed on 28 December 2021).
28. International Laboratory Accreditation Cooperation (ILAC-Proficiency Testing Consultative Group, Task Force on H&S Testing). ILAC Discussion Paper on Homogeneity and Stability Testing. ILAC, Australia, 2000. Available online: https://www.academia.edu/27733690/ILAC_Discussion_Paper_on_Homogeneity_and_Stability_Testing (accessed on 15 June 2021).
29. International Organization for Standardization (ISO-ISO/CASCO Technical Committee on Conformity Assessment). ISO/IEC 17043:2010 International Standard on Conformity Assessment-General Requirements for Proficiency Testing. ISO, Switzerland, 2010. Available online: https://www.iso.org/standard/29366.html (accessed on 28 December 2021).
30. International Organization for Standardization (ISO-ISO/TC34/SC9 Microbiology Technical Committee). NF EN ISO 22117:2019 International Standard on Microbiology of the Food Chain-Specific Requirements and Guidance for Proficiency Testing by Interlaboratory Comparison. ISO, Switzerland, 2019. Available online: https://www.iso.org/standard/67052.html (accessed on 28 December 2021).
31. International Organization for Standardization (ISO-ISO/TC34/SC9 Microbiology Technical Committee). ISO 16140-1:2016 International Standard on Microbiology of the Food Chain-Method Validation-Part 1: Vocabulary. ISO, Switzerland, 2016. Available online: https://www.iso.org/standard/54869.html (accessed on 15 June 2021).
32. Silacci, P.; Biolley, C.; Jud, C.; Charrière, J.-D.; Dainat, B. An improved DNA method to unambiguously detect small hive beetle *Aethina tumida*, an invasive pest of honeybee colonies. *Pest Manag. Sci.* **2018**, *74*, 2667–2670. [CrossRef]
33. World Organisation for Animal Health (OIE). Animal Health Surveillance. In *Terrestrial Animal Health Code*, 28th ed.; OIE: Paris, France, 2019; pp. 1–10.
34. Dufour, B.; Hendrikx, P.; Toma, B. The design and establishment of epidemiological surveillance systems for high-risk diseases in developed countries. *Rev. Sci. Et Tech. (Int. Off. Epizoot.)* **2006**, *25*, 187–198. [CrossRef] [PubMed]

35. World Organisation for Animal Health (OIE). Principles and methods of validation of diagnostic assays for infectious diseases. In *Manual of Diagnostic Tests and Vaccines for Terrestrial Animals*, 9th ed.; OIE: Paris, France, 2019; pp. 11–22.
36. Rozet, E.; Dewé, W.; Ziemons, E.; Bouklouze, A.; Boulanger, B.; Hubert, P. Methodologies for the transfer of analytical methods: A review. *J. Chromatog. B* **2009**, *877*, 2214–2223. [CrossRef]
37. Lee, S.; Hong, K.-J.; Cho, Y.S.; Choi, Y.S.; Yoo, M.-S.; Lee, S. Review of the subgenus *Aethina* Erichson s. str. (Coleoptera: Nitidulidae: Nitidulinae) in Korea, reporting recent invasion of small hive beetle, *Aethina tumida*. *J. Asia Pac. Entomol.* **2017**, *20*, 553–558. [CrossRef]
38. World Organisation for Animal Health (OIE). Collection, submission and storage of diagnostic specimens. In *Manual of Diagnostic Tests and Vaccines for Terrestrial Animals*, 9th ed.; OIE: Paris, France, 2019; pp. 72–87.
39. European Food Safety Authority (EFSA). EFSA scientific report on small hive beetle diagnosis and risk reduction options. *EFSA J.* **2015**, *13*, 4048. [CrossRef]
40. Chabirand, A.; Anthoine, G.; Pierson, O.; Hostachy, B. The organization of proficiency testing in plant pathology (qualitative methods of analysis) according to the ISO/IEC 17043: Example of the French national reference laboratory. *Accredit. Qual. Assur.* **2014**, *19*, 111–125. [CrossRef]

insects

MDPI

Article

Micronutrient Deficiency May Be Associated with the Onset of Chalkbrood Disease in Honey Bees

Ratko Pavlović [1], Robert Brodschneider [2,*], Walter Goessler [3], Ljubiša Stanisavljević [4], Zoran Vujčić [1] and Nenad M. Zarić [3,4,*]

1 Department of Biochemistry, Faculty of Chemistry, University of Belgrade, Studentski trg 12-16, 11000 Belgrade, Serbia; ratkopav@gmail.com (R.P.); zvujcic@chem.bg.ac.rs (Z.V.)
2 Department of Biology, University of Graz, Universitätsplatz 2, 8010 Graz, Austria
3 Analytical Chemistry for Health and Environment, Institute of Chemistry, University of Graz, Universitätsplatz 1, 8010 Graz, Austria; walter.goessler@uni-graz.at
4 Faculty of Biology, University of Belgrade, Studentski trg 12-16, 11000 Belgrade, Serbia; ljstanis@bio.bg.ac.rs
* Correspondence: robert.brodschneider@uni-graz.at (R.B.); nenad.zaric@bio.bg.ac.rs (N.M.Z.)

Simple Summary: Chalkbrood is a fungal honey bee disease that effects honey bee larvae. It usually does not cause colony death. However, it can weaken the colony by reducing the number of bees and thus reducing chances for colony survival. It is known that poor nutrition can be behind the onset of honey bee diseases. Until now, the mineral content of larvae and mummies and the onset of chalkbrood disease have not been linked. Here, we show that there are differences in the elemental composition of larvae from colonies with different statuses of the chalkbrood disease. Mummies had higher concentrations of macroelements in comparison to apparently typically developed larvae from the same hive, while at the same time they had much lower concentrations of microelements that have known antifungal and antimicrobial activities. It may be that the lack of these elements contributes to the onset of chalkbrood disease.

Abstract: Chalkbrood is a disease of honey bee brood caused by the fungal parasite *Ascosphaera apis*. Many factors such as genetics, temperature, humidity and nutrition influence the appearance of clinical symptoms. Poor nutrition impairs the immune system, which favors the manifestation of symptoms of many honey bee diseases. However, a direct link between dietary ingredients and the symptoms of chalkbrood disease has not yet been established. We show here that the elemental composition of chalkbrood mummies and healthy larvae from the same infected hives differ, as well as that mummies differ from larvae from healthy hives. Chalkbrood mummies had the highest concentration of macroelements such as Na, Mg, P, S, K and Ca and some microelements such as Rb and Sn, and at the same time the lowest concentration of B, As, Sr, Ag, Cd, Sb, Ba and Pb. Larvae from infected hives contained less Pb, Ba, Cs, Sb, Cd, Sr, As, Zn, Cu, Ni, Co, Mn, Cr, V and Al in contrast to healthy larvae from a disease-free apiary. This is the first study to demonstrate such differences, suggesting that an infection alters the larval nutrition or that nutrition is a predisposition for the outbreak of a chalkbrood infection. Though, based on results obtained from a case study, rather than from a controlled experiment, our findings stress the differences in elements of healthy versus diseased honey bee larvae.

Keywords: *Apis mellifera*; *Ascosphaera apis*; element composition; nutrition; feeding

Citation: Pavlović, R.; Brodschneider, R.; Goessler, W.; Stanisavljević, L.; Vujčić, Z.; Zarić, N.M. Micronutrient Deficiency May Be Associated with the Onset of Chalkbrood Disease in Honey Bees. *Insects* **2024**, *15*, 269. https://doi.org/10.3390/insects15040269

Academic Editors: Kirk E. Anderson and Silvio Erler

Received: 20 February 2024
Revised: 9 April 2024
Accepted: 11 April 2024
Published: 12 April 2024

1. Introduction

Honey bees, important pollinators, are threatened by the reduction of the availability and diversity of pollen and nectar sources due to land-use changes. Malnutrition can directly contribute to poor bee health, or indirectly effect their immunocompetence [1–3]. Honey bees consume macro- and micronutrients from nectar, pollen and water to adequately meet their nutritional requirements [4]. Feeding a diverse (multifloral) pollen diet may be a crucial factor for honey bee health [5–8]. However, the role of micronutrients in honey bee health is not yet completely understood [9]. Minerals and nutrients are vital for the reproduction and development of adult bees and larvae. It is known that some elements are essential to honey bees (Na, K, Ca, Mg, P), while others can be toxic (Al, Pb, Cd, Ba) [10,11]. Al is not a proven nutrient. It can replace other metals, especially magnesium, in protein-buried sites and induce conformational defects and changes in the protonation states of protein sidechains [12]. Pb can influence the activity of enzymes and antioxidants. It causes oxidative stress and mostly effects the nervous system [13]. Cd can have a catalytic role in the production of reactive oxygen species, increasing oxidative stress. It can interfere with DNA-repair mechanisms, ending in cell death [14]. Ba has a negative influence on K^+ accumulation inside the cells, causing hyperkalemia, which causes the depolarization of membranes [13].

Colonies can be provided with additional essential minerals through supplemental feeding. However, the definite requirements and optimal amounts that honey bees need are still mostly unexplored [4,15]. Honey bee larvae are to a certain extend buffered against the changes in the food supply of the colony [16]. The effects of larval nutrition on the susceptibility of bees to disease have not been studied [17].

Chalkbrood is a honey bee brood disease caused by the fungus *Ascosphaera apis* (Ascomycota: Eurotiomycetes: Ascosphaerales). The disease is widespread worldwide, and there is evidence that the incidence of chalkbrood may be increasing [18–21]. The typical symptoms are irregular wax capping over the brood and uncapped cells scattered over the brood frames. Mummies can often be seen in cells, at the hive entrance or found on the bottom board, where they are removed by worker bees [22]. At first, dead larvae are covered by a fluffy white mold and swollen to the hexagonal shape of the cell. Later they shrink into 'mummies', and may become grey/black if spore cysts form [23]. Around 5–37% less honey is produced globally as a result of the disease, due to decreased productivity in affected bee colonies [24]. This happens due to a reduction in workforce caused by mycosis [25]. Although it is usually not lethal to the colony, it can hinder its development by reducing its population [26]. While adult bees are not susceptible to this pathogen, they can transmit the disease within and between beehives [19]. Inherited genetic traits, such as hygienic behavior, play a role in preventing the onset of chalkbrood disease [27]. Temperature and humidity are also factors that contribute to commencement of this disease [28,29]. The chilling of brood cells 24 h before or after they are sealed is an important predisposing factor for chalkbrood disease [30]. This fungus is best regarded as an opportunistic pathogen which is efficiently dispersed and very widespread, and its presence in the larvae does not necessarily cause the disease to appear; one or more predisposing conditions must occur at the same time for the disease to develop [23].

The development of healthy honey bee colonies is supported by adequate nutrition [4], but so far there has been no direct link between poor larval nutrition and chalkbrood disease in honey bees. As suggested by recent studies, negative effects of infectious viral and fungal diseases can be increased by the poor nutrition of honey bees [31]. At the same time, the nutritional physiology of honey bees can be negatively affected by common bee pathogens and parasites [32,33]. This holds the potential for deleterious feedback loops between poor nutrition and infectious disease, which may contribute to a spiral of deteriorating bee health [31]. Here, we investigated whether mummies and larvae from infected colonies show differences in elemental composition compared to larvae from healthy honey bee colonies.

2. Materials and Methods

Honey bee larvae and chalkbrood mummies were collected from two apiaries in Vršac, Serbia. One is positioned in an urban environment (45°06′35.3″ N 21°18′27.0″ E), and another one in a rural area (45°08′14.9″ N 21°20′04.2″ E), with around 3.7 km distance between them. The urban apiary hosted up to 15 hives, while the rural apiary had more than 60 hives with at least 100 more hives nearby. The urban apiary is considered disease-free, as the chalkbrood disease has not been present in the past ten years, while chalkbrood was occasionally present in the rural apiary. At the urban apiary, we sampled 25 larvae from each of three hives on 3 June 2023. From the rural apiary, we sampled 25 mummies and larvae from each of three hives which had moderate clinical symptoms of chalkbrood, and 25 larvae from each of three hives that did not show any symptoms of disease. Mummies were easy to spot on the frame, but only a few of them could be spotted on the bottom or landing board. Sampling was conducted during nectar dearth, and infected hives had notably smaller honey and pollen reserves in contrast to healthy hives. All sampled larvae were approximately 5 days old post-hatching and appeared healthy. They were collected from the comb cells closest to capped brood, and the first signs of the capping process were visible. Mummies were collected only from frames (not from the bottom board) only from those cells that were partially opened by bees, with no regard to their color. The samples were pooled, each containing 75 larvae or mummies, leaving one pooled sample, each of the following: (1) larvae from the urban disease-free apiary, (2) larvae from rural hives that do not show symptoms of chalkbrood disease, (3) larvae from rural hives that show symptoms of chalkbrood disease and (4) mummies from rural colonies with chalkbrood disease. Each of the pooled samples were analyzed in triplicate, adding up to $n = 12$.

2.1. Chemicals and Standards

A purification system (Milli-Q, Merck Millipore, Darmstadt, Germany) was used to provide purified water (18.2 MΩ cm). Nitric acid (HNO_3) Rotipuran p. a. $\geq$ 65% (Carl Roth, Karlsruhe, Germany) was sub-boiled with an MLS duoPUR (MLS, Leutkirch, Germany) prior to its use for the preparation of samples. For internal standards and the preparation of calibration standards, we used ICP Single-Element Standards Certipur (Merck Millipore, Darmstadt, Germany) and Single Element Standards for ICP (Carl Roth, Karlsruhe, Germany). Fifteen and fifty mL Cellstar polypropylene tubes (Greiner Bio-One International GmbH, Kremsmünster, Austria) were used for the preparation of all solutions.

2.2. Sample Preparation

The sample preparation was adopted from [34]. In short, 100 mg of freeze-dried and homogenized honey bee samples was digested in an ultraCLAVE IV microwave digestion system (MLS GmbH, Leutkirch, Germany) using 5 mL of concentrated HNO_3. Each digestion was accompanied by three digestion blanks (5 mL conc. HNO_3) and three reference materials BOVM-1 "Bovine muscle powder" (NRC, Ottawa, ON, Canada). After digestion, samples were left to cool, transferred to 50 mL Cellstar tubes, and diluted with ultrapure water to a final volume of 50 mL (10% (v/v) nitric acid).

2.3. Determination of Element Concentrations

Element concentrations were determined as described [34] using inductively coupled plasma mass spectrometry—ICPMS (Agilent ICPMS 7700x, Waldbronn, Germany). We have used an external calibration curve with six points and four concentration ranges. The calibration curve was made in 10% HNO_3 to match the sample matrix. Calibration curve ranges (Text S1), the performance of the instrument (Table S1), the selected mass, the tune mode and internal standard for correction for each element analyzed and the detection limits (Table S2) are reported in the Supplementary Material.

2.4. Quality Control

Quality control was achieved through the continuous addition of Be, Ge, In and Lu (200 μg L^{-1} in 1% v/v HNO_3) and analyses of drift standards (after every 10 samples). Extraction efficiency was evaluated by subjecting BOVM-1: Bovine Muscle Certified Reference Material for Trace Metals and other Constituents (NRC, Canada) through the same digestion process as the samples (Table S3). In addition, accuracy was also evaluated using SRM 1643f Trace elements in natural water (National Institute of Standards & Technology, Gaithersburg, MD, USA) (Table S4).

2.5. Statistical Analyses

Microsoft Excel 2021, version 2108 and IBM SPSS Statistics 25 were used to process the data statistically. Descriptive statistics including mean concentrations and standard deviations were calculated (Table S5). To assess statistically significant differences between samples, an ANOVA followed by Tukey's HSD test was applied to the dataset (Table S5).

3. Results and Discussion

Chalkbrood mummies had higher concentrations of macroelements such as Na, Mg, P, S, K, Ca and some microelements, like Rb, compared to all other samples (Figure 1a,b,d, Supplementary Material Table S5). At the same time, the mummies had the lowest concentrations of B, As, Sr, Ag, Cd, Sb, Ba and Pb (Figure 1d,e). Mummies and larvae from infected hives contained statistically less Al, Cr, Mn, Co, Ni, Cu, Zn, As, Sr, Cd, Sb and Cs compared to larvae from uninfected hives in the same apiary and to larvae from the disease-free apiary (Figure 1, Table S5).

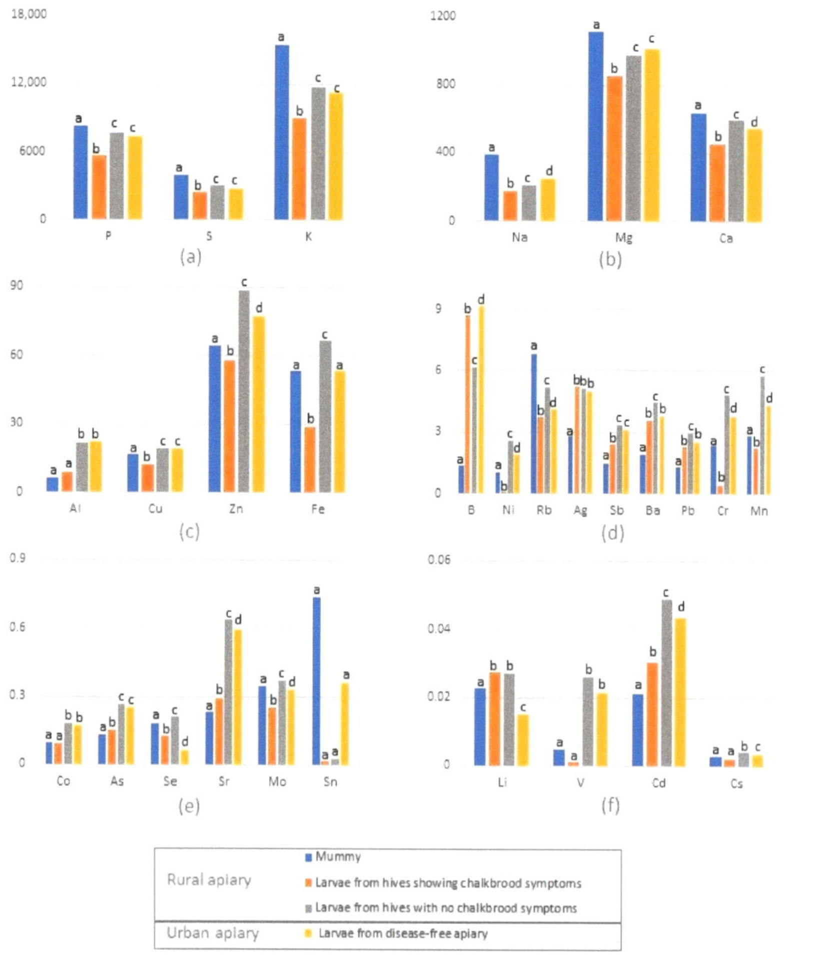

Figure 1. (**a**–**f**) Element concentrations (mg kg^{-1} dry weight) in mummies, larvae from hives showing chalkbrood symptoms and larvae from hives not showing chalkbrood symptoms from rural apiary and larvae from disease-free urban apiary (n = 12); different lower-case letters (a, b, c and d) represent statistically significant differences in element concentrations between samples.

This could indicate that the larvae from infected hives consumed a more uniform diet that is rich in macroelements, but lacking in microelements, compared to larvae from healthy hives, where the larvae are fed based on diverse food sources. Earlier studies concluded that the composition of the food is more important than the amount of food consumed [35]. It could be that honey bees from the infected hives either collected less pollen or the pollen collected was of a lower diversity. There are a number of reasons for this, including genetics (pollen-hoarding selection) [27,36], feedback between poor nutrition and infectious disease or intraspecific competition between honey bee colonies [31,37]. It is possible that the foragers from infected hives were outcompeted for pollen by honey bees from nearby, uninfected hives in the same apiary. Previous studies have shown that bees from different hives in the same apiary accumulate different amounts of elements in them. In the same study, bees from the same hive also had different element concentrations [34]. Recent studied concluded that most of the elements accumulated in bees originate from the food they eat [38–40]. This same food is processed and used to feed larvae; hence it is safe to assume that not all larvae receive the same elemental composition of food. Even in a normally functioning colony with adequate resources, some larvae are seemingly stochastically ignored by nurses for sufficient time to result in some amount of "hunger", and the deprived larva pheromonally sends a "hunger signal" to positively influence its chances of being fed [41]. This means that not all larvae are fed equally, hence not receiving the same quantity and quality of food [42,43]. There is a possibility that the infected larvae try to compensate for the nutrient deficiency caused by the nutrient consumption of *Ascosphaera apis* in the same way in which infected bees try to compensate for the energetic stress caused by *Nosema ceranae*, through eating more, by sending more "hunger signal", and in this way positively influences its chances to receive more jelly [41,44].

Based on the data presented here, we suggest that the deficiency of some elements could either be a possible cause contributing to the onset of chalkbrood disease or a consequence of the infection. Mainly, B, As, Sr, Ag, Cd, Sb, Ba and Pb are interesting elements, as we found their concentrations in mummies to be much lower than in larvae from the same colonies (Figure 1). As already mentioned, not all larvae from one colony will turn into mummies. Only the infected larvae that are mummified have lower concentrations of the mentioned elements. In addition, larvae and mummies from hives that show symptoms of chalkbrood had lower concentrations of Al, Cu, Zn, Ni, Cr, Mn, Co, Mo, V and Cs compared to larvae from hives that did not show the onset of disease and to the larvae from hives in the disease-free apiary. The fact that larvae contain more macroelements in healthy colonies in comparison to colonies with chalkbrood symptoms strengthens our hypothesis that chalkbrood is associated with poor nutrition. Larvae in healthy hives are generally better fed.

Boron concentration in the mummies is very low in contrast to the larvae from the infected hives, but also from all other hives. Several boron-containing compounds show excellent antifungal activity against important fungal pathogens *Mycosphaerella fijiensis* and *Colletotrichum sublineolum* [45]. Low B concentrations in mummies compared to larvae from the same hive suggests, as already mentioned, that not all larvae are fed food with same elemental composition. Aluminum and its salts have attracted attention in recent decades as an alternative to chemical antifungals. It could be referred to as an organic antifungal. They have been shown to be effective against pathogens from the *Ascomycota* phylum [46]. Al is mostly excreted by honey bees [47]. This could be the reason it is well tolerated and why Zaric et al. [48] found levels in live honey bees that are 10-20 times higher than in previous studies. Our results show a lower zinc content in larvae from infected hives and mummies compared to larvae from healthy colonies. Zinc supplementation increases the antioxidant defenses of honey bees [49]. A preference for zinc-treated food has been observed in *Apis mellifera*, underlining the importance of zinc for honey bee colonies [50]. Zinc oxide and silver nanoparticles inhibit *Ascosphaera apis* [51]. We observed a much lower silver concentration in mummies in contrast to larvae from infected but also from healthy colonies. Ag is well known to be antimicrobial and antibacterial [52,53]. It is possible that only larvae fed a low quantity of certain microelements are mummified while others are protected by these same elements. Larvae prior to mummification were probably fed a diet low in this element, lowering its antimicrobial activity and thus providing the opportunity for the onset of chalkbrood. In addition, most of the elements, including V, Cr, Mn and Sr, which we found in very low concentrations in mummies and larvae from infected colonies, are known for their antifungal properties [54–56]. Hence, it could be hypothesized that a lack of these elements favors the outbreak of chalkbrood; we suggest further controlled studies.

4. Conclusions

Chalkbrood is caused by *Ascosphaera apis* spores, which must be present in the colony before clinical symptoms break out. In an infected colony, some of the larvae develop into adult bees and some of them die and become mummies. Our data show that some elements have higher concentrations in mummies compared to healthy larvae from the same colony, while for other elements the opposite is true. This is possibly the outcome of the qualitative properties of the food that is fed to these larvae, which dependents on the pollen composition. Infected larvae that will turn into mummies are most likely fed a higher quantity of food that is rich in some elements (P, S, K, Na, Mg, Ca, Cu, Zn, Rb, Cr, Mn, and Mo), while lacking others (Al, B, Ag, Sb, Ba, Pb, As, Sr and Cd), some of which exhibit antifungal properties. The lack of antifungal elements in the larval diet could be the reason behind the onset of chalkbrood and the mumification of some larvae within infected hives. However, sampling more locations and different timepoints as well as different ages of larvae is needed for a more conclusive conclusion.

Supplementary Materials: The following supporting information can be downloaded at https://www.mdpi.com/article/10.3390/insects15040269/s1, Text S1: Calibration-curve ranges; Table S1. Performance of the ICPMS; Table S2. Selected mass, tune mode, internal standard and detection limits (LoD); Table S3. Certified and determined values for elements in NIST SRM 1643f Trace Elements in Natural Water; Table S4 Certified and determined values for elements in CRM BOVN-1 Bovine Muscle Powder; Table S5. Descriptive statistics and ANOVA (mg kg^{-1} dry weight ± standard deviation).

Author Contributions: R.P.: Conceptualization, Methodology, Investigation, Writing—original draft; R.B.: Investigation, Writing—review and editing; W.G.: Methodology, Resources, Writing—review and editing, Supervision; L.S.: Investigation, Writing—review and editing; Z.V.: Investigation, Writing—review and editing; N.M.Z.: Conceptualization, Data curation, Formal analysis, Investigation, Resources, Writing—original draft, Writing—review and editing, Supervision. All authors have read and agreed to the published version of the manuscript.

Funding: This work was funded by a bilateral project between Serbia and Austria, under the Ministry of Science, technological development and innovation (No. 337-00-577/2021-09/19) and OeAD-GmbH (No. RS 17/2022). The authors acknowledge the open access funding by the University of Graz. Part of the funding was provided by contracts No. 451-03-47/2023-01/200178, 451-03-47/2023-01/200168 and 451-03-65/2024-03/200178 from the Ministry of Science, Technological Development and Innovation of the Republic of Serbia, and CEEPUS network SI-0905-10-2324.

Data Availability Statement: The data are available in Supplementary Material Table S5.

Acknowledgments: This work was made possible by a bilateral project between Serbia and Austria, funded by the Ministry of Science, Technological Development and Innovation and the OeAD-GmbH. The authors gratefully acknowledge the financial support of the University of Graz. We also acknowledge financial support from the Ministry of Science, Technological Development and Innovation of the Republic of Serbia.

Conflicts of Interest: The authors declare no conflicts of interest.

References

1. Alaux, C.; Ducloz, F.; Crauser, D.; Le Conte, Y. Diet Effects on Honeybee Immunocompetence. *Biol. Lett.* **2010**, 6, 562–565. [CrossRef]

2. Goulson, D.; Nicholls, E.; Botías, C.; Rotheray, E.L. Bee Declines Driven by Combined Stress from Parasites, Pesticides, and Lack of Flowers. *Science* **2015**, *347*, 1255957. [CrossRef]
3. Danihlík, J.; Škrabišová, M.; Lenobel, R.; Šebela, M.; Omar, E.; Petřivalský, M.; Crailsheim, K.; Brodschneider, R. Does the Pollen Diet Influence the Production and Expression of Antimicrobial Peptides in Individual Honey Bees? *Insects* **2018**, *9*, 79. [CrossRef]
4. Brodschneider, R.; Crailsheim, K. Nutrition and Health in Honey Bees. *Apidologie* **2010**, *41*, 278–294. [CrossRef]
5. Di Pasquale, G.; Salignon, M.; Le Conte, Y.; Belzunces, L.P.; Decourtye, A.; Kretzschmar, A.; Suchail, S.; Brunet, J.-L.; Alaux, C. Influence of Pollen Nutrition on Honey Bee Health: Do Pollen Quality and Diversity Matter? *PLoS ONE* **2013**, *8*, e72016. [CrossRef]
6. Di Pasquale, G.; Alaux, C.; Le Conte, Y.; Odoux, J.-F.; Pioz, M.; Vaissière, B.E.; Belzunces, L.P.; Decourtye, A. Variations in the Availability of Pollen Resources Affect Honey Bee Health. *PLoS ONE* **2016**, *11*, e0162818. [CrossRef]
7. Omar, E.; Abd-Ella, A.A.; Khodairy, M.M.; Moosbeckhofer, R.; Crailsheim, K.; Brodschneider, R. Influence of Different Pollen Diets on the Development of Hypopharyngeal Glands and Size of Acid Gland Sacs in Caged Honey Bees (*Apis mellifera*). *Apidologie* **2017**, *48*, 425–436. [CrossRef]
8. Barroso-Arévalo, S.; Vicente-Rubiano, M.; Ruiz, J.A.; Bentabol, A.; Sánchez-Vizcaíno, J.M. Does Pollen Diversity Influence Honey Bee Colony Health? *Span. J. Agric. Res.* **2019**, *17*, e0504. [CrossRef]
9. Khan, K.A.; Ghramh, H.A.; Ahmad, Z.; El-Niweiri, M.A.A.; Mohammed, M.E.A. Honey Bee (*Apis mellifera*) Preference towards Micronutrients and Their Impact on Bee Colonies. *Saudi J. Biol. Sci.* **2021**, *28*, 3362–3366. [CrossRef]
10. Ilijević, K.; Vujanović, D.; Orčić, S.; Purać, J.; Kojić, D.; Zarić, N.; Gržetić, I.; Blagojević, D.P.; Čelić, T. V Anthropogenic Influence on Seasonal and Spatial Variation in Bioelements and Non-Essential Elements in Honeybees and Their Hemolymph. *Comp. Biochem. Physiol. Part C Toxicol. Pharmacol.* **2021**, *239*, 108852. [CrossRef]
11. Monchanin, C.; Blanc-Brude, A.; Drujont, E.; Negahi, M.M.; Pasquaretta, C.; Silvestre, J.; Baqué, D.; Elger, A.; Barron, A.B.; Devaud, J.-M.; et al. Chronic Exposure to Trace Lead Impairs Honey Bee Learning. *Ecotoxicol. Environ. Saf.* **2021**, *212*, 112008. [CrossRef]
12. Burge, P.S.; Scott, J.A.; McCoach, J. Occupational Asthma Caused by Aluminium. *Allergy* **2000**, *55*, 779–780. [CrossRef]
13. Al osman, M.; Yang, F.; Massey, I.Y. Exposure Routes and Health Effects of Heavy Metals on Children. *BioMetals* **2019**, *32*, 563–573. [CrossRef]
14. Rani, A.; Kumar, A.; Lal, A.; Pant, M. Cellular Mechanisms of Cadmium-Induced Toxicity: A Review. *Int. J. Environ. Health Res.* **2014**, *24*, 378–399. [CrossRef]
15. Bonoan, R.E.; Tai, T.M.; Tagle Rodriguez, M.; Feller, L.; Daddario, S.R.; Czaja, R.A.; O'Connor, L.D.; Burruss, G.; Starks, P.T. Seasonality of Salt Foraging in Honey Bees (*Apis mellifera*). *Ecol. Entomol.* **2017**, *42*, 195–201. [CrossRef]
16. Schmickl, T.; Crailsheim, K. Inner Nest Homeostasis in a Changing environment with Special Emphasis on Honey Bee Brood nursing and Pollen Supply. *Apidologie* **2004**, *35*, 249–263. [CrossRef]
17. Foley, K.; Fazio, G.; Jensen, A.B.; Hughes, W.O.H. Nutritional Limitation and Resistance to Opportunistic Aspergillus Parasites in Honey Bee Larvae. *J. Invertebr. Pathol.* **2012**, *111*, 68–73. [CrossRef]
18. Gerdts, J.R.; Roberts, J.M.K.; Simone-Finstrom, M.; Ogbourne, S.M.; Tucci, J. Genetic Variation of *Ascosphaera apis* and Colony Attributes Do Not Explain Chalkbrood Disease Outbreaks in Australian Honey Bees. *J. Invertebr. Pathol.* **2021**, *180*, 107540. [CrossRef]
19. Aronstein, K.A.; Murray, K.D. Chalkbrood Disease in Honey Bees. *J. Invertebr. Pathol.* **2010**, *103*, S20–S29. [CrossRef]
20. Rowland, B.W.; Rushton, S.P.; Shirley, M.D.F.; Brown, M.A.; Budge, G.E. Identifying the Climatic Drivers of Honey Bee Disease in England and Wales. *Sci. Rep.* **2021**, *11*, 21953. [CrossRef]
21. Sevim, A.; Akpınar, R.; Karaoğlu, Ş.A.; Bozdeveci, A.; Sevim, E. Prevalence and Phylogenetic Analysis of *Ascosphaera apis* (Maassen ex Claussen) LS Olive & Spiltoir (1955) Isolates from Honeybee Colonies in Turkey. *Biologia* **2022**, *77*, 2689–2699. [CrossRef]
22. Jensen, A.B.; Aronstein, K.; Flores, J.M.; Vojvodic, S.; Palacio, M.A.; Spivak, M. Standard Methods for Fungal Brood Disease Research. *J. Apic. Res.* **2013**, *52*, 1–20. [CrossRef]
23. Heath, L.A.F. Development of Chalk Brood in a Honeybee Colony: A Review. *Bee World* **1982**, *63*, 119–130. [CrossRef]
24. Deneke, Y.A.; Dero, B.S.; Mekonnen, A.S. Review on Chalkbrood Disease of Honey Bee. *Vet. Med. Open J.* **2023**, *8*, 47–55. [CrossRef]
25. Evison, S.E.F. Chalkbrood: Epidemiological Perspectives from the Host–Parasite Relationship. *Curr. Opin. Insect Sci.* **2015**, *10*, 65–70. [CrossRef]
26. Castagnino, G.L.B.; Mateos, A.; Meana, A.; Montejo, L.; Zamorano Iturralde, L.V.; Cutuli De Simón, M.T. Etiology, Symptoms and Prevention of Chalkbrood Disease: A Literature Review. *Rev. Bras. De Saude E Prod. Anim.* **2020**, *21*, e210332020. [CrossRef]
27. Gilliam, M.; Taber, S.; Lorenz, B.J.; Prest, D.B. Factors Affecting Development of Chalkbrood Disease in Colonies of Honey Bees, *Apis mellifera*, Fed Pollen Contaminated with *Ascosphaera apis*. *J. Invertebr. Pathol.* **1988**, *52*, 314–325. [CrossRef]
28. Flores, J.M.; Ruiz, J.A.; Ruz, J.M.; Puerta, F.; Bustos, M.; Padilla, F.; Campano, F. Effect of Temperature and Humidity of Sealed Brood on Chalkbrood Development under Controlled Conditions. *Apidologie* **1996**, *27*, 185–192. [CrossRef]
29. Yoder, J.A.; Nelson, B.W.; Main, L.R.; Lorenz, A.L.; Jajack, A.J.; Aronstein, K.A. Water Activity of the Bee Fungal Pathogen *Ascosphaera apis* in Relation to Colony Conditions. *Apidologie* **2017**, *48*, 159–167. [CrossRef]
30. Puerta, F.; Flores, J.M.; Bustos, M.; Padilla, F.; Campano, F. Chalkbrood Development in Honeybee Brood under Controlled Conditions. *Apidologie* **1994**, *25*, 540–546. [CrossRef]

31. Dolezal, A.G.; Toth, A.L. Feedbacks between Nutrition and Disease in Honey Bee Health. *Curr. Opin. Insect Sci.* **2018**, *26*, 114–119. [CrossRef]
32. DeGrandi-Hoffman, G.; Chen, Y.; Huang, E.; Huang, M.H. The Effect of Diet on Protein Concentration, Hypopharyngeal Gland Development and Virus Load in Worker Honey Bees (*Apis mellifera* L.). *J. Insect Physiol.* **2010**, *56*, 1184–1191. [CrossRef]
33. Goblirsch, M. Nosema Ceranae Disease of the Honey Bee (*Apis mellifera*). *Apidologie* **2018**, *49*, 131–150. [CrossRef]
34. Zarić, N.M.; Brodschneider, R.; Goessler, W. Honey Bees as Biomonitors—Variability in the Elemental Composition of Individual Bees. *Environ. Res.* **2022**, *204*, 112237. [CrossRef]
35. Pavlović, R.; Dojnov, B.; Šokarda Slavić, M.; Pavlović, M.; Slomo, K.; Ristović, M.; Vujčić, Z. In Pursuit of the Ultimate Pollen Substitute (Insect Larvae) for Honey Bee (*Apis mellifera*) Feed. *J. Apic Res.* **2022**, *62*, 1007–1016. [CrossRef]
36. Hellmich II, R.L.; Kulincevic, J.M.; Rothenbuhler, W.C. Selection for High and Low Pollenhoarding Honey Bees. *J. Hered.* **1985**, *76*, 155–158. [CrossRef]
37. Henry, M.; Rodet, G. Controlling the Impact of the Managed Honeybee on Wild Bees in Protected Areas. *Sci. Rep.* **2018**, *8*, 9308. [CrossRef]
38. Taylor, M.P.; Gillings, M.M.; Fry, K.L.; Barlow, C.F.; Gunkel-Grillion, P.; Gueyte, R.; Camoin, M. Tracing Nickel Smelter Emissions Using European Honey Bees. *Environ. Pollut.* **2023**, *335*, 122257. [CrossRef]
39. Gekière, A.; Vanderplanck, M.; Michez, D. Trace Metals with Heavy Consequences on Bees: A Comprehensive Review. *Sci. Total Environ.* **2023**, *895*, 165084. [CrossRef]
40. Zarić, N.M.; Brodschneider, R.; Goessler, W. Sex-Specific Element Accumulation in Honey Bees (*Apis mellifera*). *Environ. Sci. Pollut. Res.* **2024**, 1–8. [CrossRef]
41. Sagili, R.R.; Metz, B.N.; Lucas, H.M.; Chakrabarti, P.; Breece, C.R. Honey Bees Consider Larval Nutritional Status Rather than Genetic Relatedness When Selecting Larvae for Emergency Queen Rearing. *Sci. Rep.* **2018**, *8*, 7679. [CrossRef]
42. Huang, Z.Y.; Otis, G.W. Inspection and Feeding of Larvae by Worker Honey Bees (Hymenoptera: Apidae): Effect of Starvation and Food Quantity. *J. Insect Behav.* **1991**, *4*, 305–317. [CrossRef]
43. Lindauer, M.; Watkin, B. Division of Labour in the Honeybee Colony. *Bee World* **1953**, *34*, 63–73. [CrossRef]
44. Mayack, C.; Naug, D. Energetic Stress in the Honeybee *Apis mellifera* from Nosema Ceranae Infection. *J. Invertebr. Pathol.* **2009**, *100*, 185–188. [CrossRef]
45. Liu, C.; Steere, L.; McGregor, C.; Frederick, B.H.; Pastoor, T.; Zhou, Y.; Liu, C.T.; Cai, Y.; Zhou, H.; Xu, M.; et al. Exploring Boron Applications in Modern Agriculture: A Structure-Activity Relationship Study of a Novel Series of Multi-Substitution Benzoxaboroles for Identification of Potential Fungicides. *Bioorg. Med. Chem. Lett.* **2021**, *43*, 128089. [CrossRef]
46. Avis, T.J.; Rioux, D.; Simard, M.; Michaud, M.; Tweddell, R.J. Ultrastructural Alterations in Fusarium Sambucinum and Heterobasidion Annosum Treated with Aluminum Chloride and Sodium Metabisulfite. *Phytopathology* **2009**, *99*, 167–175. [CrossRef]
47. Gauthier, M.; Aras, P.; Jumarie, C.; Boily, M. Low Dietary Levels of Al, Pb and Cd May Affect the Non-Enzymatic Antioxidant Capacity in Caged Honey Bees (*Apis mellifera*). *Chemosphere* **2016**, *144*, 848–854. [CrossRef]
48. Zarić, N.M.; Ilijević, K.; Stanisavljević, L.; Gržetić, I. Metal Concentrations around Thermal Power Plants, Rural and Urban Areas Using Honeybees (*Apis mellifera* L.) as Bioindicators. *Int. J. Environ. Sci. Technol.* **2016**, *13*, 413–422. [CrossRef]
49. Zhang, G.; Zhang, W.; Cui, X.; Xu, B. Zinc Nutrition Increases the Antioxidant Defenses of Honey Bees. *Entomol. Exp. Appl.* **2015**, *156*, 201–210. [CrossRef]
50. Hýbl, M.; Šipoš, J.; Krejčová, A.; Sodomová, K.; Polák, O.; Koláčková, I.; Mráz, P. Preference of Pollinators over Various Forage Mixtures and Microelement Treatments. *Agronomy* **2022**, *12*, 370. [CrossRef]
51. Hussain, R.; Hasan, M.; Iqbal, K.J.; Zafar, A.; Tariq, T.; Saif, M.S.; Hassan, S.G.; Shu, X.; Caprioli, G.; Anjum, S.I. Nano-Managing Silver and Zinc as Bio-Conservational Approach against Pathogens of the Honey Bee. *J. Biotechnol.* **2023**, *365*, 1–10. [CrossRef]
52. Pacheco, N.I.N.; Semerad, J.; Pivokonsky, M.; Cajthaml, T.; Filip, J.; Busquets-Fité, M.; Dvorak, J.; Rico, A.; Prochazkova, P. Effects of Silver Sulfide Nanoparticles on the Earthworm Eisenia Andrei. *Comp. Biochem. Physiol. Part C Toxicol. Pharmacol.* **2022**, *257*, 109355. [CrossRef]
53. Sibiya, A.; Gopi, N.; Jeyavani, J.; Mahboob, S.; Al-Ghanim, K.A.; Sultana, S.; Mustafa, A.; Govindarajan, M.; Vaseeharan, B. Comparative Toxicity of Silver Nanoparticles and Silver Nitrate in Freshwater Fish Oreochromis Mossambicus: A Multi-Biomarker Approach. *Comp. Biochem. Physiol. Part C Toxicol. Pharmacol.* **2022**, *259*, 109391. [CrossRef]
54. Ścibior, A.; Pietrzyk, Ł.; Plewa, Z.; Skiba, A. Vanadium: Risks and Possible Benefits in the Light of a Comprehensive Overview of Its Pharmacotoxicological Mechanisms and Multi-Applications with a Summary of Further Research Trends. *J. Trace Elem. Med. Biol.* **2020**, *61*, 126508. [CrossRef]
55. Nriagu, J.O. *Production and Uses of Chromium*; Wiley-Interscience: New York, NY, USA, 1988.
56. Ali, M.; Wang, X.; Haroon, U.; Chaudhary, H.J.; Kamal, A.; Ali, Q.; Saleem, M.H.; Usman, K.; Alatawi, A.; Ali, S.; et al. Antifungal Activity of Zinc Nitrate Derived Nano Zno Fungicide Synthesized from Trachyspermum Ammi to Control Fruit Rot Disease of Grapefruit. *Ecotoxicol. Environ. Saf.* **2022**, *233*, 113311. [CrossRef]

insects

MDPI

Article

Interaction between Thiamethoxam and Deformed Wing Virus Type A on Wing Characteristics and Expression of Immune and Apoptosis Genes in *Apis mellifera*

Patcharin Phokasem [1,*], Wannapha Mookhploy [1], Sasiprapa Krongdang [2], Chainarong Sinpoo [1] and Panuwan Chantawannakul [1,3,*]

1 Bee Protection Laboratory, Department of Biology, Faculty of Science, Chiang Mai University, Chiang Mai 50200, Thailand; wannapha.mookhploy@gmail.com (W.M.); chainarong.sp@gmail.com (C.S.)
2 Faculty of Science and Social Sciences, Burapha University Sa Kaeo Campus, Sa Kaeo 27160, Thailand; sasiprapa.kr@buu.ac.th
3 Environmental Science Research Center, Faculty of Science, Chiang Mai University, Chiang Mai 50200, Thailand
* Correspondence: patcharin.phokasem@gmail.com (P.P.); panuwan@gmail.com (P.C.); Tel.: +66-612399782 (P.P.); +66-53-943346 (P.C.)

Simple Summary: Honey bees are key pollinators in agricultural crops. Today, honey bee colonies in decline are a global concern as a result of various stressors, including pesticides, pathogens, honey bee health, and parasites. A healthy honey bee colony refers to colonies that are not exposed to biotic and abiotic stressors. In this study, we examine how thiamethoxam (pesticide) and deformed wing virus type A (DWV-A) interact in effects on honey bee health. The results revealed that the honey bees were infected with DWV-A and were additionally exposed to thiamethoxam, showing effects that increased the mortality rate, and crippled wings in newly emerged adult honey bees. Moreover, the exposure to thiamethoxam and DWV-A injection resulted in induced expression of immune genes (*hymenoptaecin* gene) while downregulation of two apoptosis genes (*caspase8-like*, *caspase9-like* genes). The impact interaction of pesticide and DWV-A have on the expression of apoptosis genes can directly affect viral susceptibility in the honey bee host.

Abstract: Honey bees are economically important insects for crop pollination. They play a significant role as pollinators of wild plants and agricultural crops and produce economical products, such as honey, royal jelly, wax, pollen, propolis, and venom. Despite their ecological and economical importance, the global honey bee population is in decline due to factors including pathogens, parasites, intensive agriculture, and pesticides. Moreover, these factors may be interlinked and exacerbate the loss of honey bees. This study aimed to investigate the interaction between a pesticide, thiamethoxam, and deformed wing virus type A (DWV-A) to honey bees and the effects on survival rate, wing characteristics, and expression of immune and apoptosis genes in *Apis mellifera*. We described the potential interaction between thiamethoxam and DWV-A on honey bee wing characteristics, DWV-A loads, and the expressions of immune (*defensin*, *abaecin*, and *hymenoptaecin*) and apoptosis genes (*buffy*, *apaf1*, *caspase3-like*, *caspase8-like*, and *caspase9-like*). Honey bee larvae were fed with three different thiamethoxam doses (0.001, 1.4, and 14.3 ng/μL of the diet). Then, thiamethoxam-treated white-eyed pupae were injected with 10^7 copy numbers/honey bee of the DWV-A genome. The interaction between thiamethoxam and DWV-A caused a high mortality rate, crippled wings in newly emerged adult honey bees (100%), and resulted in induced expression of *hymenoptaecin* gene compared to the control group, while downregulation of *caspase8-like*, *caspase9-like* genes compared to the DWV injection group. Therefore, the potential interaction between thiamethoxam and DWV-A might have a deleterious effect on honey bee lifespan. The results from this study could be used as a tool to combat DWV-A infection and mitigate pesticide usage to alleviate the decrease in the honey bee population.

Keywords: apoptosis; DWV-A; immune; pathogens; pesticides; thiamethoxam

Citation: Phokasem, P.; Mookhploy, W.; Krongdang, S.; Sinpoo, C.; Chantawannakul, P. Interaction between Thiamethoxam and Deformed Wing Virus Type A on Wing Characteristics and Expression of Immune and Apoptosis Genes in *Apis mellifera*. *Insects* **2022**, *13*, 515. https://doi.org/10.3390/insects13060515

Academic Editors: Ivana Tlak Gajger, Franco Mutinelli and Luc P. Belzunces

Received: 27 April 2022
Accepted: 27 May 2022
Published: 31 May 2022

1. Introduction

The western honey bee, *Apis mellifera*, is the main pollinator of wild plants and agricultural crops. They also produce honey and other hive products such as royal jelly, wax, pollen, propolis, and venom [1–3]. The global and rapid loss of honey bee colonies has been associated with various factors, including pesticides, pathogens, honey bee health, and parasites [4]. However, these losses are thought to be largely attributed to the pesticide as well as emergent pathogens, including viruses [5]. Furthermore, viruses and pesticides can be concurrent threats to honey bee colonies, as honey bees infected with different pathogens encounter pesticides when collecting pollen and nectar [6,7].

When honey bees forage pollen, pesticide residue in crops up to 10 km away can pollute collected pollen and nectar and consequently cause pesticide contamination in colonies. Thiamethoxam (nitro-substituted neonicotinoid) is now the most commonly used insecticide in crops worldwide for seed coatings or directly sprayed on crops [8,9]. Previous studies have shown that thiamethoxam had negative effects on honey bees in the larval stage [10–13], pupal stage [14], and adult stage [15–18]. Moreover, neonicotinoid insecticides are likely to cause changes in honey bee physiology, such as hypopharyngeal gland development [19,20], honey bee behavior [21–23], colony development [24], foraging [25,26], and memory and learning [27–29].

Among honey bee pathogens, viruses have been one of the main culprits associated with honey bees' colony decline [30,31]. To date, about 26 honey bee viruses have been described, most of which are single-stranded RNA viruses, primarily belonging to the Dicistroviridae and Iflaviridae families [30]. The most common honey bee viruses include acute bee paralysis virus (ABPV), black queen cell virus (BQCV), chronic bee paralysis virus (CBPV), deformed wing virus (DWV), Israeli acute bee paralysis virus (IABPV), Kashmir bee virus (KBV), and sacbrood virus (SBV) were detected in honey bee colonies [5,32]. DWV is widespread and dominant in *A. mellifera*, positively correlated with varroa mites and tropilaelaps mites infestation [32–35]. DWV causes crippled wings and reduced body size in adult honey bees [36]. Several studies have documented that DWV has been linked to colony losses [32,37,38]. Three master variants of DWV (DWV-A, DWV-B, and DWV-C) have been discovered, with DWV-A being the most widespread variant [39–41]

Effects of co-exposure between pesticides and honey bee viruses have already been reported, resulting in an increase in DWV loads [42], BQCV loads [43], and CBPV loads [44] in honey bees. The effect of these factors has also been found to cause higher mortality rates in honey bee larvae [45]. Moreover, the change in gene expression pattern has been observed in immune and detoxification genes in honey bees [44]. Although the co-exposure of honey bees to DWV and thiamethoxam were investigated in previous studies [45], information on the relationship between crippled wings honey bees and gene pattern is still scarce. In this study, we described the effects of DWV-A infection and different concentrations of thiamethoxam treatment on the survival, viral loads, wing characteristics, and expressions of immune and apoptosis genes in newly emerged adult honey bees.

2. Materials and Methods

2.1. Honey Bee Samples

Seven *Apis mellifera* colonies maintained at Bee Protection Laboratory (BeeP) apiary, Chiang Mai University, Thailand (18°48′14.3″ N 98°57′22.2″ E) during 2018–2019 were used in this study. The crippled honey bees were collected from four colonies kept without ectoparasitic mites treatment. Three honey bee colonies, no visible clinical symptoms, and low/no ectoparasitic mites infestation were used for in vitro larval rearing.

2.2. Preparation of Deformed Wing Virus Type A (DWV-A) Lysate

The crippled adult honey bees were collected from *A. mellifera* colonies to prepare a DWV-A lysate. Five crippled adult honey bees were frozen in liquid nitrogen and crushed with a mortar. The ground crippled adult honey bees were suspended in 5 mL of phosphate buffer solution (pH 7.4) and then centrifuged at 6440× g for 10 min at 4 °C (K3 Series,

Centurion Scientific Ltd., London, UK). The supernatant was collected after centrifugation and filtered through a 0.2-micron filter (Millipore, Merck, Darmstadt, Germany) to eliminate bacteria, fungi, and Nosema. The absence of six common honey bee viruses (acute bee paralysis virus (ABPV) [46], black queen cell virus (BQCV) [46], chronic bee paralysis virus (CBPV) [46], Israeli acute bee paralysis virus (IABPV) [47], Kashmir bee virus (KBV) [46], and sacbrood virus (SBV) [46]) in the lysate was confirmed by quantitative real-time PCR (qRT-PCR). The sequence of primers used is shown in Table S1. Lysate without all six common honey bee viruses was used for this study. The lysate was kept at −80 °C until use [48]. The level of DWV genome equivalents in the lysate was measured using the same qRT-PCR technique described later in the Materials and Methods (Sections 2.6 and 2.7).

2.3. Diet and Larval Feeding

The first instar larval stage of *A. mellifera* was grafted onto an artificial diet plate. The levels of different sugar and yeast extract concentrations in food were provided for each developmental larval stage to meet the nutritional requirements. The artificial diet consisting of 50% *w*/*w* of royal jelly and 50% *w*/*w* of distilled water that contained either diet A (12% *w*/*v* glucose, 12% *w*/*v* fructose, and 2% *w*/*v* yeast extract), or diet B (15% *w*/*v* glucose, 15% *w*/*v* fructose, and 3% *w*/*v* yeast extract), or diet C (18% *w/v* glucose, 18% *w*/*v* fructose, and 4% *w*/*v* yeast extract) was refreshed every day. On the first and second days of in vitro rearing, each larva was fed with diet A, and then diet B was fed on the third day. Finally, diet C was fed on the fourth, fifth, and sixth days of the larvae developmental stage. Plates of larvae were incubated at 34 ± 1 °C and 96% RH [49].

2.4. Exposure to Thiamethoxam

Thiamethoxam was mixed in with diet C at three concentrations, including 0.001 (LT group), 1.4 (MT group), and 14.3 (HT group) ng/µL of the diet (note that the concentration of 0.001 ng/µL was the equivalent level of residues found in nectar, pollen, and beebread) [13]. The medium and high concentrations of thiamethoxam were selected according to a previous study [12], which were the lethal and sub-lethal concentrations of thiamethoxam to honey bee larvae reared in vitro. Diet C with no thiamethoxam was used in the control group (C group). The experimental groups were fed with diet C at different concentrations on the 4th day after grafting. After that, larvae received only food without the insecticide on the 5th and 6th days. On the 4th, 5th, and 6th days, each larva was fed 30, 40, and 50 ng/µL of diet C, respectively [12]. Overall, 105 honey bee larvae were subject to each treatment. Larval mortality was checked individually by observation under a stereomicroscope (Olympus, Tokyo, Japan) until they developed into the white-eyed pupae stage.

2.5. Injection of DWV-A to Honey Bee White-Eyed Pupae

The white-eyed pupae were collected and divided into 9 groups. Thiamethoxam-treated white-eyed pupae were injected laterally between the second and third tergite of the abdomen with 2 µL per honey bee of PBS containing 10^7 copy numbers/honey bee of DWV-A genome. Thiamethoxam-treated white-eyed pupae were divided into six groups: LT/V- (treated with 0.001 ng/µL thiamethoxam); LT/NC (treated with 0.001 ng/µL thiamethoxam with PBS injection); LT/V+ (treated 0.001 ng/µL thiamethoxam with DWV-A injection); MT/V- (treated 1.4 ng/µL thiamethoxam); MT/NC (treated 1.4 ng/µL thiamethoxam with PBS injection); MT/V+ (treated 1.4 ng/µL thiamethoxam with DWV-A injection). Thiamethoxam-untreated white-eyed pupae were injected with 2 µL per honey bee of 10^7 copy numbers/honey bee of DWV-A genome as a positive DWV-A control group (PC group). White-eyed pupae that were not treated with thiamethoxam and PBS injected were used as a negative control group (NC group). The PBS injection treatments were used as a control for the injection [45,50]. White-eyed pupae that were not treated with thiamethoxam and not injected were used as a handling control group (C group). All white-eyed pupae were incubated at 34 ± 1 °C and 70% RH until developing into newly emerged adult honey bees [51,52]. The honey bee survival rate was monitored during development.

2.6. RNA Extraction and cDNA Synthesis

Total RNA of adult honey bees was individually extracted by using TRIzol® (Invitrogen, Carlsbad, CA, USA) following the manufacturer's protocol. RNA concentration and quantity were determined using a BioDrop Duo spectrophotometer (BioDrop Ltd., Cambridge, UK). Reverse transcription was performed from 1 μg RNA to cDNA using the Tetro cDNA synthesis kit (Bioline, Alexandria, NSW, Australia) following the manufacturer's protocol.

2.7. Quantitative Real-Time PCR Parameters

The number of DWV-A genome copies was determined by the absolute quantification method. The standard curve was established by plotting seven 10-fold dilutions of DWV-A insert in TOPO ®TA Cloning® plasmid (Invitrogen, Carlsbad, CA, USA). The qRT-PCR was performed on BioRad iQ^{TM} 5 (Bio-Rad Crop., Hercules, CA, USA), using SensiFAST SYBRR® No-ROX Kit master mix (Bioline, Alexandria, NSW, Australia). The amplification was performed in a 20 μL reaction volume using SensiFAST SYBR® No-ROX Mix, consisting of 10 μL of 2x SensiFAST SYBR® No-ROX Mix, 0.8 μL of each 10 μM primer, 1 μL of 10-fold diluted cDNA and nuclease-free water to adjust the volume to 20 μL. For amplification step with the following profile was used: 50 °C for 30 min and 95 °C for 10 min, followed by 40 cycles of 95 °C for 30 s, 55 °C for 1 min, and 72 °C for 30 s. The melting curve was generated from 55 °C to 95 °C in 0.5 °C/s increments. The sequence of DWV-A [53], and housekeeping genes [54,55] primers is described in Table S1.

Relative quantification in real-time PCR was determined in antimicrobial peptides (AMPs), and apoptosis-related genes [50,56]. Ribosomal protein subunit 5 (RPS5) and β-actin were used as housekeeping genes for all primers shown in Table S1. qRT-PCR was performed as described above. All reactions were carried out using a thermal program of 95 °C for 30 s followed by 40 cycles of 95 °C for 30 s, 60 °C for 30 s, and 72 °C for 1 min. The final qRT-PCR amplification was confirmed by the analysis of the melting curve generated from 55 °C to 95 °C in 0.5 °C/s increments. Each experiment was performed in triplicate, and negative controls (no template) were included in each reaction. Gene expression was calculated as $2^{-\Delta\Delta CT}$ [57].

2.8. Statistical Analysis

The survival of white-eyed pupae and newly emerged adult honey bees were established using Kaplan–Meier survival statistics with the log-rank test. Log-transformed DWV-A loads and gene transcripts were analyzed using one-way ANOVA (Welch ANOVA in cases of unequal variance) followed by the Games-Howell post-hoc *t*-test. The data were analyzed using generalized linear models (GLMs) to evaluate significant variations among treatments and genes, with treatments and genes as fixed factors, and the interaction was included. *p*-values less than 0.05 were noted as significant. All statistical analyses were tested using the SPSS v 25 program (IBM Corp., Armonk, NY, USA).

3. Results

3.1. Effects of Thiamethoxam on Survival of Larvae to White-Eyed Pupae

The cumulative survival curves of *A. mellifera* white-eyed pupae were significantly different between the C group (thiamethoxam-untreated) and thiamethoxam-treated groups after 12 days post feed (Kaplan–Meier log-rank test, $x^2 = 170.826$, $p < 0.0001$; Figures 1 and S1).

The survival rate of the C group (90%) was not significantly different compared to the LT group (70%) (log-rank test, $p = 0.059$). In addition, the survival rates between MT (39%) and HT (22%) groups were not significantly different (log-rank test, $p = 0.128$; Figures 1 and S1 and Table S2). Honey bees fed with the highest thiamethoxam dose (14.3 ng/μL; HT group) showed a significantly lower survival rate than the C group (log-rank test, $p < 0.0001$), and honey bees fed with 0.001 (LT group), 1.4 (MT group) ng/μL of thiamethoxam (log-rank test, $p = 0.013$ and 0.028, respectively) (Figure 1 and Table S2).

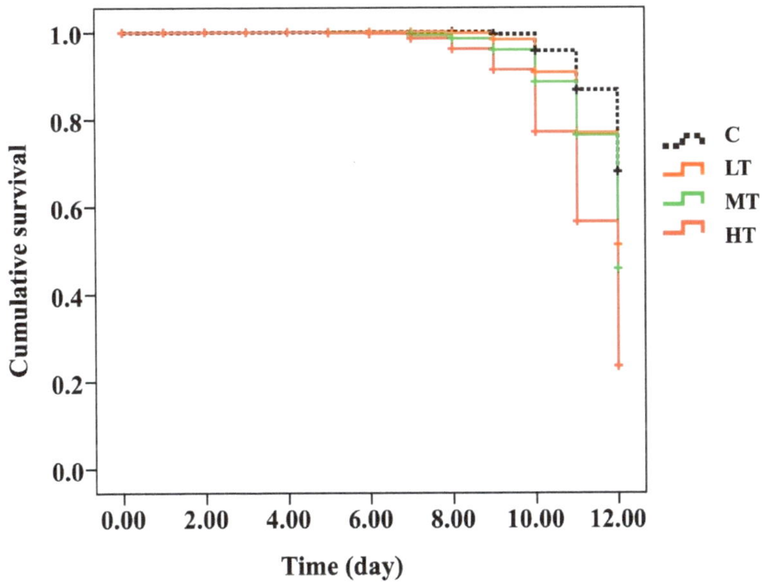

Figure 1. Kaplan–Meier survival curve of white-eyed pupae that were treated with three concentrations of thiamethoxam (0.001, 1.4, and 14.3 ng/μL) and control (untreated thiamethoxam) in the larval stage.

3.2. Effects of Co-Exposure of Thiamethoxam and DWV-A on the Survival of White-Eyed Pupae to Newly Emerged Adult Honey Bees

The cumulative survival curves of *A. mellifera* newly emerged adult honey bees were significantly different between the control and treated groups 8 days post injection (Kaplan–Meier log-rank test, $x^2 = 131.182$, $p < 0.0001$; Figure 2). There was no significant difference in cumulative survival rates among the C group (thiamethoxam-untreated with no DWV-A injection) (89%), NC (thiamethoxam-untreated with PBS injection) (87%), LT/V- (91%), LT/NC (86%), MT/V- (85%), and MT/NC (83%) (log-rank test, $p > 0.05$). However, these treatment groups showed higher cumulative survival rates when compared to groups injected with DWV-A. The injection with DWV-A-untreated thiamethoxam group (PC group) and LT/V+, MT/V+, and HT/V+ groups showed survival rates of 61%, 47%, 50%, and 13%, respectively. The HT/V- and HT/NC groups resulted in survival rates of 13% and 14%, respectively (at $p < 0.05$, Figures 2 and S2, and Table S3). Moreover, PC groups showed higher cumulative survival rate than LT/V+, MT/V+, and HT/V+ groups at p-value = 0.012, <0.0001, and <0.0001, respectively (Figure 2). A significant effect of interaction between thiamethoxam and DWV-A on mortality was found in all co-exposure groups when compared with DWV-A alone or thiamethoxam alone, except for the high dose thiamethoxam groups (HT/V- and HT/NC).

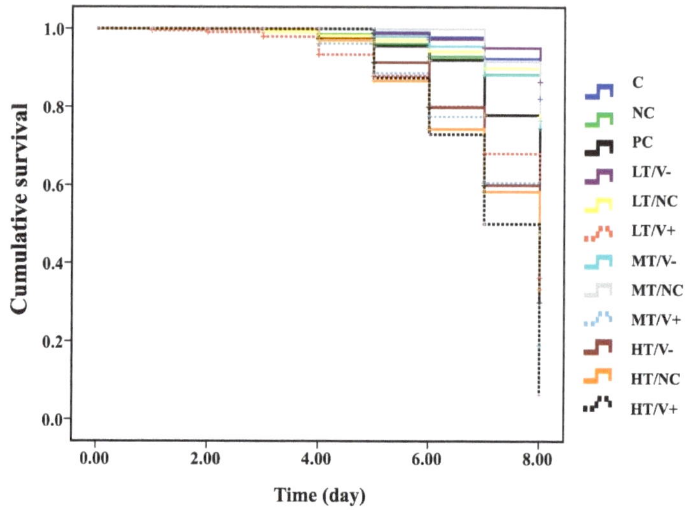

Figure 2. Kaplan–Meier survival curve of newly emerged adult honey bees treated with three concentrations of thiamethoxam (0.001, 1.4, and 14.3 ng/μL) in the larval stage that were injected with DWV-A, PBS, and control (not treated with thiamethoxam and uninfected group) in the white-eyed pupal stage.

3.3. *Effects of Co-Exposure of Thiamethoxam and DWV-A on Wing Characteristics of Newly Emerged Adult Honey Bees*

All newly emerged adult honey bees showed normal wings in the C group (100%) and the NC group (96%). Newly emerged adult honey bees that were not treated with thiamethoxam and injected with DWV-A (PC group) showed both normal and deformed wings at 5% and 95%, respectively. All thiamethoxam-treated groups were investigated for the crippled wings. The groups that were subject to 0.001 ng/μL of thiamethoxam (LT/V- group) and 0.001 ng/μL of thiamethoxam with PBS (LT/NC group) showed normal wing at 65% and 74%, respectively. The groups that were subject to 1.4 ng/μL of thiamethoxam (MT/V- group) and 1.4 ng/μL of thiamethoxam with PBS (MT/NC group) showed normal wing at 73% and 50%, respectively. The results showed that all concentrations of thiamethoxam treatments that were injected with DWV-A resulted in crippled wings in newly emerged adult honey bees (100%) (Figure 3). The survival rate of the HT/V+, HT/V-, and HT/NC groups was very low, and, therefore, the wing characteristic analysis was not performed.

3.4. *DWV-A Loads in Newly Emerged Adult Honey Bees*

Low DWV-A loads were detected in C and NC groups in newly emerged adult honey bees ($2.9 \times 10^4 \pm 8.0 \times 10^3$ and $3.8 \times 10^4 \pm 1.6 \times 10^4$ copy numbers/honey bee, respectively). The DWV-A levels of the MT/V- group showed a statistically significant difference in DWV-A levels compared to the C group ($p = 0.031$). The PC groups had higher DWV-A levels compared to the C group and all treatment groups at a p-value less than 0.05, except LT/V+ and MT/V+ groups (Figure 4 and Table S4). Crippled wings honey bees in the PC, LT/V+, and MT/V+ groups showed DWV-A loads of $1.2 \times 10^8 \pm 1.4 \times 10^7$, $1.1 \times 10^8 \pm 1.1 \times 10^7$, and $1.2 \times 10^8 \pm 1.9 \times 10^7$ copy numbers/honey bee, respectively, and there was no statistically significant difference in DWV-A loads between each other ($p > 0.05$). The interaction between thiamethoxam and DWV-A did not result in a significant modulation of DWV-A loads.

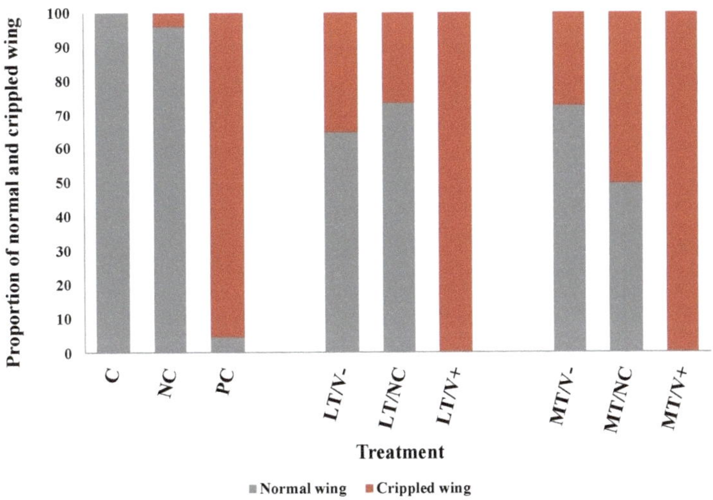

Figure 3. Percentages of the normal and crippled wings of newly emerged adult honey bees after being treated with thiamethoxam at 0.001 and 1.4 ng/µL in the larval stage and injected with DWV-A and PBS in the white-eyed pupal stage. The untreated and uninjected larvae were used as controls.

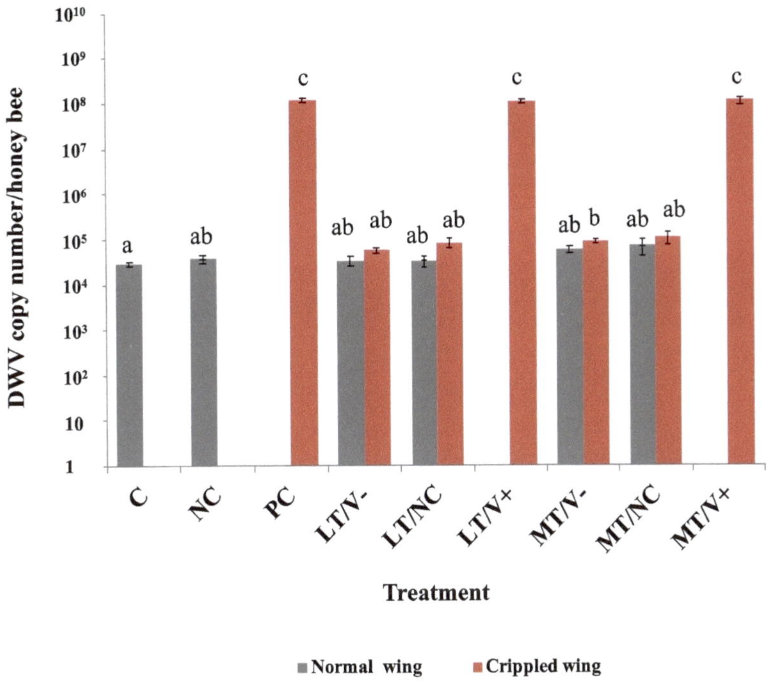

Figure 4. DWV-A loads in newly emerged adult honey bees treated with thiamethoxam at 0.001 and 1.4 ng/µL in the larval stage and injected with DWV-A and PBS in the white-eyed pupal stage. The control group was not treated with thiamethoxam and not injected. Vertical bars represent means ± SEM. One-way ANOVA with Games–Howell post-hoc test was used. The lowercase letters indicate significant differences at *p*-values less than 0.05.

3.5. Immune- and Apoptosis-Related Gene Expression in Newly Emerged Adult Honey Bees

White-eyed honey bee pupae treated **with** 0.001 and 1.4 ng/µL thiamethoxam and control (thiamethoxam-untreated) in the larval stage were injected with PBS and DWV-A. Three immune (*defensin*, *abaecin*, and *hymenoptaecin*) and five apoptosis-related genes (*buffy*, *apaf1*, *caspase3-like*, *caspase8-like*, and *caspase9-like*) were investigated at the newly emerged adult honey bee stage. The results showed no significant differences in the expressions of *buffy* and *apaf1* (Welch ANOVA, $p = 0.062$ and 0.095, respectively) in all experimental groups. In contrast, there were statistically significant differences in the expressions of six genes, including *defensin*, *abaecin*, *hymenoptaecin*, *caspase3-like*, *caspase8-like*, and *caspase9-like* (Welch ANOVA, all genes, $p < 0.01$). PC group showed an upregulation in three immune genes, including *defensin*, *abaecin*, and *hymenoptaecin*, compared to the C group (Games-Howell, $p = 0.003$, 0.034, and 0.011, respectively) (Figure 5 and Tables S5–S7). The expressions of two immune genes (*defensin* and *abaecin*) were lower in all groups treated with thiamethoxam and DWV-A injection than in the PC group, though not significantly different. Only the *hymenoptaecin* gene showed slightly higher upregulation in LT/V+ and MT/V+ groups than in the C group (Games–Howell, $p = 0.005$ and 0.006, respectively) (Figure 5 and Table S7). Honey bees that were treated with 1.4 ng/µL thiamethoxam and with DWV-A injection (MT/V+ group) showed an upregulation of the *caspase3-like* gene (Games-Howell, $p = 0.008$) compared to the PC group (Figure 5 and Table S8). The *caspase8-like* and *caspase9-like* genes showed the highest upregulation in the PC group, but only *caspase9-like* was significantly different compared to the control at $p = 0.018$ (Figure 5 and Tables S9 and S10). Moreover, the mRNA levels of the two genes were significantly suppressed in LT/V+ and MT/V+ groups compared to the PC group (Games-Howell, *caspase8-like* $p = 0.046$ and 0.031, respectively, and *caspase9-like* $p = 0.014$ and 0.026, respectively).

Gene expression was significantly influenced by treatments and genes, and the interaction was also significant (GLMs: $p < 0.001$ for treatments; $p < 0.001$ for genes; and $p < 0.001$ for the interaction).

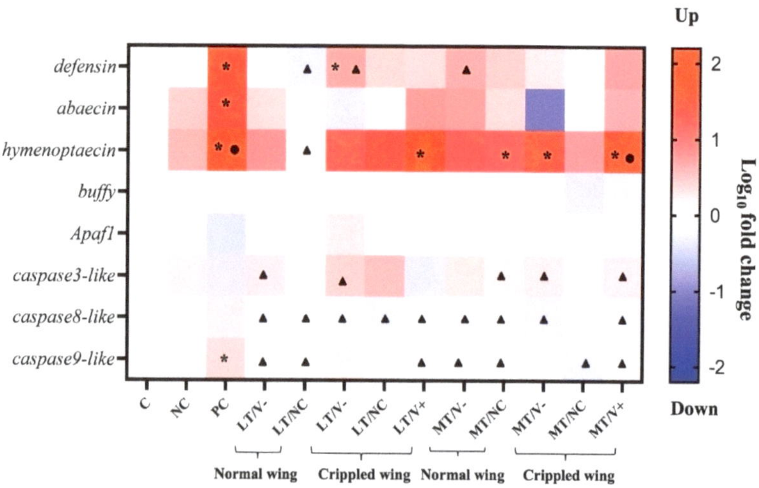

Figure 5. Heatmap of immune and apoptosis genes expression levels in newly emerged adult honey bees. A black asterisk (*) indicates a significant difference between the treatment compared to the control group. A black circle (•) indicates a significant difference between the treatment compared to the PBS group. A black triangle (▲) indicates a significant difference between the treatment compared to the DWV-A group ($p < 0.05$; Welch ANOVA and Games-Howell).

4. Discussion

Our study provides an insight into the effects on survival, DWV-A loads, wing characteristic, and expression of immune and apoptosis genes of *Apis mellifera* after exposure to different doses of thiamethoxam and DWV-A infection in newly emerged adult honey bees. Our results are consistent with previous reports, as we found that honey bees exposed to thiamethoxam in the larval stage had a significantly reduced survival rate in the white-eyed pupal stage [13,39]. Moreover, the combined effect of thiamethoxam and DWV-A further decreased the survival rate of newly emerged adult honey bees. Coulon et al. [45] also reported that a high dosage of thiamethoxam decreased the survival rate of honey bees after being injected with DWV. In this study, we showed that treatment with a low concentration of thiamethoxam (environmental dose in the colony) induces increased crippled wings in newly emerged adult honey bees. Nevertheless, there are limitations in this study that could be addressed in future research. The study used two high concentrations (1.4 and 14.3 ng/μL) that are not environmentally relevant. These concentrations not only induced high mortality but also resulted in an uneven number of individual tested.

Previous studies have shown that thiamethoxam caused changes in honey bee physiology [19,20]. Honey bees exposed to pesticides in the larval stage developed deformed physical characteristics in the adult stage, such as wing malformation, stunted bodies, and crippled legs [58]. Our study demonstrated that the effects of treatment with only thiamethoxam induce increased wing deformity in newly emerged adult honey bees. Moreover, numerous studies have also demonstrated that DWV is also the cause of crippled wings in honey bees [56,59]. Our result showed that honey bees exposed to thiamethoxam and DWV-A stressors had a high percentage of crippled wings as newly emerged adult honey bees. As a consequence, exposure to pesticides and DWV-A in honey bee colonies may impact the ability of adult honey bees to perform duties and forage effectively, leading to a decreased rate of colony survival.

In the present study, DWV-A levels of honey bees co-exposed to DWV-A and thiamethoxam were significantly higher than in treated groups, except for the PC group. Thus, the combination of neonicotinoid insecticides and DWV infection induced significantly higher DWV viral loads in honey bees [42]. These results coincide with the low survival rate of honey bees co-exposed to DWV-A and thiamethoxam, suggesting the effect between thiamethoxam and DWV-A infection on honey bee survival.

The immune-related gene expressions of honey bees co-exposed to thiamethoxam and DWV-A were upregulated in newly emerged adult honey bees. However, only the *hymenoptaecin* gene was significantly upregulated compared to the control group. The *hymenoptaecin* gene is one of the antimicrobial peptides that have been identified in honey bees to be active against microorganisms [60]. AMPs play a crucial role in the insect immune system and contribute to individual and social immunity [50,61,62]. Previous studies have indicated that the expressions of AMP genes in honey bees were upregulated after the invasion of pathogens, including microsporidian *Nosema* [63], *Paenibacillus larvae* [64], viruses [56], and ectoparasitic mites [65]. Viral infection within the host via viral entry, replication, and spreading can induce the antiviral innate immune responses [66]. Upregulation of several AMP genes, including *abaecin*, *hymenoptaecin*, and *defensin*, was also shown in other studies where honey bees were infected with DWV-A [56]. Treatment with thiamethoxam led to the downregulation of *abaecin* and *defensin* genes in crippled wings adult honey bees. Interestingly, honey bees exposed to thiamethoxam and DWV-A injection were also found to downregulate *abaecin* and *defensin* genes, implying immunological toxicity.

We found that co-exposure to thiamethoxam and DWV-A decreased the expression of apoptosis-related genes and significantly down-regulated *caspases8-like* and *caspases9-like* genes. The *caspases* gene is known to be related to programmed cell death and is associated with the final proteases in apoptosis [67]. Apoptosis is an important component of various processes, including normal cell development, embryonic development, function of the immune system, hormone-dependent atrophy, and chemical-induced cell death [68,69]. Evidence from previous studies suggested that virus infection induced apoptosis in insects and that the infection was mitigated by the elimination of the infected cells [70–72]. Honey bees injected with DWV had suppressed the expression of caspases in the pupal stage, which likely promoted the virus survival in hosts [56]. Honey bee co-exposure groups showed a strong alteration of immune gene expressions and downregulation of two apoptosis genes. Further studies are needed to investigate in greater detail the mechanisms for the viruses and pesticides that destroy immune pathways and the ability of viral replication in honey bee hosts.

5. Conclusions

This study showed the combined effect of DWV-A and thiamethoxam on *A. mellifera*, resulting in an increased mortality rate, crippled wings, and increased DWV-A loads. Our finding showed that honey bees exposed to thiamethoxam and DWV-A could intensify DWV-A infection, which could result in long-term physical deformity and decreased honey bees' life span. Data from our investigation revealed that gene expression patterns changed in each treatment group. The effect of both thiamethoxam and DWV-A results in the transcriptome imbalance, which may also have an effect on stress recovery and, subsequently, on honey bees' survival rate. Therefore, the results of our study could be explained by a negative interaction between thiamethoxam and DWV-A on honey bee lifespan in laboratory conditions. Future studies should be undertaken to examine the effects of pesticide exposure and viral infection occurring under field conditions.

Supplementary Materials: The following supporting information can be downloaded at: https://www.mdpi.com/article/10.3390/insects13060515/s1, Table S1: Primers used for qRT-PCR amplification in this study, Table S2: Statistic of survival in white-eyed pupae honey bees, Table S3: Statistic of survival in newly emerged adult honey bees, Table S4: Statistic of DWV-A load in newly emerged adult honey bees, Table S5: Statistic of *defensin* in newly emerged adult honey bees, Table S6: Statistic of *abaecin* in newly emerged adult honey bees, Table S7: Statistic of *hymenoptaecin* in newly emerged adult honey bees, Table S8: Statistic of *caspase3*-like in newly emerged adult honey bees, Table S9: Statistic of *caspase8*-like in newly emerged adult honey bees, Table S10: Statistic of *caspase9*-like in newly emerged adult honey bees, Figure S1: The percentage of survival rate of white-eyed pupae honey bees at 12 days post feed, Figure S2: The percentage of survival rates in newly emerged adult honey bees.

Author Contributions: Conceptualization, P.C. and P.P.; methodology, P.C.; software, S.K.; validation, P.P.; formal analysis, P.P., W.M., S.K. and C.S.; investigation, P.P.; resources, P.C.; data curation, P.P.; writing—original draft preparation, P.P., W.M., S.K., C.S. and P.C.; writing—review and editing, P.P., W.M., S.K., C.S. and P.C.; visualization, P.C. and P.P.; supervision, P.C.; project administration, P.C.; funding acquisition, P.C. All authors have read and agreed to the published version of the manuscript.

Funding: This research work was partially supported by Chiang Mai University Fund to P.C., and P.P. Open Access funding was covered by Mekong-Lancang Cooperation Special Fund.

Institutional Review Board Statement: Not applicable.

Informed Consent Statement: Not applicable.

Data Availability Statement: Data is contained within the article and supplementary material.

Acknowledgments: P.C. and P.P. are supported by Post-Doctoral Fellowship 2022, Office of Research Administration, Chiang Mai University, Thailand. We thank Terd Disayathanoowat for editing this paper, providing many valuable comments, and financial support. This research was supported by Chiang Mai University. The Department of Biology, Faculty of Science, Chiang Mai University, Thailand is also acknowledged.

Conflicts of Interest: The authors declared that there are no competing interest.

References

1. Gallai, N.; Salles, J.-M.; Settele, J.; Vaissière, B.E. Economic valuation of the vulnerability of world agriculture confronted with pollinator decline. *Ecol. Econ.* **2009**, *68*, 810–821. [CrossRef]
2. Liu, Z.; Chen, C.; Niu, Q.; Qi, W.; Yuan, C.; Su, S.; Liu, S.; Zhang, Y.; Zhang, X.; Ji, T. Survey results of honey bee (*Apis mellifera*) colony losses in China (2010–2013). *J. Apic. Res.* **2016**, *55*, 29–37. [CrossRef]

3. Crane, E. The past and present importance of bee products to man. In *Bee Products: Properties, Applications, and Apitherapy*; Avshalom, M., Yaacov, L., Eds.; Springer Science & Business Media: Berlin, Germany, 2013; pp. 1–14.
4. Potts, S.G.; Biesmeijer, J.C.; Kremen, C.; Neumann, P.; Schweiger, O.; Kunin, W.E. Global pollinator declines: Trends, impacts and drivers. *Trends Ecol. Evol.* **2010**, *25*, 345–353. [CrossRef]
5. Cox-Foster, D.L.; Conlan, S.; Holmes, E.C.; Palacios, G.; Evans, J.D.; Moran, N.A.; Quan, P.-L.; Briese, T.; Hornig, M.; Geiser, D.M.; et al. A metagenomic survey of microbes in honey bee colony collapse disorder. *Science* **2007**, *318*, 283–287. [CrossRef] [PubMed]
6. Neumann, P.; Carreck, N.L. Honey bee colony losses. *J. Apic. Res.* **2010**, *49*, 1–6. [CrossRef]
7. van der Sluijs, J.P.; Simon-Delso, N.; Goulson, D.; Maxim, L.; Bonmatin, J.-M.; Belzunces, L.P. Neonicotinoids, bee disorders and the sustainability of pollinator services. *Curr. Opin. Environ. Sustain.* **2013**, *5*, 293–305. [CrossRef]
8. Nauen, R.; Ebbinghaus-Kintscher, U.; Salgado, V.L.; Kaussmann, M. Thiamethoxam is a neonicotinoid precursor converted to clothianidin in insects and plants. *Pestic. Biochem. Phys.* **2003**, *76*, 55–69. [CrossRef]
9. van Sluijs, L.; Pijlman, G.P.; Kammenga, J.E. Why do Individuals differ in viral susceptibility? A story told by model organisms. *Viruses* **2017**, *9*, 284. [CrossRef]
10. Friol, P.S.; Catae, A.F.; Tavares, D.A.; Malaspina, O.; Roat, T.C. Can the exposure of Apis mellifera (Hymenoptera, Apiadae) larvae to a field concentration of thiamethoxam affect newly emerged bees? *Chemosphere* **2017**, *185*, 56–66. [CrossRef]
11. Grillone, G.; Laurino, D.; Manino, A.; Porporato, M. Toxicity of thiametoxam on in vitro reared honey bee brood. *Apidologie* **2017**, *48*, 635–643. [CrossRef]
12. Tavares, D.A.; Roat, T.C.; Carvalho, S.M.; Silva-Zacarin, E.C.M.; Malaspina, O. In vitro effects of thiamethoxam on larvae of Africanized honey bee *Apis mellifera* (Hymenoptera: Apidae). *Chemosphere* **2015**, *135*, 370–378. [CrossRef] [PubMed]
13. Tavares, D.A.; Dussaubat, C.; Kretzschmar, A.; Carvalho, S.M.; Silva-Zacarin, E.C.M.; Malaspina, O.; Bérail, G.; Brunet, J.-L.; Belzunces, L.P. Exposure of larvae to thiamethoxam affects the survival and physiology of the honey bee at post-embryonic stages. *Environ. Pollut.* **2017**, *229*, 386–393. [CrossRef] [PubMed]
14. Tesovnik, T.; Cizelj, I.; Zorc, M.; Čitar, M.; Božič, J.; Glavan, G.; Narat, M. Immune related gene expression in worker honey bee (*Apis mellifera* carnica) pupae exposed to neonicotinoid thiamethoxam and Varroa mites (*Varroa destructor*). *PLoS ONE* **2017**, *12*, e0187079. [CrossRef]
15. Badiou-Bénéteau, A.; Carvalho, S.M.; Brunet, J.-L.; Carvalho, G.A.; Buleté, A.; Giroud, B.; Belzunces, L.P. Development of biomarkers of exposure to xenobiotics in the honey bee Apis mellifera: Application to the systemic insecticide thiamethoxam. *Ecotoxicol. Environ. Saf.* **2012**, *82*, 22–31. [CrossRef] [PubMed]
16. Coulon, M.; Schurr, F.; Martel, A.C.; Cougoule, N.; Bégaud, A.; Mangoni, P.; Dalmon, A.; Alaux, C.; Le Conte, Y.; Thiéry, R.; et al. Metabolisation of thiamethoxam (a neonicotinoid pesticide) and interaction with the Chronic bee paralysis virus in honeybees. *Pestic. Biochem. Phys.* **2018**, *144*, 10–18. [CrossRef] [PubMed]
17. Henry, M.; Béguin, M.; Requier, F.; Rollin, O.; Odoux, J.-F.; Aupinel, P.; Aptel, J.; Tchamitchian, S.; Decourtye, A. A Common Pesticide Decreases Foraging Success and Survival in Honey Bees. *Science* **2012**, *336*, 348–350. [CrossRef]
18. Tosi, S.; Burgio, G.; Nieh, J.C. A common neonicotinoid pesticide, thiamethoxam, impairs honey bee flight ability. *Sci. Rep.* **2017**, *7*, 1201. [CrossRef]
19. Hatjina, F.; Papaefthimiou, C.; Charistos, L.; Dogaroglu, T.; Bouga, M.; Emmanouil, C.; Arnold, G. Sublethal doses of imidacloprid decreased size of hypopharyngeal glands and respiratory rhythm of honeybees in vivo. *Apidologie* **2013**, *44*, 467–480. [CrossRef]
20. Renzi, M.T.; Rodríguez-Gasol, N.; Medrzycki, P.; Porrini, C.; Martini, A.; Burgio, G.; Maini, S.; Sgolastra, F. Combined effect of pollen quality and thiamethoxam on hypopharyngeal gland development and protein content in *Apis mellifera*. *Apidologie* **2016**, *47*, 779–788. [CrossRef]
21. Suchail, S.; Guez, D.; Belzunces, L.P. Discrepancy between acute and chronic toxicity induced by imidacloprid and its metabolites in Apis mellifera. *Environ. Toxicol. Chem.* **2001**, *20*, 2482–2486. [CrossRef]
22. Williamson, S.; Moffat, C.; Gomersall, M.; Saranzewa, N.; Connolly, C.; Wright, G. Exposure to Acetylcholinesterase Inhibitors Alters the Physiology and Motor Function of Honeybees. *Front. Physiol.* **2013**, *4*, 13. [CrossRef] [PubMed]
23. Wright, G.A.; Softley, S.; Earnshaw, H. Low doses of neonicotinoid pesticides in food rewards impair short-term olfactory memory in foraging-age honeybees. *Sci. Rep.* **2015**, *5*, 15322. [CrossRef] [PubMed]
24. Wu, J.Y.; Anelli, C.M.; Sheppard, W.S. Sub-Lethal Effects of Pesticide Residues in Brood Comb on Worker Honey Bee (*Apis mellifera*) Development and Longevity. *PLoS ONE* **2011**, *6*, e14720. [CrossRef] [PubMed]
25. Schneider, C.W.; Tautz, J.; Grünewald, B.; Fuchs, S. RFID Tracking of Sublethal Effects of Two Neonicotinoid Insecticides on the Foraging Behavior of *Apis mellifera*. *PLoS ONE* **2012**, *7*, e30023. [CrossRef]
26. Yang, E.C.; Chuang, Y.C.; Chen, Y.L.; Chang, L.H. Abnormal Foraging Behavior Induced by Sublethal Dosage of Imidacloprid in the Honey Bee (Hymenoptera: Apidae). *J. Econ. Entomol.* **2008**, *101*, 1743–1748. [CrossRef]
27. Frost, E.H.; Shutler, D.; Hillier, N.K. Effects of fluvalinate on honey bee learning, memory, responsiveness to sucrose, and survival. *J. Exp. Biol.* **2013**, *216*, 2931–2938. [CrossRef]
28. Desneux, N.; Decourtye, A.; Delpuech, J.-M. The Sublethal Effects of Pesticides on Beneficial Arthropods. *Annu. Rev. Entomol.* **2007**, *52*, 81–106. [CrossRef]
29. Williamson, S.M.; Wright, G.A. Exposure to multiple cholinergic pesticides impairs olfactory learning and memory in honeybees. *J. Exp. Biol.* **2013**, *216*, 1799–1807. [CrossRef]
30. McMenamin, A.J.; Genersch, E. Honey bee colony losses and associated viruses. *Curr. Opin. Insect Sci.* **2015**, *8*, 121–129. [CrossRef]

31. McMahon, D.P.; Natsopoulou, M.E.; Doublet, V.; Fürst, M.; Weging, S.; Brown, M.J.; Gogol-Döring, A.; Paxton, R.J. Elevated virulence of an emerging viral genotype as a driver of honeybee loss. *Proc. R. Soc. Lond. B Biol. Sci.* **2016**, *283*, 20160811. [CrossRef]
32. Martin, S.J.; Highfield, A.C.; Brettell, L.; Villalobos, E.M.; Budge, G.E.; Powell, M.; Nikaido, S.; Schroeder, D.C. Global honey bee viral landscape altered by a parasitic mite. *Science* **2012**, *336*, 1304–1306. [CrossRef] [PubMed]
33. Bowen-Walker, P.L.; Martin, S.J.; Gunn, A. The transmission of deformed wing virus between honeybees (*Apis mellifera* L.) by the ectoparasitic mite *Varroa jacobsoni* Oud. *J. Invertebr. Pathol.* **1999**, *73*, 101–106. [CrossRef] [PubMed]
34. Highfield, A.C.; El Nagar, A.; Mackinder, L.C.M.; Noël, L.M.L.J.; Hall, M.J.; Martin, S.J.; Schroeder, D.C. Deformed wing virus implicated in overwintering honeybee colony losses. *Appl. Environ. Microbiol.* **2009**, *75*, 7212–7220. [CrossRef] [PubMed]
35. Nordström, S. Distribution of deformed wing virus within honey bee (*Apis mellifera*) brood cells infested with the ectoparasitic mite Varroa destructor. *Exp. Appl. Acarol.* **2003**, *29*, 293–302. [CrossRef]
36. Yue, C.; Genersch, E. RT-PCR analysis of Deformed wing virus in honeybees (*Apis mellifera*) and mites (*Varroa destructor*). *J. Gen. Virol.* **2005**, *86*, 3419–3424. [CrossRef]
37. Dainat, B.; Vanengelsdorp, D.; Neumann, P. Colony collapse disorder in Europe. *Environ. Microbiol. Rep.* **2012**, *4*, 123–125. [CrossRef]
38. Martin, S. A population model for the ectoparasitic mite *Varroa jacobsoni* in honey bee (*Apis mellifera*) colonies. *Ecol. Model.* **1998**, *109*, 267–281. [CrossRef]
39. Lanzi, G.; de Miranda, J.R.; Boniotti, M.B.; Cameron, C.E.; Lavazza, A.; Capucci, L.; Camazine, S.M.; Rossi, C. Molecular and biological characterization of deformed wing virus of honeybees (*Apis mellifera* L.). *Virol. J.* **2006**, *80*, 4998–5009. [CrossRef]
40. Mordecai, G.J.; Wilfert, L.; Martin, S.J.; Jones, I.M.; Schroeder, D.C. Diversity in a honey bee pathogen: First report of a third master variant of the Deformed Wing Virus quasispecies. *ISME J.* **2016**, *10*, 1264–1273. [CrossRef]
41. Ongus, J.R.; Peters, D.; Bonmatin, J.M.; Bengsch, E.; Vlak, J.M.; van Oers, M.M. Complete sequence of a picorna-like virus of the genus Iflavirus replicating in the mite *Varroa destructor*. *J. Gen. Virol.* **2004**, *85*, 3747–3755. [CrossRef]
42. Di Prisco, G.; Cavaliere, V.; Annoscia, D.; Varricchio, P.; Caprio, E.; Nazzi, F.; Gargiulo, G.; Pennacchio, F. Neonicotinoid clothianidin adversely affects insect immunity and promotes replication of a viral pathogen in honey bees. *Proc. Natl. Acad. Sci. USA* **2013**, *110*, 18466. [CrossRef] [PubMed]
43. Doublet, V.; Labarussias, M.; de Miranda, J.R.; Moritz, R.F.A.; Paxton, R.J. Bees under stress: Sublethal doses of a neonicotinoid pesticide and pathogens interact to elevate honey bee mortality across the life cycle. *Environ. Microbiol.* **2015**, *17*, 969–983. [CrossRef] [PubMed]
44. Coulon, M.; Schurr, F.; Martel, A.-C.; Cougoule, N.; Bégaud, A.; Mangoni, P.; Di Prisco, G.; Dalmon, A.; Alaux, C.; Ribière-Chabert, M.; et al. Influence of chronic exposure to thiamethoxam and chronic bee paralysis virus on winter honey bees. *PLoS ONE* **2019**, *14*, e0220703. [CrossRef] [PubMed]
45. Coulon, M.; Dalmon, A.; Di Prisco, G.; Prado, A.; Arban, F.; Dubois, E.; Ribière-Chabert, M.; Alaux, C.; Thiéry, R.; Le Conte, Y. Interactions Between Thiamethoxam and Deformed Wing Virus Can Drastically Impair Flight Behavior of Honey Bees. *Front. Microbiol.* **2020**, *11*, 766. [CrossRef]
46. Locke, B.; Forsgren, E.; Fries, I.; de Miranda, J.R. Acaricide treatment affects viral dynamics in *Varroa destructor*-infested honey bee colonies via both host physiology and mite control. *Appl. Environ. Microbiol.* **2012**, *78*, 227–235. [CrossRef]
47. de Miranda, J.R.; Cordoni, G.; Budge, G. The acute bee paralysis virus–kashmir bee virus–israeli acute paralysis virus complex. *J. Invertebr. Pathol.* **2010**, *103*, S30–S47. [CrossRef]
48. Khongphinitbunjong, K.; de Guzman, L.I.; Rinderer, T.E.; Tarver, M.R.; Frake, A.M.; Chen, Y.; Chantawannakul, P. Responses of varroa-resistant honey bees (*Apis mellifera* L.) to deformed wing virus. *J. Asia-Pac. Entomol.* **2016**, *19*, 921–927. [CrossRef]
49. Aupinel, P.; Fortin, D.; Dufour, H.; Tasei, J.-N.; Michaud, B.; Odoux, J.; Pham-Delègue, M.-H. Improvement of artificial feeding in a standard in vitro method for rearing *Apis mellifera* larvae. *Bull. Insectol.* **2005**, *58*, 107–111.
50. Evans, J.D.; Aronstein, K.; Chen, Y.P.; Hetru, C.; Imler, J.L.; Jiang, H.; Kanost, M.; Thompson, G.J.; Zou, Z.; Hultmark, D. Immune pathways and defence mechanisms in honey bees *Apis mellifera*. *Insect Mol. Biol.* **2006**, *15*, 645–656. [CrossRef]
51. Gätschenberger, H.; Azzami, K.; Tautz, J.; Beier, H. Antibacterial immune competence of honey bees (*Apis mellifera*) is adapted to different life stages and environmental risks. *PLoS ONE* **2013**, *8*, e66415. [CrossRef]
52. Azzami, K.; Ritter, W.; Tautz, J.; Beier, H. Infection of honey bees with acute bee paralysis virus does not trigger humoral or cellular immune responses. *Arch. Virol.* **2012**, *157*, 689–702. [CrossRef] [PubMed]
53. DeGrandi-Hoffman, G.; Chen, Y.; Huang, E.; Huang, M.H. The effect of diet on protein concentration, hypopharyngeal gland development and virus load in worker honey bees (*Apis mellifera L.*). *J. Insect Physiol.* **2010**, *56*, 1184–1191. [CrossRef] [PubMed]
54. Evans, J.D.; Pettis, J.S. Colony-level impacts of immune responsiveness in honey bees, *Apis mellifera*. *EVO* **2005**, *59*, 2270–2274. [CrossRef] [PubMed]
55. Simone, M.; Evans, J.D.; Spivak, M. Resin collection and social immunity in honey bees. *EVO* **2009**, *63*, 3016–3022. [CrossRef]
56. Mookhploy, W.; Krongdang, S.; Chantawannakul, P. Effects of Deformed Wing Virus Infection on Expressions of Immune- and Apoptosis-Related Genes in Western Honeybees (*Apis mellifera*). *Insects* **2021**, *12*, 82. [CrossRef]
57. Livak, K.J.; Schmittgen, T.D. Analysis of relative gene expression data using real-time quantitative PCR and the $2^{-\Delta\Delta CT}$ method. *Methods* **2001**, *25*, 402–408. [CrossRef]
58. Atkins, E.L.; Kellum, D. Comparative Morphogenic and toxicity Studies on the Effect of Pesticides on Honeybee Brood. *J. Apic. Res.* **1986**, *25*, 242–255. [CrossRef]

59. de Miranda, J.R.; Genersch, E. Deformed wing virus. *J. Invertebr. Pathol.* **2010**, *103* (Suppl. 1), S48–S61. [CrossRef]
60. Casteels, P.; Ampe, C.; Jacobs, F.; Tempst, P. Functional and chemical characterization of hymenoptaecin, an antibacterial polypeptide that is infection-inducible in the honeybee (*Apis mellifera*). *J. Biol. Chem.* **1993**, *268*, 7044–7054. [CrossRef]
61. Evans, J.D. Transcriptional immune responses by honey bee larvae during invasion by the bacterial pathogen, *Paenibacillus larvae*. *J. Invertebr. Pathol.* **2004**, *85*, 105–111. [CrossRef]
62. Ilyasov, R.; Gaifullina, L.; Saltykova, E.; Poskryakov, A.; Nikolenko, A. Review of the expression of antimicrobial peptide defensin in honey bees *Apis mellifera* L. *J. Apic. Sci* **2012**, *56*, 115–124. [CrossRef]
63. Sinpoo, C.; Paxton, R.J.; Disayathanoowat, T.; Krongdang, S.; Chantawannakul, P. Impact of *Nosema ceranae* and *Nosema apis* on individual worker bees of the two host species (*Apis cerana* and *Apis mellifera*) and regulation of host immune response. *J. Insect Physiol.* **2018**, *105*, 1–8. [CrossRef] [PubMed]
64. Krongdang, S.; Evans, J.D.; Chen, Y.; Mookhploy, W.; Chantawannakul, P. Comparative susceptibility and immune responses of Asian and European honey bees to the American foulbrood pathogen, *Paenibacillus larvae*. *Insect Sci.* **2018**, *26*, 831–842. [CrossRef] [PubMed]
65. Khongphinitbunjong, K.; de Guzman, L.I.; Tarver, M.R.; Rinderer, T.E.; Chantawannakul, P. Interactions of *Tropilaelaps mercedesae*, honey bee viruses and immune response in *Apis mellifera*. *J. Apic. Res.* **2015**, *54*, 40–47. [CrossRef]
66. Mueller, S.N.; Rouse, B.T. Immune responses to viruses. In *Clinical Immunology*, 3rd ed.; Mosby: Maryland Heights, MO, USA, 2008; pp. 421–431. [CrossRef]
67. McIlwain, D.R.; Berger, T.; Mak, T.W. Caspase functions in cell death and disease. *Cold Spring Harb. Perspect. Biol.* **2013**, *5*, a008656. [CrossRef]
68. Elmore, S. Apoptosis: A review of programmed cell death. *Toxicol. Pathol.* **2007**, *35*, 495–516. [CrossRef]
69. Kerr, J.F.; Wyllie, A.H.; Currie, A.R. Apoptosis: A basic biological phenomenon with wide-ranging implications in tissue kinetics. *Br. J. Cancer* **1972**, *26*, 239–257. [CrossRef] [PubMed]
70. Esch, H.; Esch, I.; Kerr, W.E. Sound: An element common to communication of stingless bees and to dances of the honey bee. *Science* **1965**, *149*, 320–321. [CrossRef]
71. Settles, E.W.; Friesen, P.D. Flock House Virus Induces Apoptosis by Depletion of Drosophila Inhibitor-of-Apoptosis Protein DIAP1. *Virol. J.* **2008**, *82*, 1378–1388. [CrossRef]
72. Marques, J.T.; Imler, J.-L. The diversity of insect antiviral immunity: Insights from viruses. *Curr. Opin.* **2016**, *32*, 71–76. [CrossRef]

insects

Article

Suitable Areas for Apiculture Expansion Determined by Antioxidant Power, Chemical Profiles, and Pesticide Residues in *Caldcluvia paniculata* Honey and Beeswax Samples

Enrique Mejías [1,*], Carlos Gómez [2] and Tatiana Garrido [2]

1 Centro de Tecnologías Nucleares en Ecosistemas Vulnerables, División de Investigación y Aplicaciones Nucleares—Comisión Chilena de Energía Nuclear, Nueva Bilbao 12501, Santiago 7600713, Chile
2 Departamento de Química Inorgánica y Analítica, Facultad de Ciencias Químicas y Farmacéuticas, Universidad de Chile, Dr. Carlos Lorca Tobar 964, Santiago 8391063, Chile; cvgomez@gmail.com (C.G.); tgarrido@ciq.uchile.cl (T.G.)
* Correspondence: enrique.mejias@cchen.cl; Tel.: +56-9-8750-00-05

Simple Summary: Honey is biologically desirable for antioxidant powers and antiradical capacities. However, pesticide use in farming means that any nearby beehives might become contaminated with undesirable and often harmful compounds. Apart from considerations for bee and human health, producing pesticide-free honey is economically important for Chile, the primary export market of which is the regulation-strict European Union. In the present study, honey and beeswax samples were collected from the Los Lagos Region of Chile and subjected to chemical profiling (phenol contents via Folin–Ciocalteu method; antioxidant power via Ferric Reducing Antioxidant Power Assay (FRAP) antiradical activity via 2,2-Diphenyl-1-picrylhydrazyl Assay (DPPH) and evaluations for pesticide residues (via HPLC-MS/MS and GS-MS).

Citation: Mejías, E.; Gómez, C.; Garrido, T. Suitable Areas for Apiculture Expansion Determined by Antioxidant Power, Chemical Profiles, and Pesticide Residues in *Caldcluvia paniculata* Honey and Beeswax Samples. *Insects* **2022**, *13*, 31. https://doi.org/10.3390/insects13010031

Academic Editors: Ivana Tlak Gajger and Franco Mutinelli

Received: 26 October 2021
Accepted: 12 December 2021
Published: 28 December 2021

Abstract: Forty-two samples of Tiaca Honey (*Caldcluvia paniculata*) obtained from beehives belonging to 14 apiaries (three honey samples per apiary) were collected at the end of January near Osorno (40°34′ S, 73°8′ W), Puyehue (40°40′ S, 72°37′ W) and Frutillar 41°7′ S, 72°59′ W) covering an area of 1240 km^2. They presented the highest phenol contents (0.36 mg gallic acid equivalent/kg) and antioxidant power (1.27 mM equivalent of Fe^{+2}/g of sample), and were among the highest for antiradical activity. Phenol contents and antioxidant power (r = 0.72, p-value < 0.01) and total phenol contents and antiradical activity (r = 0.69; p-value < 0.01) displayed linear correlations. Only two beeswax samples showed residues of the pesticide fenhexamid. The respective sites (Purranque [40°55′ S, 73°10′ W] and Coligual [40°49′ S, 72°54′ W]) were the only areas located near active farms. Additionally, the m/z value 163.1091 was found as an element to identify honeys. Data were used to construct a mapped suitability index ranking for pesticide-free areas with high biological quality. The provided chemical profiles will aid local beekeepers in obtaining international certifications, particularly for the EU market. In turn, the constructed maps indicate suitable areas for apiculture expansion, while differentiated pesticide detection in honey and beeswax requires further comparative research.

Keywords: honey; beeswax; apiculture; pesticides; antioxidants; phenols

1. Introduction

Pesticides play an important role in farming, with primary positive benefits, such as pest control, helping to improve crop yields [1,2]. However, pesticides also present several widely described negative effects for the environment and human health [3–6]. Estimates indicate that more than 98% of insecticides do not reach their destination [7], which is why these products are frequently found in the water, soil, atmosphere, and farmed crops [8].

Insecticide use has shifted over the last 20 years from organophosphates and carbamates to neonicotinoids [9]. Neonicotinoids are acetylcholine agonists that bind to nicotinic acetylcholine receptors, thereby triggering continuous signaling and causing insect death [10]. Due to the high persistence of pesticides in the environment, these compounds can be transferred to honey and other apicultural products either directly or indirectly by bees during production [11–14]. A lack of regulations and appropriate oversight have resulted in an indiscriminate use of pesticides, thus potentiating the lethal effects for populations that should not be exterminated, such as bees [15]. For this pollinating insect, the median lethal dose (LD50) of neonicotinoids is 0.003–0.006 μg/bee via oral ingestion [16]. Furthermore, damage is induced in proportion to the amount of insecticide [17]. Sublethal neonicotinoid doses consequently provoke nervous system disorders in bees that result in disorientation, memory loss, behavioral changes, communication difficulties, and an inability to carry out pollinating functions [9,18,19]. Neonicotinoids can also cause immunodeficiency, which is one of the causes for colony collapse disorder [20]. In the same way, pesticides residues detected in bee product samples offer a wide spectrum of risk for health of consumers, from slight allergenic reaction after exposure to carcinogenic effect [21]. Although, there are several sources of pollution, in many instances the presence of varroacides compounds is related to beekeeping regular activities, and thus, it is up to appropriate application of products. This explains the differences among honeys produced in the same apiary and samples from one country to another [22,23].

The European Union (EU) is one of the primary worldwide importers of honey and is the main market for Chilean honey, with 96% of national honey exports destined for EU countries [24]. The EU is one of the most stringent markets in terms of sustainable production and is particularly concerned about the negative consequences of pesticides, among other farming practices [25]. In 2013, the European Food Safety Authority identified at least three high-risk neonicotinoids for bees, especially regarding colony survival and development: clothianidin, imidacloprid, and thiamethoxam. As such, use of these insecticides is restricted by the EU [26–28].

It is worth highlighting that in addition to neonicotinoids, other pesticides used in farming, such as organophosphates, can also affect bees and, consequently, honey production [29]. In the interest of sustainable apicultural practices, current research interests in the area include determining the possible routes through pesticides affect honey, beeswax, bee pollen, and propolis, as well as establishing if there is a relationship of any effects with farming activities near hives [30]. Therefore, the objectives of this study were to (i) characterize the phenolic, antioxidant, antiradical profiles, calories and content of total carbohydrates, and ashes of honey samples. Likewise, pesticides residues in honey and beeswax samples and (ii) use the obtained data to establish suitable geographical areas in Chile for the pesticide-free production of honey and other biologically valuable apicultural products.

2. Materials and Methods

2.1. Honey and Beeswax Samples

The sampled beehives (n = 14) were in the Los Lagos Region of southern Chile (39°16′ S to 44°04′ S). Three samples of honey and a single sample of beeswax were collected from each beehive. All samples were collected during the summer (January–February 2016). Immediately after collection, the samples were transported to the laboratory for posterior analyses. Information regarding bee deaths in the colonies was gathered through interviews with local beekeepers.

2.2. Mellisopalynological Analysis for Determining the Botanical Origin of Honey Samples

The botanical composition of honey samples was quantitatively counted following methods described by Louveaux, Maurizio, and Vorwohl (1978) [31]. Briefly, honey (20 g) was placed on acetolyzed slides (Montenegro, Gómez, Díaz-Forestier, and Pizarro, 2008) [32]. Then, a sample aliquot was diluted with warm distilled water (20 mL at 40 °C), and the solution was transferred to an appropriate tube and centrifuged at 3500 rpm for 10 min. The supernatant was discarded, and the pollen residue was deposited at the bottom of the tube for resuspension in distilled water (100 μL). An aliquot (20 μL) was then taken and added to a slide together with Calberla's solution (10 μL), which was either basic fuchsine or diamond. The slide was gently dried. Finally, melted glycerinated gelatin (15 μL) was added to the mixture. For each sample, pollen grain residues were identified using an optical microscope at 400 and 1000× magnifications.

2.3. Preparation of Honey Solutions

First, honey samples (50 g) were mixed with distilled water (100 mL) acidified with HCl (pH = 2). The mixture was placed in a volumetric flask, and water was added until reaching a final volume of 250 mL. The extract was then filtered with cotton. Phenolic compounds were separated by column chromatography using the Amberlite XAD-2 resin (250 mm height, 20 mm diameter, 2 mL/min drop speed; Sigma-Aldrich, St. Louis, MO, USA). The column was washed with acid water (100 mL, pH = 2) and, subsequently, neutral distilled water (200 mL). Phenolic compounds were eluted with methanol p.a EMSURE® Merck Darmstadt, Germany (300 mL) The phenolic extract was collected and concentrated in vacuo to dryness at 45 °C. The dry residue was resuspended in distilled water (5 mL). The suspension was put in a decantation funnel, and diethyl ether (5 mL) was added. The organic phase was collected and washed twice with diethyl ether p.a EMSURE ® Merck (5 mL). The solution was concentrated to dryness in vacuo at 45 °C. The residual was resuspended in 2 mL of high-performance liquid chromatography (HPLC) grade methanol (Merck LiChrosolv Darmstadt, Germany), filtered (0.45 μm pore size), and stored at −20 °C until analysis. The extract was weighed prior to storage.

2.4. Colorimetric Assays for Determining Total Phenolic Compounds

The procedure described by Singleton and Rossi [33] and Buratti et al. [34] was used with minor modifications. Briefly, honey solution (200 μL) was mixed with the Folin–Ciocalteu reagent (50 μL—Merck KGaA, Darmstadt, Germany) and, subsequently, 20% Na_2CO_3 (150 μL). Distilled water was then added to a total volume of 1 mL. Absorbance was read at 765 nm after 30 min in a DLab SP—UV 1000 spectrophotometer (Beijing, China). Gallic acid (Sigma-Aldrich, St. Louis, MO, USA) was used as a standard to derive the calibration curve (0–150 mg/mL). The results defined the phenolic contents, which were expressed as the g equivalent of gallic acid/kg of sample.

2.5. Ferric Reducing/Antioxidant Power (FRAP) Assays for Determining Antioxidant Power

FRAP assays were performed according to Bertoncelj et al. [35]. The FRAP reagent was prepared by mixing 2,4,6-Tris(2-pyridyl)-s-triazine Sigma-Aldrich (2.5 mL; 10 mM 2,4,6-Tris(2-pyridyl)-s-triazine/40 mM of HCl) with 20 mM $FeCl_3$ (2.5 mL). Finally, 0.3 M acetate buffer (25 mL, pH = 3.6) was added to the mixture. The FRAP reagent was prepared just prior to each assay run. Antioxidant power was determined by mixing honey solution (0.2 mL) with the FRAP reagent (1.8 mL). Absorbance was read at 593 nm after 10 min. $FeSO_4$ $7H_2O$ was used as a standard to derive the calibration curve (50–1000 mM). Values were expressed as the mM equivalent of Fe^{+2}/g of sample. Assays were performed at room temperature.

2.6. Determinations of Antiradical Activity

The procedure described by Meda et al. [36] and modified by Mejías and Montenegro [37] was followed to determine antiradical activity. The 1,1-diphenyl-2-picrylhydrazyl radical DPPH (Calbiochem, Darmstadt, Germany) assay was used to determine the antiradical properties of the chemical compounds in honey by inhibiting or decreasing the oxidant activity of DPPH. For this, honey solutions (750 μL) were mixed with the DPPH radical (1.5 mL) in methanol (0.02 mg DPPH/mL MeOH). Absorbance was read at 517 nm after 15 min. A blank sample was prepared with methanol. Ascorbic acid (Calbiochem, Darmstadt, Germany) was used as a standard to derive the calibration curve (1–10 mg/mL). The values for antiradical activity were expressed as mg equivalent of ascorbic acid/g of sample.

2.7. Determination of Total Carbohydrate Content in Honey Samples

The total carbohydrate percentage of each honey was previously measured by refractometry (w/w percentage). Next, 15 g of honey were weighed and mixed with 10 mL of water. The pH of the resulting solution was adjusted to 1.0 by adding HCl 1.2 M from an automatic titrator, provided by a combined pH electrode. The total carbohydrate percentage was then reduced to a final value of 40.0%, w/w, by dilution with acidified water at pH 1.5. The total carbohydrate percentage of pure honeys is in average 80%, w/w [38].

2.8. Determinations of Total Ash Content in Honey Samples

The ash content was determined according to the methods of AOAC 2000 [39] with modifications. First, 10 g of honey were placed in combustion pots, which required preheating to darkness with a gas flame to prevent honey foaming. Thereafter, the samples were incinerated at 600 °C in a burning muffle for 5 h. After cooling at room temperature, the obtained ash was weighed.

2.9. Direct Sample Analysis-Time of Flight-Mass Spectrometry (DSA-TOF-MS) for Determining Chromatography Profiles

For direct sample analysis-time of flight-mass spectrometry, HPLC grade water containing 0.1 M NaOH was added to honey (500 mg). The liquid sample (10 μL) was placed in a mesh holder for analysis. Assays were run on a direct sample analysis-time of flight-mass spectrometry (DSA-TOF-MS, Perkin Elmer, Waltham, MA, USA) using the following conditions: corona current = 30 μA; heater temperature = 300 °C; auxiliary gas (N_2) flow = 4 L/min; nebulizer gas (N_2) pressure = 80 psi; drying gas (N_2) flow = 3 L/min; and drying gas (N_2) temperature = 25 °C. The DSA-TOF-MS was run in positive ionization mode with a flight tube voltage of −10,000 V. The capillary exit voltage was set to 100 V for normal MS analysis and 155 V for collision induced dissociation analysis. Mass spectra were acquired with a mass range of 100–3000 m/z and acquisition rate of 1 spectra/s. To maintain mass accuracy, five lock mass ions were used (m/z 121.0509, m/z 622.0299, m/z 922.0119, m/z 1521.9771, and m/z 2121.9405). All samples were analyzed for only 10 s.

2.10. Extraction Methodologies

2.10.1. Quick, Easy, Cheap, Effective, Rugged, and Safe (QuEChERS)

QuEChERS extraction was performed following methodology proposed by Barganska et al. [11], with certain modifications. Briefly, honey (2 g) samples were dissolved with 15 mL of a solution of 1% acetic acid in acetonitrile. This mixture was transferred to the Extraction Tube containing the salt kit provided with the Dispersive QuEChERS (DisQuE) was added (Cat. No. 176001903; Waters, Milford, MA, USA). The composition of this kit included 4 g MgSO4, 1 g NaCl, 1 g trisodium citrate dehydrate, and 0.5 g disodium hydrogen citrate sesquihydrate. Internal standards (50 μL; triphenyl phosphate 100 μg/mL) were subsequently added to the mixture. The samples were shaken vigorously for 1 min and centrifuged at 4400 rpm for 5 min. Samples were further cleaned by transferring the obtained supernatant (4 mL) to a dispersive sample preparation extraction tube, which was then shaken for 45 s. Thereafter, the tube was centrifuged at 5000 rpm for 2 min. The resulting supernatant was used for chromatographic analysis. Samples were cleaned using $MgSO_4$ (150 mg), primary–secondary amine (PSA; 25 mg), and a C18 (PSA) sorbent (25 mg). The above methodology was also applied to beeswax samples, excepting the dissolution of beeswax (1 g) with chloroform (5 mL) in the initial steps.

2.10.2. Solid-Phase Extraction

Pesticides were extracted and identified from honey samples using the methodology proposed by Bohm et al. [40], with modifications. Briefly, honey samples (2 g) were homogenized with a citrate buffer solution (10 mL, pH 4.0). After agitation for 15 min, the mixture was centrifuged (4000 × RPM, 5 min, 5 °C) and filtered. The entire supernatant was transferred to the Oasis HLB 3 cc Vac Cartridge (Cat. No. WAT 094226; Waters, Milford,

MA, USA) for sample preparation extraction on a vacuum station previously conditioned with MeOH (6 mL) and water (6 mL). The extracts were rinsed and then eluted with a solution (5 mL) containing 3% formic acid in MeOH. Finally, the elutions were concentrated to dryness. Dry residues were reconstituted in a mobile phase solution A (750 µL; see below for details) and filtered for further analysis via liquid chromatography tandem-mass spectrometry (LC-MS/MS). Beeswax samples (2 g) were dissolved by vigorous agitation with 3:1 chloroform/MeOH (5 mL). The homogenized solution was mixed with 96% MeOH (5 mL) and centrifuged. The entire supernatant was transferred onto an OASIS HLB cartridge. All subsequent steps for beeswax processing were the same as with honey processing.

2.11. *Chromatography*

A total of 242 pesticides from the following groups were analyzed: organochlorines, organophosphates, carbamates, thiocarbamates, pyrethroids, and neonicotinoids. Regarding neonicotinoids, acetamiprid, imidacloprid, thiamethoxam, and thiacloprid were considered (Supporting Material Table S1).

2.11.1. LC-MS/MS Analysis

Honey and beeswax were analyzed by UPLC–MS/MS using a XEVO Triple Quadrupole Tandem Mass Spectrometer (ACQUITY UPLC H-Class System; Waters Corp., Milford, MA, USA). Separation was facilitated by using an Acquity- Ethylene Bridged Hybrid (BEH) C18 column (1.7 lm, 2.1 9 50 mm; Waters Corp., Milford, MA, USA). A mobile phase gradient was composed of solutions A and B. Solution A was comprised by 5 mM of 10% ammonium formate in 10% methanol and 90% HPLC grade water. Solution B was comprised by 5 mM ammonium formate in 90% methanol and 10% HPLC grade water. The oven temperature was 30 °C, with an injection volume of 10 µL. The following MS/MS parameters were used: ionization mode = positive; scan type = MRM; dwell-time = 20 ms; ion spray voltage = 5.500 V; source = 300 °C; and analysis time = 21 min.

2.11.2. Gas Chromatography–Mass Spectrometry (GC-MS) Analysis

Chromatography analyses were conducted in an Agilent 7890A GC-MS (Santa Clara, CA, USA) with solvent vent-mode injection using a programmable temperature vaporization inlet with the 5975C Mass Selective Detector (Agilent, Santa Clara, CA, USA). Chromatography conditions were as follows: injector temperature = 250 °C; column temperature = 40 °C for 5 min, then increased to 240 °C at a speed of 3 °C/min, and finally 240 °C for 10 min; carrier gas = helium at 20 mL/min flow rate; and column = Zebron ZB-5ms 30 m × 0.24 mm × 0.25 mm (Phenomenex, Torrance, CA, USA). Mass detector conditions were as follows: transference line temperature = 260 °C; ionization trap temperature = 17 °C; ion impact energy = 70 eV; and analysis time = 37.5 min. The used pesticide standards were from Dr. Ehrenstorfer GmBH (Augsburg, Germany). Stock solutions were prepared at a concentration of 400 µg mL^{-1} in ethyl acetate and were stored at −20 °C in amber vials. All solvents and reagents used were HPLC grade (Merck Millipore, Darmstadt, Germany).

2.12. *Suitability Index*

The values obtained for each analyzed honey sample in relation to total phenols, antioxidant power, and antiradical activity were separately calculated in decreasing order for each georeferenced location. From this, an increasing scale of 1 to 3 was used to rank each variable from lowest to highest. The remaining numbers in the series were calculated proportional to the following parameterization (e.g., for total phenols [TP]):

$$TP_{\text{new scale}} = \frac{TP_{\text{original scale}} - \min\left(TP_{\text{original scale}}\right)}{\max\left(TP_{\text{original scale}}\right) - \min\left(TP_{\text{original scale}}\right)} \times (3-1) + 1$$

where $TP_{new\ scale}$ is expressed in the original units of measurement for total phenols, i.e., g equivalent of gallic acid/kg and min ($TP_{original\ scale}$) was defined as the smallest Total Phenolic value observed among all honeys.

This new parameterization methodology was applied to obtain values for total phenols (TP), antioxidant power (AP), antiradical activity (AA), Total Carbohydrate Content (TC) and Total Ash (TA). Additionally, there is a set of Energy values obtained from TC. TC may be alternatively replaced by Energy when these data are available. To construct a suitability index for each location, the respective scores were added in such a way that sample XX met the following:

Sample XX: [(X(TP); Z(AP); Y(AA); W(TC);T(TA)] where $1 < T, W, X, Y, Z < 3$

Finally, the suitability index (SI) for location i was calculated as follows:

$$SI_i = \sum(T_i + W_i + X_i + Y_i + Z_i) \qquad 5 \leq SI_i \leq 15$$

2.13. Statistical Analysis

All assays for each honey and beeswax sample were performed in triplicate. An exploratory analysis of the data was conducted to evaluate assumptions of normality and to select appropriate statistical methodologies. All calculations and map constructions were performed in the R v.3.2.5. software (2016) with the ggmap and ggplot2 packages. Furthermore, all statistical analyses followed methodological guidelines for reproducible research using the knitr library. The source code in R can be requested from the corresponding author.

For the comparison between the different honeys, an analysis of variance (ANOVA) was carried out for each of the variables studied. Tukey's multiple comparisons test was carried out to evaluate the statistical significance of the differences between honeys. The assumptions of the ANOVA test were corroborated by residual analysis: normality, independence and homoscedasticity. The statistical significance of the correlations was evaluated using Pearson's correlation test. The normality of the data was assessed using Shapiro's test and QQ normality plots.

3. Results

3.1. Botanical Origin and Chemical Analyses of Honey Samples

The botanical origins of the studied honey samples are indicated in Table 1, which also shows the percentage of the three most predominant botanical species found in analyses.

Table 1. Predominant botanical species (%) found in each honey sample.

Apiary (*)-Total Pollen Grains	*Caldcluvia paniculata*	*Luma/Myrceugenia*	*Weinmannia trichosperma*	Other Species
A (1.851)	35 ± 0.02 h	0	0	65
B (1.908)	72 ± 0.02 c	5 ± 0.01 b	6 ± 0.01 b	17
C (2.232)	52 ± 0.03 f	2 ± 0.01 c	0	46
D (2.241)	33 ± 0.02 h	0	0	67
E (1.836)	85 ± 0.04 a	6 ± 0.02 b	6 ± 0.01 b	3
F (1.986)	81 ± 0.04 a	7 ± 0.01 a	5 ± 0.01 b	7
G (1.968)	28 ± 0.03 i	0	0	72
H (1.953)	58 ± 0.02 e	2 ± 0.02 c	2 ± 0.01 c	38
I (2.049)	75 ± 0.01 b	5 ± 0.02 b	5 ± 0.02 b	15
J (2.169)	63 ± 0.02 d	2 ± 0.01 c	2 ± 0.01 c	33
K (1.965)	70 ± 0.03 c	6 ± 0.01 b	7 ± 0.02 a	17
L (2.073)	71 ± 0.02 c	5 ± 0.02 b	6 ± 0.01 b	18
M (2.001)	48 ± 0.03 f	0	0	52
N (1.827)	43 ± 0.03 g	0	0	57

Values represent the mean of triplicate samples. The means reported in the same column are significantly different according to Tukey's test. ($p < 0.05$) if denoted by these letters. (*). Three honey samples were taken from each apiary. The number of pollen grains corresponds to the sum of the total number of grains of the 3 honeys from each apiary.

Overall results for total phenol contents, antioxidant power, and antiradical activity are shown in Table 2. The index of each component was also calculated for posterior suitability index determinations for the assessed areas. More specifically, total phenol contents were established based on a gallic acid standard (i.e., mg gallic acid equivalent/kg of sample; Figure 1). Honey sample E, collected in proximity to Puyehue (40°40′ S, 72°37′ W), presented the highest phenol contents (i.e., 0.36 mg/kg). Similarly, honey sample E presented the highest antioxidant power (1.27 mm equivalent of Fe^{+2}/g of sample; Figure 2), as established by FRAP analyses. Finally, antiradical activity was measured as the ability to inhibit or decrease the oxidizing effect of DPPH (Figure 3). Honey samples E, F, and L, respectively located in proximity to Puyehue (40°40′ S, 72°37′ W), Purranque (40°55′ S, 73°10′ W), and Fresia (41°09′ S, 73°27′ W), had the highest antiradical activities.

Table 2. DSA-TOF-MS signals for analyzed honeys. ND: Not Detected.

Sample	*m/z*					
A	105.0708	ND	121.0497	163.1091	207.1680	322.0545
B	105.0709	120.0804	ND	163.0613	ND	322.0544
C	105.0389	120.0804	121.0499	163.0617	207.1767	322.0553
D	105.0399	120.0807	121.0529	163.0618	207.1399	322.0501
E	105.0709	120.0804	121.0618	163.1095	207.1445	322.0500
F	105.0422	ND	ND	163.1094	207.1399	322.0494
G	105.0710	ND	121.0520	163.1096	ND	322.0549
H	105.0708	ND	121.0499	163.0614	207.1667	322.0502
I	105.0709	ND	121.0475	163.0704	207.1666	322.0501
J	105.0708	120.0807	121.1037	163.1093	207.1635	322.0506
K	105.0709	ND	121.0501	163.0615	ND	322.0506
L	105.0709	ND	121.0500	163.0615	207.1569	322.0509
M	105.0709	ND	121.0499	163.0616	207.1666	322.0508
N	105.0709	ND	ND	163.0616	ND	322.0511
Z (CONTROL)	ND	ND	ND	163.0465	ND	ND

Notably, results for phenol contents, antioxidant power, and antiradical activity coincided for honey sample E (Table 3, Figures 1–3). Furthermore, positive linear correlations were found among all evaluated honeys for total phenol contents and antioxidant power ($r = 0.72$, p-value < 0.01), as well as for total phenol contents and antiradical activity ($r = 69$; p-value < 0.01). These results suggest that biological antioxidant activity primarily depends on the phenol contents of the honey sample, which would be inherited from the predominant nectar-containing plants, as established through botanical origin analyses (Table 1).

Table 3. Phenol, antioxidant power, and antiradical activity results obtained for honey samples from the Los Lagos Region (Chile).

Sample	Phenols † ± SD	Antioxidant Power ‡ ± SD	Antiradical Activity ≡ ± SD	Total Carbohydrates ⊥ ± SD	Energy #	Total Ash *	Phenol Index	Antioxidant Index	Antiradical Index	Carbohydrate Index	Energy Index	Ash Index	Suitability Index
A	0.222 ± 0.011 a	0.65 ± 0.09 a	485.5 ± 0.01 a	82.1 ± 0.5 a	329 ± 4 a	0.08 ± 0.008 i	1.1	1.3	1.0	2.1	2.1	1.0	6.4
B	0.320 ± 0.008 b	1.15 ± 0.06 b	490.1 ± 0.01 a	83.4 ± 0.9 a	332 ± 4 a	0.19 ± 0.006 b	2.4	2.7	1.0	2.9	2.5	2.8	11.8
C	0.303 ± 0.010 c	0.64 ± 0.06 a	558.5 ± 0.06 b	80.4 ± 0.5 a	321 ± 6 a	0.21 ± 0.008 a	2.2	1.2	1.4	1.0	1.0	3.0	8.8
D	0.272 ± 0.013 d	0.61 ± 0.07 a	492.4 ± 0.02 a	81.5 ± 0.6 a	326 ± 5 a	0.12 ± 0.007 g	1.7	1.2	1.0	1.7	1.7	1.7	7.3
E	0.366 ± 0.011 e	1.27 ± 0.04 b	681.8 ± 0.06 c	83.6 ± 0.6 a	334 ± 4 a	0.15 ± 0.006 d	3.0	3.0	2.2	3.0	2.7	2.2	13.3
F	0.227 ± 0.008 f	0.71 ± 0.07 d	737.5 ± 0.07 d	80.7 ± 0.1 a	323 ± 6 a	0.22 ± 0.008 a	1.1	1.4	2.5	1.2	1.3	3.0	9.3
G	0.239 ± 0.005 g	0.62 ± 0.05 e	546.0 ± 0.03 b	82.5 ± 0.5 a	331 ± 7 a	0.09 ± 0.006 j	1.3	1.2	1.4	2.3	2.3	1.2	7.3
H	0.282 ± 0.010 d	0.79 ± 0.01 d	539.2 ± 0.06 b	83.6 ± 0.8 a	330 ± 6 a	0.17 ± 0.005 c	1.9	1.6	1.3	3.0	2.2	2.5	10.3
I	0.263 ± 0.010 d	0.67 ± 0.08 f	678.5 ± 0.01 c	82.2 ± 0.3 a	327 ± 5 a	0.12 ± 0.007 f	1.6	1.3	2.2	2.1	1.8	1.7	8.9
J	0.308 ± 0.004 h	0.70 ± 0.07 f	536.8 ± 0.01 a	81.8 ± 0.4 a	327 ± 4 a	0.14 ± 0.006 e	2.2	1.4	1.3	1.9	1.8	2.0	8.8
K	0.284 ± 0.005 d	0.77 ± 0.08 g	624.1 ± 0.04 c	81.6 ± 0.7 a	326 ± 7 a	0.16 ± 0.005 d	1.9	1.6	1.8	1.8	1.7	2.3	9.4
L	0.273 ± 0.009 i	0.95 ± 0.06 h	817.5 ± 0.03 e	81.2 ± 0.9 a	323 ± 5 a	0.09 ± 0.007 i	1.8	2.0	3.0	1.5	1.3	1.2	8.4
M	0.265 ± 0.008 d	0.79 ± 0.06 g	560.1 ± 0.06 b	83.5 ± 0.7 a	336 ± 8 a	0.18 ± 0.006 b	1.6	1.7	1.4	2.9	3.0	2.7	10.3
N	0.217 ± 0.004 j	0.66 ± 0.09 c	654.7 ± 0.05 c	81.6 ± 0.5 a	326 ± 5 a	0.12 ± 0.007 h	1.0	1.3	2.0	1.8	1.7	1.7	7.7

† Phenols = equivalent g of gallic acid/kg of sample. ‡ Antioxidant Power = equivalent mM of Fe^{+2}/g of sample. ≡ Antiradical Activity = equivalent mg of ascorbic acid/g of sample. ⊥ Total Carbohydrates = g/100 g of sample. # Energy = kcal/100 g of sample. * Total Ash = g/100 g of sample. SD = Standard Deviation. Values represent the mean of triplicate samples. a, b, c, d, e, f, g, h, i, j. The means reported in the same column are significantly different according to Tukey's test ($p < 0.05$) if denoted by these letters.

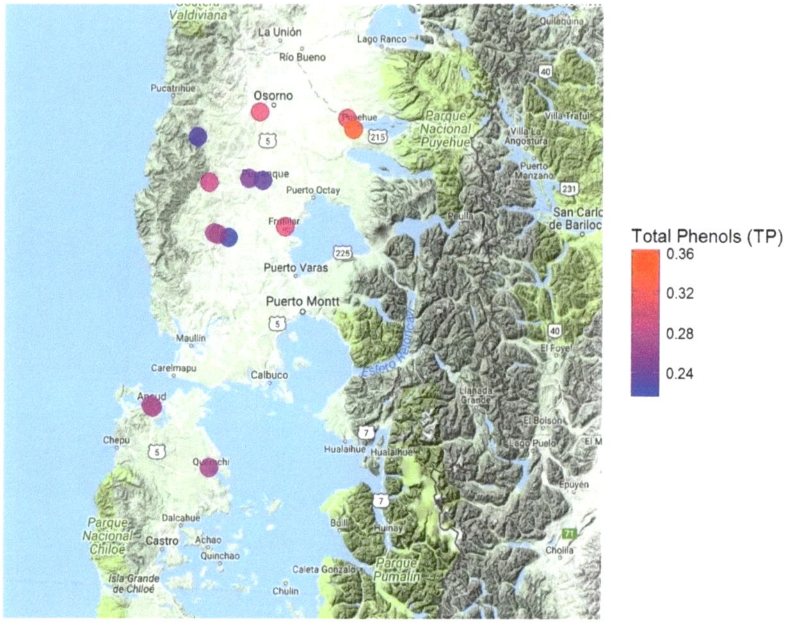

Figure 1. Map of the Los Lagos Region (Chile, 39°16′ S to 44°04′ S) indicating the sampled hive locations and respective phenol contents in honey samples. Contents concentration is indicated by a color scale going from blue (lower contents) to red (higher contents).

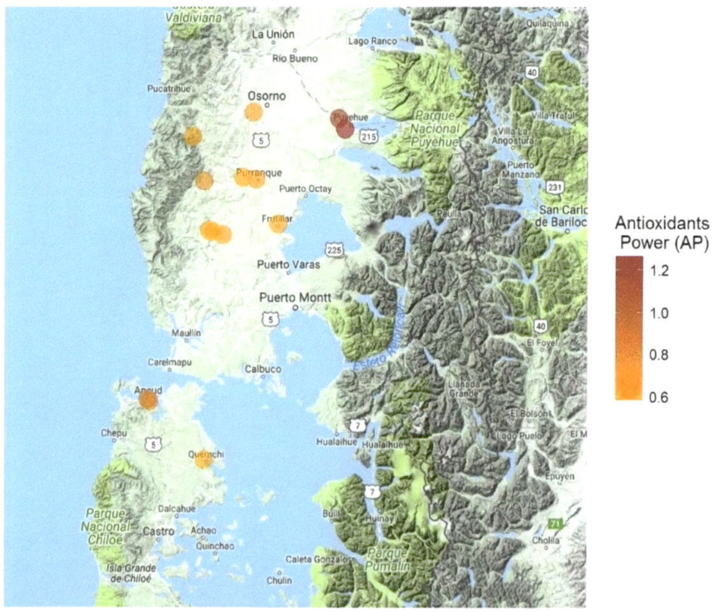

Figure 2. Map of the Los Lagos Region (Chile, 39°16′ S to 44°04′ S) indicating the sampled hive locations and respective antioxidant power of honey samples (mg of ascorbic acid equivalents/g of sample). Antioxidant power is indicated by a color scale going from light orange (lower power) to dark orange (higher power).

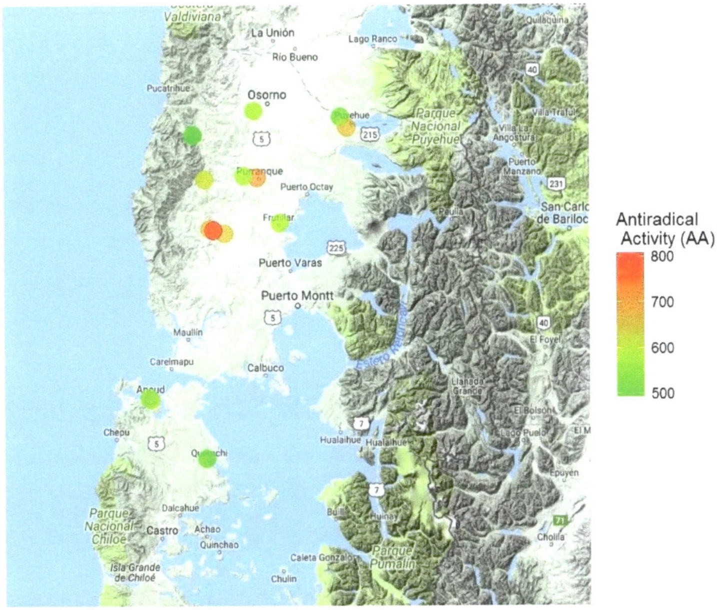

Figure 3. Map of the Los Lagos Region (Chile, 39°16′ S to 44°04′ S) indicating the sampled hive locations and respective antiradical activity of honey samples. Antiradical activity is indicated by a color scale going from green (lower activity) to red (higher activity).

3.2. MS Analysis of Honey Samples

Direct sample analysis-time of flight-mass spectrometry for the collected honey samples was performed between 100 and 3000 m/z. Signals were principally distributed between 100 and 350 m/z, and the calibration process resulted in residues with less than 0.0006 m/z. The calibrating volume used during analysis was 20 μL. Direct sample analysis is a source of ambient ionization. Ambient mass spectrometry can sample and ionize analyte molecules directly from surfaces with little to no preparation. Direct sample analysis operates on the principles of atmospheric pressure chemical ionization. Therefore, m/z values obtained via direct sample analysis-time of flight-mass spectrometry can be interpreted as honey fingerprints showing a distribution pattern that should be related to botanical origin.

In this study, six signals were regularly detected for the honeys analyzed. However, three of these were observed without exception in all honeys. Likewise, the obtained mass spectrometry signals for the control sample differed from honey samples A–N. For example, an m/z value of 105.0708 was detected in honey samples A–N, but not in honey sample Z (control). The same case was noted for m/z values of 322.0545. In addition to this, the signal corresponding to the m/z value 163.1093 was observed both in the selected honeys and in the control sample. This suggests the presence of a signal that could be an element to identify honeys independently of their geographical origin (Table 2).

For example, honey samples A-N originated from the same region and harvested in the same period, and all contained the evergreen species *Caldcluvia paniculata* (Figure 4). The m/z values also indicated distribution similarities among certain samples, such as honey samples C-D-E-J, which were collected from proximal beehives (Figure 4). These samples showed a presence of *C. paniculata* (Table 1) and similar composition percentages for the other two identified botanical species (i.e., *Luma/Myrceugenia and Weinmannia trichosperma*).

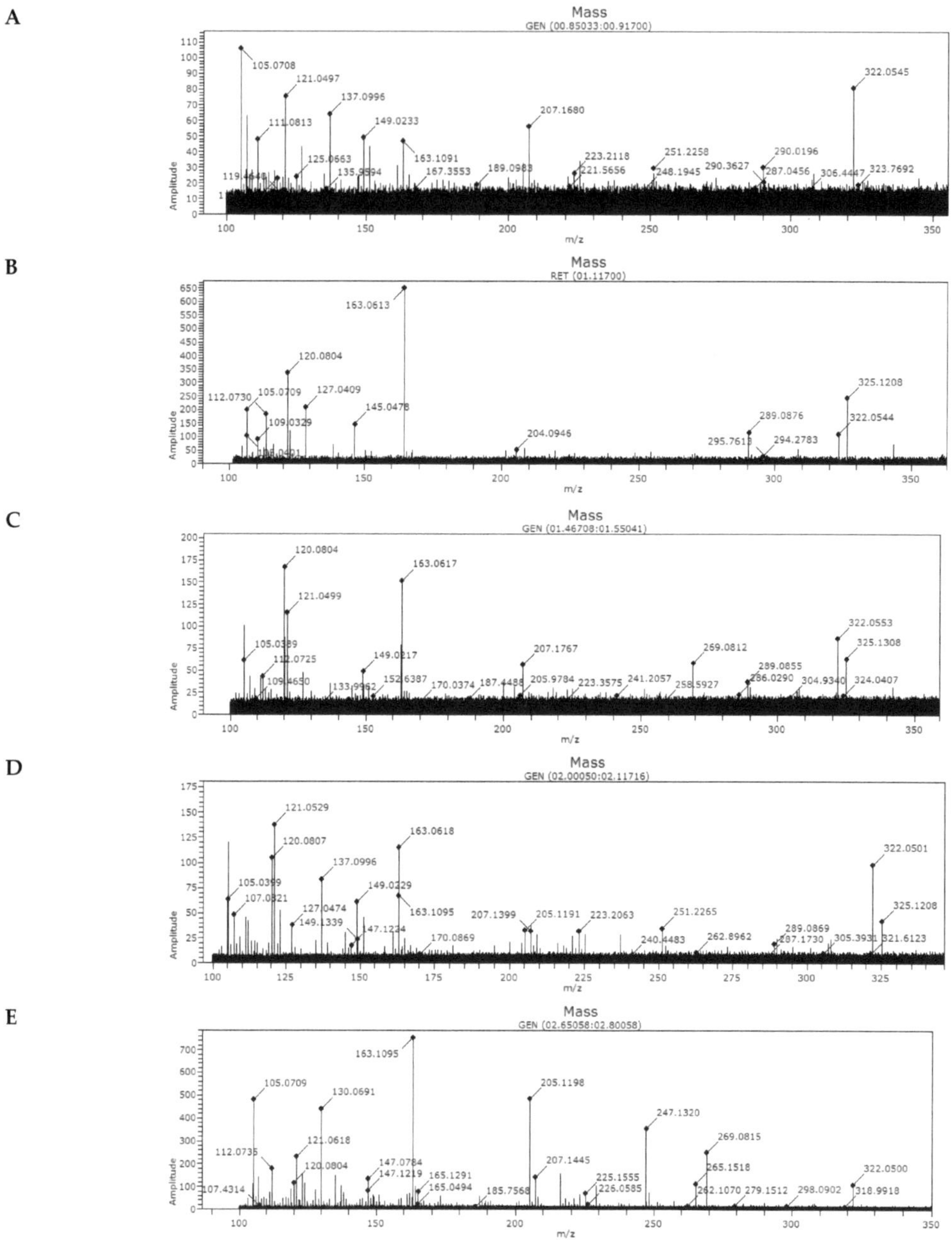

Figure 4. *Cont.*

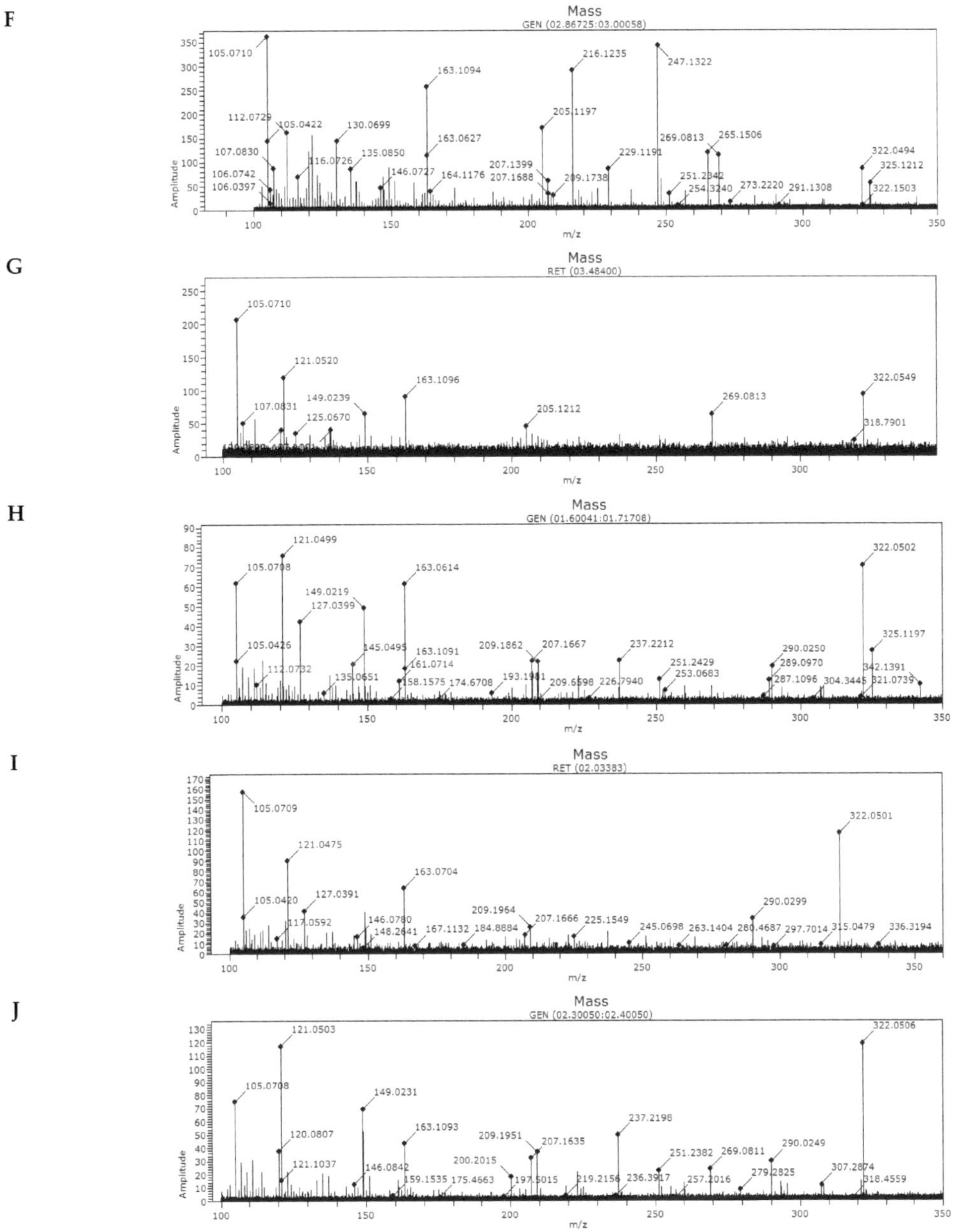

Figure 4. *Cont.*

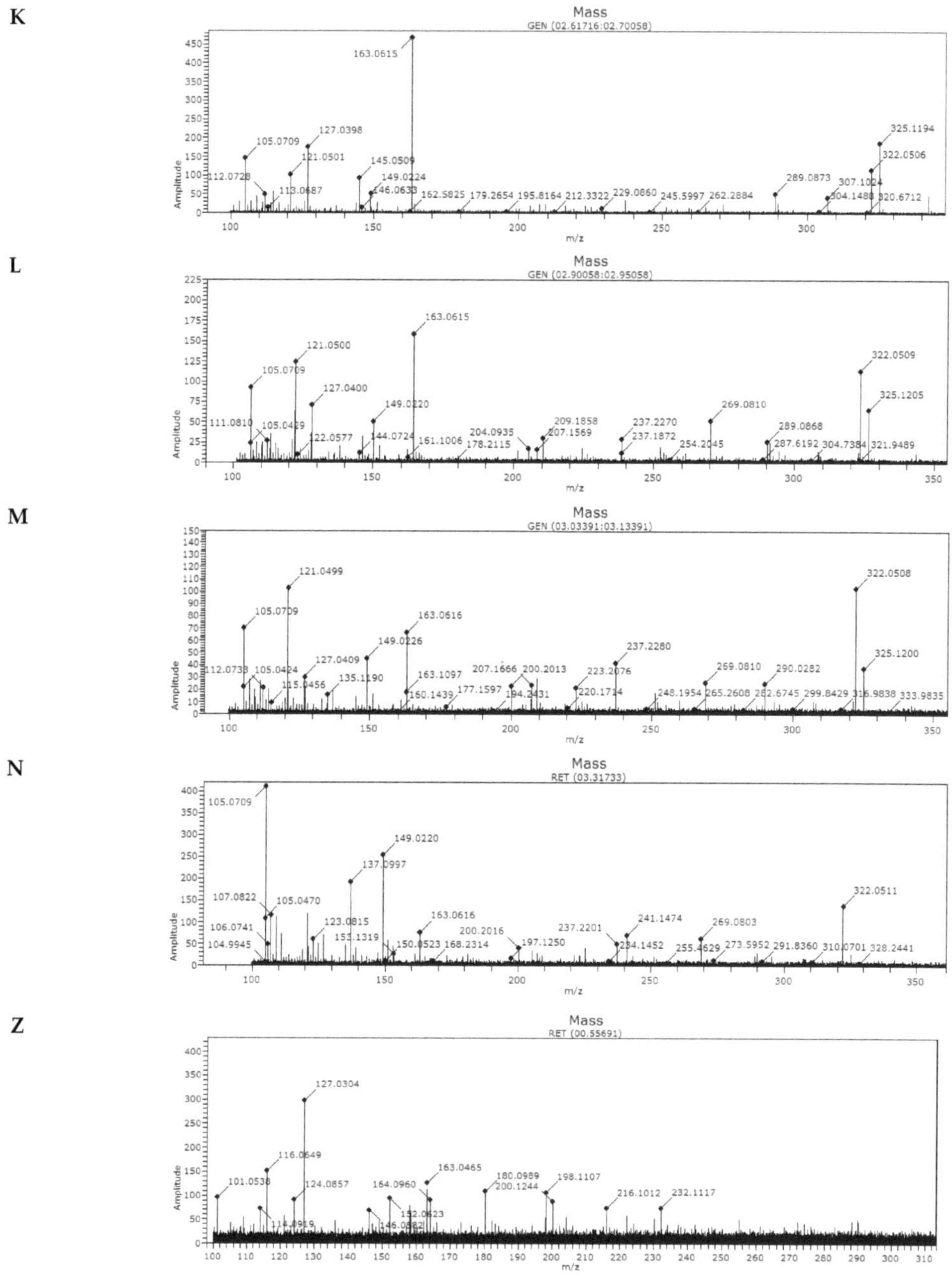

Figure 4. Spectroscopic profiles for phenolic compounds in 14 honey samples (**A–N**) obtained from the Las Lagos Region (Chile, 40°15′ S to 44°14′ S), as well as in 1 sample (**Z**) obtained from the Araucanía Region (Chile, 37°35′ S to 39°37′ S).

In turn, honey sample Z was collected as a control from the Araucanía Region (38°45′ S 72°40′ W), located approximately 310 km from the other sampled beehives. The obtained mass spectrometry signals for the control sample differed from honey samples A-N.

3.3. Presence of Pesticides in Honey/Beeswax Samples

No residues of the 242 evaluated pesticides were found for 14 of the 16 assessed beehives. However, fenhexamid was detected in beeswax samples collected in proximity to Purranque (40°55′ S, 73°10′ W) and Coligual (40°49′ S, 72°54′ W) (Figure 5). These two localities were the only sites with nearby farming activities (e.g., berries and raps).

Figure 5. Map of the Los Lagos Region (Chile, 39°16′ S to 44°04′ S) indicating the two hive locations where beeswax samples tested positive for the pesticide fenhexamid.

3.4. Suitable Areas for the Pesticide-Free Production of Honey and Beeswax

From the chemical profile and pesticide residue results, as well as information provided by beekeepers, a map was constructed indicating the most suitable zones for apicultural activities in the Los Lagos Region of Chile (Figure 6). Area suitability for honey and beeswax production was established based on biological attributes, indicators of quality, and the absence of pesticides (Table 3). The area proximal to Puyehue (40°40′ S, 72°37′ W) presented a suitability index value of 7.95, which was significantly higher than the other assessed areas. Honey and beeswax samples from this area also had the highest total phenol contents and antiradical activities (Figures 1 and 2), as well as a lack of pesticide residues (Figure 5).

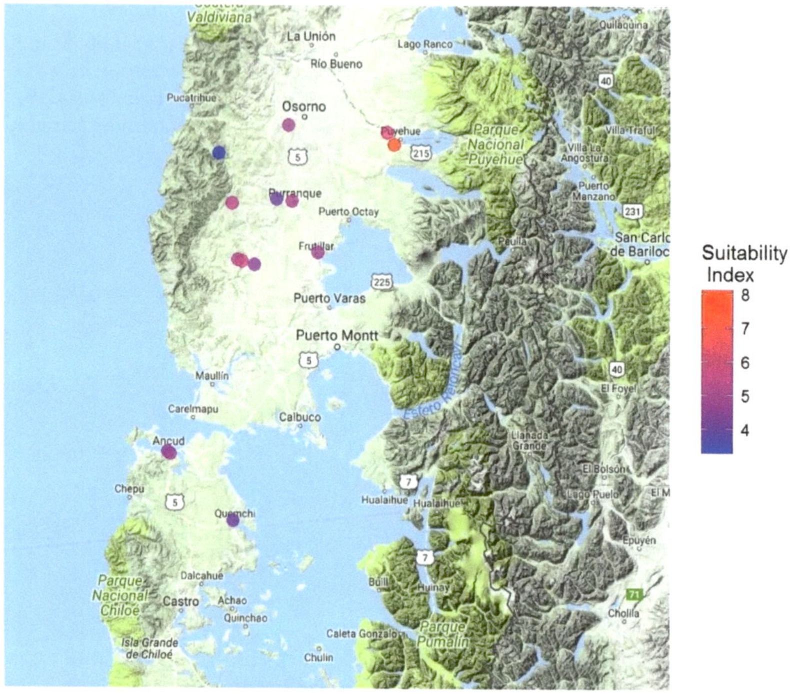

Figure 6. Map of the Los Lagos Region (Chile, 39°16′ S to 44°04′ S) indicating the most suitable areas for apiculture. The suitability index is indicated by a color scale going from blue (more suitable) to red (less suitable).

4. Discussion

The highly variable environments of the Los Lagos Region (Chile) give rise to wide floral diversity. Distinguishable among this diversity are two large plant formations, temperate laurel forests (Wintero-Nothofagetea) and sub-Antarctic deciduous forests (Nothofagetea pumilionis antarcticae) [41]. Laurel forests are composed of three forest subtypes: Valdivian or evergreen forests (38–43° S), north-Patagonian forests (43–47° S), and sub-Antarctic forests (47–55° S) [42]. Environmental floral components include species of liverworts, mosses, ferns, and gymnosperms, such as *Pilgerodendron uviferum* (Guaitecas cypres), *Podocarpus nubigenus* (Chilean podocarp), *Saxegothaea conspicua* (female maniu), *Drimys winteri* (winter's bark), *Eucryphia cordifolia* (Ulmo), *Gevuina avellana* (Chilean hazel), *Laureliopsis philippiana* (tepa), *Luma apiculata* (Chilean myrtle), *Nothofagus* sp. (coihue), *Berberis buxifolia* (Magellan barberry), *Pernettya* sp. (chaura), and *Ugni molinae* (Chilean guava), among others; as well as monocotyledon species such as *Philesia magellanica* (Chilean bellflower), *Chusquea quila* (colihue cane), *Luzuriaga* sp. (coralito), *Carex* sp. (sedges), *Codonorchis lessonii* (field lily), *Juncus* sp. (rushes), and *Uncinia* sp. (clin-clín), among others [43]. Nearly all of these species are endemic to the Los Lagos Region, meaning unique representation in south-central Chilean Patagonia and the extreme south of Argentina [42]. Other species found in the humid woodlands of the Chilean and Argentinean Mountain ranges are the native evergreen *C. paniculata* (tiaca) and endemic Cunoniaceae evergreen *W. trichosperma* (tineo), both of which serve as an important source of nectar for honey production [44]. All the assessed samples showed a presence of *C. paniculata*, which is consistent with the geographical origin of the collected honey samples. Similarly, this result coincided with the harvest date (January), which aligned with the peak flowering period for this species in the region where the sampled colonies were located [45].

The chemical composition of honey varies according to the floral origin from which bees collect nectar. Consequently, the properties of honey, such as antibacterial, antioxidant, or antidiabetic activities depend on geographical location and respective flora [37,45,46]. Different chemical compounds have been found in honey and can be related to floral origin, such as volatile aromatic compounds, derived from carotenes; amino acids and respective derivatives; aromatic acids and respective esters; aromatic aldehydes; and phenolic compounds [47]. In fact, many of these phenolic and aromatic compounds are used as markers of floral origin for honey and other apicultural products [44,48–50]. This would explain the similarity in spectroscopic profiles obtained for the 14 honey samples collected from the Los Lagos Region and the calibrating variations detected for the control honey sample obtained 310 km from the other colonies (Figure 4).

The antioxidant and antiradical abilities of honey are highly related to the presence and types of phenolic compounds [35,51–53]. Furthermore, antioxidant power depends on the number and position of OH- groups present on flavonoid structures [54]. In turn, antiradical ability, evaluated through a FRAP assay, is based on the capacity to reduce Fe^{3+} to Fe^{2+} in the presence of 2,4,6-Tris(2-pyridyl)-s-triazine [55]. Regarding the presently obtained results for these traits, honey samples collected in proximity to Puyehue showed significantly greater total phenol contents (Figure 1) and higher antioxidant power (Figure 2). Furthermore, linear and positive correlations were found between total phenols and antioxidant power ($r = 0.72$; p-value < 0.01) and between total phenols and antiradical activity ($r = 0.69$; p-value < 0.01). These results are indicative of honey quality recommended for human consumption, particularly as oxidative stress would induce cell-level damage, such as lipoperoxidation, protein damage, and nucleic acid, all of which would give rise to biological complications such as carcinogenesis, mutagenesis, aging, and arteriosclerosis [56].

Honey and beeswax samples were tested for 242 pesticides, including organochlorines, organophosphates, carbamates, thiocarbamates, pyrethroids, and neonicotinoids. No traces of these pesticide groups were found in 14 of the evaluated honey samples. However, fenhexamid was detected in two beeswax samples originating from near Purranque and Coligual (Figure 6). Fenhexamid is a widely used fungicide with site-specific actions that inhibit the 3-ketoreductase enzyme, which is involved in C-4 demethylation during the biosynthesis of ergosterol, a cellular membrane component of fungi [57]. This fungicide has an LD50 > 215 μg/bee depending on exposure contact [58]. Fenhexamid is frequently used to control *Botrytis cinerea*, translating into a commonplace presence of this fungicide in farming sectors [57,59–61]. This finding aligns with some reports provided by beekeepers with colonies located near sites with a known presence of fenhexamid. Namely, berry farms affected by *B. cinerea* existed in the same area as the bee colonies with detected pesticide residues. Interestingly, although some local beekeepers have reported decreased colony populations, the present results may indicate that this phenomenon might be due to disease in the bees (e.g., varroasis or nosema) rather than an improper use of pesticides; however, bee population can decrease and still not find pesticide residues in honey.

Chilean honey exports have increased in recent years. The primary buyer of this national product is the EU, accounting for 96% of honey exports. This market is very strict regarding sustainable and environmentally friendly productive practices [25]. In the EU, pesticide risks for bees are evaluated according to European and Mediterranean Plant Protection Organization guidelines. Consequently, the use of various pesticides is restricted because of risks to the environment and human health [58]. These economic and regulatory factors highlight the need to identify adequate areas for apiculture in Chile. In that way, the absence of pesticides in all honey samples despite the nearby crops gives good confirmation about good agricultural practices fulfillment [62]. Although in the south of Chile, a great number of native melliferous species are found [63], beekeeping is involved in pollination of fruits, as it occurs in other places along the country and the continent [64]. For this reason, the risk of exposure to one toxic compound or a mixture of them increase concerns of beekeepers [15]. Additionally, changes observed on the original and natural properties of honeys are detected when pesticides are present at the same time in the final content of these samples [65].

5. Conclusions

To this end, the present study is the first to provide a suitability index map for apiculture sites in the Los Lagos Region (Figure 6). This map establishes areas free of 242 pesticides and with honey of interesting biological quality. The developed maps and calculated data will aid local beekeepers in obtaining certifications as to the quality and safety of their products. Finally, the differentiated concentration of pesticides in honey and beeswax highlights the need for further comparative studies in order to apply this model to other regions of the country.

Supplementary Materials: The following are available online at https://www.mdpi.com/article/10.3390/insects13010031/s1, supporting Table S1: List of pesticides analyzed in honey and beeswax samples.

Author Contributions: Conceptualization, E.M.; Formal analysis, E.M., C.G. and T.G.; Funding acquisition, E.M.; Investigation, E.M. and T.G.; Methodology, E.M., C.G. and T.G.; Validation, C.G.; Writing—review & editing, E.M. and T.G. All authors have read and agreed to the published version of the manuscript.

Funding: Funding by ANID—PAI/Inserción sector productivo, 1era conv. 2019, Grant number I7819010001.

Institutional Review Board Statement: Not applicable.

Informed Consent Statement: Not applicable.

Conflicts of Interest: The authors declare no conflict of interest.

References

1. Carvalho, F.P. Pesticides, environment, and food safety. *Food Energy Secur.* **2017**, *6*, 48–60. [CrossRef]
2. Cooper, J.; Dobson, H. The benefits of pesticides to mankind and the environment. *Crop. Prot.* **2007**, *26*, 1337–1348. [CrossRef]
3. Garrido-Reyes, T.I.; Mendoza-Crisosto, J.E.; Varela-Echeverría, P.S.; Mejias-Barrios, E.G.; Alvarez-Salgado, X.A. Interaction between polychlorinated biphenyls and dissolved organic matter of different molecular weights from natural and anthropic sources. *J. Environ. Manag.* **2021**, *299*, 113645. [CrossRef]
4. Bonmatin, J.M.; Giorio, C.; Girolami, V.; Goulson, D.; Kreutzweiser, D.P.; Krupke, C. Environmental fate and exposure; neonicotinoids and fipronil. *Environ. Sci. Pollut. Res.* **2015**, *22*, 35–67. [CrossRef]
5. Cimino, A.M.; Boyles, A.L.; Thayer, K.A.; Perry, M.J. Effects of Neonicotinoid Pesticide Exposure on Human Health: A Systematic Review. *Environ. Health Perspect.* **2015**, *125*, 155–162. [CrossRef]
6. Hernández, A.F.; Parrón, T.; Tsatsakis, A.M.; Requena, M.; Alarcón, R.; López-Guarnido, O. Toxic effects of pesticide mixtures at a molecular level: Their relevance to human health. *Toxicology* **2012**, *307*, 136–145. [CrossRef] [PubMed]
7. Llorent-Martínez, E.J.; Ortega-Barrales, P.; Fernández-de Córdova, M.L.; Ruiz-Medina, A. Trends in flow-based analytical methods applied to pesticide detection: A review. *Anal. Chim. Acta* **2011**, *684*, 30–39. [CrossRef] [PubMed]
8. Souza Tette, P.A.; Guidi, L.R.; De Abreu Glória, M.B.; Fernandes, C. Pesticides in honey: A review on chromatographic analytical methods. *Talanta* **2016**, *149*, 124–141. [CrossRef]
9. Williamson, S.; Wright, A. Exposure to multiple cholinergic pesticides impairs olfatory learning and memory in honeybees. *J. Exp. Biol.* **2013**, *216*, 1799–1807.
10. Tomizawa, M.; Casida, J.E. Neonicotinoid Insecticide Toxicology: Mechanisms of Selective Action. *Annu. Rev. Pharm. Toxicol.* **2005**, *45*, 247–268. [CrossRef]
11. Barganska, Z.; Slebioda, M.; Namiesnik, J. Pesticide residues levels in honey from apiaries located of Northern Poland. *Food Control* **2013**, *31*, 196–201. [CrossRef]
12. Codling, G.; Al Naggar, Y.; Giesy, J.P.; Robertson, A.J. Concentrations of neonicotinoid insecticides in honey, pollen and honey bees (*Apis mellifera* L.) in central Saskatchewan, Canada. *Chemosphere* **2016**, *144*, 2321–2328. [CrossRef]
13. Kasiotis, K.M.; Anagnostopoulos, C.; Anastasiadou, P.; Machera, K. Pesticide residues in honeybees, honey and bee pollen by LC-MS/MS screening: Reported death incidents in honeybees. *Sci. Total Environ.* **2014**, *485–486*, 633–642. [CrossRef] [PubMed]
14. Mitchell, E.A.D.; Mulhauser, B.; Mulot, M.; Mutabazi, A.; Glauser, G.; Aebi, A. A worldwide survey of neonicotinoids in honey. *Science* **2017**, *358*, 109–111. [CrossRef] [PubMed]
15. Rodríguez López, D.; Ahumada, D.A.; Díaz, A.C.; Guerrero, J.A. Evaluation of pesticide residues in honey from different geographic regions of Colombia. *Food Control* **2014**, *37*, 33–40. [CrossRef]

16. Boily, M.; Sarrasin, B.; DeBlois, C.; Aras, P.; Chagnon, M. Acetylcholinesterase in honey bees (*Apis mellifera*) exposed to neonicotinoids, atrazine and glyphosate: Laboratory and field experiments. *Environ. Sci. Pollut. Res.* **2013**, *20*, 5603–5614. [CrossRef] [PubMed]
17. Tosi, S.; Demares, F.; Nicolson, S.; Medrzycki, P.; Pirk, C.; Human, H. Effects of a neonicotinoid pesticide on thermoregulation of African honey bees (*Apis mellifera scutellata*). *J. Insect Physiol.* **2016**, *93–94*, 56–63. [CrossRef]
18. Aliouane, Y.; El Hassani, A.K.; Gary, V.; Armengaud, C.; Lambin, M.; Gauthier, M. Subchronic Exposure of Honeybees to Sublethal Doses of Pesticides: Effects on Behavior. *Environ. Toxicol. Chem.* **2009**, *28*, 113. [CrossRef]
19. El Hassani, A.K.; Dacher, M.; Gary, V.; Lambin, M.; Gauthier, M.; Armengaud, C. Effects of sublethal doses of acetamiprid and thiamethoxam on the behavior of the honeybee (*Apis mellifera*). *Arch. Environ. Contam. Toxicol.* **2008**, *54*, 653–661. [CrossRef]
20. vanEngelsdorp, D.; Evans, J.D.; Saegerman, C.; Mullin, C.; Haubruge, E.; Nguyen, B.K. Colony collapse disorder: A descriptive study. *PLoS ONE* **2009**, *4*, e6481. [CrossRef]
21. Al-Waili, N.; Salom, K.; Al-Ghamdi, A.; Javed Ansari, M. Antibiotic, Pesticide, and Microbial Contaminants of Honey: Human Health Hazards. *Sci. World J.* **2012**, *11*, 930849. [CrossRef] [PubMed]
22. Bogdanov, S. Contaminants of bee products. *Apidologie* **2006**, *37*, 1–18. [CrossRef]
23. Choudhary, A.; Sharma, D.C. Pesticide Residues in Honey Samples from Himachal Pradesh (India). *Bull. Environ. Contam. Toxicol.* **2008**, *80*, 417–422. [CrossRef]
24. Oficina de Estudios y Políticas Agrarias; Ministerio de Agricultura de Chile. Apicultura Chilena: Actualización de Mercado y Estadísticas Sectoriales. 2018. Available online: https://www.odepa.gob.cl/publicaciones/apicultura-chilena-actualizacion-de-mercado-y-estadisticas-sectoriales-octubre-de-2018 (accessed on 26 November 2021).
25. Lamichhane, J.R.; Dachbrodt-Saaydeh, S.; Kudsk, P.; Messéan, A. Toward a reduce reliance on conventional pesticides in European agriculture. *Plant Dis.* **2015**, *100*, 10–24. [CrossRef]
26. EFSA. Conclusion on the peer review of the pesticide risk assessment for bees for the active substance imidacloprid. *EFSA J.* **2013**, *11*, 1–55.
27. EFSA. Conclusion on the peer review of the pesticide risk assessment for bees for the active substance thiamethoxam. *EFSA J.* **2013**, *11*, 12–68.
28. EFSA. Conclusion on the peer review of the pesticide risk assessment for bees for Clothianidin. *EFSA J.* **2013**, *11*, 1–58.
29. Eissa, F.; El-Sawi, S.; Zidan, N.E.-H. Determining Pesticide Residues in Honey and their Potential Risk to Consumers. *Pol. J. Environ. Stud.* **2014**, *23*, 1573–1580.
30. Krupke, C.H.; Hunt, G.J.; Eitzer, B.D.; Andino, G.; Given, K. Multiple routes of pesticide exposure for honey bees living near ag-ricultural fields. *PLoS ONE* **2012**, *7*, e29268.
31. Louveaux, J.; Maurizio, A.; Vorwohl, G. Methods of melissopalynology. *Bee World* **1978**, *59*, 139–157. [CrossRef]
32. Montenegro, G.; Gómez, M.; Díaz-Forestier, J.; Pizarro, R. Aplicación de la Norma Chilena Oficial de denominación de origen botánico de la miel para la caracterización de la producción apícola. *Cienc. Investig. Agrar.* **2008**, *35*, 181–190. [CrossRef]
33. Singleton, V.L.; Rossi, J.A. Colorimetry of total phenolics with phosphomolybdic-phosphotungstic acid reagents. *Am. J. Enol. Vitic.* **1965**, *16*, 144–158.
34. Buratti, S.; Benedetti, S.; Cosio, M.S. Evaluation of the antioxidant power of honey, propolis and royal jelly by amperometric flow injection analysis. *Talanta* **2007**, *71*, 1387–1392. [CrossRef] [PubMed]
35. Bertoncelj, J.; Doberšek, U.; Jamnik, M.; Golob, T. Evaluation of the phenolic content, antioxidant activity and colour of Slovenian honey. *Food Chem.* **2007**, *105*, 822–828. [CrossRef]
36. Meda, A.; Lamien, C.E.; Romito, M.; Millogo, J.; Nacoulma, O.G. Determination of the total phenolic, flavonoid and proline contents in Burkina Fasan honey, as well as their radical scavenging activity. *Food Chem.* **2005**, *91*, 571–577. [CrossRef]
37. Mejías, E.; Montenegro, G. The Antioxidant Activity of Chilean Honey and Bee Pollen Produced in the Llaima Volcano's Zones. *J. Food Qual.* **2012**, *35*, 315–322. [CrossRef]
38. del Campo, G.; Zuriarrain, J.; Zuriarrain, A.; Berregi, I. Quantitative determination of carboxylic acids, amino acids, carbohydrates, ethanol and hydroxymethylfurfural in honey by 1H NMR. *Food Chem.* **2016**, *196*, 1031–1039. [CrossRef]
39. Association of Official Analytical Chemists. *Official Methods of Analysis of the AOAC*, 17th ed.; Association of Official Analytical Chemists: Gaithersburgs, MD, USA, 2000.
40. Bohm, D.A.; Stachel, C.S.; Gowik, P. Confirmatory method for the determination of streptomycin and dihydrostreptomycin in honey by LC–MS/MS. *Food Addit. Contam. Part A* **2012**, *29*, 189–196. [CrossRef]
41. Borsdorf, A.; Marchant, C.; Rovira, A.; Sánchez, R. Chile cambiando. In *Revisitando la Geografía Regional de Wolfgang Weischet*; Serie GEOlibros N°36; Instituto de Geografía, Pontificia Universidad Católica de Chile/Instituto de Ciencias Ambientales y Evolutivas, Facultad de Ciencias, Universidad Austral de Chile: Santiago, Chile, 2020; pp. 123–144.
42. Gayoso, J.; Iroume, A. Impacto del manejo de plantaciones sobre el ambiente físico. *Bosque* **1995**, *16*, 3–12. [CrossRef]
43. Meneses, M.; Gayoso, J. *Estudio de Impacto Ambiental Proyecto Forestal de los Predios Tepuhueico y El Canelo*; Golden Spring Forestal: Santiago, Chile, 1995.
44. Montenegro, G.; Díaz-Forestier, J.; Fredes, C.; Rodríguez, S. Phenolic profiles of nectar and honey of Quillaja saponaria Mol. (Quillajaceae) as potential chemical markers. *Biol. Res.* **2013**, *46*, 177–182. [CrossRef]
45. Viteri, R.; Zacconi, F.; Montenegro, G.; Giordano, A. Bioactive compounds in *Apis mellifera* monofloral honeys. *J. Food Sci.* **2021**, *86*, 1552–1582. [CrossRef]

46. Cuevas-Glory, L.F.; Pino, J.A.; Santiago, L.S.; Sauri-Duch, E. A review of volatile analytical methods for determining the botanical origin of honey. *Food Chem.* **2007**, *103*, 1032–1043. [CrossRef]
47. Tomás-Barberán, F.A.; Martos, I.; Ferreres, F.; Radovic, B.S.; Anklam, E. HPLC flavonoid profiles as markers for the botanical origin of European unifloral honeys. *J. Sci. Food Agric.* **2001**, *81*, 485–496. [CrossRef]
48. Cabras, P.; Angioni, A.; Tuberoso, C.; Floris, I.; Reniero, F.; Guillou, C. Homogentisic acid: A phenolic acid as a marker of straw-berry-tree (*Arbutus unedo*) honey. *J. Agric. Food Chem.* **1999**, *47*, 4064–4067. [CrossRef]
49. Ferreres, F.; García-Viguera, C.; Tomás-Lorente, F.; Tomás-Barberán, F.A. Hesperetin: A marker of the floral origin of citrus honey. *J. Sci. Food Agric.* **1993**, *61*, 121–123. [CrossRef]
50. Kaškoniene, V.; Venskutonis, P.R. Floral Markers in Honey of Various Botanical and Geographic Origins: A Review. *Compr. Rev. Food Sci. Food Saf.* **2010**, *9*, 620–634. [CrossRef]
51. Baltrušaityte, V.; Venskutonis, P.R.; Čeksteryte, V. Radical scavenging activity of different floral origin honey and beebread phenolic extracts. *Food Chem.* **2007**, *101*, 502–514. [CrossRef]
52. Ferreira, I.C.F.R.; Aires, E.; Barreira, J.C.M.; Estevinho, L.M. Antioxidant activity of Portuguese honey samples: Different contribu-tions of the entire honey and phenolic extract. *Food Chem.* **2009**, *114*, 1438–1443. [CrossRef]
53. Serem, J.C.; Bester, M.J. Physicochemical properties, antioxidant activity and cellular protective effects of honeys from southern Africa. *Food Chem.* **2012**, *133*, 1544–1550. [CrossRef]
54. Perna, A.; Simonetti, A.; Intaglietta, I.; Sofo, A.; Gambacorta, E. Metal content of southern Italy honey of different botanical origins and its correlation with polyphenol content and antioxidant activity. *Int. J. Food Sci. Technol.* **2012**, *47*, 1909–1917. [CrossRef]
55. Ulusoy, E.; Kolayli, S. Phenolic composition and antioxidant properties of anzer bee pollen. *J. Food Biochem.* **2014**, *38*, 73–82. [CrossRef]
56. Kishore, R.K.; Halim, A.S.; Syazana, M.S.N.; Sirajudeen, K.N.S. Tualang honey has higher phenolic content and greater radical scavenging activity compared with other honey sources. *Nutr. Res.* **2011**, *31*, 322–325. [CrossRef]
57. Grabke, A.; Fernández-Ortuño, D.; Schnabel, G. Fenhexamid Resistance in Botrytis cinerea from Strawberry Fields in the Carolinas Is Associated with Four Target Gene Mutations. *Plant Dis.* **2013**, *97*, 271–276. [CrossRef] [PubMed]
58. Stoner, K.A.; Eitzer, B.D. Using a Hazard Quotient to Evaluate Pesticide Residues Detected in Pollen Trapped from Honey Bees (Apis mellifera) in Connecticut. *PLoS ONE* **2013**, *8*, e10. [CrossRef]
59. Malhat, F.M.; El Sharkawi, H.M.; Loutfy, N.M.; Ahmed, M.T. Field dissipation and health hazard assessment of Fenhexamid on Egyptian grapes. *Toxicol. Environ. Chem.* **2014**, *96*, 722–729. [CrossRef]
60. Ravoet, J.; Reybroeck, W.; De Graaf, D.C. Pesticides for apicultural and/or agricultural application found in Belgian honey bee wax combs. *Bull. Environ. Contam. Toxicol.* **2015**, *94*, 543–548. [CrossRef] [PubMed]
61. Pacífico da Silva, I.; Oliveira, F.A.S.; Pedroza, H.P.; Gadelha, I.C.N.; Melo, M.M.; Soto-Blanco, B. Pesticide exposure of honeybees (Apis mellifera) pollinating melon crops. *Apidologie* **2015**, *46*, 703–715. [CrossRef]
62. Heard, M.S.; Baas, J.; Dorne, J.L.; Lahive, E.; Robinson, A.G. Comparative toxicity of pesticides and environmental contaminants in bees: Are honey bees a useful proxy for wild bee species? *Sci. Total Environ.* **2017**, *578*, 357–365. [CrossRef]
63. Bastias, A. *Biodiversidad y Plantas Nativas de Chile*, 1st ed.; Universidad Autónoma de Chile: Santiago, Chile, 2020; pp. 61–90.
64. Harriet, J.; Campá, J.P.; Grajales, M.; Lhéritier, C.; Gómez-Pajuelo, A. Agricultural pesticides and veterinary substances in Uruguayan beeswax. *Chemosphere* **2017**, *177*, 77.e83. [CrossRef]
65. Mejias, E.; Gómez, C.; Garrido, T.; Godoy, P.; Gómez, M.; Montenegro, G. Natural attributes of Chilean honeys modified by the presence of neonicotinoids residues. *Agrofor. Syst.* **2019**, *93*, 2257–2266. [CrossRef]

insects

MDPI

Article

Polyphenols as Food Supplement Improved Food Consumption and Longevity of Honey Bees (*Apis mellifera*) Intoxicated by Pesticide Thiacloprid

Marian Hýbl [1], Petr Mráz [2], Jan Šipoš [1], Irena Hoštičková [2], Andrea Bohatá [3], Vladislav Čurn [2] and Tomáš Kopec [4,*]

1 Department of Zoology, Fisheries, Hydrobiology and Apiculture, Faculty of AgriSciences, Mendel University in Brno, Zemedelska 1, 613 00 Brno, Czech Republic; mario.eko@seznam.cz (M.H.); jan.sipos@mendelu.cz (J.Š.)

2 Department of Genetics and Agricultural Biotechnology, Faculty of Agriculture, University of South Bohemia in Ceske Budejovice, Studentska 1668, 370 05 Ceske Budejovice, Czech Republic; mrazpe01@zf.jcu.cz (P.M.); jelini00@zf.jcu.cz (I.H.); curn@zf.jcu.cz (V.Č.)

3 Department of Crop Production, Faculty of Agriculture, University of South Bohemia in Ceske Budejovice, Studentska 1668, 370 05 Ceske Budejovice, Czech Republic; bohata@zf.jcu.cz

4 Department of Animal Breeding, Faculty of AgriSciences, Mendel University in Brno, Zemedelska 1, 613 00 Brno, Czech Republic

* Correspondence: tomas.kopec@mendelu.cz

Simple Summary: Worldwide, mass losses of honey bee colonies are being observed more frequently. Poor nutrition may cause honey bees to be more susceptible to pesticides and more vulnerable to diseases, and as a direct result of this, honey bee colonies can collapse. Another cause of mass bee colony collapse that is no less important is the use of pesticides. The level of toxicity of most pesticides is greatly affected by nutrient uptake. In addition, the honey bee genome is known to be specific for a significantly lower number of genes associated with detoxification compared with other insect species. Intake of phenolic and flavonoid substances in food can lead to increased expression of genes encoding detoxification enzymes in bees. Therefore, in this study, we evaluated in vitro the effect of phenolic and flavonoid substances on bee mortality and food consumption in the case of intoxication by pesticide thiacloprid. The results of this study showed a significant positive effect on honey bee survival rate as well as increased food intake. In addition, the expression level of genes encoding detoxification enzymes was determined.

Abstract: Malnutrition is one of the main problems related to the global mass collapse of honey bee colonies, because in honey bees, malnutrition is associated with deterioration of the immune system and increased pesticide susceptibility. Another important cause of mass bee colonies losses is the use of pesticides. Therefore, the goal of this study was to verify the influence of polyphenols on longevity, food consumption, and cytochrome P450 gene expression in worker bees intoxicated by thiacloprid. The tests were carried out in vitro under artificial conditions (caged bees). A conclusively lower mortality rate and, in parallel, a higher average food intake, were observed in intoxicated bees treated using a mixture of phenolic acids and flavonoids compared to untreated intoxicated bees. This was probably caused by increased detoxification capacity caused by increased expression level of genes encoding the cytochrome P450 enzyme in the bees. Therefore, the addition of polyphenols into bee nutrition is probably able to positively affect the detoxification capacity of bees, which is often reduced by the impact of malnutrition resulting from degradation of the environment and common beekeeping management.

Keywords: cage experiments; cytochrome P450; detoxification; food intake; mortality rate

Citation: Hýbl, M.; Mráz, P.; Šipoš, J.; Hoštičková, I.; Bohatá, A.; Čurn, V.; Kopec, T. Polyphenols as Food Supplement Improved Food Consumption and Longevity of Honey Bees (*Apis mellifera*) Intoxicated by Pesticide Thiacloprid. *Insects* **2021**, *12*, 572. https://doi.org/10.3390/insects12070572

Academic Editors: Ivana Tlak Gajger and Franco Mutinelli

Received: 19 May 2021
Accepted: 16 June 2021
Published: 23 June 2021

1. Introduction

One of the most worrying phenomena is the global mass losses of honey bee colonies, including in Europe and the USA [1,2]. Along with diseases, nutrition stress and malnutrition appears to be one of the main causes of bee mortality [3–5]. The healthy development and survival of bee colonies depends to a large extent on the availability and quality of nutrients in the environment [3,6]. However, the availability and diversity of bee food resources are steadily declining due to the ever-increasing intensification of agriculture and the associated changes in the landscape, leading to a decrease in environmental sustainability [7]. As a result, there has been a decrease in the diversity of flowering plants, and low species diversity of blooming plants means reduced availability and diversity of macro- and microelements in bee nutrition [4,8], which, in the end, negatively affects bee populations [7,9]. The lack of nutrients is also the result of inefficient beekeeping practices; when replenishing winter supplies, bees are often provided only with a solution of sugar and artificial pollen substitutes. These food supplements usually lack nutrients that are naturally occurring in bees' natural diets [10]. Consequently, bee colonies are not provided with full-value nutrition [11]. Poor nutrition may cause greater susceptibility to pesticides [12], more vulnerability to diseases [13], and, as a direct result of this, the number of honey bee colonies may be decreasing [14].

Another cause of mass bee colony collapse that is no less important is the use of pesticides [15,16], which can act synergistically with other pesticides [17] or with pathogens [18,19]. The level of toxicity of most pesticides varies depending on many factors, including the means of exposure, the age of the bees, the fitness of the colonies or bee subspecies [20,21], and the optimal nutrient distribution [4,22]. In addition, the degree of toxicity of different pesticides may vary depending on whether they are tested on individual bees or on whole colonies. In vitro tests often show high pesticide toxicity and associated negative effects on bees [17,18,23]; in contrast, entire colonies appear to be relatively less susceptible to pesticides [24]. A similar trend can be observed in other social bees such as bumblebees [25]. In a broader context, some bee species may even be advantaged in anthropogenic areas such as agricultural land or urban areas [26]. However, the results of the study by Alburaka et al. [27] suggest that, while neonicotinoids do not directly affect the health and strength of bee colonies, they indirectly weaken bee health by inducing physiological stress and increasing the burden of pathogens.

However, the bee genome is known to be specific for a significantly lower number of genes associated with detoxification compared to other insect species. Where the honey bee has only 46 genes encoding the cytochrome P450 enzyme, which is thought to be the major enzyme responsible for detoxification, other insect species have around 80 or more genes encoding the cytochrome P450 enzyme [28]. There are several honey bee cytochrome P450 genes that have defined functions, including CYP9Q1, CYP9Q2, and CYP9Q3. These genes metabolize both natural and synthetic xenobiotics [29].

The intake of phenolic and flavonoid substances, which are commonly found in honey and, to a greater extent, in pollen, via food can lead to an increased expression of genes encoding cytochrome P450 enzyme in bees. The amount and proportion can vary significantly depending on food sources [30]. Of these, the highest efficacies have been observed for p-coumaric acid and quercetin [31]. The natural diet of bees usually contains a large amount and great diversity of phenolic acids, flavonoids and their derivatives [32,33] and it is their different amounts and proportions that influence the detoxifying effects [31].

The first goal of this study was to determine the real effect of phenolic acids and flavonoids in vitro on the mortality of bees intoxicated by thiacloprid, one of the most widely used neonicotinoids. The second aim was to determine the effect of phenolic substances on the rate of food intake by bees; and the last target, although no less important, was to determine the expression level of several genes potentially responsible for detoxification via the enzyme cytochrome P450.

2. Materials and Methods

The experiment was carried out at the beginning of the summer of 2019 in Brno (South Moravia, Czech Republic)

2.1. Bees

The honey bees used in this study were obtained from the experimental apiary of Mendel University in Brno. Honey bees from four colonies were used (one frame with hatching bees per colony). The colonies were maintained following standard beekeeping practices. In all the bee colonies, inseminated queens belonging to *Apis mellifera carnica* were used. As a result, the genetic variability of bees in individual colonies was reduced so that the average coefficient of relatedness between workers from one colony was r = 0.5.

The brood frames with hatching bees (one from each colony) were incubated at 35 °C and 65–80% relative humidity for 12 h. This allowed bees of the same age ± 12 h to be obtained. Then the frames were brushed, and all bees were mixed together and divided into four groups according to the treatment with three replications (three cages each). There were 40 bees of the same age in each cage. The cages were maintained for 2 weeks in the thermostat with conditions 30 °C and 65–70% relative humidity [34]. The bee mortality and food consumption were noted down every day and dead bees were continuously removed from the cages.

2.2. Chemicals

The sucrose solution consisted of 50% (*w/v*) sucrose and distilled water. A dosage of thiacloprid was mixed with the sucrose solution in two different concentrations, 35 mL/L or 70 mg/L, depending on the treatment [19].

The mixture of phenolic compounds consisted of 200 mg/kg of phenolic acids and 10 mg/kg of flavonoids in proportions based on the real concentrations found in common honey [33]. The concentration of p-Coumaric acid was scaled up on the basis of Mao et al. [30]. The final contain of phenolic compounds is in Table 1. The thiacloprid and sucrose were purchased from Sigma Aldrich (Schnelldorf, BO, Germany), phenolic acids and flavonoids were purchased from Alfa Aestar (Kandel, RP, Germany).

Table 1. Content of phenolic acids and flavonoids used in the phenolic mixture.

Phenolic Substance Classification	Phenolic Substance Name	Amount (%)	Amount (mg/kg)
Phenolic acids	caffeic acid	10	20
	benzoic acid	20	40
	gallic acid	7.5	15
	ferulic acid	20	40
	p-Coumaric acid	35	70
	vanillic acid	7.5	15
Flavonoids	rutin	25	2.5
	quercetin	25	2.5
	naringin	25	2.5
	hesperidin	25	2.5

2.3. Design of the Experiment

The bees in cages were fed with two top feeders with scales (*ad libitum*) per cage, enabling measurement of the daily food consumption. The rate of consumption of a prequantified amount by a set number of live bees was evaluated over a set time period. The experimental groups were set up as follows:

1. Treatment TL—sucrose solution with a low dosage of Thiacloprid (35 mg/L).

2. Treatment FTL—sucrose solution (50% *w/v*) with a mixture of phenolic compounds and low dosage of Thiacloprid (35 mg/L).
3. Treatment TH—sucrose solution (50% *w/v*) with high dosage of Thiacloprid (70 mg/L).
4. Treatment FTH—sucrose solution (50% *w/v*) with a mixture of phenolic compounds and a high dosage of Thiacloprid (70 mg/L).
5. Treatment F—sucrose solution (50% *w/v*) and a mixture of phenolic compounds.
6. Treatment C—sucrose solution (50% *w/v*).

2.4. RNA Isolation and RT-qPCR

Samples for studying gene expression were collected as a bulk of three bees and frozen immediately in liquid nitrogen, and were stored at −80 °C. Total RNA was extracted using the TRI Reagent (MRC, Montgomery, OH, USA) according to manufacturer's instructions. Contaminating DNA was removed using the DNA-freeTMKit (Ambion, supplied by ThermoFisher scientific, Loughborough, UK). BioSpec Nano (Shimadzu, Nakagyo-ku, Kyoto, Japan) was used to quantify RNA (OD260) and to assess sufficient quality (OD260/280 ratio and OD260/230 ratio). cDNA templates were prepared using a Standard Reverse Transcription Protocol (Promega, Madison, WI, USA) and stored at −20 °C until use.

The RT-qPCR was performed on the QuantStudio™ 6 Flex Real-Time PCR System (Applied Biosystems, supplied by ThermoFisher scientific, Loughborough, UK) using Power SYBR® Green PCR Master Mix (Applied Biosystems, supplied by ThermoFisher scientific, Loughborough, UK) in a 96-well reaction plate using parameters recommended by the manufacturer (2 min at 50 °C, 10 min at 95 °C and 40 cycles of 15 s 95 °C, 1 min of 60 °C, 15 s at 95 °C, 1 min at 60 °C and 15 s at 95 °C). The three replicates and no-template controls were included. The specificity of amplification was determined by dissociation curve analyses. A comparative threshold cycle method was applied to determine relative concentrations of mRNA. The primers used are shown in Table 2. All the gene expression levels were normalized to Am Rp49 gene expression, as a reference gene [35], and the obtained data were normalized to Am Rp49 using the ΔΔCT method according to Livak, Schmittgen [36].

Table 2. Primers for qPCR analysis.

Gene	Sequences 5′-3′	Reference
Cyp9q1	F: TCGAGAAGTTTTTCCACCG R: CTCTTTCCTCCTCGATTG	Mao et al. [37]
Cyp9q2	F: GATTATCGCCTATTATTA R: GTTCTCCTTCCCTCTGAT	Mao et al. [37]
Cyp4g11	F: AATGCGAGAAGTGTCGTCGA R: AGCGGTTTCCAGAAGGATGT	Calla et al. [38]
AmRp49	F: CGTCATATGTTGCCAACTGGT R: TTGAGCACGTTCAACAATGG	Tesovnik et al. [39]

2.5. Data Analyses

The survival curves were fitted by the Kaplan-Meier method. On the basis of this method, the survival probability for each tested treatment during 14 days of observation was estimated [40]. The conclusive difference between each survival curve was evaluated by log-rank test [41]. The log-rank test compares a monitored case number with the case number that would have been expected under the null hypothesis (i.e., identical survival curves). All data were analyzed using the R statistical program (R Core Team, 2017).

Daily food intake was analyzed using the statistical program Statistica 12. The effect of fed substances on the rate of diet consumption was tested by the analysis of variance procedure ANOVA (post hoc analysis using Tukey test), preceded by a normality test. Statistical significance was tested at a level of significance $\alpha = 0.05$.

3. Results

The bee survival rate corresponding fed treatment is presented in Table 3 and Figure 1. The FTH group exhibited a significantly lower mortality rate than group TH ($p < 0.001$), but a higher mortality rate than the control groups C and F ($p < 0.001$ for both). Comparatively lower mortality rates were observed in the treatment FTL than in the treatment TL ($p < 0.001$), although the mortality rate was higher than in the control group C ($p = 0.03$), but no significant differences were observed in comparison with group F ($p = 0.17$). Additionally, no significant differences were registered between control groups C and F ($p = 0.44$). Significant differences were also observed between TH and C, TH and F, TL and C, and TL and F ($p < 0.001$ in all cases).

Table 3. The results of the log-rank test, which was used to compare different treatment groups of bees treated using various chemical substances.

Treatment	Degrees of Freedom	Chi-Square Statistic	*p*-Value
TH/C	1	310	<0.001
TH/F	1	270	<0.001
FTH/TH	1	72	<0.001
FTH/C	1	62.9	<0.001
FTH/F	1	51.2	<0.001
FTL/TL	1	6	0.01
TL/C	1	5.7	0.01
FTL/C	1	4.6	0.03
FTL/F	1	1.8	0.17
F/C	1	0.6	0.44

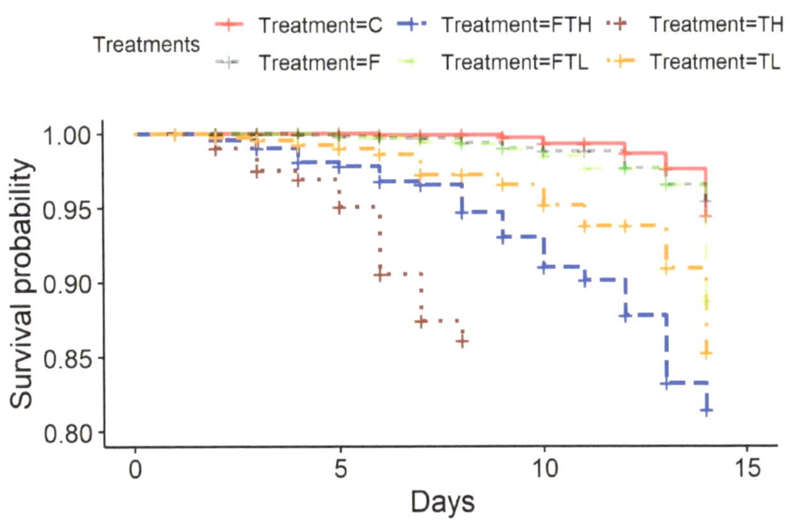

Figure 1. The relationship between the bee mortality and the treatments over 14 days (Kaplan-Maier survival analyses). Legend: TH—sucrose solution (50% *w/v*) with high dosage of Thiacloprid (70 mg/L), FTH—sucrose solution (50% *w/v*) with a mixture of phenolic compounds and a high dosage of Thiacloprid (70 mg/L), TL—sucrose solution with a low dosage of Thiacloprid (35 mg/L), FTL—sucrose solution (50 *w/v*) with a mixture of phenolic compounds and low dosage of Thiacloprid (35 mg/L), F—sucrose solution (50% *w/v*) and a mixture of phenolic compounds, C—sucrose solution (50% *w/v*).

The food intake was dependent on the treatment (Figure 2). The amount of diet consumed was higher in groups C and F than in any of the other groups, whereas the food consumption was higher in group F than in group C. In group FTL, higher food intake was observed than in group TL. The same trend was observed in the case of the FTH and TH groups. The lowest food consumption was observed in groups TH and TL.

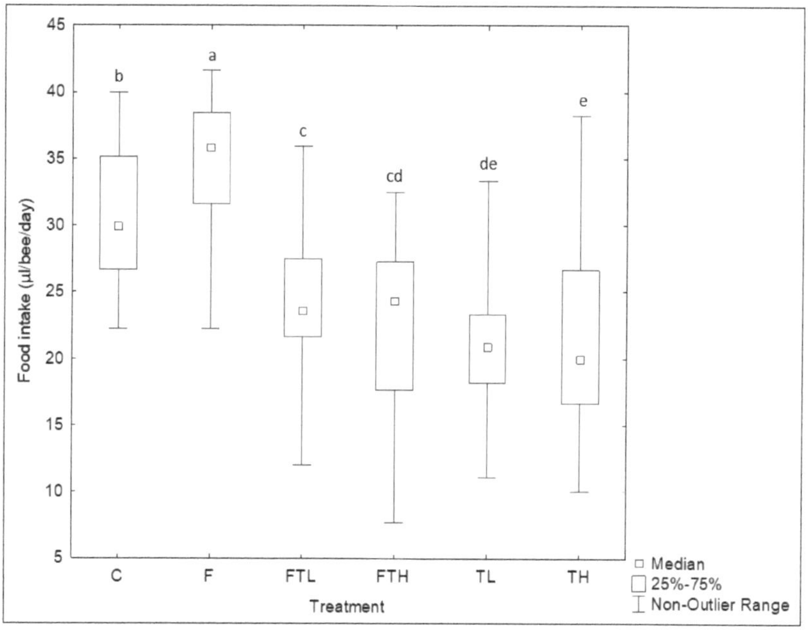

Figure 2. The daily average of the food intake per 1 bee over 14 days. Significant differences ($p < 0.05$) in food consumption between treatments are indicated by different letters.

The expression of CYP9Q1 (Figure 3a), CYP9Q2 (Figure 3b), CYP9Q3 (Figure 3c) and CYP4G11 (Figure 3d) genes was analyzed using RT-qPCR. Differences in gene expression between the testing groups were not statistically significant. However, despite that, some trends in the levels of expression were noted between the testing groups. The relative expression of the CYP9Q1 gene decreased in bees fed with sucrose solution enriched by phenolic compounds, irrespective of thiacloprid intoxification (F, FTL, FTH) in comparison with the C group after 7 days of treatment. In bees from groups TL and TH, the relative gene expression of this gene was comparable with its expression in the C group after 7 days. After 14 days of treatment, the relative expression of CYP9Q1 was increased in bees from group TH. In other groups, the relative gene expression of this gene was comparable with its expression in the C group.

After 7 days of treatment, the relative expression of CYP9Q2 in bees from groups F and FTL was comparable with the C group. In groups FTH, TL and TH, it was slightly increased in comparison with the C group. After 14 days of treatment, the relative expression of this gene was increased in groups F, FTL and FTH. In groups TH and TL it was comparable with group C.

The relative expression of CYP9Q3 was higher in the TL group after 7 days of treatment and also in the FTH and TL groups after 14 days of treatment. In other groups, the relative expression of this gene was comparable with group C.

After 7 days of treatment, the relative expression of the CYP4G11 gene was comparable in bees fed with sucrose solution enriched by phenolic compounds regardless of whether they were intoxicated with thiacloprid (F, FTL, FTH) and in bees from C group. In groups TL and TH, it was increased in comparison with C. After 14 days of treatment, the relative expression of this gene was increased in the FTH, TL and TH groups in comparison to the C group. In the F and FTL groups it was comparable with the C group.

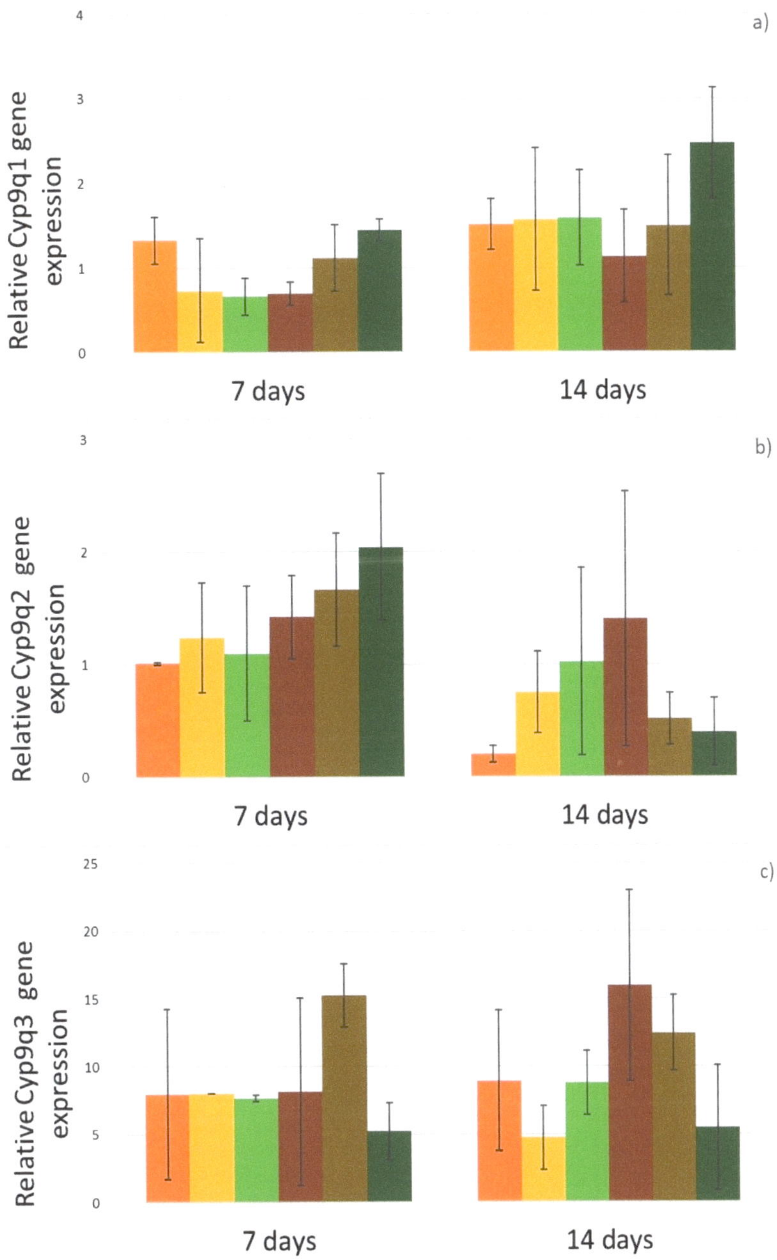

Figure 3. *Cont.*

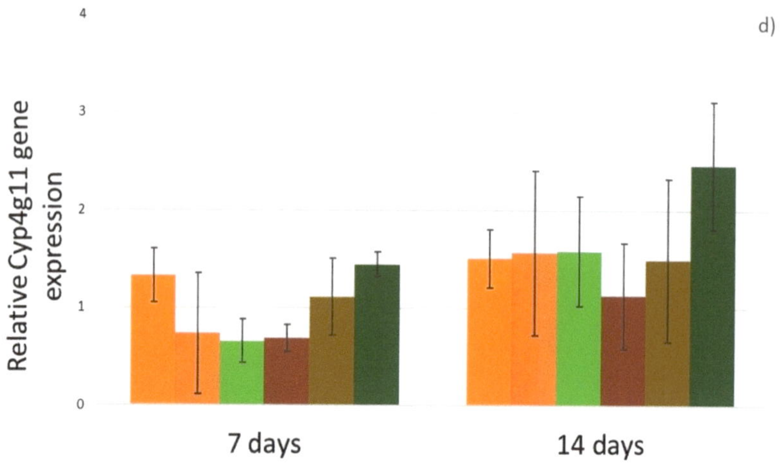

Figure 3. Relative expressions of four cytochrome P450 genes in bee workers depending on treatment at the 7th and 14th days of the experiment. Error bars denote two technical replications of three samples. (**a**) CYP9Q1 gene, (**b**)CYP9Q2 gene, (**c**) CYP9Q3 gene, (**d**) CYP4G11 genes.

4. Discussion

As expected, significantly higher mortality was observed in the treatment groups containing thiaclopride (TL, TH) compared to thiacloprid-free groups (C, F). This is consistent with the findings reported by Retschnig et al. [19]. However, Retschnig et al. [19] observed significantly lower mortality levels in bees intoxicated with high doses of thiacloprid than those observed in this study (TH). This difference may be due to different levels of sensitivity between bee subtaxons [20]. On the other hand, a low level of bee mortality was observed in group F, which did not differ from group C, indicating the safety of phenolic compounds for bees, which is in accordance with the results of Liao et al. [31]. Conversely, a statistically significant decrease was observed in the mortality rate of the group FTL in comparison with group TL, as well as in FTH compared to TH, which was probably caused by the increased detoxification capacity [30] and antioxidant activity [33] of the experimental bees due to phenolic-enriched diets [31]. The fact that with the addition of phenolic compounds (FTL), the mortality of intoxicated bees decreased significantly compared to the TL group, but did not reach the same level as in non-intoxicated bees (C), indicates the limited detoxification capacity in honey bees [28,42]. This trend was even more pronounced in the groups with a high dose of thiacloprid. A very significant reduction in mortality was observed with the addition of phenolic compounds (FTH) in comparison with the group without phenolic compounds (TH), but losses were still higher than in all other groups. The relationship between the experimental groups FTL, F and C seems to be an interesting phenomenon. No statistically significant difference in mortality was observed between the FTL and F groups, but there was a difference between the FTL and C groups. This can be explained by the probable increase of metabolic load caused by increased flavonoid levels [43].

The food consumption in the group containing phenolic compounds (F) was higher than in the group fed only with sucrose solution (C). Similar results were obtained by Porrini et al. [44]. They observed increased food intake when feeding bees with the addition of essential oils that contained phenolic compounds as their main components. Nevertheless, in their study, the rate of food consumption was lower in the control group, as well as in the experimental groups, compared to our study. This difference was probably caused by the difference in the carbohydrate concentration of the feed solution. A higher concentration of sugars in the feed leads to lower feed consumption, and vice versa [45].

On the other hand, in groups C and F, the food consumption rate was significantly higher than the groups with the addition of pesticide (TL and TH). This is consistent with the findings of Gregorc et al. [46] and Tosi et al. [47]. This was probably caused by increased levels of stress as a result of the addition of pesticides in the food [45]. However, in the case of intoxicated groups fed food enriched with phenolic compounds, the rate of food intake was significantly increased compared with groups without phenolic compounds, both in the case of low amounts of pesticide (FTL) and the case of high amounts of pesticide (FTH). This trend may be explained by the increased detoxification capacity [30] and antioxidant activity [33] caused by phenolic compounds in the food [31]. Differences in food consumption between TL and TH were not observed, nor were they observed between FTL and FTH. Therefore, the amount of pesticide in the food did not affect the level of consumption, which is consistent with the results of Retschnig et al. [19].

The relative expression of CYP9Q1, CYP9Q2, CYP9Q3 and CYP4G11 genes was analyzed using RT-qPCR. The cytochrome P450 enzyme group was chosen as the main endpoint in the detoxification process because it is responsible for the activity of the detoxification pathways of neonicotinoids [17]. The gene expression of four genes responsible for detoxication was analyzed after 7 and 14 days of treatment. In previous studies [30,37,48], bees were fed once with pesticide at the beginning of the experiment, and then the mortality and gene expression were analyzed in the first days after treatment. Conversely, in this study, a long-term experiment with long-term exposition to the tested substances was performed, with bees being fed continuously throughout the whole experiment. The expression levels of detoxification genes are highly dependent on time after pesticide treatment [49]. Therefore, the time of collection of genetic material could be the main reason why differences in detoxification gene expressions between experimental groups were not conclusive, and that the expression levels did not differ significantly between groups. Our results suggest a trend in which the expression in the F, FTL and C groups was comparable, and the gene expression in other intoxicated groups was increased. This could indicate that increased expression probably took place at the beginning of experiment, and that in the first days of bee sampling, the level of enzymes cytochrome P450 were already increased. However, better explanation of this issue could be provided by quantification of expressed protein. It would be suitable to carry out this investigation in future experiments.

Wheeler and Robinson [50] point out the problem that beekeepers use artificial bee food for bees, which, however, usually does not contain certain ingredients with high nutritional value and importance that are natural components of honey and pollen. Thus, it is clear that in addition to macronutrients (carbohydrates and proteins), the bee diet should also contain other elements (such as phenolic compounds) that have a conclusive impact on their detoxification capacity [30,31]. Based on the results of this study, we suggest that the addition of phenolic compounds to bee nutrition could to some extent increase the detoxification capacity of bees [30,31], which is often reduced due to malnutrition caused by degradation of the environment and the associated loss and contamination of food resources, as well as factors related to routine beekeeping management [4,50]. In addition, according to Mao et al. [29], some phenolic substances have an effect on the suppression of ovarian development, suggesting that phenolic substances could be used in the future to solve other problems in beekeeping practice.

5. Conclusions

Phenolic compounds, as natural components of the bee diet, have been demonstrated to have a positive impact on the longevity of honey bees intoxicated by thiacloprid, as well as their food intake.

The results of the experiments suggest that by adding phenolic substances to bee nutrition, the risks associated with the intoxication of bees can be reduced.

The expression levels of detoxification genes alone, depending on the treatment, may not be sufficient, and it is appropriate to support this with quantification of expressed proteins.

Author Contributions: Conceptualization, M.H.; Data curation, J.Š. and I.H.; Formal analysis, P.M., J.Š. and T.K.; Funding acquisition, V.Č. and A.B.; Investigation, P.M. and A.B.; Methodology, M.H. and I.H.; Project administration, A.B. and T.K.; Supervision, V.Č.; Validation, J.Š.; Writing—original draft, M.H., P.M., I.H. and T.K. All authors have read and agreed to the published version of the manuscript.

Funding: This research was funded by Mendel University in Brno, grant number AF–IGA2019–IP007 and Ministry of agriculture, grant number QK1910356.

Institutional Review Board Statement: Not applicable.

Informed Consent Statement: Not applicable.

Data Availability Statement: The study did not report any data.

Acknowledgments: We thank Antonín Přidal (Department of Zoology, Fisheries, Hydrobiology and Apiculture, Mendel University in Brno) for providing us with the used chemicals, the background of the university apiary, and apicultural consultations.

Conflicts of Interest: The authors declare no conflict of interest.

References

1. Aizen, M.A.; Garibaldi, L.A.; Cunningham, S.A.; Klein, A.M. How much does agriculture depend on pollinators? Lessons from long-term trends in crop production. *Ann. Bot.* **2009**, *103*, 1579–1588. [CrossRef]
2. Stokstad, E. The Case of the Empty Hives. *Science* **2007**, *316*, 970–972. [CrossRef]
3. Brodschneider, R.; Crailsheim, K. Nutrition and Health in Honey Bees. *Apidologie* **2010**, *41*, 278–294. [CrossRef]
4. Di Pasquale, G.; Salignon, M.; Le Conte, Y.; Belzunces, L.P.; Decourtye, A.; Kretzschmar, A.; Suchail, S.; Brunet, J.-L.; Alaux, C. Influence of Pollen Nutrition on Honey Bee Health: Do Pollen Quality and Diversity Matter? *PLoS ONE* **2013**, *8*, e72016. [CrossRef] [PubMed]
5. Sharpe, R.J.; Heyden, L.C. Honey Bee Colony Collapse Disorder Is Possibly Caused by a Dietary Pyrethrum Deficiency. *Biosci. Hypotheses* **2009**, *2*, 439–440. [CrossRef]
6. Haydak, M.H. Honey Bee Nutrition. *Annu. Rev. Entomol.* **1970**, *15*, 143–156. [CrossRef]
7. Naug, D. Nutritional Stress Due to Habitat Loss May Explain Recent Honeybee Colony Collapses. *Biol. Conserv.* **2009**, *142*, 2369–2372. [CrossRef]
8. Johnson, R.M.; Ellis, M.D.; Mullin, C.A.; Frazier, M. Pesticides and Honey Bee Toxicity—USA. *Apidologie* **2010**, *41*, 312–331. [CrossRef]
9. Decourtye, A.; Mader, E.; Desneux, N. Landscape Enhancement of Floral Resources for Honey Bees in Agro-Ecosystems. *Apidologie* **2010**, *41*, 264–277. [CrossRef]
10. Manning, R.; Rutkay, A.; Eaton, L.; Dell, B. Lipid-enhanced Pollen and Lipid-reduced Flour Diets and Their Effect on the Longevity of Honey Bees (*Apis mellifera* L.). *Aust. J. Entomol.* **2007**, *46*, 251–257. [CrossRef]
11. Hayes, J., Jr.; Underwood, R.M.; Pettis, J. A Survey of Honey Bee Colony Losses in the US, Fall 2007 to Spring 2008. *PLoS ONE* **2008**, *3*, e4071.
12. Wahl, O.; Ulm, K. Influence of Pollen Feeding and Physiological Condition on Pesticide Sensitivity of the Honey Bee *Apis mellifera* Carnica. *Oecologia* **1983**, *59*, 106–128. [CrossRef]
13. Alaux, C.; Ducloz, F.; Crauser, D.; Le Conte, Y. Diet Effects on Honeybee Immunocompetence. *Biol. Lett.* **2010**, *6*, 562–565. [CrossRef] [PubMed]
14. Keller, I.; Fluri, P.; Imdorf, A. Pollen Nutrition and Colony Development in Honey Bees—Part II. *Bee World* **2005**, *86*, 27–34. [CrossRef]
15. Frazier, M.; Mullin, C.; Frazier, J.; Ashcraft, S. What Have Pesticides Got to Do with It? *Am. Bee J.* **2008**, *148*, 521–524.
16. Medrzycki, P.; Sgolastra, F.; Bortolotti, L.; Bogo, G.; Tosi, S.; Padovani, E.; Porrini, C.; Sabatini, A.G. Influence of Brood Rearing Temperature on Honey Bee Development and Susceptibility to Poisoning by Pesticides. *J. Apic. Res.* **2010**, *49*, 52–59. [CrossRef]
17. Iwasa, T.; Motoyama, N.; Ambrose, J.T.; Roe, R.M. Mechanism for the Differential Toxicity of Neonicotinoid Insecticides in the Honey Bee, *Apis mellifera*. *Crop Prot.* **2004**, *23*, 371–378. [CrossRef]
18. Alaux, C.; Brunet, J.; Dussaubat, C.; Mondet, F.; Tchamitchan, S.; Cousin, M.; Brillard, J.; Baldy, A.; Belzunces, L.P.; Le Conte, Y. Interactions between Nosema Microspores and a Neonicotinoid Weaken Honeybees (*Apis mellifera*). *Environ. Microbiol.* **2010**, *12*, 774–782. [CrossRef]
19. Retschnig, G.; Neumann, P.; Williams, G.R. Thiacloprid–Nosema Ceranae Interactions in Honey Bees: Host Survivorship but Not Parasite Reproduction Is Dependent on Pesticide Dose. *J. Invertebr. Pathol.* **2014**, *118*, 18–19. [CrossRef]
20. Suchail, S.; Guez, D.; Belzunces, L.P. Characteristics of Imidacloprid Toxicity in Two *Apis mellifera* Subspecies. *Environ. Toxicol. Chem.* **2000**, *19*, 1901–1905. [CrossRef]
21. Nauen, R.; Ebbinghaus-Kintscher, U.; Schmuck, R. Toxicity and Nicotinic Acetylcholine Receptor Interaction of Imidacloprid and Its Metabolites in *Apis mellifera* (Hymenoptera: Apidae). *Pest Manag. Sci.* **2001**, *57*, 577–586. [CrossRef]

22. Wehling, M.; Ohe, W.; Brasse, D.; Forster, R. Colony Losses-Interactions of Plant Protection Products and Other Factors. *Jul. Kühn Arch.* **2009**, *423*, 153–154.
23. Grillone, G.; Laurino, D.; Manino, A.; Porporato, M. Toxicity of Thiametoxam on in vitro Reared Honey Bee Brood. *Apidologie* **2017**, *48*, 635–643. [CrossRef]
24. Osterman, J.; Wintermantel, D.; Locke, B.; Jonsson, O.; Semberg, E.; Onorati, P.; Forsgren, E.; Rosenkranz, P.; Rahbek-Pedersen, T.; Bommarco, R. Clothianidin Seed-Treatment Has No Detectable Negative Impact on Honeybee Colonies and Their Pathogens. *Nat. Commun.* **2019**, *10*, 1–13. [CrossRef] [PubMed]
25. Rundlöf, M.; Lundin, O. Can Costs of Pesticide Exposure for Bumblebees Be Balanced by Benefits from a Mass-Flowering Crop? *Environ. Sci. Technol.* **2019**, *53*, 14144–14151. [CrossRef]
26. Ghazoul, J. Buzziness as Usual? Questioning the Global Pollination Crisis. *Trends Ecol. Evol.* **2005**, *20*, 367–373. [CrossRef] [PubMed]
27. Alburaki, M.; Boutin, S.; Mercier, P.-L.; Loublier, Y.; Chagnon, M.; Derome, N. Neonicotinoid-Coated Zea Mays Seeds Indirectly Affect Honeybee Performance and Pathogen Susceptibility in Field Trials. *PLoS ONE* **2015**, *10*, e0125790. [CrossRef]
28. Claudianos, C.; Ranson, H.; Johnson, R.; Biswas, S.; Schuler, M.; Berenbaum, M.; Feyereisen, R.; Oakeshott, J.G. A Deficit of Detoxification Enzymes: Pesticide Sensitivity and Environmental Response in the Honeybee. *Insect Mol. Biol.* **2006**, *15*, 615–636. [CrossRef]
29. Mao, W.; Schuler, M.; Berenbaum, M. Task-related Differential Expression of Four Cytochrome P450 Genes in Honeybee Appendages. *Insect Mol. Biol.* **2015**, *24*, 582–588. [CrossRef] [PubMed]
30. Mao, W.; Schuler, M.A.; Berenbaum, M.R. Honey Constituents Up-Regulate Detoxification and Immunity Genes in the Western Honey Bee *Apis mellifera*. *Proc. Natl. Acad. Sci. USA* **2013**, *110*, 8842–8846. [CrossRef]
31. Liao, L.-H.; Wu, W.-Y.; Berenbaum, M.R. Impacts of Dietary Phytochemicals in the Presence and Absence of Pesticides on Longevity of Honey Bees (*Apis mellifera*). *Insects* **2017**, *8*, 22. [CrossRef]
32. Cianciosi, D.; Forbes-Hernández, T.Y.; Afrin, S.; Gasparrini, M.; Reboredo-Rodriguez, P.; Manna, P.P.; Zhang, J.; Bravo Lamas, L.; Martínez Flórez, S.; Agudo Toyos, P. Phenolic Compounds in Honey and Their Associated Health Benefits: A Review. *Molecules* **2018**, *23*, 2322. [CrossRef]
33. Moniruzzaman, M.; Yung An, C.; Rao, P.V.; Hawlader, M.N.I.; Azlan, S.A.B.M.; Sulaiman, S.A.; Gan, S.H. Identification of Phenolic Acids and Flavonoids in Monofloral Honey from Bangladesh by High Performance Liquid Chromatography: Determination of Antioxidant Capacity. *BioMed Res. Int.* **2014**, *2014*, 737490. [CrossRef]
34. Williams, G.R.; Alaux, C.; Costa, C.; Csaki, T.; Doublet, V.; Eisenhardt, D.; Brodschneider, R. Standard Methods for Maintaining Adult *Apis mellifera* in Cages under in Vitro Laboratory Conditions. *J. Apic. Res.* **2013**, *52*, 1–36. [CrossRef]
35. Lourenço, A.P.; Mackert, A.; dos Santos Cristino, A.; Simões, Z.L.P. Validation of Reference Genes for Gene Expression Studies in the Honey Bee, *Apis mellifera*, by Quantitative Real-Time RT-PCR. *Apidologie* **2008**, *39*, 372–385. [CrossRef]
36. Livak, K.J.; Schmittgen, T.D. Analysis of Relative Gene Expression Data Using Real-Time Quantitative PCR and the 2-ΔΔCT Method. *Methods* **2001**, *25*, 402–408. [CrossRef] [PubMed]
37. Mao, W.; Schuler, M.A.; Berenbaum, M.R. CYP9Q-Mediated Detoxification of Acaricides in the Honey Bee (*Apis mellifera*). *Proc. Natl. Acad. Sci. USA* **2011**, *108*, 12657–12662. [CrossRef] [PubMed]
38. Calla, B.; MacLean, M.; Liao, L.; Dhanjal, I.; Tittiger, C.; Blomquist, G.J.; Berenbaum, M.R. Functional Characterization of CYP4G11—a Highly Conserved Enzyme in the Western Honey Bee *Apis mellifera*. *Insect Mol. Biol.* **2018**, *27*, 661–674. [CrossRef]
39. Tesovnik, T.; Cizelj, I.; Zorc, M.; Čitar, M.; Božič, J.; Glavan, G.; Narat, M. Immune Related Gene Expression in Worker Honey Bee (*Apis mellifera* Carnica) Pupae Exposed to Neonicotinoid Thiamethoxam and Varroa Mites (Varroa Destructor). *PLoS ONE* **2017**, *12*, e0187079. [CrossRef] [PubMed]
40. Kaplan, E.L.; Meier, P. Nonparametric Estimation from Incomplete Observations. *J. Am. Stat. Assoc.* **1958**, *53*, 457–481. [CrossRef]
41. Therneau, T.M.; Grambsch, P.M. The cox model. In *Modeling Survival Data: Extending the Cox Model*; Springer: New York, NY, USA, 2000; pp. 39–77.
42. Yu, S.J. Interactions of Allelochemicals with Detoxication Enzymes of Insecticide-Susceptible and Resistant Fall Armyworms. *Pestic. Biochem. Physiol.* **1984**, *22*, 60–68. [CrossRef]
43. Mao, W.; Schuler, M.A.; Berenbaum, M.R. Disruption of Quercetin Metabolism by Fungicide Affects Energy Production in Honey Bees (*Apis mellifera*). *Proc. Natl. Acad. Sci. USA* **2017**, *114*, 2538–2543. [CrossRef] [PubMed]
44. Porrini, M.P.; Fernández, N.J.; Garrido, P.M.; Gende, L.B.; Medici, S.K.; Eguaras, M.J. In Vivo Evaluation of Antiparasitic Activity of Plant Extracts on Nosema Ceranae (Microsporidia). *Apidologie* **2011**, *42*, 700–707. [CrossRef]
45. Tong, L.; Nieh, J.C.; Tosi, S. Combined Nutritional Stress and a New Systemic Pesticide (Flupyradifurone, Sivanto®) Reduce Bee Survival, Food Consumption, Flight Success, and Thermoregulation. *Chemosphere* **2019**, *237*, 124408. [CrossRef]
46. Gregorc, A.; Alburaki, M.; Rinderer, N.; Sampson, B.; Knight, P.R.; Karim, S.; Adamczyk, J. Effects of Coumaphos and Imidacloprid on Honey Bee (Hymenoptera: Apidae) Lifespan and Antioxidant Gene Regulations in Laboratory Experiments. *Sci. Rep.* **2018**, *8*, 15003. [CrossRef]
47. Tosi, S.; Nieh, J.C.; Sgolastra, F.; Cabbri, R.; Medrzycki, P. Neonicotinoid Pesticides and Nutritional Stress Synergistically Reduce Survival in Honey Bees. *Proc. R. Soc. B* **2017**, *284*, 20171711. [CrossRef] [PubMed]
48. Manjon, C.; Troczka, B.J.; Zaworra, M.; Beadle, K.; Randall, E.; Hertlein, G.; Kumar, S.S.; Nauen, R. Unravelling the Molecular Determinants of Bee Sensitivity to Neonicotinoid Insecticides. *Curr. Biol.* **2018**, *28*, 1137–1143. [CrossRef]

49. Alptekin, S.; Bass, C.; Nicholls, C.; Paine, M.J.; Clark, S.J.; Field, L.; Moores, G.D. Induced Thiacloprid Insensitivity in Honeybees (*Apis mellifera* L.) Is Assoc. with Up-regulation of Detoxification Genes. *Insect Mol. Biol.* **2016**, *25*, 171–180. [CrossRef]
50. Wheeler, M.M.; Robinson, G.E. Diet-Dependent Gene Expression in Honey Bees: Honey vs. Sucrose or High Fructose Corn Syrup. *Sci. Rep.* **2014**, *4*, 5726. [CrossRef] [PubMed]

insects

MDPI

Article

Translocation of Tebuconazole between Bee Matrices and Its Potential Threat on Honey Bee (*Apis mellifera* Linnaeus) Queens

Risto Raimets [1,*], Sigmar Naudi [1], Marika Mänd [1], Vadims Bartkevičs [2], Guy Smagghe [3] and Reet Karise [1]

1 Institute of Agricultural and Environmental Sciences, Estonian University of Life Sciences, F. R. Kreutzwaldi 1, 51006 Tartu, Estonia; sigmar.naudi@emu.ee (S.N.); marika.mand@emu.ee (M.M.); reet.karise@emu.ee (R.K.)
2 Institute of Food Safety, Animal Health and Environment "BIOR", Lejupes Street 3, LV-1076 Riga, Latvia; vadims.bartkevics@bior.lv
3 Department of Plants and Crops, Ghent University, B-9000 Ghent, Belgium; Guy.Smagghe@ugent.be
* Correspondence: ristorai@gmail.com or risto.raimets@emu.ee; Tel.: +372-58453095

Simple Summary: Numerous pesticide residues have been found in bee products. It is unclear whether and to what degree pesticides migrate between different bee matrices. Even though the use of many common insecticides is strictly regulated, fungicide residues are still ubiquitous in bee matrices and data regarding this problem are still insufficient. The aim of this work was to determine the migration of fungicide tebuconazole between bee matrices and to assess its potential risk to honey bee queens. We found that tebuconazole mixed into wax has the potential to migrate into royal jelly (RJ), but no residues were found in honey bee queen larvae and newly emerged queens. The residues of tebuconazole found in queen cell cups and RJ decreased over time and probably posed no direct lethal threat to queens. Nevertheless, sub-lethal effects of tebuconazole on honey bee queens might occur even at low concentrations.

Abstract: Various pesticide residues can be found in different bee colony components. The queen larvae of honey bee (*Apis mellifera* L.) receive non-contaminated food from nurse bees. However, there is little knowledge about how pesticide residues affect developing bees. Additionally, little is known about the migration of lipophilic pesticides between bee matrices. While wax, royal jelly (RJ), and bee larvae are chemically distinct, they all contain lipids and we expected the lipophilic fungicide tebuconazole to be absorbed by different contacting materials. Our aim was to analyze the translocation of tebuconazole residues from queen cell wax to RJ, queen larvae, and newly emerged queens and to evaluate its potential risk to queens. We demonstrated the potential for the migration of tebuconazole from wax to RJ, with a strong dilution effect from the original contamination source. No residues were detected in queen bee larvae and newly emerged queens, indicating that the migration of tebuconazole probably did not directly endanger the queen bee, but there was some risk that tebuconazole might still affect the homeostasis of developing bees.

Keywords: translocation of pesticides; wax; hazard quotient; tebuconazole; honey bee queen

Citation: Raimets, R.; Naudi, S.; Mänd, M.; Bartkevičs, V.; Smagghe, G.; Karise, R. Translocation of Tebuconazole between Bee Matrices and Its Potential Threat on Honey Bee (*Apis mellifera* Linnaeus) Queens. *Insects* **2022**, *13*, 45. https://doi.org/10.3390/insects13010045

Academic Editors: Ivana Tlak Gajger, Franco Mutinelli and David Ward Roubik

Received: 28 October 2021
Accepted: 29 December 2021
Published: 31 December 2021

1. Introduction

Pesticides are known to be amplify all stress elements related to honey bee colony losses [1]. Various pesticides accumulate in different bee products [2,3] and often beeswax is the most contaminated bee product [3–5]. Most of the conducted studies have been focused on insecticide effects on bees, while fungicides have gained little attention. Nevertheless, it has been shown that at least azole type fungicides can potentiate the toxicity of insecticides [6–8].

Azole fungicides are among the highest volume fungicides used against fungal diseases of agricultural crops [9]. Among other azole fungicides, tebuconazole is very commonly used and its residues have been found in bee matrices [2,10,11]. Tebuconazole has synergistic negative effects on honey bees together with insecticides [10]. Additionally, it can separately affect microbial communities in different environments, including those in insect guts [11,12].

The large majority of bee toxicology studies have been focused on the impact of pesticide residues on worker bees [13–15], while queens have received little attention. The main task of a mated queen is to sustain colony development and survival via laying eggs [16]. During their lifetime, queens are fed pure royal jelly (RJ) secreted by nurse bees [17] that feed on nectar, pollen, and beebread, which can be contaminated by various pesticides [3,18–20]. Contaminated pollen and nectar can lead to contamination of beeswax, which absorbs lipophilic compounds well [3,5,21,22]. This may pose long-term risk to the viability of a beehive. Although the queen bee is protected from xenobiotic compounds due to feeding solely on RJ, it spends most of its life on or in wax combs. It has been shown that honey bee larvae can suffer morphological changes when grown on contaminated wax [23]. The same changes may occur in queen larvae, because the ambient conditions are the same during queen development.

Long-term contact with pesticide residues exposes bees to a certain risk. One way to quantify the potential risk is to calculate a hazard quotient (HQ) [21]. HQ is calculated by dividing the concentration of contaminant in wax with the LD_{50} value (the concentration that would kill 50% of the test group individuals) of the substance. HQ simply evaluates the risk of individual contaminants to bees. However, depending on their specific tasks in colony, the individual honey bees are in contact with different materials [3]. In the case of multiple contaminants, the calculation of HQ still helps to understand the contribution of individual contaminants to the overall risk.

The contaminating molecules are able to migrate between materials inside the hive [22]. The migration ability depends on the chemical nature of the contaminant, on the chemical and physical characteristics of the material.

The information of pesticide migration among bee products is still scarce. For instance, Böhme et al. showed that RJ is not contaminated [24], despite the contaminated pollen they were fed with. Similar results were shown by Johnson and Percel, where nurse bees that were fed contaminated pollen secreted pure RJ [25]. However, contamination of RJ can occur via exposure to wax comb cells.

The purpose of this study was to test whether tebuconazole residues mixed into honey bee wax migrate from one bee product to another (RJ; queen bee larvae; newly emerged queens). The aim thereafter was to evaluate the potential hazard to queens.

2. Materials and Methods

2.1. Honey Bees Used

The experiments were conducted in summer 2020 at a single apiary (OÜ R-honey) located in the Eastern region of Estonia. Queens (*Apis mellifera ligustica*) were bred in the experiment from one-day old larvae originating from a single queen. Queenright normal-sized honey bee colony (50 Langstroth frames) was used as a cell builder. An egg laying queen was allocated to the first Langstroth box on the hive bottom and separated with queen excluder. In addition, an extra flight entrance was installed above the queen excluder to encourage building up the queen cells by nurse bees. Honey supers were allocated on the extra flight entrance and the grafted queen cells with open brood frames were all inserted into the top box.

2.2. Exposure to Tebuconazole

Wax obtained from a local organic beekeeping operation was used for making queen cell cups. The active ingredient tebuconazole (purity 99.3%) was purchased from Sigma Aldrich. Tebuconazole was dissolved in acetone and incorporated into molten wax. The tebuconazole concentration mixed into wax was 412 $\mu g\ kg^{-1}$. Field-realistic pesticide concentration was selected based on the analysis of Estonian bee products by Raimets et al. [26]. In addition, tebuconazole has been found in honey bee wax in other studies [2,3,6].

Immediately after the mixing of tebuconazole into molten wax, queen cell cups were prepared using special wooden dowel to shape the cup. Separate dowels were used for the control and test group cups. It is a common procedure in beekeeping to make queen cell cups from molten wax [26]. One-day-old honey bee worker larvae were grafted into the newly made cups. A frame with the grafted cups (control cups (n = 20)) and tebuconazole spiked cups (n = 20) was inserted into queenright cell builder colony. After 24 h, the queen cell acceptance by nurse bees was controlled visually.

The closest conventionally managed agricultural crops were located 5.4 km away. The experiment was conducted in July, which is the main honey flow period in Estonia. At that time, there is abundance of plants blooming in the wild. Considering the distance between the apiary and conventional farming fields, it is very unlikely that the bees visited conventional farming fields and were exposed to tebuconazole from outside environment.

2.3. Pesticide Residues in Bee Matrices

In order to investigate whether tebuconazole can migrate from one bee product to another, we collected queen cell cups, RJ, accepted queen larvae, and newly emerged queens. RJ and larvae were collected from cups 3 days after the acceptance. RJ was collected using micropipette and larvae were taken from the cells using a special spatula. Adult queens and built-up queen cell cups were taken for pesticide residue analysis on the day of their emergence. All the samples were put into freezer (−20 °C) immediately after the collection. While large volume experiments with honey bee queens are difficult, we could not split the individually treated cells into several groups to obtain the required minimal sample mass (2 g) for pesticide residue analyses. Therefore, the samples were pooled according to the types of bee matrices. The pooled samples were sent to a laboratory (Institute of Food Safety, Animal Health and Environment "BIOR", Riga, Latvia) for pesticide analysis.

2.4. Pesticide Residue Analyses from Bee Matrices

Tebuconazole residues from honey bee matrices were analyzed at the Institute of Food Safety, Animal Health and Environment "BIOR". The UHPLC-MS/MS assay was performed using an Ultimate 3000 high performance liquid chromatograph (Thermo Scientific, MA, USA) coupled to TSQ Quantiva tandem mass spectrometer (Thermo Scientific, MA, USA).

2.5. Chemicals and Materials Used in Pesticide Residue Analyses

Pesticide reference standards were supplied by Dr. Ehrenstorfer (Germany). Methanol and HPLC grade acetonitrile were acquired from Merck Millipore (Darmstadt, Germany). Ammonium formate (99%), acetic acid (glacial, >99.85%), and ACS grade formic acid (>96.0%) were obtained from Sigma-Aldrich (St. Louis, MO, USA). Ultrapure deionized water was prepared by using a Millipore Milli-Q™ system (Billerica, MA, USA). The following substances were purchased from Phenomenex (Torrance, CA, USA): buffer salt mixture (1 g trisodium citrate dihydrate, 1 g sodium chloride, 0.5 g disodium hydrogen citrate sesquihydrate, and 4 g of anhydrous magnesium sulphate) and a mixture of dSPE (900 mg anhydrous magnesium sulphate, 150 mg PSA, and 150 mg C18E). Stock solutions (approximately 1000 mg L^{-1}) were made by weighing 10 mg of standard in a 10 mL graduated flask and then dissolving in acetonitrile. While preparing the standard solutions of final concentration, the purity of standard was taken into account. In order to prepare a working standard solution with a concentration of 0.01 mg L^{-1}, the appropriate volume of stock solution was diluted with acetonitrile. The prepared solutions were stored at 20 °C.

2.5.1. Sample Preparation

The sample (2.0 ± 0.1 g) was weighed into a 50 mL centrifuge tube. The standard solutions were added at the appropriate spiking level in order to prepare calibration and quality control samples. Acetonitrile (10 mL) and deionized water (10 mL) were both added and the tubes were shaken vigorously by hand for 1 min. Then, a mixture of trisodium citrate dihydrate (1 g), sodium chloride (1 g), disodium hydrogen citrate sesquihydrate (0.5 g), and anhydrous magnesium sulphate (4 g) was added, the tubes were closed, shaken for 10 min, and centrifuged for 10 min at 3500 rpm.

The supernatant was transferred into a 15 mL PP centrifuge tube and frozen out at −70 °C for 30 min using a Heto Ultra freeze (Thermo Fisher Scientific, MA, USA), followed by centrifugation of the resulting organic sample fraction for 10 min at 3500 rpm. For further clean-up procedure, 6 mL of the extract was transferred into 15 mL PP tubes, each containing anhydrous magnesium sulphate (900 mg), PSA (150 mg), and C_{18} sorbent (150 mg). The tubes were shaken vigorously by hand for 30 s and then centrifuged for 10 min at 3500 rpm. A 250 µL aliquot of the purified extract was mixed with 500 µL of the mobile phase A, consisting of 5 mM ammonium formate and 0.1% formic in water and filtered through a 0.22 µm PVDF membrane centrifuge filter. An aliquot of the extract was transferred to an autosampler vial for UHPLC-MS/MS analysis.

2.5.2. UHPLC-MS/MS Analysis

For pesticide analyses in samples, an UltiMate 3000 high performance liquid chromatograph coupled to a TSQ Quantiva tandem mass spectrometer equipped with an electrospray ionization source was applied. The parameters of the ion source were the following: vaporizer temperature was adjusted at 450 °C and ion transfer capillary at 320 °C, ion spray voltage 3.5 kV (positive mode), sheath gas 45 arbitrary units (arb), auxiliary gas 25 arb, and sweep gas 4 arb. The analysis was performed by multiple reaction monitoring (MRM) in the positive ionization mode. Table 1 lists the analyte-dependent parameters: MRM transitions and collision energies (CE).

Table 1. Instrumental parameters of the applied method.

Pesticide	Molecular Ion, Da	Daughter Ion, Da	Collision Energy, eV
Tebuconazole	308	70	21
	308	125	34

Chromatographic separation was performed on a Kinetex C18 analytical column (50 × 3.0 mm, 1.7 mm) from Phenomenex. The mobile phase A consisting of 5 mM ammonium formate and 0.1% formic in water and acetonitrile (mobile phase B) were delivered at the flow rate of 0.4 mL min^{-1}. A gradient program was used: 20% of mobile phase B was used from 0 to 1.0 min, 20% (B) to 90% (B) from 1.0 to 10.0 min, maintained at 90% (B) for 1 min, then decreased back to 20% (B) at 11.0 min and finally the column was re-equilibrated with 20% (B) from 11.0 to 15.0 min. A 10 µL aliquot of the extract was injected. The column and autosampler were maintained at 30 °C and 10 °C, respectively.

2.6. Statistical Analysis

For calculating the hazard quotient (HQ), the wax toxicity calculation tool (Bee Tox Wax) was used [21]. The aim of HQ was to evaluate the potential exposure to a pesticide and to set a level at which no adverse effect is expected. When using the "Bee Tox Wax" tool, the median lethal dose (LD_{50}) is a quantitative indicator of the pesticide toxicity. In this case, HQ values under 250 were shown to be slightly toxic to bees and the wax can be re-used.

3. Results

3.1. The Residues Detected

UHPLC-MS/MS assay showed that tebuconazole was present in certain bee matrices (Table 2). Tebuconazole residues were found in the entire queen cell wax (0.19 mg kg^{-1}), indicating that this substance was incorporated into wax during the experiment. The residues were also found in RJ samples but already at lower concentrations. The tebuconazole concentration found in RJ was 0.08 mg kg^{-1} (0.08 µg per bee), which was 2.4 times lower than found in wax and probably should not pose a risk to the survival of honey bee queens. Despite the fact that tebuconazole was present in queen cell cups and RJ, its residues were not found in honey bee queen larvae and newly emerged queens. It could be assumed that queen larvae and adult queens metabolised tebuconazole and thus no residues were found. Nevertheless, the results show the potential for tebuconazole migration from wax to RJ.

Table 2. Tebuconazole residues detected in bee matrices after spiking queen cell cup wax with tebuconazole at the concentration of 0.412 mg kg^{-1}. Additionally, the HQ values of tebuconazole residue in wax are shown. According to the "Bee Tox Wax" tool data, a HQ value under 250 indicates low wax contamination rate [21].

Bee Matrix.	Control (Tebuco-Nazole Found (mg kg^{-1})	Tebuco-Nazole Found (mg kg^{-1})	% Left from Spiking Concentra-Tion	Tebucona-Zole Contact LD_{50} to Honey Bees 48 h (mg kg^{-1})	Dose (µg) per Bee Found	HQ Value
Queen cells	<0.01	0.19 ± 0.09	46.1	0.2	0.19	1
Royal jelly	<0.01	0.08 ± 0.04	19.4	0.2	0.08	
Larvae	<0.01	<0.01	0	0.2	0	
Queens	<0.01	<0.01	0	0.2	0	

3.2. *Hazard Quotient*

According to the findings of tebuconazole residues in wax, the Hazard Quotient (HQ) was used to determine its potential toxicity level to honey bee queens. The HQ value of initial concentration mixed into wax was 1.

The tebuconazole concentration found in queen cells was 0.19 mg kg^{-1} (0.19 µg per bee), which means that it should not cause direct mortality of bees, but this sub-lethal concentration may have significant negative effect on queen homeostasis. In the case of tebuconazole, 200 µg per bee has been shown to be the acute lethal contact dose after 48 h. The HQ value in our case was 1, which means that the wax was slightly contaminated and it can be used in beekeeping and recycling. According to the Bee Tox Wax tool, any HQ value under 250 is considered to be low-polluted and the wax could be used and recycled in beekeeping operations [21]. At the same time, it should be taken into account that the risk for bees will increase with increasing number and levels of different contaminants.

4. Discussion

This study has demonstrated the migration of tebuconazole from wax to RJ. The only source of tebuconazole was the addition of this xenobiotic compound to molten wax. However, further migration from RJ to queen larvae or newly emerged queens did not occur. Despite the high initial concentration, the content of tebuconazole detected from the entire queen cell wax was lower.

Lipophilic pesticides like tebuconazole can accumulate in wax [3,27] and, due to continuous agricultural application, the amount of residues in bee matrices can be expected to increase with time. Little is known about the effects of certain pesticides on queen bee performance. In addition, various studies show that different pesticides can cause synergistic effects in bees. Raimets et al. showed that azole fungicide imazalil had a synergistic effect with three insecticides on the mortality of bumble bee (*Bombus terrestris*) [6]. Besides, Vandame and Belzunces showed that insecticide deltamethrin had a synergistic effect with azole fungicides, causing changes in bee thermoregulation [28]. Azole fungicides inhibit cytochrome P450 detoxification system in bees [7] and thus the bees are more vulnerable to other pesticides.

The octanol-water partition coefficient of tebuconazole (Log P_{OW}) has been shown to be 3.7 [29], which means that this compound has lipophilic properties and can concentrate in lipophilic bee matrices. In addition, tebuconazole is quite stable at elevated temperatures and its melting point is 102.4 °C [29], while beeswax melting point is about 63–65 °C [30].

This also confirms that tebuconazole incorporation into molten wax is possible without degradation.

It is vital to understand whether pesticide residues from wax can be taken up by developing honey bees. Medici et al. showed that the presence of insecticides in wax negatively affected honey bee brood survival [31]. There have been only a few studies focused on the migration of pesticides from wax to bees, especially to queen bees. A study conducted by Böhme et al. showed that some pesticide residues (with different Log p values) can be found in RJ, though in apparently negligible concentrations (the highest concentration found was 0.016% of the original concentration fed to the nurse bees) [24]. In another experimental study, hives were treated with known amounts of tau-fluvalinate via contaminated plywood inserts, and no residues were detected in RJ [32]. On the contrary, our study revealed that fungicide tebuconazole was translocated from polluted wax to RJ. Nevertheless, tebuconazole concentration found in RJ was 2.4 times lower than in wax. Another study showed that nurse bees fed on contaminated pollen and nectar produced uncontaminated RJ for honey bee queens [33]. Our study revealed that RJ can be contaminated via exposure to residues in wax. Our results are in accordance with Milone and Tarpy who showed that different pesticides mixed into queen cell cups successfully migrated to RJ [34].

Larval feeding pattern performed by nurse bees is different among honey bee castes and sexes (workers, drones, queens). In the case of queen larvae, feeding was performed by nurse bees within relatively short visits and the content of RJ remained almost constant, while worker larvae received longer feeding periods after 48 h [35]. In addition, Dietz and Lambremont showed that honey bee queen larvae consume 13% more food than worker larvae within first the 3 days of larval development [36]. Thus, the higher food consumption by queen bee larvae may simultaneously increase the amount of ingested pesticide residues in the case if RJ is contaminated.

Tebuconazole concentrations found in bee matrices are related to the lipophilicity of this compound, which tends to accumulate in nonpolar media. This tendency was demonstrated in our experiments where the highest concentration was revealed in wax (very non-polar matrix), lower concentration was detected in RJ (up to 6% of fats), and the lowest content in larvae (mostly water-protein sample). Due to the chemical and physical properties of tebuconazole, it preferentially remained in the samples with the highest lipid content. Additionally, tebuconazole has been shown to be persistent in soil (aerobic metabolism $T_{1/2}$ in soils is 796 days) [37]. Even under hydrolytic conditions, it was stable for >28 days. Therefore, the most probable reason for the occurrence of tebuconazole in certain bee products is related to its tendency to stay in a lipophilic environment.

It is noteworthy that there was no conventional farming within 5.4 km radius in our study, which excludes the possibility of bee exposure from agricultural use of tebuconazole. Morales et al. showed that pesticides used on the fields, as well as in apiculture, can move from beeswax to honey bee brood [38]. Interestingly, our results did not indicate tebuconazole residues in the sampled honey bee queen larvae and newly emerged queens. We saw rapid dilution of tebuconazole through different matrices. Compared to the cell cup wax, the whole queen cells contained less than one half the concentration of tebuconazole. Due to the dilution with newly added wax, the initial concentration mixed into queen cell cups had decreased 2.2 times. In addition, the concentration of tebuconazole decreased 2.4 times when comparing RJ to wax. However, if the apiary had been located next to conventional farming lands, the amount of tebuconazole residues in bee matrices could be considerably higher due to the agricultural use of tebuconazole.

Besides chemical decomposition of pesticide molecules over time, the individual bee organisms are also capable of detoxification [39]. In addition, Berenbaum and Johnson have proposed that detoxification of xenobiotic compounds among eusocial honey bees may be complemented by a "social detoxification system", which includes colony food processing via microbial fermentation, dilution by pollen mixing, and worker discrimination [40]. Without detoxification, the effects of pesticides on honey bee health could be more severe.

Some studies have shown that pesticides may have effects on the longevity, olfactory functions, and water consumption by bees during long-term exposure [41,42]. Nevertheless, in our case, the concentration of tebuconazole after 3 days of exposure was not observed, probably due to the inability of this lipophilic contaminant to migrate from wax to biological tissues of larvae and queen bees.

Tebuconazole concentrations found in bee products in our study probably do not possess any direct lethal effect to developing queens. Still, pesticides even at low concentrations have been shown to cause sub-lethal effects in bees [43]. Fungicide tebuconazole is considered to have low toxicity to bees [44]. Despite the fact that tebuconazole probably does not kill bees directly, it might cause changes in bee homeostasis. As it is known, there are eight different dominant bacterial species in honey bee gut and they tend to exhibit strain diversity according to differences in tolerance of pesticide exposure and metabolic capability [45]. Tebuconazole at high concentrations (5, 50, and 500 mg kg^{-1}) has been shown to decrease soil microbial biomass and activity but, interestingly, no clear effect of different concentrations was found [11]. In addition, tebuconazole reduced gut fungal diversity in brown planthopper (*Nilaparvata lugens* Stål). Interestingly, a study conducted by Powell et al. showed that newly emerged honey bees are lacking gut bacteria and their gut is colonized 4–6 days after the emergence via contacts with other workers [46]. Even though tebuconazole does not spread from RJ to bee larvae, a possibility of tebuconazole exposure inside the hive (contaminated pollen, beebread, etc.) cannot be completely excluded. Changes in bee gut microbiota are shown to increase the susceptibility to diseases [47] and bee mortality may be increased due to different sub-lethal effects [45].

5. Conclusions

The current pilot study demonstrated the migration of a single pesticide from one bee product to another. Further studies are needed to investigate the potential for the migration of multiple pesticides among bee products, with the goal to identify and quantify the impact on bees associated with possible synergistic and sub-lethal effects.

Author Contributions: Conceptualization, R.R., R.K., M.M., S.N., V.B. and G.S.; Methodology, R.R., R.K. and S.N.; Formal analysis, R.R., R.K. and V.B.; Investigation, R.R.; Resources, M.M. and G.S.; Writing—original draft preparation, R.R., R.K., S.N. and M.M.; Writing—review and editing, R.K., M.M., G.S. and V.B.; Visualization, R.R. and R.K.; Supervision, M.M., R.K. and G.S.; Project administration, M.M., R.K. and G.S.; Funding acquisition, M.M., R.K. and G.S. All authors have read and agreed to the published version of the manuscript.

Funding: This research was supported by Baseline funding for supporting the activities related to the project "Honey bee (*Apis mellifera*) sensitivity to food quality and diseases: explained by collected field data and subsequent individual lab testing (P200192PKTE)" of the Estonian Ministry of Education and RITA Programme project ForBee (L180250) of the Estonian Research Council and the European Union, European Regional Development Fund (Estonian University of Life Sciences ASTRA project "Value-chain based bioeconomy").

Institutional Review Board Statement: Not applicable.

Data Availability Statement: Data sharing not applicable.

Acknowledgments: Acknowledgements go to Jonathan Martin Willow and Jānis Jaunbergs for English editing.

Conflicts of Interest: The authors declare that they have no potential conflict of interest in relation to the study in this paper.

References

1. vanEngelsdorp, D.; Meixner, M.D. A Historical Review of Managed Honey Bee Populations in Europe and the United States and the Factors That May Affect Them. *J. Invertebr. Pathol.* **2010**, *103*, S80–S95. [CrossRef] [PubMed]
2. Chauzat, M.-P.; Martel, A.-C.; Cougoule, N.; Porta, P.; Lachaize, J.; Zeggane, S.; Aubert, M.; Carpentier, P.; Faucon, J.-P. An Assessment of Honeybee Colony Matrices, *Apis mellifera* (Hymenoptera: Apidae) to Monitor Pesticide Presence in Continental France. *Environ. Toxicol. Chem.* **2011**, *30*, 103–111. [CrossRef] [PubMed]

3. Mullin, C.A.; Frazier, M.; Frazier, J.L.; Ashcraft, S.; Simonds, R.; vanEngelsdorp, D.; Pettis, J.S. High Levels of Miticides and Agrochemicals in North American Apiaries: Implications for Honey Bee Health. *PLoS ONE* **2010**, *5*, e9754. [CrossRef] [PubMed]
4. El Agrebi, N.; Traynor, K.; Wilmart, O.; Tosi, S.; Leinartz, L.; Danneels, E.; de Graaf, D.C.; Saegerman, C. Pesticide and Veterinary Drug Residues in Belgian Beeswax: Occurrence, Toxicity, and Risk to Honey Bees. *Sci. Total Environ.* **2020**, *745*, 141036. [CrossRef] [PubMed]
5. Ravoet, J.; Reybroeck, W.; de Graaf, D.C. Pesticides for Apicultural and/or Agricultural Application Found in Belgian Honey Bee Wax Combs. *Bull. Environ. Contam. Toxicol.* **2015**, *94*, 543–548. [CrossRef]
6. Raimets, R.; Karise, R.; Mand, M.; Kaart, T.; Ponting, S.; Song, J.; Cresswell, J.E. Synergistic Interactions between a Variety of Insecticides and an Ergosterol Biosynthesis Inhibitor Fungicide in Dietary Exposures of Bumble Bees (*Bombus terrestris* L.). *Pest Manag. Sci.* **2018**, *74*, 541–546. [CrossRef] [PubMed]
7. Pilling, E.D.; Jepson, P.C. Synergism between EBI Fungicides and a Pyrethroid Insecticide in the Honeybee (*Apis mellifera*). *Pestic. Sci.* **1993**, *39*, 293–297. [CrossRef]
8. Johnson, R.M.; Dahlgren, L.; Siegfried, B.D.; Ellis, M.D. Acaricide, Fungicide and Drug Interactions in Honey Bees (*Apis mellifera*). *PLoS ONE* **2013**, *8*, e54092. [CrossRef]
9. Sanchez-Bayo, F.; Goka, K. Pesticide Residues and Bees—A Risk Assessment. *PLoS ONE* **2014**, *9*, e94482. [CrossRef]
10. Schmuck, R.; Stadler, T.; Schmidt, H.-W. Field Relevance of a Synergistic Effect Observed in the Laboratory between an EBI Fungicide and a Chloronicotinyl Insecticide in the Honeybee (*Apis mellifera* L, Hymenoptera). *Pest Manag. Sci.* **2003**, *59*, 279–286. [CrossRef]
11. Muñoz-Leoz, B.; Ruiz-Romera, E.; Antigüedad, I.; Garbisu, C. Tebuconazole Application Decreases Soil Microbial Biomass and Activity. *Soil Biol. Biochem.* **2011**, *43*, 2176–2183. [CrossRef]
12. Song, Y.; Shi, J.; Xiong, Z.; Shentu, X.; Yu, X. Three Antimicrobials Alter Gut Microbial Communities and Causing Different Mortality of Brown Planthopper, Nilaparvata Lugens Stål. *Pestic. Biochem. Physiol.* **2021**, *174*, 104806. [CrossRef]
13. Xavier, V.M.; Message, D.; Picanco, M.C.; Chediak, M.; Santana Junior, P.A.; Ramos, R.S.; Martins, J.C. Acute Toxicity and Sublethal Effects of Botanical Insecticides to Honey Bees. *J. Insect Sci.* **2015**, *15*, 137. [CrossRef]
14. Doublet, V.; Labarussias, M.; de Miranda, J.R.; Moritz, R.F.A.; Paxton, R.J. Bees under Stress: Sublethal Doses of a Neonicotinoid Pesticide and Pathogens Interact to Elevate Honey Bee Mortality across the Life Cycle. *Environ. Microbiol.* **2015**, *17*, 969–983. [CrossRef] [PubMed]
15. Oruc, H.H.; Hranitz, J.M.; Sorucu, A.; Duell, M.; Cakmak, I.; Aydin, L.; Orman, A. Determination of Acute Oral Toxicity of Flumethrin in Honey Bees. *J. Econ. Entomol.* **2012**, *105*, 1890–1894. [CrossRef] [PubMed]
16. Milchreit, K.; Ruhnke, H.; Wegener, J.; Bienefeld, K. Effects of an Insect Growth Regulator and a Solvent on Honeybee (*Apis mellifera* L.) Brood Development and Queen Viability. *Ecotoxicology* **2016**, *25*, 530–537. [CrossRef]
17. Haydak, M.H. Honey Bee Nutrition. *Annu. Rev. Entomol.* **1970**, *15*, 143–156. [CrossRef]
18. Škerl, M.I.S.; Bolta, Š.V.; Česnik, H.B.; Gregorc, A. Residues of Pesticides in Honeybee (*Apis mellifera* Carnica) Bee Bread and in Pollen Loads from Treated Apple Orchards. *Bull. Environ. Contam. Toxicol.* **2009**, *83*, 374–377. [CrossRef] [PubMed]
19. Chauzat, M.-P.; Faucon, J.-P.; Martel, A.-C.; Lachaize, J.; Cougoule, N.; Aubert, M. A Survey of Pesticide Residues in Pollen Loads Collected by Honey Bees in France. *J. Econ. Entomol.* **2006**, *99*, 253–262. [CrossRef]
20. Karise, R.; Raimets, R.; Bartkevics, V.; Pugajeva, I.; Pihlik, P.; Keres, I.; Williams, I.H.; Viinalass, H.; Mand, M. Are Pesticide Residues in Honey Related to Oilseed Rape Treatments? *Chemosphere* **2017**, *188*, 389–396. [CrossRef]
21. Bee Tox Wax: Outil de Calcul de Toxicité Des Cires (FMV ULiège). Available online: https://www.beetools.uliege.be/beetoxwax/?langue=en (accessed on 12 October 2021).
22. Tremolada, P.; Bernardinelli, I.; Colombo, M.; Spreafico, M.; Vighi, M. Coumaphos Distribution in the Hive Ecosystem: Case Study for Modeling Applications. *Ecotoxicology* **2004**, *13*, 589–601. [CrossRef]
23. Atkins, E.L.; Kellum, D. Comparative Morphogenic and Toxicity Studies on the Effect of Pesticides on Honeybee Brood. *J. Apic. Res.* **1986**, *25*, 242–255. [CrossRef]
24. Böhme, F.; Bischoff, G.; Zebitz, C.P.W.; Rosenkranz, P.; Wallner, K. From Field to Food—Will Pesticide-Contaminated Pollen Diet Lead to a Contamination of Royal Jelly? *Apidologie* **2018**, *49*, 112–119. [CrossRef]
25. Johnson, R.M.; Percel, E.G. Effect of a Fungicide and Spray Adjuvant on Queen-Rearing Success in Honey Bees (Hymenoptera: Apidae). *J. Econ. Entomol.* **2013**, *106*, 1952–1957. [CrossRef] [PubMed]
26. Raimets, R.; Bontsutsnaja, A.; Bartkevics, V.; Pugajeva, I.; Kaart, T.; Puusepp, L.; Pihlik, P.; Keres, I.; Viinalass, H.; Mand, M.; et al. Pesticide Residues in Beehive Matrices Are Dependent on Collection Time and Matrix Type but Independent of Proportion of Foraged Oilseed Rape and Agricultural Land in Foraging Territory. *Chemosphere* **2020**, *238*, 124555. [CrossRef]
27. García, M.D.G.; Duque, S.U.; Fernández, A.B.L.; Sosa, A.; Fernández-Alba, A.R. Multiresidue Method for Trace Pesticide Analysis in Honeybee Wax Comb by GC-QqQ-MS. *Talanta* **2017**, *163*, 54–64. [CrossRef] [PubMed]
28. Vandame, R.; Belzunces, L.P. Joint Actions of Deltamethrin and Azole Fungicides on Honey Bee Thermoregulation. *Neurosci. Lett.* **1998**, *251*, 57–60. [CrossRef]
29. PubChem Tebuconazole. Available online: https://pubchem.ncbi.nlm.nih.gov/compound/86102 (accessed on 7 October 2021).
30. Tulloch, A.P. Beeswax—Composition and Analysis. *Bee World* **1980**, *61*, 47–62. [CrossRef]
31. Medici, S.K.; Castro, A.; Sarlo, E.G.; Marioli, J.M.; Eguaras, M.J. The Concentration Effect of Selected Acaricides Present in Beeswax Foundation on the Survival of *Apis mellifera* Colonies. *J. Apic. Res.* **2012**, *51*, 164–168. [CrossRef]

32. Tsigouri, A.D.; Menkissoglu-Spiroudi, U.; Thrasyvoulou, A.; Diamantidis, G. Fluvalinate Residues in Honey and Beeswax after Different Colony Treatments. *Bull. Environ. Contam. Toxicol.* **2004**, *72*, 975–982. [CrossRef]
33. Chauzat, M.-P.; Faucon, J.-P. Pesticide Residues in Beeswax Samples Collected from Honey Bee Colonies (*Apis mellifera* L.) in France. *Pest Manag. Sci.* **2007**, *63*, 1100–1106. [CrossRef]
34. Milone, J.P.; Tarpy, D.R. Effects of Developmental Exposure to Pesticides in Wax and Pollen on Honey Bee (*Apis mellifera*) Queen Reproductive Phenotypes. *Sci. Rep.* **2021**, *11*, 1020. [CrossRef]
35. Brouwers, E.V.M.; Ebert, R.; Beetsma, J. Behavioural and Physiological Aspects of Nurse Bees in Relation to the Composition of Larval Food During Caste Differentiation in the Honeybee. *J. Apic. Res.* **1987**, *26*, 11–23. [CrossRef]
36. Dietz, A.; Lambremont, E.N. Caste Determination in Honey Bees.1 II. Food Consumption of Individual Honey Bee Larvae, Determined with 32P-Labeled Royal Jelly2. *Ann. Entomol. Soc. Am.* **1970**, *63*, 1342–1345. [CrossRef]
37. United States Environmental Protection Agency. Available online: https://archive.epa.gov/pesticides/chemicalsearch/chemical/foia/web/pdf/128997/128997-2006-12-05a.pdf (accessed on 10 December 2021).
38. Murcia Morales, M.; Gómez Ramos, M.J.; Parrilla Vázquez, P.; Díaz Galiano, F.J.; García Valverde, M.; Gámiz López, V.; Manuel Flores, J.; Fernández-Alba, A.R. Distribution of Chemical Residues in the Beehive Compartments and Their Transfer to the Honeybee Brood. *Sci. Total Environ.* **2020**, *710*, 136288. [CrossRef]
39. Ardalani, H.; Vidkjær, N.H.; Kryger, P.; Fiehn, O.; Fomsgaard, I.S. Metabolomics Unveils the Influence of Dietary Phytochemicals on Residual Pesticide Concentrations in Honey Bees. *Environ. Int.* **2021**, *152*, 106503. [CrossRef]
40. Berenbaum, M.R.; Johnson, R.M. Xenobiotic Detoxification Pathways in Honey Bees. *Curr. Opin. Insect Sci.* **2015**, *10*, 51–58. [CrossRef] [PubMed]
41. Moncharmont, F.-X.D.; Decourtye, A.; Hennequet-Hantier, C.; Pons, O.; Pham-Delègue, M.-H. Statistical Analysis of Honeybee Survival after Chronic Exposure to Insecticides. *Environ. Toxicol. Chem.* **2003**, *22*, 3088–3094. [CrossRef] [PubMed]
42. Aliouane, Y.; el Hassani, A.K.; Gary, V.; Armengaud, C.; Lambin, M.; Gauthier, M. Subchronic Exposure of Honeybees to Sublethal Doses of Pesticides: Effects on Behavior. *Environ. Toxicol. Chem.* **2009**, *28*, 113–122. [CrossRef] [PubMed]
43. Mänd, M.; Karise, R. Recent Insights into Sublethal Effects of Pesticides on Insect Respiratory Physiology. *Open Access Insect Physiol.* **2015**, *5*, 31–39. [CrossRef]
44. Harriet, J.; Campá, J.P.; Grajales, M.; Lhéritier, C.; Gómez Pajuelo, A.; Mendoza-Spina, Y.; Carrasco-Letelier, L. Agricultural Pesticides and Veterinary Substances in Uruguayan Beeswax. *Chemosphere* **2017**, *177*, 77–83. [CrossRef] [PubMed]
45. Motta, E.V.S.; Raymann, K.; Moran, N.A. Glyphosate Perturbs the Gut Microbiota of Honey Bees. *Proc. Natl. Acad. Sci. USA* **2018**, *115*, 10305–10310. [CrossRef] [PubMed]
46. Powell, J.E.; Martinson, V.G.; Urban-Mead, K.; Moran, N.A. Routes of Acquisition of the Gut Microbiota of the Honey Bee *Apis mellifera*. *Appl. Environ. Microbiol.* **2014**, *80*, 7378–7387. [CrossRef]
47. Koch, H.; Schmid-Hempel, P. Socially Transmitted Gut Microbiota Protect Bumble Bees against an Intestinal Parasite. *Proc. Natl. Acad. Sci. USA* **2011**, *108*, 19288–19292. [CrossRef] [PubMed]

insects

Article

Sensitivity of Buff-Tailed Bumblebee (*Bombus terrestris* L.) to Insecticides with Different Mode of Action

Guillermo Cabezas and Gema P. Farinós *

Laboratory of Applied Entomology for Human and Plant Health, Centro de Investigaciones Biológicas Margarita Salas, Ramiro de Maeztu 9, 28040 Madrid, Spain; gcabezastorrero@gmail.com

* Correspondence: gpfarinos@cib.csic.es

Simple Summary: Several neonicotinoid insecticides that were once widely used for pest control are currently banned for outdoor use in the European Union (EU) because they pose a risk to bees. This restriction meant that farmers had to look for alternatives for pest management and use known insecticides or new substances with supposedly more bee-friendly characteristics. We evaluated the toxicity of six insecticides on buff-tailed bumblebee workers (*Bombus terrestris*): two banned neonicotinoids (imidacloprid, thiacloprid), two pyrethroids (deltamethrin, esfenvalerate), one sulfoximine (sulfoxaflor) and a microbial insecticide based on *Bacillus thuringiensis* toxins, which are present in genetically modified (Bt) maize. The results obtained show that certain insecticides in use have higher acute toxicity to *B. terrestris* than some of the banned neonicotinoids.

Abstract: Systemic insecticides are recognized as one of the drivers of the worldwide bee decline as they are exposed to them through multiple pathways. Specifically, neonicotinoids, some of which are banned for outdoor use in the European Union (EU), have been pointed out as a major cause of bee collapse. Thus, farmers have had to look for alternatives for pest control and use known insecticides or new substances reportedly less harmful to bees. We evaluated the oral acute toxicity of six insecticides (three of them systemic: imidacloprid, thiacloprid and sulfoxaflor) with four different modes of action on buff-tailed bumblebee workers (*Bombus terrestris*): two banned neonicotinoids (imidacloprid, thiacloprid), two pyrethroids (deltamethrin, esfenvalerate), one sulfoximine (sulfoxaflor) and a microbial insecticide based on *Bacillus thuringiensis* toxins, present in genetically modified (Bt) maize. The microbial insecticide only caused mortality to bumblebee workers at extremely high concentrations, so it is expected that Bt maize does not pose a risk to them. The toxicity of the other five insecticides on bumblebees was, from highest to lowest: imidacloprid, sulfoxaflor, deltamethrin, esfenvalerate and thiacloprid. This outcome suggests that certain insecticides in use are more toxic to *B. terrestris* than some banned neonicotinoids. Further chronic toxicity studies, under realistic conditions, are necessary for a proper risk assessment.

Keywords: pollinators; neonicotinoids; imidacloprid; thiacloprid; pyrethroids; deltamethrin; esfenvalerate; sulfoxaflor; *Bacillus thuringiensis*; Cry1Ab; acute toxicity; bioassays; commercial hives

Citation: Cabezas, G.; Farinós, G.P. Sensitivity of Buff-Tailed Bumblebee (*Bombus terrestris* L.) to Insecticides with Different Mode of Action. *Insects* **2022**, *13*, 184. https://doi.org/10.3390/insects13020184

Academic Editors: Ivana Tlak Gajger and Franco Mutinelli

Received: 13 December 2021
Accepted: 6 February 2022
Published: 10 February 2022

1. Introduction

The genus *Bombus* Latreille, with a wide world distribution that includes Europe, Asia, North America, parts of South America and North Africa [1], is highly efficient for pollinating not only a wide variety of wild plants but also crops [2,3]. Within this genus, the *Bombus terrestris* L./*Bombus lucorum* L. complex is the first in the ranking list of the most important crop pollinators in Europe, with a mean contributed value of USD 425 per hectare [4]. The different subspecies of *B. terrestris* are originally distributed throughout Eurasia and North Africa [5], and their characteristics as pollinators, together with the relative ease of rearing them in captivity, have led to their domestication for use in the pollination of horticultural, fruit and seed crops, with particular importance in tomato and berry greenhouses [2,3].

Since the mid-1980s, a severe decline in managed honey bee populations has been observed in Europe and the US [6–9]. Furthermore, where evidence is available, such as the UK, Netherlands, Belgium or the US, it has been demonstrated that the abundance and diversity of non-*Apis* bees, including bumblebees, were also being affected [6,10–14]. Although there is no single cause responsible for this decline, agricultural intensification has been cited as one of the most important threats to pollinators [15,16]. Pesticides (and particularly insecticides) have been the subject of debate ever since the collapse in bee populations was observed [17], although the neonicotinoids are the ones that have mainly been in the focus for years. Since their launch in the 1990s, neonicotinoids have become one of the most widespread classes of insecticides globally, to such an extent that in 2014 they accounted for 25% of the insecticide market [18]. This success has been mainly due to their effectiveness in controlling a number of important pests and their versatility in application (soil and seed treatments and foliar spray) [19]. One of their most valuable characteristics is their systemic nature, which allows them to reach all plant tissues and organs. However, it is precisely their systemicity, together with their toxicity on invertebrates, persistence and environmental impact, that makes them potentially harmful to a wide range of non-target organisms [20].

Various laboratory and semi-field trials have reported that dietary neonicotinoids produce harmful sublethal effects on honey bees and non-*Apis* bees [21–24]. These negative effects have also been reported in different studies conducted under field conditions, in which bumblebees and solitary bees were exposed to several neonicotinoids by different routes, leading to important sublethal effects, such as reduced solitary bee nesting and reproductive success and bumblebee colony growth and reproduction [25–27]. Nowadays, systemic insecticides are recognized as one of the drivers of worldwide honey and wild bee declines as they are exposed to them when they collect crop pollen and nectar [28–30]. Despite the large number of studies conducted to assess the effects of neonicotinoids on bees, numerous gaps have been identified in the methodology used and the subjects studied in relation to this issue, and further research using species other than the honey bee *Apis mellifera* L. has been recommended to improve laboratory, semi-field and field tests [31–33].

In 2013, in the light of scientific and technical evidences, the European Commission considered that the active substances clothianidin, thiamethoxam and imidacloprid posed a high risk to bees, so restrictions on their use were imposed in the European Union (EU), mainly affecting maize, sunflower and oilseed rape [34]. Due to extensive evidence from scientific research on the detrimental effects of neonicotinoids on bees, in May 2018 the Commission definitively banned the outdoor use of these three neonicotinoids based on risk assessment and scientific data [35–37]). This ban meant that farmers had to look for alternatives for pest management. The first option was to replace them with other neonicotinoids with a similar mode of action, such as acetamiprid or thiacloprid, the latter mainly applied to seed in the case of maize and sunflower, or by foliar spray in sunflower and rapeseed [38]. However, an application to renew the approval of thiacloprid was rejected by the European Commission in January 2020. Another option was to replace the banned neonicotinoids with other broad-spectrum and widely used chemical insecticides, mainly pyrethroids [38,39]. A third avenue for the replacement of neonicotinoids was to explore promising new active substances that are supposedly less aggressive towards non-target pests. One example is the active substance sulfoxaflor, belonging to the sulfoximine family, which was authorized in 2015 in the EU. It is a systemic insecticide with translaminar movement and acts mainly by ingestion and contact. It has been proposed as an alternative to neonicotinoids given its lower toxicity to bees and better ecological and human health profile, among other reasons [40,41].

In addition to insecticides sprayed on crops, bumblebees may also be exposed to other insecticidal substances, such as the toxin Cry1Ab from *Bacillus thuringiensis*, expressed in genetically modified maize varieties derived from the MON810 event (Bt maize). It has been found that maize pollen can be attractive for honey bees during dry growing seasons or periods when more favorable protein sources may not be available [42,43]. Furthermore, in agricultural landscapes with large-scale monocultures, pollen foragers might be forced to almost exclusively collect pollen from a single source, even from wind pollinated crops such as maize [44]. Thus, in areas where its level of adoption is high, exposure to Cry1Ab protein could represent a risk to bees. It has been demonstrated that the pollen from Bt crops can contain significant amounts of Cry toxin, although the toxin levels found differed depending on the study [45–47]. Information on the toxic effects of Cry1Ab toxin on bees is scarce and sometimes contradictory. Some studies indicate that it does not cause acute or chronic toxic effects on honey bees (*A. mellifera*) or bumblebees [46,48]. However, it has been shown that it can cause sublethal effects on honey bees by affecting their foraging behavior or feeding [49,50].

The aim of this study is to evaluate the oral acute toxicity of six insecticides with four different modes of action on the bumblebee *Bombus terrestris*: two neonicotinoids (imidacloprid and thiacloprid), two pyrethroids (deltamethrin and esfenvalerate), one sulfoximine (sulfoxaflor) and a microbial insecticide based on Cry toxins from *Bacillus thuringiensis*. The possible implications of the results obtained, following the restriction of the use of neonicotinoids in the EU, are discussed.

2. Materials and Methods

2.1. Maintenance of the Hives

Workers of the buff-tailed bumblebee *B. terrestris* were used to carry out the bioassays with insecticides. The workers came from hives supplied by the company Agrobío S.L (Almería, Spain), consisting of a fertilized queen, 20–40 workers and a nest with larvae and eggs. Twenty-one commercial hives were used to perform the bioassays (Table S1). Once in the laboratory, they were kept in a chamber at a constant temperature of 25 ± 1 °C, relative humidity of 60 ± 5% and in conditions of complete darkness, illuminating the chamber with red light only when it was necessary for handling or for picking up individuals. The bees were provided ad libitum with syrup (Api 65®, Agrobío S.L., Almería, Spain) and food balls, prepared with dry pollen, artificial syrup and distilled water. Workers for the bioassays were extracted at the moment of exponential growth of the hive.

2.2. Insecticides

We selected six different active ingredients for bioassays: (i) two neonicotinoids, imidacloprid (Confidor® SL 20% *W/V*, Bayer S.A.S., Lyon, France) and thiacloprid (Calypso® SC, 48% *W/V*, Bayer CropScience, S.L., Valencia, Spain), both currently restricted in the EU; (ii) two pyrethroids, deltamethrin (Decis Protech® EW, 1.5% *W/V*, Bayer CropScience S.L., Valencia, Spain) and esfenvalerate (Sumifive® Plus EW, 5% *W/V*, Kenogard, S.A., Madrid, Spain), belonging to a new generation of pyrethroids, which show a higher effectiveness than the previous ones and widely used on a large number of crops for the control of different pests; (iii) the sulfoximine-based pesticide sulfoxaflor (Closer® SC EW 11.43% *W/W*, Dow Agrosciences Iberica S.A., Seville, Spain) for being a relatively new systemic insecticide, proposed as a candidate for replacement of the neonicotinoids [51]; (iv) the selective biological insecticide *Bacillus thuringiensis* ssp. *kurstaki* (Btk) (DiPel® DF WG, 54% *W/W*, strain ABTS-351 (32 mill. CLU/g), Kenogard, S.A., Valent BioScience LLC, Libertyville, IL, USA), containing four insecticide Cry toxins, one of which (Cry1Ab) is expressed in the genetically modified maize cultivated in the European Union. All insecticide concentrations were prepared in Api 65® artificial syrup with the exception of the insecticide DiPel® DF, which was previously diluted in 0.1% Triton® X-100 (Sigma-Aldrich, St. Louis, MO, USA).

2.3. Bioassays

The methodology used to carry out the bioassays was almost entirely in line with that recommended for studying the acute oral toxicity of insecticides in bumblebees, with some modifications [52]. Concentration–response bioassays were conducted on individualized *B. terrestris* workers. Each bioassay consisted of treating individual workers with 5–7 increasing nominal test concentrations of insecticide diluted in 2 mL of artificial nectar. The range of nominal test concentrations of each of the insecticides used in the bioassays was decided after preliminary trials in order to obtain a response (mortality) between 10% and 90%. Once the extreme concentrations were defined, intermediate concentrations were determined on a logarithmic scale. A negative control was added consisting of 2 mL of artificial nectar without insecticide treatment and a positive control consisting of 2 mL of artificial nectar treated with 80 ppm (*W/V*) of the pyrethroid lambda-cyhalothrin (Karate Zeon® 10% *W/V* CS Syngenta S.A., Madrid, Spain), previously known to be toxic to bumblebees [53].

Bees between 1 and 4 days old were selected for the bioassays. Newly emerged workers, distinguishable by the absence of pigmentation and the shape of the wings yet to unfold, were removed from the hives and separated into groups of five bees of similar size and same hive of origin. They were placed in round transparent plastic boxes (11.5 cm diameter, 5 cm high and 450 mL capacity) with a lid fitted with a ventilation grid and the base covered with filter paper. Each box contained a jar (2.5 cm in diameter, 1.3 cm high and 3 mL capacity) with pollen similar to that used to feed the hives, and another jar with absorbent cotton impregnated with Apis 65® artificial nectar (Agrobío S.L., Almería, Spain). The bees were kept in these conditions for 24 h. After this time, their condition was evaluated to assess their inclusion in the bioassay, using only those that were in optimal conditions. We evaluated whether there was a relationship between the weight of the bees and the volume of artificial nectar consumed, which could affect the outcome of the bioassays. For this purpose, 40 bees from 10 different hives were selected and fed for 48 h with untreated nectar. After this time, the bees were weighed and the weight of food consumed per bee was calculated. Once the weights of the bees were known, the weight of the workers used in the bioassays was standardized for each of the hives. Thus, those bees whose weight was not within 2 standard deviations of the mean weight were discarded.

The bioassay was performed by placing individual bees, previously anaesthetized with CO_2 for 10 s, in queen cages to which an empty 10 mL plastic syringe was attached. The bees were starved for 3 h. After this period they were weighed with a balance (LPW-523i, VWR, Milan, Italy) to determine their initial weight and the empty syringes were replaced with pre-weighed syringes containing the different treatments. To enable the treatment to be taken, the tip of the syringes was cut off, and to ensure that the liquid was accessible throughout the bioassay, they were placed with a slight inclination. Once the treatment syringe was attached, each bee was checked for feeding. In each bioassay, between 2 and 6 replicates were carried out, where a replicate is defined as the set of treated bees from each commercial hive. At least 4 bees from the same source hive were assigned to each of the concentrations and controls.

The bioassays lasted for 48 h or until the bees died. The condition of each worker bee at 2, 4, 24 and 48 h after the start of the bioassay was assessed according to the responses to mechanical stimuli (gentle blows on the cage containing the bee) and light stimuli (illuminating each bee with white light for a few seconds). The bees were given a numerical value according to the following scale: 0 (dead bee that does not react to any stimulus), 1 (bee in ataxia state that does not move, but abdominal or limb movements are observed), 2 (bee that moves awkwardly but can access the food source) and 3 (bee that is in an optimal state and is able to move without difficulty through the cylinder and reach the food source). After 48 h, or when the bees died, they were reweighed. For mortality purposes, bees whose status was 0 and 1 at the end of the bioassay were taken into account.

To determine how much the workers had consumed in each of the treatments, the syringes with the unconsumed treatment inside were reweighed when the bees died or at the end of the bioassay. In this way, the weight of the consumed treatment was calculated by subtracting the final weight of the syringes from the initial weight. The value obtained was corrected for the weight loss of the treatments caused by evaporation. To calculate this weight loss, 3 to 10 syringes containing 2 mL of each insecticide concentration, as well as both controls in each bioassay, were attached to empty queen cages and kept in the same conditions and for the same period of time as the bioassay syringes, and the mean value of weight lost due to evaporation was subtracted from the consumption value. To obtain the daily consumption/bee, the corrected weight of treated nectar consumed was divided by the number of days the bee was able to access the food source (stages 2 and 3 of the numerical scale).

2.4. Maximum Exposure of Bumblebee Workers in the Field

Numerous studies carried out with different methodologies have identified a large number of pesticide residues in pollen, nectar and honey samples. Taking into account the available information, we estimated the maximum dose to which bumblebee workers could be exposed in the field in a worst case scenario. For this purpose, we used values of maximum daily nectar consumption of bumblebee workers in our bioassays (554 mg/bee/day) and values of the maximum concentration detected in nectar or honey of each of the active substances evaluated in the bioassays [54,55]. In the case of esfenvalerate, as no published data on residues in nectar or honey were found, the value of fenvalerate residue in honey was taken, as esfenvalerate is an isomer (2S, αS) of fenvalerate.

2.5. Statistical Analyses

To test whether there was a relationship between the weight of the workers and the amount of food ingested in *B. terrestris*, a Pearson correlation was performed with the values of the weight of the workers after feeding on nectar for 48 h and the amount of nectar consumed by each one of them. SPSS (IBM© SPSS© Statistics, Version 25.0, 2017) was used to perform this analysis, and a significance level of $\alpha = 0.05$ was set.

Mortality data from the concentration–response bioassays for each of the insecticides were analyzed by probit analysis. The PoloPlus program (LeOra Software, 2002–2019) was used, which calculates a concentration–response regression line on a logarithmic scale. To determine the susceptibility of *B. terrestris* workers to the different insecticides, the LC_{50} and LC_{90} values (concentration that causes mortality at 50% and 90% of the population, respectively), as well as their 95% confidence intervals, were calculated for each of the insecticides. The dose (μg active ingredient/bee) was calculated taking into account the daily amount of insecticide-treated nectar consumed by each bee. Once the insecticide doses were obtained, the LD_{50} and LD_{90} values were calculated (dose that causes 50% and 90% mortality in the population, respectively) together with their 95% confidence intervals. Significant differences in the susceptibility of workers to the different insecticides were tested using the 95% confidence intervals of the lethal concentration ratio (LCR) and lethal dose ratio (LDR) at the LC_{50} and LD_{50} point, respectively. Confidence intervals including value 1 indicate that there are no significant differences between the values compared [56].

The survival rate of the workers at the different applied concentrations of each insecticide was calculated using the Kaplan–Meier method, and the differences between the survival probabilities at each concentration were analyzed using the Mantel–Cox log-rank test, setting a significance level of $\alpha = 0.05$. The probability of survival (%) was defined as the probability that an individual survives longer than time "t". These analyses were carried out with GraphPad Prism 5.01 software.

3. Results

3.1. Relationship between Worker Weight and Nectar Consumed

Forty workers with a mean ($\pm$SE) weight of 247.4 $\pm$ 12.8 mg (range: 60–418 mg) were used to evaluate the relationship between the bee weight and the amount of food consumed. They consumed an average of 302.6 $\pm$ 15.2 mg of artificial nectar in 48 h.

Pearson's correlation showed a moderate positive relationship between both variables ($r = 0.474$; $p = 0.002$), indicating that nectar consumption was higher with increasing bee size (Figure 1). Therefore, the acute toxicity of the insecticides was evaluated by selecting workers of similar size within each hive.

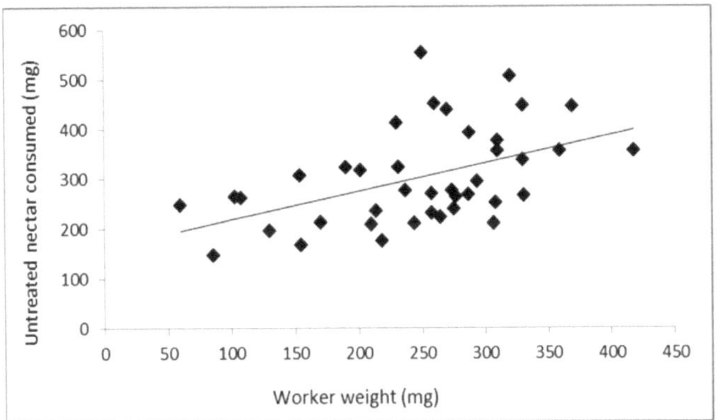

Figure 1. Correlation between the weight of *B. terrestris* workers fed with untreated artificial nectar and the amount of food consumed.

3.2. Intake of Treated Artificial Nectar

We observed a hive effect in the amount of nectar consumed by worker bumblebees. The values of untreated nectar consumption of the controls ranged from 150 to 250 mg, except for the workers used in the deltamethrin bioassays, whose consumption was almost double (467 ± 31) (Figure 2). In general, the intake of insecticide-treated nectar was irregular, although with the pyrethroids and sulfoxaflor a trend of decreasing consumption was observed as the concentration of insecticide in the nectar increased (Figure 2). Workers exposed to nectar containing thiacloprid showed a drastic reduction in nectar ingestion compared to controls. Likewise, a very sharp decrease in insecticide consumption of artificial nectar treated with *B. thuringiensis* was observed at the three highest concentrations of the four tested (Figure 2), which in this case corresponded to those used in the preliminary trial.

3.3. Toxicity of Insecticides on Bumblebee Workers

In all bioassays of the six insecticides tested, the mortality of the positive controls was 100%, while no bees died in the negative controls, except in the case of sulfoxaflor where there were two dead bees (9.1% mortality). A total of 149 bees weighing 240.5 ± 8.9 mg from five commercial hives were fed for 48 h with concentrations of the neonicotinoid imidacloprid ranging between 0.05 and 3 ppm. Fitting the mortality data to the probit model resulted in an LC_{50} of 0.38 (0.22–0.76) ppm and LD_{50} of 0.13 (0.08–0.24) μg a.i./bee (Table 1). For thiacloprid, 95 workers weighing 225.2 ± 7.9 mg extracted from two hives were treated with increasing concentrations of thiacloprid, from 150 to 800 ppm. The toxicity of this insecticide to *B. terrestris* workers gave values of LC_{50} of 424 (296–815) ppm and LD_{50} of 90.5 (58.8–172.7) μg a.i./bee (Table 1). The deltamethrin bioassay was conducted with 99 workers with an average weight of 251.8 ± 8.1 mg from two commercial hives. The concentrations used ranged between 2.04 and 73.53 ppm, and mortality resulted in LC_{50} of 7.1 (3.3–11.9) ppm and LD_{50} of 3.65 (2.19–5.11) μg a.i./bee (Table 1). Seven commercial hives were used to carry out the bioassay with the insecticide esfenvalerate, with a total of 191 bees weighing on average 205.4 ± 5.6 mg. The concentrations used for this insecticide were between 5.6 and 56.2 ppm, and its toxicity gave values of LC_{50} of 17.8 (14.4–22.4) ppm and LD_{50} of 5.52 (4.55–6.70) μg a.i./bee (Table 1). A total of 126 workers with a mean weight of 245.2 ± 5.3 mg from five hives were treated with increasing concentrations of sulfoxaflor ranging between 0.9 and 3 ppm. Fitting the mortality data to the probit model resulted in an LC_{50} of 2.22 (1.76–3.85) ppm and LD_{50} of 0.71 (0.56–1.01) µg a.i./bee (Table 1). In all cases with these five insecticides, the χ^2 values of the probit analyses performed for the calculation of the LC and LD values did not exceed the tabular values corresponding to the degrees of freedom in each case, indicating a good fit of the models.

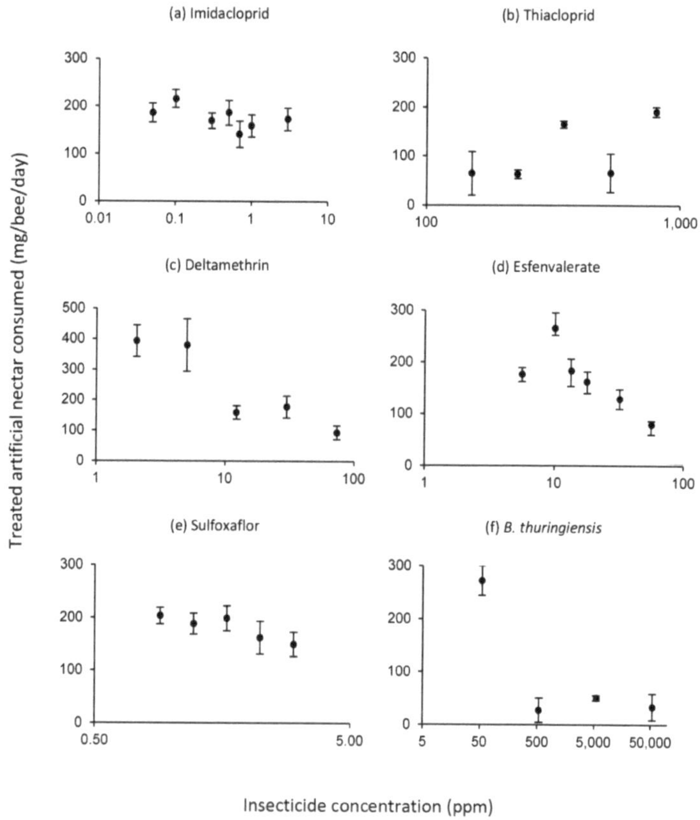

Figure 2. Average daily consumption per bee of artificial nectar treated with different concentrations of the insecticides imidacloprid (**a**), thiacloprid (**b**), deltamethrin (**c**), esfenvalerate (**d**), sulfoxaflor (**e**) and *B. thuringiensis* (**f**). Values expressed in mg of nectar consumed/bee/day. Average values of untreated nectar consumed daily by control bees in each bioassay are: 153.1 ± 20.3 (imidacloprid), 234.7 ± 44.3 (thiacloprid), 466.8 ± 31.5 (deltamethrin), 204.9 ± 13.9 (esfenvalerate), 177.6 ± 15.5 (sulfoxaflor) and 225.1 ± 17.3 (*B. thuringiensis*).

Four commercial hives were used to carry out the bioassays with the microbial insecticide *B. thuringiensis*, with a total of 94 bumblebee workers weighing 267.9 ± 5.3 mg. The range of concentrations used in the preliminary bioassay was between 54 and 54,000 ppm, the latter being the maximum concentration that could be prepared with the commercial insecticide DiPel® DF. It was not possible to fit the mortality data to the probit model because responses between 10% and 90% were not achieved. Mortality was only recorded in the group of workers treated with the 54,000 ppm concentration, which caused 73.7% mortality.

Table 1. Toxicity of insecticides with different mode of action on *B. terrestris* workers after 48 h oral exposure. Data expressed as lethal concentration (LC) or lethal dose (LD).

			Lethal Concentration					Lethal Dose				
Active Ingredient [a]	**N** [b]	**d.f.**	**Slope (SE)**	χ^2	**LC_{50} (CI 95%)** [c]	**LC_{90} (CI 95%)** [c]	**LCR (LC_{50}) (CI 95%)** [d]	**Slope (SE)**	χ^2	**LD_{50} (CI 95%)** [e]	**LD_{90} (CI 95%)** [e]	**LDR (LC_{50}) (CI 95%)** [d]
Imidacloprid (N)	149	23	1.15 (0.23)	28.9	0.38 (0.22–0.76)	4.96 (1.85–58.6)	1103 (608–2003) *	1.2 (0.2)	28.4	0.13 (0.08–0.24)	1.31 (0.55–10.78)	687.3 (352.5–1340.4) *
Thiacloprid (N)	95	8	2.52 (0.64)	9.2	424 (296–815)	1366 (744–15338)	1	1.8 (0.4)	7.1	90.5 (58.8–172.7)	554 (252–4330)	1
Deltamethrin (*p*)	99	8	1.33 (0.32)	5.1	7.1 (3.3–11.9)	64.5 (31.2–382.9)	60.1 (29.9–121) *	1.8 (0.4)	5.4	3.65 (2.19–5.11)	15.5 (9.6–52.8)	24.85 (12.86–47.65) *
Esfenvalerate (P)	191	32	2.68 (0.38)	36.5	17.8 (14.4–22.4)	53.5 (38.3–96.1)	23.8 (16.4–34.6) *	3.2 (0.5)	43.4	5.52 (4.55–6.70)	12.5 (9.4–23.2)	44.21 (12.7–155) *
Sulfoxaflor (S)	126	19	3.73 (0.97)	26.4	2.22 (1.66–3.85)	4.90 (3.15–30.1)	191 (162–289) *	5.6 (1.5)	22.8	0.71 (0.56–1.01)	1.28 (0.93–5.30)	123.3 (72.6–209) *

[a] (N): neonicotinoid; (P): pyrethroid; (S): sulfoximine. [b] Number of *B. terrestris* workers used in the bioassay including the positive and negative controls. [c] LC_{50} and LC_{90} and their 95% confidence intervals (CI) expressed in ppm (*W/V*). [d] LCR: lethal concentration ratio; LDR: lethal dose ratio. LC_{50} and LD_{50} are significantly different (* $p < 0.05$) if the 95% CI of the of the LCR or LDR do not include the value 1 [56]. [e] LD_{50} and LD_{90} and their 95% CI expressed in µg of active ingredient/bee.

3.4. Survival Probability after Insecticide Treatment

The Kaplan–Meier survival test was performed for the six insecticides tested. The probability of survival was below 20% by 24 h after starting treatment with the highest concentrations tested, except for sulfoxaflor and *B. thuringiensis*. In these two cases, the survival probability with all concentrations was higher than 20% at the end of the treatment (48 h) (Figure 3).

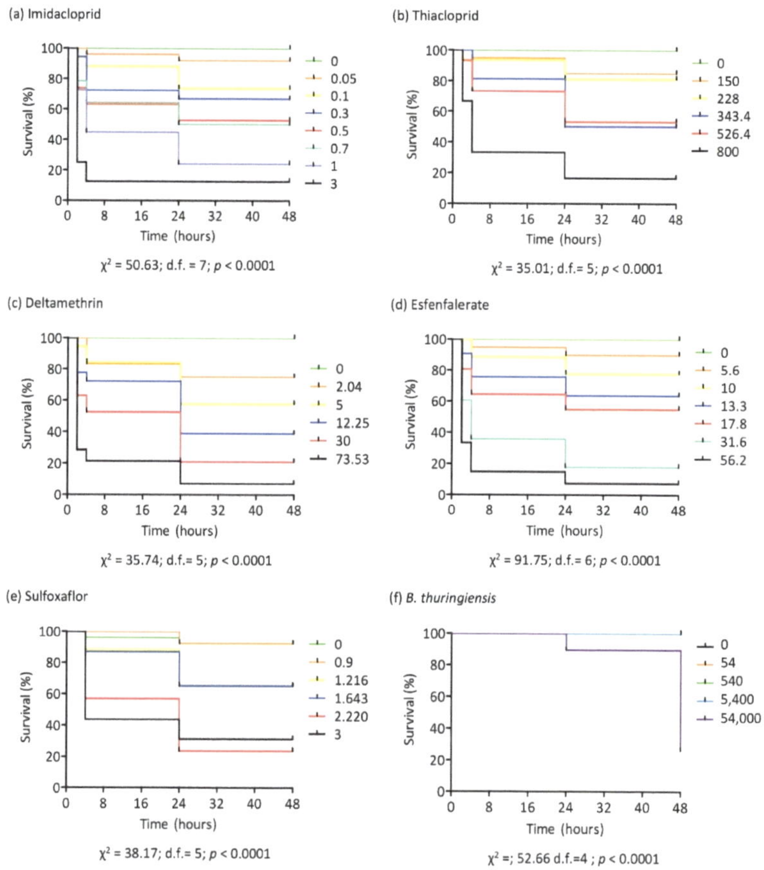

Figure 3. Kaplan–Meier survival analyses for *B. terrestris* workers when they were exposed for 48 h to different concentrations of the insecticides imidacloprid, thiacloprid, deltamethrin, esfenvalerate, sulfoxaflor and *B. thuringiensis*. Concentrations are represented by different colors and are expressed in ppm (*W/V*). The probability of survival (%) is defined as the probability that an individual survives longer than time "t".

3.5. Maximum Exposure of Bumblebee Workers in the Field

Of the five insecticides studied, neonicotinoids were found to be those to which *B. terrestris* workers were most exposed (first thiacloprid and secondly imidacloprid, with 115.7 and 40.3 ng a.i./bee/day, respectively), taking into account the residues found in nectar and the amount of nectar they can ingest (Table 2), followed by sulfoxaflor and deltamethrin. Taking into account the LD values obtained in our bioassays, with imidacloprid a bumblebee worker would need only 3.2 days of maximum ingestion of the highest concentration of insecticide found in the residue (worst case scenario) to reach the LD_{50} value (0.13 µg a.i./bee). This result contrasts with 92.9 days to reach the corresponding LD_{50} value in the case of sulfoxaflor, and almost 800 days for the other neonicotinoid, thiacloprid. Pyrethroids showed the longest time to reach the LD_{50} values in a worst case scenario. (Table 2).

Table 2. Maximum exposure dose of *B. terrestris* workers to insecticides based on the maximum insecticide residues found in nectar or honey.

	Max. Residue (ppb) [a]	LD_{50} (μg a.i./Bee)	Max. Exposure Dose (ng a.i./Bee/Day) [c]	No. Days a Worker Needs to Reach the LD_{50} (Worst Case Scenario) [d]
Imidacloprid	72.8	0.13	40.3	3.2
Thiacloprid	208.8	90.5	115.7	782.3
Deltamethrin	6.7	3.65	3.7	983.8
Esfenvalerate	0.7 [b]	5.52	0.39	14,234.1
Sulfoxaflor	13.8	0.71	7.6	92.9

[a] Maximum insecticide residues detected in nectar or honey for imidacloprid, thiacloprid, deltamethrin, fenvalerate [54] and sulfoxaflor [55]. [b] This value corresponds to maximum fenvalerate residue found in honey, as no published data on esfenvalerate residue were found. [c] Maximum dose of insecticide consumed per bee and day, assuming that each bee consumes 554 mg of nectar per day (maximum consumption value). [d] Number of days a worker needs to reach the LD_{50} after maximum consumption of the highest concentration of insecticide found in residue.

4. Discussion

We have evaluated the oral acute toxicity of six insecticide active substances in the buff-tailed bumblebee *B. terrestris*, one of most economically important pollinator species. Of the insecticides evaluated, both neonicotinoids are currently restricted for use in the EU (imidacloprid since 2018 and thiacloprid since 2021), and the rest (two pyrethroids, sulfoxaflor and *B. thuringiensis*) are used on many crops in the EU for the control of a wide range of pests.

The intake of insecticide-treated nectar did not show a dependence on the concentration used, except in the cases of deltamethrin and sulfoxaflor. With the two pyrethroids evaluated, deltamethrin and esfenvalerate, a trend towards reduced feeding with increasing insecticide concentration was observed. This could be an indication of a repellency effect, such as it was observed with deltamethrin in *B. terrestris* [57], or with alpha-cypermethrin and lambda-cyhalothrin in *A. mellifera* [58]. However, apart from a possible deterrent effect, it could also be an indication that the insecticide in the diet reduces the bees' ability or need to forage [59], which cannot be ruled out. Some studies have suggested that some neonicotinoids, such as imidacloprid, may have an antifeedant effect on *B. terrestris* [22,60]. On the contrary, two-choice feeding assays showed that *B. terrestris* preferred to eat more of sucrose solutions laced with the neonicotinoids imidacloprid and thiametoxam than sucrose alone [61]. Our results with the insecticides imidacloprid and thiacloprid do not match either of these two cases, as no concentration-dependent reduction or increase in consumption was observed.

The results obtained with both LC_{50} and LD_{50} reveal that the acute toxicity levels of the different insecticides that caused mortality to bumblebees followed the following sequence: imidacloprid > sulfoxaflor > deltamethrin > esfenvalerate > thiacloprid. Particularly remarkable is the large difference in toxicity found between the two neonicotinoids. The differential toxicity observed in these insecticides within the same class (almost 700 times more toxic imidacloprid than thiacloprid in terms of LD_{50} and 275 times in terms of LC_{50}) were likewise noticed for buff-tailed bumblebees in other studies where LD_{50} of imidacloprid resulted in being 518 times lower than that of thiacloprid [62,63]. Similar results had also been obtained previously in *A. mellifera*. In this species, the acute toxicity of imidacloprid expressed in LD_{50} was 884 [64] or 815 [65] times higher than thiacloprid, or six times more toxic considering LC_{50} values [66]. The differences observed in bee toxicity between the two neonicotinoids tested seem to be caused by the different chemical structure of these two compounds, despite having the same mode of action. Thus, the nitro-substituted neonicotinoids (clothianidin, dinotefuran, imidacloprid and its metabolites, thiamethoxam, nitenpyram) are much more toxic to bees than the cyano-substituted neonicotinoids (acetamiprid and thiacloprid), which exhibit a much lower toxicity [67]. In addition, in both honey bees and bumblebees, it has been found that sensitivity to nitro- and cyano-substituted neonicotinoids are associated to differences in their detoxification by P450 enzymes, demonstrating that both pollinators have biochemical defense systems that define their sensitivity to insecticides [62]. The results of our experiment with imidacloprid showed a somewhat lower survival probability than previously observed in *B. terrestris* workers, which was 75% after four days of oral exposure at 0.1 ppm [60]. However, in our study, the probability of survival was about the same only two days after an exposure at that concentration, so a lower survival would be expected if the treatment had been continued for two more days.

Typically, acute toxicity risk is considered for direct oral or contact exposure of bees to insecticides due to spray drift. However, an important part of the toxicity to which bees are subjected in crop fields comes from the insecticide residues to which they are exposed through multiple pathways, not only within the crop but also in the surrounding area [68–70]. In general, residues can be found in pollen and nectar after spray treatments [54]. Bees may also be exposed to residues through guttation droplets, as is the case with neonicotinoids in maize crops [71,72], and by dust drift from sowing treated seeds, especially significant in many neonicotinoids [73]. Neonicotinoids, as systemic insecticides, can be present in nectar and pollen, so it is easy for pollinators to be exposed to them [74,75]. Thiacloprid is in residues much more than imidacloprid [54,76]. Our results show that the maximum exposure dose of bumblebee workers to thiacloprid in a worst case scenario is almost three times higher than that of imidacloprid. Despite this, as thiacloprid has a much lower acute oral toxicity, it can be observed that in this scenario, a worker would need about 800 days to reach the LD_{50} value after maximum ingestion of contaminated nectar, compared to 3 days for imidacloprid.

Despite the differential toxicity shown by imidacloprid and thiacloprid, as well as their different residual presence in pollen and nectar, there were various reasons related to their environmental risk that led to the banning of both in the EU. Following their withdrawal, it is likely that other insecticides, particularly pyrethroids, will take their place in agricultural pest control [38,39]. We have found that two of them, deltamethrin and esfenvalerate, have intermediate toxicity on bumblebee workers, between the two neonicotinoids, in terms of LC_{50} and LD_{50} values, although our deltamethrin LD_{50} result was higher (lower toxicity) than a previous observation with the same product [57]. All pyrethroids have been found to be potentially toxic towards *B. terrestris* [77]. On the other hand, pyrethroids are lipophilic insecticides and very easily degraded in the natural environment. This characteristic means that many of them, such as esfenvalerate, are not normally present in nectar or honey after field treatments at recommended dosages [54,76]. We used deltamethrin and fenvalerate to evaluate the maximum exposure to these insecticides in a worst case scenario, and we found that it was very low for bumblebee workers, indicating that the nectar residue levels of these pyrethroids in nectar would be below the acute toxicity levels for *B. terrestris*. This result coincides with that found in *A. mellifera*, where it was observed that despite the presence of insecticide residues in honey and nectar, these were below the acute oral toxicity for this species [76]. If pyrethroids are finally the insecticides that will mainly replace neonicotinoids, it is important that acute oral toxicity studies on buff-tailed bumblebee are complemented by other studies of contact toxicity, chronic exposure and effects on larvae, as well as possible synergies and/or additive effects with other substances. Chronic treatment with the pyrethroid λ-cyhalothrin resulted in *B. terrestris* workers with a significantly lower body mass, although this did not affect the reproductive output of colonies or survival [78]. However, chronic dietary λ-cyhalothrin exposure caused severe decreases in survival, food consumption and reproduction of the same species [79]. Moreover, chronic exposure of bumblebees to imidacloprid and λ-cyhalothrin at concentrations that could approximate field-level exposure affected foraging behaviour and increased worker mortality, leading to significant reductions in brood development and colony success [53]. Available information indicates much variability in the toxicity of pyrethroids, which depends on the study subject, the active ingredient used and the type of bioassay performed to assess toxicity, so that risk assessment should be analyzed on a case-by-case basis. In addition, from the point of view of environmental and health risks, it should be considered that, sometimes, pyrethroid metabolites or photodegradates are just as or more harmful than parent compounds for non-target organisms, including mammals and the environment [80].

The sulfoximine sulfoxaflor was the second most toxic insecticide to buff-tailed bumblebee workers of the six tested in terms of LC_{50} and LD_{50}, behind only imidacloprid. It belongs to the group of competitive modulators of the nicotinic receptor of acetylcholine, and it is a systemic insecticide that acts mainly by ingestion and contact but with lower toxicity to bees, so it was proposed as an alternative to neonicotinoids [40]. The data available so far on its effects on *B. terrestris* seem to indicate that chronic exposure of colonies to sulfoxaflor may have important sublethal effects, such as reduction in egg laying and larval and worker production, leading to fewer reproductive offspring [51,81], which may also be aggravated by different stressors [82]. A recent study found sulfoxaflor to be about ten times more toxic after 48 h treatment than we observed [83]. However, our results are consistent with those of this study, which suggests that sulfoxaflor was less toxic than imidacloprid for *B. terrestris* but more toxic than the recently banned thiacloprid. Sulfoxaflor has also been found to appear residually in nectar or honey after field treatments, although in smaller quantities, in general, than neonicotinoids [54,55], but it is highly soluble, allowing it to reach wild plants near the treated crop, which could pose a risk factor for visiting pollinators [84]. We have found that the maximum exposure dose of sulfoxaflor for bumblebees in a worst case scenario is higher that the two pyrethroids but lower than the neonicotinoids. It could then be considered that it is a relatively safe insecticide for pollinators as far as its presence in nectar residue is concerned, but its sublethal effects after chronic exposures should be carefully considered. Despite this, in the USA, it is used in a large number of crops for the control of different pests, and, recently, the Environmental Protection Agency has announced that it poses a lower risk to non-target wildlife, including pollinators, than other registered alternative products if used according to the label [41].

The microbial insecticide formulated from the Cry insecticidal toxins of *B. thuringiensis* was the only one of the six insecticides tested with which no acute toxicity was observed in buff-tailed bumblebee workers. Nectar consumption was 5–7 times lower in the three highest concentrations tested (540, 5,400 and 54,000 ppm) than in the control, most likely due to the texture resulting from the dilution of the insecticide, which considerably thickened the treated artificial nectar, and this low consumption could have affected the toxicity results. Even so, workers were only affected at the highest concentration. Although the quantity of nectar consumed was small, there was nectar intake in all three concentrations, comparable to the amount of treated nectar consumed at certain concentrations in thiacloprid, deltamethrin or esfenvalerate bioassays, where the probability of the survival of workers was much below 50% after 48 h (Figure 3). Thus, if Bt were toxic, it would be at significantly lower levels than the other insecticides because we did not find differences in the probability of survival with respect to the control. Given the results of nectar intake in this bioassay, the commercial formulation of Bt used is not appropriate, so it would be convenient to use other sources of Cry toxin to make a more accurate assessment of its toxicity. Spain is the European leading country in the cultivation of maize expressing the insecticidal Cry1Ab toxin of *B. thuringiensis* (MON810 Bt maize), with 98,152 ha grown in 2020 (95% of the total in the EU). This area of Bt crop could pose a problem for pollinators visiting it. Even though maize does not require pollinating insects, they may occasionally use this pollen as a resource [42–44], thus being exposed to the insecticidal protein. However, it has been determined that pollen is the maize tissue in which the lowest levels of the Cry1Ab insecticidal toxin is found [45,47]. On the basis of the levels of Cry1Ab measured by these authors, it follows that the maximum concentration we have tested is about 10^5 times more concentrated than the maximum concentration of toxin detected in the pollen of MON810 maize grown in the EU. Thus, Bt maize would not be expected to be a source of toxicity for *B. terrestris*. Toxic effects have also not been found in bumblebee workers when eating Bt maize pollen [50] or following chronic exposure to Cry1Ab toxin [85]. Nevertheless, Bt maize could represent a cumulative/synergistic stressor since there are evidences that Bt toxins can pose sublethal effects to honey bees by affecting their foraging behavior [49], so further studies under realistic Bt maize field conditions are needed to confirm our findings.

5. Conclusions

The withdrawal of neonicotinoids in the EU has been the subject of debate and controversy [86,87]. Our results support that these systemic insecticides show differential toxicity to buff-tailed bumblebees. The neonicotinoid imidacloprid, banned for use in the EU since 2018, caused high mortality in *B. terrestris* workers, while thiacloprid, also currently banned, showed significantly less acute toxicity to this species. Pyrethroids, which are highly probable candidates to replace the banned neonicotinoids, showed intermediate toxicity between the two neonicotinoids, and there was a tendency for bumblebees to decrease their consumption with increasing concentrations. Sulfoxaflor, also a systemic insecticide that has been on the market for only a few years, was found to be the second most toxic insecticide tested, behind only imidacloprid. Taking into account the presence of residues in nectar/honey of these five insecticides, under worst case scenario conditions imidacloprid would have by far the highest risk for this species, followed by sulfoxaflor. Finally, the microbial insecticide formulated from insecticidal toxins of *B. thuringiensis* showed no significant toxicity to this species.

Therefore, the replacement of neonicotinoids by other types of insecticides for pest control may also have an impact on bee populations as some of the alternatives have already been shown to be toxic not only to *B. terrestris*, but also to other bees. It is vital to strive for a more comprehensive approach to ensure bee health in the broadest sense. To this end, laboratory studies on acute and chronic toxicity need to be complemented by studies under realistic natural conditions to allow a proper risk assessment of insecticides in adults and immatures in managed and wild species [31–33].

Supplementary Materials: The following supporting information can be downloaded at: https://www.mdpi.com/article/10.3390/insects13020184/s1, Table S1: Number of workers included in the bioassays with insecticides, from each of the commercial colonies used. Only the dose-response bioassays used for subsequent probit analysis are included.

Author Contributions: Conceptualization, G.P.F. and G.C.; laboratory work, G.C.; data analysis, G.P.F. and G.C.; writing—original draft preparation, G.P.F.; writing—review and editing, G.P.F. and G.C.; project administration and funding acquisition, G.P.F. All authors have read and agreed to the published version of the manuscript.

Funding: This research was funded by the Spanish Ministry of Science and Innovation (grant number PID2019-104578RB-I00) and the Spanish Ministry of Economy and Competitiveness (grant number AGL2015-64825-R). G.C. was funded by a predoc fellowship (BES-2016-077782) by the Spanish Ministry of Economy, Industry and Competitiveness.

Institutional Review Board Statement: Not applicable.

Informed Consent Statement: Not applicable.

Data Availability Statement: Not applicable.

Acknowledgments: We are grateful to Nuria Arranz de Pablo, Clara Benavent Celma, Carlos García Benítez and Andoni Jiménez Bravo for their help in maintaining the hives and for their support during the bioassays.

Conflicts of Interest: The authors declare no conflict of interest.

References

1. Hines, H.M. Historical Biogeography, Divergence Times, and Diversification Patterns of Bumble Bees (*Hymenoptera: Apidae: Bombus*). *Syst. Biol.* **2008**, *57*, 58–75. [CrossRef] [PubMed]
2. Velthuis, H.H.W.; Van Doorn, A. A century of advances in bumblebee domestication and the economic and environmental aspects of its commercialization for pollination. *Apidologie* **2006**, *37*, 421–451. [CrossRef]
3. Trillo, A.; Montero-Castaño, A.; Vilà, M. Seasonality of bumblebee spillover between strawberry crops and adjacent pinewoods. *Apidologie* **2020**, *51*, 1051–1061. [CrossRef]
4. Kleijn, D.; Winfree, R.; Bartomeus, I.; Carvalheiro, L.G.; Henry, M.; Isaacs, R.; Klein, A.; Kremen, C.; M'Gonigle, L.K.; Rader, R.; et al. Delivery of crop pollination services is an insufficient argument for wild pollinator conservation. *Nat. Commun.* **2015**, *6*, 7414. [CrossRef] [PubMed]
5. Rasmont, P.; Coppee, A.; Michez, D.; De Meulemeester, T. An overview of the Bombus terrestris (L. 1758) subspecies (Hymenoptera: Apidae). *Annales de la Société entomologique de France (N.S.)* **2008**, *44*, 243–250. [CrossRef]
6. Biesmeijer, J.C.; Roberts, S.P.M.; Reemer, M.; Ohlemüller, R.; Edwards, M.; Peeters, T.; Schaffers, A.P.; Potts, S.G.; Kleukers, R.; Thomas, C.D.; et al. Parallel Declines in Pollinators and Insect-Pollinated Plants in Britain and the Netherlands. *Science* **2006**, *313*, 351–354. [CrossRef]
7. Van Engelsdorp, D.; Hayes, J.; Underwood, R.M.; Pettis, J. A Survey of Honey Bee Colony Losses in the U.S., Fall 2007 to Spring 2008. *PLoS ONE* **2008**, *3*, e4071. [CrossRef]
8. Potts, S.G.; Biesmeijer, J.C.; Kremen, C.; Neumann, P.; Schweiger, O.; Kunin, W.E. Global pollinator declines: Trends, impacts and drivers. *Trends Ecol. Evol.* **2010**, *25*, 345–353. [CrossRef]
9. Potts, S.G.; Imperatriz-Fonseca, V.; Ngo, H.T.; Aizen, M.; Biesmeijer, J.C.; Breeze, T.; Dicks, L.; Garibaldi, L.A.; Hill, R.; Settele, J.; et al. Safeguarding pollinators and their values to human well-being. *Nature* **2016**, *540*, 220–229. [CrossRef] [PubMed]
10. Bartomeus, I.; Ascher, J.S.; Gibbs, J.; Danforth, B.; Wagner, D.L.; Hedtke, S.M.; Winfree, R. Historical changes in northeastern US bee pollinators related to shared ecological traits. *Proc. Natl. Acad. Sci. USA* **2013**, *110*, 4656–4660. [CrossRef]
11. Bommarco, R.; Lundin, O.; Smith, H.G.; Rundlöf, M. Drastic historic shifts in bumble-bee community composition in Sweden. *Proc. Royal Soc. B* **2011**, *279*, 309–315. [CrossRef] [PubMed]
12. Cameron, S.A.; Lozier, J.D.; Strange, J.P.; Koch, J.B.; Cordes, N.; Solter, L.F.; Griswold, T.L. Patterns of widespread decline in North American bumble bees. *Proc. Natl. Acad. Sci. USA* **2011**, *108*, 662–667. [CrossRef]
13. Goulson, D.; Lye, G.C.; Darvill, B. Decline and Conservation of Bumble Bees. *Annu. Rev. Èntomol.* **2008**, *53*, 191–208. [CrossRef] [PubMed]
14. Powney, G.D.; Carvell, C.; Edwards, M.; Morris, R.K.A.; Roy, H.E.; Woodcock, B.A.; Isaac, N.J.B. Widespread losses of pollinating insects in Britain. *Nat. Commun.* **2019**, *10*, 1–6. [CrossRef] [PubMed]
15. González-Varo, J.P.; Biesmeijer, J.C.; Bommarco, R.; Potts, S.G.; Schweiger, O.; Smith, H.G.; Steffan-Dewenter, I.; Szentgyörgyi, H.; Woyciechowski, M.; Vilà, M. Combined effects of global change pressures on animal-mediated pollination. *Trends Ecol. Evol.* **2013**, *28*, 524–530. [CrossRef] [PubMed]
16. Vanbergen, A.J.; the Insect Pollinators Initiative. Threats to an ecosystem service: Pressures on pollinators. *Front. Ecol. Environ.* **2013**, *11*, 251–259. [CrossRef]
17. Vanengelsdorp, D.; Evans, J.D.; Saegerman, C.; Mullin, C.; Haubruge, E.; Nguyen, B.K.; Frazier, M.; Frazier, J.; Cox-Foster, D.; Chen, Y.; et al. Colony Collapse Disorder: A Descriptive Study. *PLoS ONE* **2009**, *4*, e6481. [CrossRef] [PubMed]
18. Bass, C.; Denholm, I.; Williamson, M.S.; Nauen, R. The global status of insect resistance to neonicotinoid insecticides. *Pestic. Biochem. Physiol.* **2015**, *121*, 78–87. [CrossRef] [PubMed]
19. Jeschke, P.; Nauen, R.; Schindler, M.; Elbert, A. Overview of the Status and Global Strategy for Neonicotinoids. *J. Agric. Food Chem.* **2011**, *59*, 2897–2908. [CrossRef]
20. Simon-Delso, N.; Amaralrogers, V.; Belzunces, L.P.; Bonmatin, J.M.; Chagnon, M.; Downs, C.; Furlan, L.; Gibbons, D.W.; Giorio, C.; Girolami, V.; et al. Systemic insecticides (neonicotinoids and fipronil): Trends, uses, mode of action and metabolites. *Environ. Sci. Pollut. Res.* **2015**, *22*, 5–34. [CrossRef] [PubMed]
21. Blacquière, T.; Smagghe, G.; van Gestel, K.; Mommaerts, V. Neonicotinoids in bees: A review on concentrations, side-effects and risk assessment. *Ecotoxicol.* **2012**, *21*, 973–992. [CrossRef]
22. Cresswell, J.E.; Page, C.J.; Uygun, M.B.; Holmbergh, M.; Li, Y.; Wheeler, J.G.; Laycock, I.; Pook, C.J.; de Ibarra, N.H.; Smirnoff, N.; et al. Differential sensitivity of honey bees and bumble bees to a dietary insecticide (imidacloprid). *Zoology* **2012**, *115*, 365–371. [CrossRef]
23. Henry, M.; Béguin, M.; Requier, F.; Rollin, O.; Odoux, J.-F.; Aupinel, P.; Aptel, J.; Tchamitchian, S.; Decourtye, A. A Common Pesticide Decreases Foraging Success and Survival in Honey Bees. *Science* **2012**, *336*, 348–350. [CrossRef]
24. Sandrock, C.; Tanadini, L.G.; Pettis, J.S.; Biesmeijer, J.C.; Potts, S.G.; Neumann, P. Sublethal neonicotinoid insecticide exposure reduces solitary bee reproductive success. *Agric. For. Èntomol.* **2014**, *16*, 119–128. [CrossRef]
25. Rundlöf, M.; Andersson, G.; Bommarco, R.; Fries, I.; Hederström, V.; Herbertsson, L.; Jonsson, O.; Klatt, B.; Pedersen, T.R.; Yourstone, J.; et al. Seed coating with a neonicotinoid insecticide negatively affects wild bees. *Nature* **2015**, *521*, 77–80. [CrossRef]
26. Woodcock, B.A.; Bullock, J.M.; Shore, R.F.; Heard, M.S.; Pereira, M.G.; Redhead, J.W.; Ridding, L.; Dean, H.; Sleep, D.; Henrys, P.A.; et al. Country-specific effects of neonicotinoid pesticides on honey bees and wild bees. *Science* **2017**, *356*, 1393–1395. [CrossRef] [PubMed]

27. Chan, D.S.W.; Raine, N.E. Population decline in a ground-nesting solitary squash bee (Eucera pruinosa) following exposure to a neonicotinoid insecticide treated crop (Cucurbita pepo). *Sci. Rep.* **2021**, *11*, 1–11. [CrossRef]
28. Goulson, D. REVIEW: An overview of the environmental risks posed by neonicotinoid insecticides. *J. Appl. Ecol.* **2013**, *50*, 977–987. [CrossRef]
29. Van der Sluijs, J.P.; Simon-Delso, N.; Goulson, D.; Maxim, L.; Bonmatin, J.-M.; Belzunces, L.P. Neonicotinoids, bee disorders and the sustainability of pollinator services. *Curr. Opin. Environ. Sustain.* **2013**, *5*, 293–305. [CrossRef]
30. Sgolastra, F.; Medrzycki, P.; Bortolotti, L.; Maini, S.; Porrini, C.; Simon-Delso, N.; Bosch, J. Bees and pesticide regulation: Lessons from the neonicotinoid experience. *Biol. Conserv.* **2020**, *241*, 108356. [CrossRef]
31. EFSA Panel on Plant Protection Products and their Residues (PPR). Scientific Opinion on the science behind the development of a risk assessment of Plant Protection Products on bees (*Apis mellifera*, Bombus spp. and solitary bees). *EFSA J.* **2012**, *10*, 2668. [CrossRef]
32. Rortais, A.; Arnold, G.; Dorne, J.-L.; More, S.J.; Sperandio, G.; Streissl, F.; Szentes, C.; Verdonck, F. Risk assessment of pesticides and other stressors in bees: Principles, data gaps and perspectives from the European Food Safety Authority. *Sci. Total Environ.* **2017**, *587-588*, 524–537. [CrossRef] [PubMed]
33. Franklin, E.L.; Raine, N.E. Moving beyond honeybee-centric pesticide risk assessments to protect all pollinators. *Nat. Ecol. Evol.* **2019**, *3*, 1373–1375. [CrossRef]
34. European Commission. Commission Implementing Regulation (EU) No 485/2013 of 24 May 2013 amending Implementing Regulation (EU) No 540/2011, as regards the conditions of approval of the active substances clothianidin, thiamethoxam and imidacloprid, and prohibiting the use and sale of seeds treated with plant protection products containing those active substances. *Off. J.* **2013**, *139*, 12–26. Available online: http://data.europa.eu/eli/reg_impl/2013/485/oj (accessed on 13 November 2021).
35. European Commission. Commission Implementing Regulation (EU) 2018/783 of 29 May 2018, amending Implementing Regulation (EU) No 540/2011 as Regards the Conditions of Approval of the Active Substance Imidacloprid. *Off. J.* **2018**, *132*, 31–34. Available online: http://data.europa.eu/eli/reg_impl/2018/783/oj (accessed on 13 November 2021).
36. European Commission. Commission Implementing Regulation (EU) 2018/784 of 29 May 2018, amending Implementing Regulation (EU) No 540/2011, as regards the conditions of approval of the active substance clothianidin). *Off. J.* **2018**, *132*, 35–39. Available online: http://data.europa.eu/eli/reg_impl/2018/784/oj (accessed on 13 November 2021).
37. European Commission. Commission Implementing Regulation (EU) 2018/785 of 29 May 2018 amending Implementing Regulation (EU) No 540/2011 as regards the conditions of approval of the active substance thiamethoxam.). *Off. J.* **2018**, *132*, 40–44. Available online: http://data.europa.eu/eli/reg_impl/2018/785/oj (accessed on 13 November 2021).
38. Kathage, J.; Castañera, P.; Alonso-Prados, J.L.; Gómez-Barbero, M.; Rodríguez-Cerezo, E. The impact of restrictions on neonicotinoid and fipronil insecticides on pest management in maize, oilseed rape and sunflower in eight European Union regions. *Pest Manag. Sci.* **2018**, *74*, 88–99. [CrossRef]
39. Jactel, H.; Verheggen, F.; Thiéryc, D.; Escobar-Gutiérrez, A.J.; Gachete, E.; Desneuxf, N. Alternatives to neonicotinoids. *Environ. Int.* **2019**, *129*, 423–429. [CrossRef]
40. Brown, M.J.; Dicks, L.V.; Paxton, R.J.; Baldock, K.C.; Barron, A.B.; Chauzat, M.-P.; Freitas, B.M.; Goulson, D.; Jepsen, S.; Kremen, C.; et al. A horizon scan of future threats and opportunities for pollinators and pollination. *PeerJ* **2016**, *4*, e2249. [CrossRef]
41. EPA. Decision Memorandum Supporting the Registration Decision for New Uses of the Active Ingredient Sulfoxaflor on Alfalfa, Cacao, Citrus, Corn, Cotton, Cucurbits, Grains, Pineapple, Sorghum, Soybeans, Strawberries and Tree Plantations and Amendments to the Labels. 2019, Docket Number EPA-HQ-OPP-2010-0889-0570. Available online: https://www.regulations.gov/document/EPA-HQ-OPP-2010-0889-0570 (accessed on 3 December 2021).
42. Bailey, J.; Scott-Dupree, C.; Harris, R.; Tolman, J.; Harris, B. Contact and oral toxicity to honey bees (*Apis mellifera*) of agents registered for use for sweet corn insect control in Ontario, Canada. *Apidologie* **2005**, *36*, 623–633. [CrossRef]
43. Nguyen, B.K.; Saegerman, C.; Pirard, C.; Mignon, J.; Widart, J.; Thirionet, B.; Verheggen, F.J.; Berkvens, D.; De Pauw, E.; Haubruge, E. Does Imidacloprid Seed-Treated Maize Have an Impact on Honey Bee Mortality? *J. Econ. Èntomol.* **2009**, *102*, 616–623. [CrossRef] [PubMed]
44. Hendriksma, H.; Küting, M.; Härtel, S.; Näther, A.; Dohrmann, A.B.; Steffan-Dewenter, I.; Tebbe, C.C. Effect of Stacked Insecticidal Cry Proteins from Maize Pollen on Nurse Bees (Apis mellifera carnica) and Their Gut Bacteria. *PLoS ONE* **2013**, *8*, e59589. [CrossRef]
45. Nguyen, H.T.; Jehle, J.A. Quantitative analysis of the seasonal and tissue-specific expression of Cry1Ab in transgenic maize Mon810. *J. Plant Dis. Prot.* **2007**, *114*, 82–87. [CrossRef]
46. Duan, J.J.; Marvier, M.; Huesing, J.; Dively, G.; Huang, Z.Y. A Meta-Analysis of Effects of Bt Crops on Honey Bees (Hymenoptera: Apidae). *PLoS ONE* **2008**, *3*, e1415. [CrossRef] [PubMed]
47. Székács, A.; Weiss, G.; Quist, D.; Takács, E.; Darvas, B.; Meier, M.; Swain, T.; Hilbeck, A. Inter-laboratory comparison of Cry1Ab toxin quantification inMON 810maize by enzyme-immunoassay. *Food Agric. Immunol.* **2012**, *23*, 99–121. [CrossRef]
48. Morandin, L.A.; Winston, M.L. Effects of Novel Pesticides on Bumble Bee (Hymenoptera: Apidae) Colony Health and Foraging Ability. *Environ. Èntomol.* **2003**, *32*, 555–563. [CrossRef]
49. Ramirez-Romero, R.; Desneux, N.; Decourtye, A.; Chaffiol, A.; Pham-Delègue, M. Does Cry1Ab protein affect learning performances of the honey bee *Apis mellifera* L. (Hymenoptera, Apidae)? *Ecotoxicol. Environ. Saf.* **2008**, *70*, 327–333. [CrossRef] [PubMed]

50. Mommaerts, V.; Jans, K.; Smagghe, G. Impact ofBacillus thuringiensisstrains on survival, reproduction and foraging behaviour in bumblebees (Bombus terrestris). *Pest Manag. Sci.* **2009**, *66*, 520–525. [CrossRef] [PubMed]
51. Siviter, H.; Brown, M.J.F.; Leadbeater, E. Sulfoxaflor exposure reduces bumblebee reproductive success. *Nature* **2018**, *561*, 109–112. [CrossRef]
52. OECD. *Test No. 247: Bumblebee, Acute Oral Toxicity Test, OECD Guidelines for the Testing of Chemicals, Section 2, No. 247*; OECD Publishing: Paris, France, 2017. [CrossRef]
53. Gill, R.J.; Ramos-Rodriguez, O.; Raine, N.E. Combined pesticide exposure severely affects individual- and colony-level traits in bees. *Nature* **2012**, *491*, 105–108. [CrossRef] [PubMed]
54. Sanchez-Bayo, F.; Goka, K. Pesticide Residues and Bees—A Risk Assessment. *PLoS ONE* **2014**, *9*, e94482. [CrossRef] [PubMed]
55. Jiang, H.; Chen, J.; Zhao, C.; Tian, Y.; Zhang, Z.; Xu, H. Sulfoxaflor Residues in Pollen and Nectar of Cotton Applied through Drip Irrigation and Their Potential Exposure to *Apis mellifera* L. *Insects* **2020**, *11*, 114. [CrossRef]
56. Robertson, J.L.; Russell, R.M.; Preisler, H.K.; Savin, N.E. *Pesticide Bioassays with Arthropods*, 2nd ed.; CRC Press: Boca Raton, FL, USA, 2007.
57. Thompson, H.M. Assessing the exposure and toxicity of pesticides to bumblebees (Bombus sp.). *Apidologie* **2001**, *32*, 305–321. [CrossRef]
58. Thompson, H.; Wilkins, S. Assessment of the synergy and repellency of pyrethroid/fungicide mixtures. *Bull. Insectology* **2003**, *56*, 131–134.
59. Laycock, I.; Lenthall, K.M.; Barratt, A.T.; Cresswell, J.E. Effects of imidacloprid, a neonicotinoid pesticide, on reproduction in worker bumble bees (Bombus terrestris). *Ecotoxicology* **2012**, *21*, 1937–1945. [CrossRef] [PubMed]
60. Thompson, H.M.; Wilkins, S.; Harkin, S.; Milner, S.; Walters, K.F. Neonicotinoids and bumblebees (Bombus terrestris): Effects on nectar consumption in individual workers. *Pest Manag. Sci.* **2015**, *71*, 946–950. [CrossRef] [PubMed]
61. Kessler, S.C.; Tiedeken, E.J.; Simcock, K.L.; Derveau, S.; Mitchell, J.; Softley, S.; Radcliffe, A.; Stout, J.; Wright, G.A. Bees prefer foods containing neonicotinoid pesticides. *Nature* **2015**, *521*, 74–76. [CrossRef] [PubMed]
62. Manjon, C.; Troczka, B.J.; Zaworra, M.; Beadle, K.; Randall, E.; Hertlein, G.; Singh, K.S.; Zimmer, C.T.; Homem, R.A.; Lueke, B.; et al. Unravelling the Molecular Determinants of Bee Sensitivity to Neonicotinoid Insecticides. *Curr. Biol.* **2018**, *28*, 1137–1143.e5. [CrossRef]
63. Reid, R.J.; Troczka, B.J.; Kor, L.; Randall, E.; Williamson, M.S.; Field, L.; Nauen, R.; Bass, C.; Davies, T.E. Assessing the acute toxicity of insecticides to the buff-tailed bumblebee (Bombus terrestris audax). *Pestic. Biochem. Physiol.* **2020**, *166*, 104562. [CrossRef] [PubMed]
64. Krupke, C.H.; Long, E.Y. Intersections between neonicotinoid seed treatments and honey bees. *Curr. Opin. Insect Sci.* **2015**, *10*, 8–13. [CrossRef]
65. Iwasa, T.; Motoyama, N.; Ambrose, J.T.; Roe, R. Mechanism for the differential toxicity of neonicotinoid insecticides in the honey bee, *Apis mellifera*. *Crop Prot.* **2004**, *23*, 371–378. [CrossRef]
66. Jacob, C.R.O.; Malaquias, J.B.; Zanardi, O.Z.; Silva, C.A.S.; Jacob, J.F.O.; Yamamoto, P. Oral acute toxicity and impact of neonicotinoids on *Apis mellifera* L. and Scaptotrigona postica Latreille (Hymenoptera: Apidae). *Ecotoxicology* **2019**, *28*, 744–753. [CrossRef]
67. Decourtye, A.; Devillers, J. Ecotoxicity of Neonicotinoid Insecticides to Bees. *Adv. Exp. Med. Biol.* **2010**, *683*, 85–95. [CrossRef] [PubMed]
68. Botías, C.; David, A.; Horwood, J.; Abdulsada, A.K.; Nicholls, E.; Hill, E.M.; Goulson, D. Neonicotinoid Residues in Wildflowers, a Potential Route of Chronic Exposure for Bees. *Environ. Sci. Technol.* **2015**, *49*, 12731–12740. [CrossRef]
69. Uhl, P.; Brühl, C.A. The Impact of Pesticides on Flower-Visiting Insects: A Review with Regard to European Risk Assessment. *Environ. Toxicol. Chem.* **2019**, *38*, 2355–2370. [CrossRef] [PubMed]
70. Chan, D.S.W.; Prosser, R.S.; Rodríguez-Gil, J.L.; Raine, N.E. Assessment of risk to hoary squash bees (Peponapis pruinosa) and other ground-nesting bees from systemic insecticides in agricultural soil. *Sci. Rep.* **2019**, *9*, 1–13. [CrossRef]
71. Girolami, V.; Mazzon, L.; Squartini, A.; Mori, N.; Marzaro, M.; Di Bernardo, A.; Greatti, M.; Giorio, C.; Tapparo, A. Translocation of Neonicotinoid Insecticides From Coated Seeds to Seedling Guttation Drops: A Novel Way of Intoxication for Bees. *J. Econ. Èntomol.* **2009**, *102*, 1808–1815. [CrossRef] [PubMed]
72. Mörtl, M.; Takács, E.; Klátyik, S.; Székács, A. Appearance of Thiacloprid in the Guttation Liquid of Coated Maize Seeds. *Int. J. Environ. Res. Public Heal.* **2020**, *17*, 3290. [CrossRef] [PubMed]
73. Girolami, V.; Marzaro, M.; Vivan, L.; Mazzon, L.; Greatti, M.; Giorio, C.; Marton, D.; Tapparo, A. Fatal powdering of bees in flight with particulates of neonicotinoids seed coating and humidity implication. *J. Appl. Èntomol.* **2011**, *136*, 17–26. [CrossRef]
74. Bonmatin, J.-M.; Giorio, C.; Girolami, V.; Goulson, D.; Kreutzweiser, D.P.; Krupke, C.; Liess, M.; Long, E.; Marzaro, M.; Mitchell, E.A.D.; et al. Environmental fate and exposure; neonicotinoids and fipronil. *Environ. Sci. Pollut. Res.* **2015**, *22*, 35–67. [CrossRef] [PubMed]
75. Zioga, E.; Kelly, R.; White, B.; Stout, J.C. Plant protection product residues in plant pollen and nectar: A review of current knowledge. *Environ. Res.* **2020**, *189*, 109873. [CrossRef] [PubMed]
76. Kaila, L.; Ketola, J.; Toivonen, M.; Loukola, O.; Hakala, K.; Raiskio, S.; Hurme, T.; Jalli, M. Pesticide residues in honeybee-collected pollen: Does the EU regulation protect honeybees from pesticides? *Environ. Sci. Pollut. Res.* **2021**, 1–20. [CrossRef] [PubMed]

77. Mommaerts, V.; Smagghe, G. Side-effects of pesticides on the pollinator bombus: An overview. In *Pesticides in the Modern World-Pests Control and Pesticides Exposure and Toxicity Assessment*; Stoytcheva, M., Ed.; In-Tech: Rijeka, Croatia, 2011; Volume 5, pp. 507–552.
78. Baron, G.L.; Raine, N.; Brown, M.J.F. Impact of chronic exposure to a pyrethroid pesticide on bumblebees and interactions with a trypanosome parasite. *J. Appl. Ecol.* **2014**, *51*, 460–469. [CrossRef]
79. Ceuppens, B.; Eeraerts, M.; Vleugels, T.; Cnops, G.; Roldan-Ruiz, I.; Smagghe, G. Effects of dietary lambda-cyhalothrin exposure on bumblebee survival, reproduction, and foraging behavior in laboratory and greenhouse. *J. Pest Sci.* **2015**, *88*, 777–783. [CrossRef]
80. Zhu, Q.; Yang, Y.; Zhong, Y.; Lao, Z.; O'Neill, P.; Hong, D.; Zhang, K.; Zhao, S. Synthesis, insecticidal activity, resistance, photodegradation and toxicity of pyrethroids (A review). *Chemosphere* **2020**, *254*, 126779. [CrossRef] [PubMed]
81. Siviter, H.; Horner, J.; Brown, M.J.F.; Leadbeater, E. Sulfoxaflor exposure reduces egg laying in bumblebees Bombus terrestris. *J. Appl. Ecol.* **2019**, *57*, 160–169. [CrossRef] [PubMed]
82. Linguadoca, A.; Rizzi, C.; Villa, S.; Brown, M.J. Sulfoxaflor and nutritional deficiency synergistically reduce survival and fecundity in bumblebees. *Sci. Total Environ.* **2021**, *795*, 148680. [CrossRef] [PubMed]
83. Azpiazu, C.; Bosch, J.; Bortolotti, L.; Medrzycki, P.; Teper, D.; Molowny-Horas, R.; Sgolastra, F. Toxicity of the insecticide sulfoxaflor alone and in combination with the fungicide fluxapyroxad in three bee species. *Sci. Rep.* **2021**, *11*, 1–9. [CrossRef]
84. Mundy-Heisz, K.A.; Prosser, R.S.; Raine, N.E. Acute oral toxicity and risks of exposure to the neonicotinoid thiamethoxam, and other classes of systemic insecticide, for the Common Eastern Bumblebee (*Bombus impatiens*). *bioRxiv* **2020**. [CrossRef]
85. Babendreier, D.; Reichhart, B.; Romeis, J.; Bigler, F. Impact of insecticidal proteins expressed in transgenic plants on bumblebee microcolonies. *Èntomol. Exp. Appl.* **2008**, *126*, 148–157. [CrossRef]
86. Klatt, B.K.; Rundlöf, M.; Smith, H.G. Maintaining the Restriction on Neonicotinoids in the European Union—Benefits and Risks to Bees and Pollination Services. *Front. Ecol. Evol.* **2016**, *4*, 4. [CrossRef]
87. Carreck, N.L. A beekeeper's perspective on the neonicotinoid ban. *Pest Manag. Sci.* **2017**, *73*, 1295–1298. [CrossRef] [PubMed]

insects MDPI

Article

Oxidative Stress Response of Honey Bee Colonies (*Apis mellifera* L.) during Long-Term Exposure at a Frequency of 900 MHz under Field Conditions

Marinko Vilić [1], Ivona Žura Žaja [1,*], Mirta Tkalec [2], Perica Tucak [3], Krešimir Malarić [4], Nato Popara [5], Nikolino Žura [6], Selim Pašić [5] and Ivana Tlak Gajger [7]

1 Department of Physiology and Radiobiology, Faculty of Veterinary Medicine, University of Zagreb, 10000 Zagreb, Croatia; mvilic@vef.unizg.hr
2 Department of Biology, Faculty of Science, University of Zagreb, 10000 Zagreb, Croatia; mirta.tkalec@biol.pmf.hr
3 State Inspectorate of Republic of Croatia, 10000 Zagreb, Croatia; perica.tucak@gmail.com
4 Department of Communication and Space Technologies, Faculty of Electrical Engineering and Computing, University of Zagreb, 10000 Zagreb, Croatia; kresimir.malaric@fer.hr
5 Department of Physics, Faculty of Veterinary Medicine, University of Zagreb, 10000 Zagreb, Croatia; npopara@vef.unizg.hr (N.P.); spasic@vef.unizg.hr (S.P.)
6 Faculty of Veterinary Medicine, University of Zagreb, 10000 Zagreb, Croatia; nikolino.zura@zvu.hr
7 Department for Biology and Pathology of Fish and Bees, Faculty of Veterinary Medicine, University of Zagreb, 10000 Zagreb, Croatia; itlak@vef.unizg.hr
* Correspondence: izzaja@vef.unizg.hr; Tel.: +385-12390173

Simple Summary: Most research on radiofrequency electromagnetic fields (RF-EMFs) in honey bees has studied adult bees under controlled laboratory conditions after direct exposure to devices that generate RF-EMF. To date, there are many uncertainties about the effects of RF-EMFs on different developmental stages of honey bees under field conditions. We investigated oxidative stress in larvae, pupae and adult honey bees after exposure to RF-EMFs originating from phone base station towers under field conditions. The study was conducted on a total of fifteen honey bee colonies exposed to RF-EMFs at a frequency of 900 MHz, divided into three locations with different electric field levels. All the honey bee colonies at the three locations were exposed for one year. Antioxidant parameters such as glutathione-S-transferase, catalase and superoxide dismutase, as well as the formation of thiobarbituric acid reactive substances, were measured in samples of larvae, pupae and the midguts of adult honey bees. Our results show that the activity of antioxidant enzymes changes and that the level of the analyzed parameters depends on the developmental stage of the honey bee, the level of the electric field and the exposure time.

Abstract: In this study, oxidative stress and lipid peroxidation in honey bee larvae, pupae and the midguts of adult bees were investigated during a one-year exposure to radiofrequency electromagnetic fields (RF-EMFs) at a frequency of 900 MHz under field conditions. The experiment was carried out on honey bee colonies at three locations with electric field levels of 30 mV m^{-1}, 70 mV m^{-1} and 1000 mV m^{-1}. Antioxidant enzymes, glutathione-S-transferase (GST), catalase (CAT) and superoxide dismutase (SOD) and thiobarbituric acid reactive substances (TBARS) as indicators of lipid peroxidation were measured spectrophotometrically. The GST activity within the same developmental stage showed no significant differences regardless of electric field level or sampling time. The highest GST activity was found in the pupae, followed by activity in the larvae and midguts. Both CAT activity and TBARS concentration were the highest in the midguts, regardless of field level and sampling time. The larvae showed a significantly higher TBARS concentration at the location with an electric field level of 1000 mV m^{-1} compared to the locations with lower levels. Our results show that RF-EMFs at a frequency of 900 MHz can cause oxidative stress in honey bees, with the larval stage being more sensitive than the pupal stage, but there was no linear relationship between electric field level and effect in any of the developmental stages.

Citation: Vilić, M.; Žura Žaja, I.; Tkalec, M.; Tucak, P.; Malarić, K.; Popara, N.; Žura, N.; Pašić, S.; Gajger, I.T. Oxidative Stress Response of Honey Bee Colonies (*Apis mellifera* L.) during Long-Term Exposure at a Frequency of 900 MHz under Field Conditions. *Insects* **2024**, *15*, 372. https://doi.org/10.3390/insects15050372

Academic Editor: Alison Gray

Received: 19 April 2024
Revised: 12 May 2024
Accepted: 18 May 2024
Published: 20 May 2024

Keywords: radiofrequency radiation; honey bee; antioxidants; field conditions

1. Introduction

The modern way of life is undoubtedly conditioned by the use of various achievements of scientific and technological development, among which the application of radiofrequency communication (RFC) is indispensable. Most sources of RFC were already known and used in the 20th century; however, their real expansion occurred in the early 1990s, with the use of mobile phones and wireless communication technologies. In 2022, the number of mobile–cellular subscriptions exceeded the total world population, and there were a reported 8.6 billion mobile phone subscriptions worldwide, with the penetration rate of mobile–broadband subscriptions growing by an enormous 14.8% per year over the last 10 years [1], and the trend continues to grow. The intensity level of radiofrequency electromagnetic radiation fields (RF-EMFs) around a frequency of 1 GHz has increased by about 10^{18} times compared to natural levels [2].

In addition to the wide use of mobile phones and other sources that generate RF-EMFs, public concern about harmful effects on human and animal health is also growing. Indeed, the results of numerous studies have shown that exposure to RF-EMFs at the operating frequency of wireless devices (mobile phones, routers, base stations) may cause various non-thermal biological effects such as oxidative stress [3,4], immune system dysfunction [5,6], genotoxic effects [7–9], effects on reproduction [10] and adverse effects on male fertility [11,12]. The mentioned biological effects have been proven in vitro and in vivo on different species of animals including mammals and insects, especially honey bees [13–15].

The honey bee is one of the most important insects for maintaining balance in natural ecosystems. It is assumed that the honey bee (*Apis mellifera*) plays the most important role in the pollination of all insect species from the order Hymenoptera, as it is involved in almost 80–85% of the pollination of the world's crops [16–18]. In addition to pollination, honey bees are also important for the production of apian products (honey, propolis, pollen, wax, royal jelly and bee venom), which are widely used in human nutrition, medicine and the pharmaceutical industry.

Previous studies on the effects of RF-EMFs on honey bee colonies mostly examined adult honey bees in the laboratory or under unnatural conditions, i.e., after direct exposure to devices emitting radiofrequency electromagnetic radiation such as mobile phones or Wi-Fi networks [19–22]. Most results showed reduced colony strength and queen laying rate [22,23] as well as the initiation of "worker piping" [19,24,25] associated with swarming. The results of some experiments suggest that RF-EMFs generated by mobile phone base stations or devices equivalent to commercial Wi-Fi devices could actually alter honey bees' navigational abilities, i.e., reduce foraging success and prevent them from returning to their hives [26,27]. Behavioral changes and biological activities, as well as oxidative stress and short-term memory under laboratory conditions in honey bees, have been reported previously [20,21,28,29].

To date, there remain many uncertainties about the effects of radiofrequency radiation on honey bee colonies, as there is a lack of studies under field conditions. In view of this, Vanbergen et al. [30] recommended more field-realistic studies on the exposure of pollinators to RF-EMFs, and Panagopoulos et al. [31] emphasized that exposure to RF-EMFs in field conditions is an important aspect of the study of biological effects.

How frequency, electric field strength, modulation and exposure duration, as well as the developmental stage, can significantly influence biological responses after RF-EMF exposure [13,32–34], and because currently there is a lack of research on the impact of RF-EMFs on oxidative stress in different developmental stages of honey bees after being exposed to RF-EMFs under field conditions, we wanted to investigate the antioxidant parameters (indicators of antioxidant defenses) in honey bee larvae, pupae and adult bees exposed to RF-EMFs under field conditions. In this study, honey bee colonies were exposed to three electric field levels at a frequency of 900 MHz originating from mobile phone base stations: (i) electric field level of 1000 mV m^{-1}, the highest value found in the environment; (ii) 30 mV m^{-1}, the lowest value of electric field found in the environment and (iii) 70 mV m^{-1}, the value measured at the location of the stationary apiary, to which the honey bees were continuously exposed for a long time. We would also like to emphasize that all the electric field levels used in this study correspond to the values found in the natural environment but with a lower probability of the colonies being exposed to the chosen highest measured field level, which is almost 30 times higher than the lowest level.

Since changes in the activity of antioxidant enzymes such as catalase (CAT), glutathione S-transferase (GST) and superoxide dismutase (SOD), as well as the formation of thiobarbituric acid reactive substances (TBARS), as an indicator of oxidative damage were documented in our previous study [29] on honey bees after exposure to RF-EMR under laboratory conditions, the aim of this study was to answer the following questions: (a) Could RF-EMF cause lipid peroxidation and changes in three vital antioxidant enzymes (CAT, GST and SOD) at different developmental stages of honey bee workers under field conditions? (b) Is there a possibility of a long-term effect on oxidative stress in honey bees after one year of exposure?

2. Materials and Methods

2.1. Honey Bee Colonies and Exposure Conditions

The study was conducted on Carniolan honey bees (*Apis mellifera carnica*, Pollmann, 1879) exposed to RF-EMFs from mobile phone base stations in their natural environment. A total of fifteen (15) honey bee colonies (Langstroth–Root-type hives) were randomly selected and equalized after a clinical inspection.

All the experimental colonies had one-year-old queens of the same genetic origin produced by a registered queen breeder. The experiment was carried out in three different locations: (I) Five (5) honey bee colonies were located near mobile phone base station towers with a frequency of 900 MHz and an average environmental electric field level of 1000 mV m^{-1}, named high intensity (HI). The site determinants were: 43°23′7.20″ N; 17°13′16.97″ E; 534 m above sea level; and 67.24 m and 159.76 m distance from the mobile phone base stations (Figure 1a). (II) Five (5) honey bee colonies located at a site with a frequency of 900 MHz and an average environmental electric field level of 30 mV m^{-1}, named low intensity (LI), with the site determinants being 43°22′43.55″ N; 17°13′34.55″ E; 450 m above sea level; and 790 and 700 m distance from mobile phone base stations. (III) Five (5) honey bee colonies located at a site with a frequency of 900 MHz and an average environmental electric field level of 70 mV m^{-1}, named stationary location or medium intensity (MI). The site determinants were 43°23′717″ N; 17°14′30.05″ E; 277 m above sea level; and 1630 m and 1520 m distance from the mobile phone base stations. The honey bee colonies located at MI were used as a control as these 5 colonies were already stationed at that site before the experiment was carried out, and the other 10 experimental colonies were formed from colonies located at this site.

The distance between the three locations (honey bee colonies), i.e., HI and LI, HI and MI and MI and LI was 829.77 m, 1.64 km and 1.44 km, respectively (Figure 1b). All the honey bee colonies at the three mentioned locations were exposed to the radiofrequency electromagnetic field continuously for one year. The average electric field level and frequency were measured at each site using the portable spectrum analyzer, NARDA SRM 3000 (Narda Safety Test Solutions GmbH, Pfullingen, Germany). All the collected samples originated from the worker brood and worker adults. Samples were taken on three occasions (at 2 weeks, 5 months and 1 year) after the beginning of observation (Table 1). During the experiment, all the honey bee colonies were treated with the same authorized and registered veterinary medicine against varroosis at the same time.

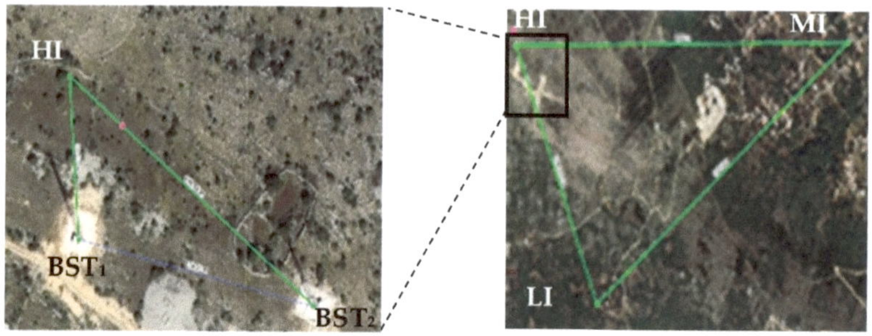

(a) (b)

Figure 1. Honey bee colonies (*Apis mellifera*) were exposed to RF-EMFs from mobile phone base stations (BST_1 and BST_2) in their natural environment: (**a**) overview of the location of honey bee colonies at the site of the high electric field (HI) and the base station towers (BST_1 and BST_2), (**b**) overview of the locations between the honey bee colonies situated at the three sites with different electric field levels: the high electric field (HI), low electric field (LI) and medium electric field (MI).

Table 1. Meteorological conditions and external temperatures during sampling of worker bee larvae, pupae and midguts of adult workers.

Sampling Time	Weather Conditions	Temperature on Sampling Day
2 weeks (April)	Cloudy	18 °C
5 months (September)	Clear	24 °C
1 year (April)	Clear	17 °C

2.2. Sample Preparation and Assays of Oxidative Stress Parameters

Five- to six-day-old larvae, pupae in the stage of purple eyes and the midguts of adult honey bees were collected from each honey bee colony and placed in five Eppendorf tubes (6 larvae/tube; 4 pupae/tube; and 10 midguts/tube). The adult bees were forager bees older than 21 days and were collected by shaking the periphery hive frames. Before the midgut of an adult bee was dissected, the adult bee was briefly exposed to ice. Dissection: we pulled out the posterior segment of the abdomen with fine forceps to remove the intestines from the abdomen. Then, we cut the midgut with a scalpel on a clean surface and placed it in an Eppendorf tube. The Eppendorf tubes containing the samples were frozen in liquid nitrogen (−196 °C) and delivered to the laboratory where they were stored at −80 °C until biochemical analysis. Immediately prior to biochemical analysis, the collected tissues were homogenized in a cold potassium phosphate buffer (50 mM, pH 7.0) containing 0.5 mM EDTA, using QIAGEN's TissueLyser II apparatus (60 s at 15 Hz) for sample disruption; the resulting homogenate was centrifuged two times (each at 15,000× *g* for 15 min, at 4 °C). The supernatants collected in this process were then used for further biochemical investigations.

The glutathione S-transferase (GST), catalase (CAT) and superoxide dismutase (SOD) activity, as well as the level of lipid peroxidation (TBARS), were determined in accordance with our previously published study [29,35].

In brief, the GST activity (EC 1.8.1.7) was measured via the modified method of [36] using 1-chloro-2,4-dinitrobenzene (CDNB) at 340 nm (ε = 9.6 mM^{-1} cm^{-1}). The reaction mixture consisted of 50 µL of the sample, 100 mM of potassium phosphate buffer at a pH of 6.5, 2 mM of CDNB and 2.5 mM of GSH. The results were expressed in units per mg of protein, where one unit was defined as the amount of enzyme that produces 1 µmol of GS-DNB of conjugate per minute under the assay conditions. CAT (EC 1.11.1.6), was measured using the absorbance decrease at 240 nm (ε = 39.4 M^{-1} cm^{-1}) as per Aebi [37]. The reaction mixture consisted of 25 µL of the sample and 50 mM of potassium phosphate buffer, at a pH of 7.0, with the addition of 10 mM H_2O_2. The CAT activity was presented in units per mg of protein. One unit was defined as the amount of enzyme that hydrolyzes 1 µmol of H_2O_2 per minute at 25 °C and a pH of 7.0. For the determination of SOD activity (EC 1.15.1.1), the reaction mixture consisted of 0.5 mM of xanthine and 0.01 mM of cytochrome c in 50 mM of potassium phosphate buffer, at a pH of 7.8, containing 0.01 mM of EDTA. This was used in a modified xanthine oxidase/cytochrome c method, according to McCord and Fridovich [38], after enough xanthine oxidase was added to cause a change in absorbance of 0.025 per min. The results were expressed in arbitrary units, where one unit was defined as inhibiting the rate of cytochrome reduction in the coupled xanthine–xanthine oxidase system by 50% at a pH of 7.8 and 25 °C. The level of lipid peroxidation was measured as the formation of thiobarbituric acid reactive substances (TBARS), a byproduct of lipid peroxidation that reacts with thiobarbituric acid [39]. The supernatants (300 µL) were mixed with 200 µL of cold 20% (*w*/*v*) trichloroacetic acid to precipitate the proteins. The precipitate was pelleted by centrifugation (10,000× *g* for 15 min at 4 °C), and the supernatant obtained was reacted with 400 µL of 1% (*w*/*v*) thiobarbituric acid prepared in 20% TCA. After heating at 95 °C for 30 min, the mixture was cooled in an ice bath. Supernatant absorbance at 532 nm was measured, and the results were corrected for nonspecific turbidity by subtracting the absorbance at 600 nm. The TBARS content, expressed per mg of protein, was obtained using an extinction coefficient of 155 mM^{-1} cm^{-1}.

The concentration of protein in the supernatants was determined according to Bradford [40], taking bovine serum albumin as a standard.

2.3. Statistical Analysis

All the results were expressed as the mean (M) and standard error (SEM) of 30 larvae, 20 pupae and 50 midguts. A 12 Software package (StatSoft, Inc., Tulsa, OK, USA) was used for the statistical analysis. Normality was tested with the Kolmogorov–Smirnov test. After the Kolmogorov–Smirnov test, the results were tested using an analysis of variance (ANOVA) to determine differences between groups. Multiple comparisons between means were determined using the Tukey HSD test. A statistical difference at $p < 0.05$ was considered significant.

3. Results

Oxidative Stress Parameters

The activity of antioxidant parameters as well as TBARS concentration in the larvae, pupae and midguts of honey bees exposed to RF-EMFs at a frequency of 900 MHz and electric field levels of 70 mV m^{-1}, 30 mV m^{-1} and 1000 mV m^{-1} for 2 weeks, 5 months and 1 year are presented in Tables 2–5.

Table 2. Glutathione S-transferase (GST) activity in the larvae, pupae and midguts of honey bees exposed to RF-EMFs at a frequency of 900 MHz and electric field levels of 70 mV m^{-1} (medium intensity—MI), 30 mV m^{-1} (low intensity—LI) and 1000 mV m^{-1} (high intensity—HI) for 2 weeks, 5 months and 1 year.

Sampling Time	Location	Electric Field Level (mV m^{-1})	GST (Unit mg^{-1} $_{proteins}$)		
			Larvae (N = 30)	Pupae (N = 20)	Midguts (N = 50)
2 weeks	MI	70	0.16 ± 0.02 ^{ab}	0.18 ± 0.004 ^{b}	0.04 ± 0.003 ^{ab}
	LI	30	0.18 ± 0.004 ^{ab}	0.26 ± 0.03 ^{ab}	0.04 ± 0.01 ^{ab}
	HI	1000	0.17 ± 0.02 ^{ab}	0.25 ± 0.01 ^{ab}	0.05 ± 0.004 ^{a}
5 months	MI	70	0.15 ± 0.01 ^{a}	0.24 ± 0.02 ^{ab}	0.04 ± 0.003 ^{ab}
	LI	30	0.16 ± 0.01 ^{ab}	0.29 ± 0.02 ^{a}	0.04 ± 0.002 ^{b}
	HI	1000	0.19 ± 0.01 ^{ab}	0.27 ± 0.03 ^{a}	0.04 ± 0.001 ^{ab}

Table 2. *Cont.*

Sampling Time	Location	Electric Field Level (mV m^{-1})	GST (Unit mg^{-1} $_{proteins}$)		
			Larvae (N = 30)	Pupae (N = 20)	Midguts (N = 50)
1 year	MI	70	0.18 ± 0.02^{ab}	0.21 ± 0.01^{ab}	0.03 ± 0.003^{b}
	LI	30	0.18 ± 0.01^{ab}	0.26 ± 0.01^{ab}	0.04 ± 0.002^{b}
	HI	1000	0.21 ± 0.02^{b}	0.23 ± 0.02^{ab}	0.03 ± 0.003^{ab}

Results are presented as the mean (M) ± standard error of the mean (SEM). Values with different letters in the same column are significantly different according to the Tukey HSD test at $p < 0.05$.

Table 3. Catalase (CAT) activity in the larvae, pupae and midguts of honey bees exposed to RF-EMFs at a frequency of 900 MHz and electric field level of 70 mV m^{-1} (medium intensity—MI), 30 mV m^{-1} (low intensity—LI) and 1000 mV m^{-1} (high intensity—HI) at 2 weeks, 5 months and 1 year.

Sampling Time	Location	Electric Field Level (mV m^{-1})	CAT (Unit mg^{-1} $_{proteins}$)		
			Larvae (N = 30)	Pupae (N = 20)	Midguts (N = 50)
2 weeks	MI	70	26.06 ± 2.54^{abc}	22.76 ± 2.22^{b}	85.43 ± 6.49^{c}
	LI	30	33.46 ± 1.30^{ab}	24.19 ± 1.89^{b}	113.50 ± 5.69^{abc}
	HI	1000	26.35 ± 1.40^{abc}	22.97 ± 1.33^{b}	119.62 ± 12.02^{ab}
5 months	MI	70	20.98 ± 1.17^{c}	38.23 ± 3.72^{a}	109.65 ± 3.67^{abc}
	LI	30	22.57 ± 2.63^{bc}	39.59 ± 4.1^{a}	122.68 ± 4.70^{ab}
	HI	1000	36.10 ± 3.78^{a}	28.22 ± 3.09^{ab}	89.83 ± 6.29^{c}
1 year	MI	70	24.18 ± 1.63^{bc}	24.26 ± 1.53^{b}	90.54 ± 2.00^{bc}
	LI	30	36.07 ± 3.17^{a}	28.52 ± 1.29^{ab}	101.39 ± 5.09^{abc}
	HI	1000	34.16 ± 3.39^{ab}	24.13 ± 0.82^{b}	130.87 ± 10.18^{a}

Results are presented as the mean (M) ± standard error of the mean (SEM). Values with different letters in the same column are significantly different according to the Tukey HSD test at $p < 0.05$.

Table 4. Superoxide dismutase (SOD) activity in the larvae, pupae and midguts of honey bees exposed to RF-EMFs at a frequency of 900 MHz and electric field levels of 70 mV m^{-1} (medium electric field—MEF), 30 mV m^{-1} (low intensity—LI) and 1000 mV m^{-1} (high intensity—HI) at 2 weeks, 5 months and 1 year.

Sampling Time	Location	Electric Field Level (mV m^{-1})	SOD (Unit mg^{-1} $_{proteins}$)		
			Larvae (N = 30)	Pupae (N = 20)	Midguts (N = 50)
2 weeks	MI	70	3.97 ± 0.51	3.70 ± 0.63^{ab}	3.07 ± 0.30^{bc}
	LI	30	4.45 ± 0.49	5.87 ± 1.23^{ab}	4.13 ± 0.50^{bc}
	HI	1000	5.29 ± 0.32	4.99 ± 1.03^{ab}	3.63 ± 0.38^{bc}
5 months	MI	70	3.61 ± 0.58	2.73 ± 0.63^{ab}	4.69 ± 0.50^{bc}
	LI	30	5.46 ± 0.37	2.11 ± 0.25^{b}	3.71 ± 0.30^{bc}
	HI	1000	4.64 ± 0.31	3.38 ± 0.50^{ab}	6.52 ± 0.58^{ab}
1 year	MI	70	4.14 ± 0.41	4.85 ± 0.77^{ab}	2.56 ± 0.25^{c}
	LI	30	5.41 ± 0.58	4.13 ± 0.66^{ab}	6.08 ± 1.63^{abc}
	HI	1000	3.75 ± 0.39	7.54 ± 1.75^{ab}	9.34 ± 1.31^{a}

Results are presented as the mean (M) ± standard error of the mean (SEM). Values with different letters in the same column are significantly different according to the Tukey HSD test at $p < 0.05$.

Table 5. Thiobarbituric acid reactive substance (TBARS) concentration in the larvae, pupae and midguts of honey bees exposed to RF-EMFs at a frequency of 900 MHz and electric field levels of 70 mV m^{-1} (medium intensity—MI), 30 mV m^{-1} (low intensity—LI) and 1000 mV m^{-1} (high intensity—HI) at 2 weeks, 5 months and 1 year.

Sampling Time	Location	Electric Field Level (mV m^{-1})	TBARS (Unit mg^{-1} $_{proteins}$)		
			Larvae (N = 30)	Pupae (N = 20)	Midguts (N = 50)
2 weeks	MI	70	0.41 ± 0.07 ^{c}	0.28 ± 0.02	1.78 ± 0.16 ^{b}
	LI	30	0.40 ± 0.07 ^{c}	0.27 ± 0.02	2.28 ± 0.10 ^{ab}
	HI	1000	0.72 ± 0.16 ^{ab}	0.18 ± 0.04	3.15 ± 0.40 ^{a}
5 months	MI	70	0.69 ± 0.11 ^{abc}	0.29 ± 0.02	1.70 ± 0.16 ^{b}
	LI	30	0.88 ± 0.14 ^{abc}	0.22 ± 0.07	1.72 ± 0.15 ^{b}
	HI	1000	1.02 ± 0.15 ^{ab}	0.19 ± 0.04	2.44 ± 0.24 ^{ab}
1 year	MI	70	0.47 ± 0.08 ^{bc}	0.21 ± 0.03	2.50 ± 0.30 ^{ab}
	LI	30	1.09 ± 0.16 ^{a}	0.21 ± 0.05	2.63 ± 0.08 ^{ab}
	HI	1000	1.26 ± 0.12 ^{a}	0.15 ± 0.02	3.00 ± 0.15 ^{a}

Results are presented as the mean (M) ± standard error of the mean (SEM). Values with different letters in the same column are significantly different according to the Tukey HSD test at $p < 0.05$.

The GST activities in all the honey bee samples during exposure to RF-EMFs at a frequency of 900 MHz at locations with different electric field levels (30, 70 and 1000 mV m^{-1}) were not statistically different when comparing the results between the different locations at the same observation periods (2 weeks, 5 months or 1 year) or when comparing the results within the same locations (electric field level) at the different observation periods (between 2 weeks, 5 months or 1 year).

The CAT activity in the larvae was statistically increased ($p < 0.05$) at the HI location (1000 mV m^{-1}) compared to the MI and LI locations with lower electric fields, i.e., 70 mV m^{-1} and 30 mV m^{-1}, at the fifth month of exposure, while it was increased at the LI (30 mV m^{-1}) compared to the MI (70 mV m^{-1}) and HI (1000 mV m^{-1}) locations at 1 year of exposure. The CAT activity in honey bee larvae at the LI location (30 mV m^{-1}) was statistically increased ($p < 0.05$) at 1 year of honey bee colony exposure compared to 5 months of exposure. The CAT activity in the pupae was not statistically different when comparing different locations at the same times of observation. The highest CAT activity in the pupae was measured over 5 months of exposure at locations LI (30 mV m^{-1}) and MI (70 mV m^{-1}), and the values at both locations were significantly higher ($p < 0.05$) when compared with those obtained at 2 weeks of exposure, while at the MI location, it was also higher compared to that at 1 year of exposure. The CAT activity in the midguts of the adult honey bee colony was significantly higher ($p < 0.05$) at the HI (1000 mV m^{-1}) compared to the MI (70 mV m^{-1}) location after the 2-week exposure as well as the 1-year exposure. On the other hand, it decreased at the HI location (1000 mV m^{-1}) compared to the LI (30 mV m^{-1}) at the 5-month exposure. The CAT activity in the guts of the bee colony at the location with the highest electric field (1000 mV m^{-1}) was the lowest at the 5-month exposure compared to the 2-week and 1-year exposures.

The SOD activities in all the honey bee samples (larvae, pupae and midgut) during exposure to RF-EMF at a frequency of 900 MHz and from locations with different electric field levels (30, 70 and 1000 mV m^{-1}) did not significantly differ when the results were compared between the different locations over the same observation period. The exception was the SOD activity in the midguts of the honey bee adults at one year of exposure, which significantly increased ($p < 0.05$) at the HI location, with the highest electric field level (1000 mV m^{-1}), compared to the MI location (70 mV m^{-1}). Regarding different times of observation, the activity of SOD in the midguts of honey bees at the HI location (1000 mV m^{-1}) was significantly lower ($p < 0.05$) at 2 weeks of exposure than at one year of exposure.

In the larvae, the TBARS concentration was significantly enhanced at the HI location (1000 mV m^{-1}) compared to the other two locations, MI (70 mV m^{-1}) and LI (30 mV m^{-1}) after the 2-week exposure, while after the 1-year exposure, the TBARS were also higher at the HI as well as at the LI locations compared to the value obtained at the MI (70 mV m^{-1}). The TBARS in larvae of honey bees from the LI location was significantly higher after 1 year of exposure than after 2 weeks of exposure. In pupae, there were no statistical differences between the results when comparing the different locations and the different exposure times. The TBARS concentration in the midguts was statistically increased at the HI location (1000 mV m^{-1}) compared to the MI (70 mV m^{-1}) after 2 weeks of exposure.

4. Discussion

To date, there are very few published studies on the effects of RF-EMFs on insects under field conditions and long-term exposure. The majority of studies were conducted under laboratory conditions and over short-term exposure. Lazaro et al. [41] showed for the first time that different wild pollinator species differ in their abundance during exposure to RF-EMFs from telecommunication antennas in natural habitats on two different Mediterranean islands (Lesvos and Limnos) in the Northeastern Aegean (Greece), with a range of electric field levels from 10 mV m^{-1} to a maximum value of 670 mV m^{-1}. The same authors concluded that the result obtained could be due to different sensitivities to RF-EMFs, especially in the larval stage. Currently, there is also a lack of data on the induction of oxidative stress in honey bees during exposure to RF-EMFs.

In this study, honey bee colonies were exposed to an electric field level of 30 mV m^{-1} as the lowest value of electric field found in the environment, then 70 mV m^{-1} and 1000 mV m^{-1}, the highest value found in the environment

We have previously reported that RF-EMFs at 900 MHz affect the antioxidant system of honey bee larvae after short-term exposure under laboratory conditions [29,35]. These previous results showed that the effects strongly depend on the physical properties of the radiofrequency electromagnetic radiation, the measured antioxidant parameters and the experimental setup, as well as there being no linear field-response relationship. The results obtained in this study also confirm the similarity with the previous study. This means that the results obtained in this study, i.e., the activity of the antioxidant enzymes and the concentration of lipid peroxidation products, depend on the developmental stage of the honey bees, the environmental electric field levels and the duration of exposure. Indeed, almost all the significant changes in the observed parameters occurred at the highest field level in the larval stage or adults (midguts of adult workers).

The results of numerous studies on the effects of RF-EMFs after short-term exposure indicate that electromagnetic radiation at the frequency of mobile telephony can cause the increased formation of reactive compounds, even at low intensity, thus inducing oxidative stress [42–44] as well as suppression of the immune system and antioxidant defense mechanisms [45].

The overproduction of reactive oxygen species (ROS) following exposure to RF-EMFs is scavenged by SOD, CAT and GST, the most important ROS-scavenging enzymes in honey bees [46–49]. One of the reasons for the observed antioxidant enzyme activity at certain developmental stages could be the physiological developmental profile of the antioxidant enzymes studied and their physiological function in honey bees. Namely, it is known that under physiological conditions, the activities of SOD, CAT and GST in a larva increase slightly from the first to the sixth day, and then decrease until the end of honey bee development, with CAT activity decreasing the most [50]. According to the same research, young worker bees have lower antioxidant enzyme activity than pupae, and CAT is the main antioxidant enzyme responsible for the degradation of hydrogen peroxide in the honey bee brood. Although, in this study, the results of enzyme activity were not compared between the different developmental stages, different levels of the enzymes tested were also found, i.e., the highest GST activity was found in the pupae, followed by the larvae and midguts, regardless of field level and time of sampling. In addition, both the

CAT activity and TBARS concentration were highest in the midguts, and GST activity was lowest, regardless of field level and time of sampling, and they followed in the larvae and pupae. In addition, Weirich et al. [46] studied the activity of antioxidant enzymes (CAT, SOD and GST) in the hemolymph, midgut, pectoral muscle and sperm in spermatheca and found that the activity of the antioxidants studied was differently distributed in the tissues of worker, unfertilized and fertilized queen bees. Indeed, the same authors found that the activity of CAT and GST was significantly higher in the midgut than in the muscle and hemolymph in all members (castes) of the honey bee colony, while the activity of SOD did not show significant differences between the tissues and bee groups studied.

Based on our results, where the levels of TBARS (at two weeks and one-year exposure) and CAT (at 5 months and one-year exposures) were significantly increased in the larvae, we hypothesize that the larvae are more sensitive to RF-EMF exposure than the pupae. The higher TBARS content could be explained by the lipid content of the larvae, which have a significantly higher fat content than the pupae [51] and therefore react more sensitively to oxidative stress. In contrast, in the developmental stage of the pupae, no statistical difference could be detected for any of the indicators examined between different locations with different electric field levels at all three sampling times. One possible reason for this is the fact that the pupa stage is able to maintain an oxidation–reduction equilibrium at this stage of frame development due to the higher physiological activity of defense enzymes, thus overcoming the possible cell damage caused by oxidative stress. The CAT, SOD and GST activity showed no linearity in relation to the field level and the time of sampling. However, if we compare the parameters obtained only in the midguts of the adult bees, we find that the greatest changes were recorded at the location with the highest electric field level (1000 mV m^{-1}). These findings indicate that the honey bees located at a higher field level are exposed to higher oxidative stress than the honey bee colonies located at a lower electric field level, which can also explain the results of our previous work [28], in which the authors demonstrated increased aggressiveness and restlessness of adult bees and an accelerated process of silent queen replacement in the colonies located in a field level of 1000 mV m^{-1}.

5. Conclusions

In conclusion, our results indicated that exposure to RF-EMFs at a frequency of 900 MHz in field conditions may cause oxidative stress in certain developmental stages of honey bees. Most of the significant changes occurred in the second week of the exposure at the location with the highest electric field level, and there was no linear relationship between field level and effect in any of the developmental stages. Among antioxidative enzymes, CAT showed the greatest change and SOD the least change in activity. In addition, the antioxidant profile could serve as a bioindicator of oxidative stress in bees during exposure to RF-EMFs, especially CAT activity and TBARS concentration in the larvae and midguts of the adult honey bees. Since RF-EMF mechanisms are not yet well known, it is difficult to say what consequences such radiation could have on their physiological characteristics. Therefore, our results show the need for further research in the developmental stages of honey bees, including not only oxidative stress parameters but, as far as possible, expression of the gene for the antioxidant enzymes, genotoxic parameters and a greater number of variations of RF-EMF sources in the natural conditions.

Author Contributions: Conceptualization, M.V. and I.T.G.; methodology, M.T., P.T., K.M., S.P. and I.Ž.Ž.; formal analysis, S.P. and N.P.; investigation, M.V., M.T., I.T.G. and P.T.; resources, P.T. and N.P.; writing—original draft preparation, M.V., S.P., N.Ž. and I.Ž.Ž.; writing—review and editing, I.T.G., K.M., S.P., M.T. and N.Ž. All authors have read and agreed to the published version of the manuscript.

Funding: This research received no external funding.

Institutional Review Board Statement: The study was reviewed and approved by the Ethics Committee of the Faculty of Veterinary Medicine, University of Zagreb (Class: 640-01/13-17/76; File number: 251/61-01/139-13-9).

Informed Consent Statement: Informed consent was obtained from the owner of the honey bee colonies.

Data Availability Statement: The data presented in this study are available on reasonable request from the corresponding author.

Conflicts of Interest: The authors declare no conflicts of interest.

References

1. ITU. Measuring Digital Development: Facts and Figures. 2022. Available online: https://www.itu.int/hub/publication/d-ind-ict_mdd-2022/ (accessed on 15 February 2024).
2. Bandara, P.; Carpenter, D.O. Planetary Electromagnetic Pollution: It Is Time to Assess Its Impact. *Lancet Planet. Health* **2018**, *2*, e512–e514. [CrossRef] [PubMed]
3. Alkis, M.E.; Bilgin, H.M.; Akpolat, V.; Dasdag, S.; Yegin, K.; Yavas, M.C.; Akdag, M.Z. Effect of 900-, 1800-, and 2100-MHz Radiofrequency Radiation on DNA and Oxidative Stress in Brain. *Electromagn. Biol. Med.* **2019**, *38*, 32–47. [CrossRef] [PubMed]
4. Schuermann, D.; Mevissen, M. Manmade Electromagnetic Fields and Oxidative Stress-Biological Effects and Consequences for Health. *Int. J. Mol. Sci.* **2021**, *22*, 3772. [CrossRef]
5. Mina, D.; Sagonas, K.; Fragopoulou, A.F.; Pafilis, P.; Skouroliakou, A.; Margaritis, L.H.; Tsitsilonis, O.E.; Valakos, E.D. Immune Responses of a Wall Lizard to Whole-Body Exposure to Radiofrequency Electromagnetic Radiation. *Int. J. Radiat. Biol.* **2016**, *92*, 162–168. [CrossRef] [PubMed]
6. Yadav, H.; Sharma, R.S.; Singh, R. Immunotoxicity of Radiofrequency Radiation. *Environ. Pollut.* **2022**, *309*, 119793. [CrossRef] [PubMed]
7. Smith-Roe, S.L.; Wyde, M.E.; Stout, M.D.; Winters, J.W.; Hobbs, C.A.; Shepard, K.G.; Green, A.S.; Kissling, G.E.; Shockley, K.R.; Tice, R.R.; et al. Evaluation of the Genotoxicity of Cell Phone Radiofrequency Radiation in Male and Female Rats and Mice Following Subchronic Exposure. *Environ. Mol. Mutag.* **2020**, *61*, 276–290. [CrossRef] [PubMed]
8. Ruediger, H.W. Genotoxic Effects of Radiofrequency Electromagnetic Fields. *Pathophysiology* **2009**, *16*, 89–102. [CrossRef] [PubMed]
9. Verschaeve, L.; Juutilainen, J.; Lagroye, I.; Miyakoshi, J.; Saunders, R.; de Seze, R.; Tenforde, T.; van Rongen, E.; Veyret, B.; Xu, Z. In Vitro and in Vivo Genotoxicity of Radiofrequency Fields. *Mut. Res./Rev. Mut. Res.* **2010**, *705*, 252–268. [CrossRef] [PubMed]
10. Singh, R.; Nath, R.; Mathur, A.K.; Sharma, R.S. Effect of Radiofrequency Radiation on Reproductive Health. *Indian J. Med. Res.* **2018**, *148*, S92–S99. [CrossRef]
11. Yadav, H.; Rai, U.; Singh, R. Radiofrequency Radiation: A Possible Threat to Male Fertility. *Reprod. Toxicol.* **2021**, *100*, 90–100. [CrossRef]
12. Jaffar, F.H.F.; Osman, K.; Ismail, N.H.; Chin, K.-Y.; Ibrahim, S.F. Adverse Effects of Wi-Fi Radiation on Male Reproductive System: A Systematic Review. *Tohoku J. Exp. Med.* **2019**, *248*, 169–179. [CrossRef] [PubMed]
13. Cucurachi, S.; Tamis, W.L.M.; Vijver, M.G.; Peijnenburg, W.J.G.M.; Bolte, J.F.B.; de Snoo, G.R. A Review of the Ecological Effects of Radiofrequency Electromagnetic Fields (RF-EMF). *Environ. Int.* **2013**, *51*, 116–140. [CrossRef] [PubMed]
14. Balmori, A. Electromagnetic Radiation as an Emerging Driver Factor for the Decline of Insects. *Sci. Total Environ.* **2021**, *767*, 144913. [CrossRef] [PubMed]
15. Thill, A.; Cammaerts, M.-C.; Balmori, A. Biological Effects of Electromagnetic Fields on Insects: A Systematic Review and Meta-Analysis. *Rev. Environ. Health* **2023**. [CrossRef] [PubMed]
16. Fontaine, C.; Dajoz, I.; Meriguet, J.; Loreau, M. Functional Diversity of Plant–Pollinator Interaction Webs Enhances the Persistence of Plant Communities. *PLoS Biol.* **2006**, *4*, e1. [CrossRef] [PubMed]
17. Klein, A.-M.; Vaissière, B.E.; Cane, J.H.; Steffan-Dewenter, I.; Cunningham, S.A.; Kremen, C.; Tscharntke, T. Importance of Pollinators in Changing Landscapes for World Crops. *Proc. Biol. Sci.* **2007**, *274*, 303–313. [CrossRef]
18. Rollin, O.; Benelli, G.; Benvenuti, S.; Decourtye, A.; Wratten, S.D.; Canale, A.; Desneux, N. Weed-Insect Pollinator Networks as Bio-Indicators of Ecological Sustainability in Agriculture. A Review. *Agron. Sustain. Dev.* **2016**, *36*, 8. [CrossRef]
19. Favre, D. Mobile Phone-Induced Honeybee Worker Piping. *Apidologie* **2011**, *42*, 270–279. [CrossRef]
20. Lopatina, N.G.; Zachepilo, T.G.; Kamyshev, N.G.; Dyuzhikova, N.A.; Serov, I.N. Effect of Non-Ionizing Electromagnetic Radiation on Behavior of the Honeybee, *Apis mellifera* L. (Hymenoptera, Apidae). *Entmol. Rev.* **2019**, *99*, 24–29. [CrossRef]
21. Odemer, R.; Odemer, F. Effects of Radiofrequency Electromagnetic Radiation (RF-EMF) on Honey Bee Queen Development and Mating Success. *Sci. Total Environ.* **2019**, *661*, 553–562. [CrossRef]
22. Sharma, V.P.; Kumar, N.R. Changes in Honeybee Behaviour and Biology under the Influence of Cellphone Radiations. *Curr. Sci.* **2010**, *98*, 1376–1378.
23. Harst, W.; Kuhn, J.; Stever, H. Can Electromagnetic Exposure Cause a Change in Behaviour? Studying Possible Non-Thermal Influences on Honey Bees—An Approach within the Framework of Educational Informatics. *Acta Syst.-IIAS Int. J.* **2006**, *6*, 1–6. Available online: https://api.semanticscholar.org/CorpusID:204874037 (accessed on 5 April 2024).
24. Favre, D. Disturbing Honeybees' Behavior with Electromagnetic Waves: A Methodology. *J. Behav.* **2017**, *2*, 1010.

25. Favre, D.; Johansson, O. Does enhanced electromagnetic radiation disturb honeybees' behaviour? observations during new year's eve 2019. *Int. J. Res.-Granthaalayah* **2020**, *8*, 7–14. [CrossRef]
26. Taye, R.R.; Deka, M.K.; Rahman, A.; Bathari, M. Effect of Electromagnetic Radiation of Cell Phone Tower on Foraging Behaviour of Asiatic Honey Bee, *Apis cerana* F. (Hymenoptera: Apidae). *J. Entomol. Zool. Stud.* **2017**, *5*, 1527–1529.
27. Treder, M.; Müller, M.; Fellner, L.; Traynor, K.; Rosenkranz, P. Defined Exposure of Honey Bee Colonies to Simulated Radiofrequency Electromagnetic Fields (RF-EMF): Negative Effects on the Homing Ability, but Not on Brood Development or Longevity. *Sci. Total Environ.* **2023**, *896*, 165211. [CrossRef]
28. Tlak Gajger, I.; Vilić, M.; Tucak, P.; Malarić, K. Effect of Electromagnetic Field on Some Behaviour Modality of Honeybee Colonies (*Apis mellifera*) in Field Conditions. *J. Anim. Vet. Adv.* **2001**, *18*, 61–64. [CrossRef]
29. Vilić, M.; Tlak Gajger, I.; Tucak, P.; Štambuk, A.; Šrut, M.; Klobučar, G.; Malarić, K.; Žura Žaja, I.; Pavelić, A.; Manger, M.; et al. Effects of Short-Term Exposure to Mobile Phone Radiofrequency (900 MHz) on the Oxidative Response and Genotoxicity in Honey Bee Larvae. *J. Apic. Res.* **2017**, *56*, 430–438. [CrossRef]
30. Vanbergen, A.J.; Potts, S.G.; Vian, A.; Malkemper, E.P.; Young, J.; Tscheulin, T. Risk to Pollinators from Anthropogenic Electro-Magnetic Radiation (EMR): Evidence and Knowledge Gaps. *Sci. Total Environ.* **2019**, *695*, 133833. [CrossRef]
31. Panagopoulos, D.J.; Johansson, O.; Carlo, G.L. Real versus Simulated Mobile Phone Exposures in Experimental Studies. *Biomed. Res. Int.* **2015**, *2015*, 607053. [CrossRef]
32. Luukkonen, J.; Hakulinen, P.; Mäki-Paakkanen, J.; Juutilainen, J.; Naarala, J. Enhancement of Chemically Induced Reactive Oxygen Species Production and DNA Damage in Human SH-SY5Y Neuroblastoma Cells by 872 MHz Radiofrequency Radiation. *Mutat. Res. /Fundam. Mol. Mech. Mutagen.* **2009**, *662*, 54–58. [CrossRef] [PubMed]
33. Kwee, S.; Raskmark, P. Changes in Cell Proliferation Due to Environmental Non-Ionizing Radiation: 2. Microwave Radiation. *Bioelectrochem. Bioenerg.* **1998**, *44*, 251–255. [CrossRef]
34. Sarimov, R.; Malmgren, L.O.G.; Markova, E.; Persson, B.R.R.; Belyaev, I.Y. Nonthermal GSM Microwaves Affect Chromatin Conformation in Human Lymphocytes Similar to Heat Shock. *IEEE Trans. Plasma Sci.* **2004**, *32*, 1600–1608. [CrossRef]
35. Vilić, M.; Žura Žaja, I.; Tkalec, M.; Štambuk, A.; Šrut, M.; Klobučar, G.; Malarić, K.; Tucak, P.; Pašić, S.; Tlak Gajger, I. Effects of a Radio Frequency Electromagnetic Field on Honey Bee Larvae (*Apis mellifera*) Differ in Relation to the Experimental Study Design. *Vet. Arhiv* **2021**, *91*, 427–435. [CrossRef]
36. Bocchetti, R.; Regoli, F. Seasonal Variability of Oxidative Biomarkers, Lysosomal Parameters, Metallothioneins and Peroxisomal Enzymes in the Mediterranean Mussel Mytilus Galloprovincialis from Adriatic Sea. *Chemosphere* **2006**, *65*, 913–921. [CrossRef] [PubMed]
37. Aebi, H. Catalase In Vitro. *Methods Enzymol.* **1984**, *105*, 121–126. [CrossRef]
38. McCord, J.M.; Fridovich, I. Superoxide Dismutase: An enzymic function for erythrocuprein (hemocuprein). *J. Biol. Chem.* **1969**, *244*, 6049–6055. [CrossRef]
39. Legeay, A.; Achard-Joris, M.; Baudrimont, M.; Massabuau, J.-C.; Bourdineaud, J.-P. Impact of Cadmium Contamination and Oxygenation Levels on Biochemical Responses in the Asiatic Clam *Corbicula fluminea*. *Aquat. Toxicol.* **2005**, *74*, 242–253. [CrossRef] [PubMed]
40. Bradford, M.M. A Rapid and Sensitive Method for the Quantitation of Microgram Quantities of Protein Utilizing the Principle of Protein-Dye Binding. *Anal. Biochem.* **1976**, *72*, 248–254. [CrossRef]
41. Lázaro, A.; Chroni, A.; Tscheulin, T.; Devalez, J.; Matsoukas, C.; Petanidou, T. Electromagnetic Radiation of Mobile Telecommunication Antennas Affects the Abundance and Composition of Wild Pollinators. *J. Insect Conserv.* **2016**, *20*, 315–324. [CrossRef]
42. Balci, M.; Devrim, E.; Durak, I. Effects of Mobile Phones on Oxidant/Antioxidant Balance in Cornea and Lens of Rats. *Curr. Eye Res.* **2007**, *32*, 21–25. [CrossRef]
43. Burlaka, A.; Tsybulin, O.; Sidorik, E.; Lukin, S.; Polishuk, V.; Tsehmistrenko, S.; Yakymenko, I. Overproduction of Free Radical Species in Embryonal Cells Exposed to Low Intensity Radiofrequency Radiation. *Exp. Oncol.* **2013**, *35*, 219–225. [PubMed]
44. Kıvrak, E.G.; Yurt, K.K.; Kaplan, A.A.; Alkan, I.; Altun, G. Effects of Electromagnetic Fields Exposure on the Antioxidant Defense System. *J. Microsc. Ultrastruct.* **2017**, *5*, 167–176. [CrossRef] [PubMed]
45. Szymański, Ł.; Sobiczewska, E.; Cios, A.; Szymanski, P.; Ciepielak, M.; Stankiewicz, W. Immunotropic Effects in Cultured Human Blood Mononuclear Cells Exposed to a 900 MHz Pulse-Modulated Microwave Field. *J. Radiat. Res.* **2020**, *61*, 27–33. [CrossRef] [PubMed]
46. Weirich, G.F.; Collins, A.M.; Williams, V.P. Antioxidant Enzymes in the Honey Bee, *Apis mellifera*. *Apidologie* **2002**, *33*, 3–14. [CrossRef]
47. Tawfik, A.I.; Ahmed, Z.H.; Abdel-Rahman, M.F.; Moustafa, A.M. Effect of Some Bee Bread Quality on Protein Content and Antioxidant System of Honeybee Workers. *Int. J. Trop. Insect. Sci.* **2023**, *43*, 93–105. [CrossRef]
48. Tawfik, A.I.; Ahmed, Z.H.; Abdel-Rahman, M.F.; Moustafa, A.M. Influence of Winter Feeding on Colony Development and the Antioxidant System of the Honey Bee, *Apis mellifera*. *J. Apic. Res.* **2020**, *59*, 752–763. [CrossRef]
49. Li, Z.; Hou, M.; Qiu, Y.; Zhao, B.; Nie, H.; Su, S. Changes in Antioxidant Enzymes Activity and Metabolomic Profiles in the Guts of Honey Bee (*Apis mellifera*) Larvae Infected with Ascosphaera Apis. *Insects* **2020**, *11*, 419. [CrossRef] [PubMed]

50. Farjan, M.; Dmitryjuk, M.; Lipiński, Z.; Biernat-Łopieńska, E.; Żółtowska, K. Supplementation of the Honey Bee Diet with Vitamin C: The Effect on the Antioxidative System of *Apis mellifera* Carnica Brood at Different Stages. *J. Apic. Res.* **2012**, *51*, 263–270. [CrossRef]
51. Haber, M.; Mishyna, M.; Itzhak Martinez, J.J.; Benjamin, O. Edible Larvae and Pupae of Honey Bee (*Apis mellifera*): Odor and Nutritional Characterization as a Function of Diet. *Food Chem.* **2019**, *292*, 197–203. [CrossRef]

insects

MDPI

Article

High Rates of Honey Bee Colony Losses and Regional Variability in Ethiopia Based on the Standardised COLOSS 2023 Survey

Teweldemedhn Gebretinsae Hailu [1,2], Alem Tadesse Atsbeha [2], Kibebew Wakjira [3] and Alison Gray [4,*]

1 Department of Livestock Population Genomics, Institute of Animal Science, University of Hohenheim, Garbenstr 17, 70599 Stuttgart, Germany; teweldemedhng.hailu@uni-hohenheim.de
2 Department of Animal Sciences, Aksum University, Shire P.O. Box 314, Ethiopia; lemtt2008@gmail.com
3 Holeta Bee Research Centre, Holeta P.O. Box 22, Ethiopia; wkibebew@gmail.com
4 Department of Mathematics and Statistics, University of Strathclyde, 26 Richmond Street, Glasgow G1 1XH, UK
* Correspondence: a.j.gray@strath.ac.uk

Simple Summary: In addition to invaluable ecosystem services, beekeeping offers opportunities for job creation, income generation and food security. Beekeepers have been experiencing economic losses due to high rates of honey bee colony losses driven by various factors worldwide. We conducted this study using the COLOSS monitoring survey tools for the first time in Ethiopia to assess honey bee colony loss rates, annual colony development, beekeeping practices, and to determine the role of management practices (varroa monitoring, varroa treatment, colony splitting, feed supplementation, use of natural comb, and purchase of beeswax from external sources), as well as region, on colony losses. For this, data were collected by interviewing beekeepers from two major beekeeping regions in the country—Oromia and Tigray. Our results showed a high rate of colony losses in Ethiopia, which significantly varied between the regions. The main drivers of honey bee colony losses are related to natural disasters (particularly war), beekeeping husbandry practices, and pest management. Therefore, it is important to promote the capacity of smallholder beekeepers to implement improved beekeeping practices such as feed supplementation, queen replacement and pest management that would lead to reduced losses, increased profitability, and improved food security and livelihood.

Abstract: The COLOSS research association has been assessing honey bee colony losses, associated risk factors and management, focusing on Western countries but with a progressive international expansion. Here, we report the first survey on the loss rates of colonies in 2022/2023 in Ethiopia using COLOSS monitoring survey tools. A face-to-face interview questionnaire survey was conducted on 64 beekeepers selected from Oromia and Tigray regions. This covered 1713 honey bee colonies distributed in 68 apiaries. The percentages of colonies lost were significantly different between Oromia (24.1%) and Tigray (66.4%) regions. Colony losses were attributed as unsolvable queen problems (8% in Oromia; 10% in Tigray), natural disaster (32%; 82%), and empty hives or dead colonies (60%; 8%). The loss rate was significantly affected by queen replacement ($p < 0.0001$), use of natural comb ($p < 0.0001$), feed supplementation ($p < 0.0001$), region ($p < 0.0001$), varroa treatment ($p < 0.0001$), colony splitting ($p < 0.01$), and merging ($p < 0.01$). Beekeepers in Oromia managed more colonies and implemented improved practices compared to those in Tigray. However, all beekeepers in Oromia detected at least some bees with signs of deformed wing virus, compared to 76% of beekeepers in Tigray. In conclusion, the colony loss rate was significantly different between Oromia and Tigray regions due to differences in natural disasters, management, environment and health factors.

Keywords: colony loss; varroa; beekeeping; natural disaster; honey bee

Citation: Hailu, T.G.; Atsbeha, A.T.; Wakjira, K.; Gray, A. High Rates of Honey Bee Colony Losses and Regional Variability in Ethiopia Based on the Standardised COLOSS 2023 Survey. *Insects* **2024**, *15*, 376. https://doi.org/10.3390/insects15060376

Academic Editor: Silvio Erler

Received: 25 April 2024
Revised: 17 May 2024
Accepted: 20 May 2024
Published: 22 May 2024

1. Introduction

Following reports on high rates of colony losses and the recognition of colony collapse disorder [1,2], assessing colony losses, their driving factors and management strategies have been among major research agendas worldwide since the last decade. International survey reports from COLOSS showed overall annual loss rates of 16% to 20.9%, which greatly varied from 2% to about 36.5% between countries [3–6]. A recent study from the Bee Informed Partnership in the United States indicates up to 50.8% average loss rate of colonies [7]. The ectoparasite *Varroa destructor* ("varroa" below) is one of the major causes of colony losses [8–11]. However, some honey bee populations have been identified as tolerant against this pest and able to survive even without anti-varroa treatment [12–15]. Since the detection of varroa for the first time in northern Ethiopia's Tigray region [16], several studies have been conducted on its distribution and prevalence in the country [17–19]. A national survey conducted in Ethiopia, which covered all regions except Tigray, showed a wide distribution and up to 95.8% prevalence in the country with a significant variation between regions [18]. The prevalence of varroa in Ethiopia was reported to be influenced by several factors including agroecology and management systems [19].

Other factors that cause honey bee colony losses include pesticides, natural disasters and climatic factors globally [4,20,21]. Recently, it has been discussed that climate change could significantly affect honey bees and beekeeping by causing extreme weather, flooding, wildfire, increased pest infestation and reduced forage availability [22,23]. Climate change affects vegetation distribution and the annual flowering calendar of plants, which varies between habitat elevations [24]. This can directly affect survival and performance of honey bees, as they depend on flowering plants. African honey bees are known for strong migratory behaviour, following forage seasonality [25]. In northern Ethiopia, a high rate of annual fluctuation in the numbers of managed honey bee colonies has recently been reported. Colony selling and losses (including absconding, death and seasonal migration) were identified as the reasons for colony outflow in the areas, while purchase, trapping, splitting and swarming are the mechanisms of inflow [26]. Honey bee colonies are also valuable trading commodities in northern Ethiopia where beekeepers exchange colonies in central markets [27]. A high rate of mobility of honey bee colonies under the influence of anthropogenic activities such as colony marketing, as well as natural circumstances (such as migratory behaviour, habitat fragmentation or rehabilitation) could exacerbate the distribution of varroa and other pests and pathogens [28]. Honey bee colony transportation across agroecological zones is a common practice in northern Ethiopia's Tigray region [27,29].

Tigray region is known as one of the major beekeeping centres in Ethiopia, where there has been a pronounced transformation of beekeeping. Data from the Central Statistical Agency of Ethiopia [30] showed the management of more than 340,000 honey bee colonies in the region. Over the last two decades, the percentage of frame hives has grown from 1% to 23% and the number of managed colonies has increased by 90% [31]. However, in recent years, Ethiopia has been facing difficulties due to war, COVID-19, climate change and outbreaks of *Desert locust* [32–34]. In particular, the two-year war from November 2020 to November 2022 in Tigray region was reported to have wiped out decades of progress in all sectors [35–38]. Considering beekeeping specifically, 70% of the managed honey bee colonies were lost due to the war [39]. This loss may threaten both the livelihoods of the beekeepers and the overall ecosystem.

A survey conducted during the pre-war period in northern Ethiopia showed that there was no difference in colony loss rates based on the beekeeping system, beekeeping experience, number of colonies per household and local areas within Tigray [26]. This differs from international surveys, which have reported that loss rates are significantly affected by the scale of beekeeping operations, geographic location and climate, although Ethiopia was not included in these studies [3–6,20,21]. Therefore, this study was conducted to assess the loss rates in different regions using the COLOSS survey protocol for the first time. The hypothesis is that honey bee colony loss rate significantly varies between regions depending on the vulnerability to natural disasters, infestation and prevalence of varroa, and beekeeping practices, as reported by the international surveys. The alternative hypothesis would be that there is no difference in honey bee colony loss rates between different regions in Ethiopia, as previously found for different local areas within Tigray. The specific objectives of this survey were to assess honey bee colony loss rates in the Oromia and Tigray regions of Ethiopia and to evaluate the degree of association between honey bee colony losses with colony management practices (varroa monitoring, varroa treatment, colony splitting, feed supplementation, operation size, use of natural comb, and purchase of beeswax from external sources). The results are discussed with reference to regional and global figures for colony losses, based on the literature.

2. Materials and Methods

2.1. Study Regions and Description

This survey was conducted in Ethiopia, namely Oromia and Tigray Regional States (Figure 1). Tigray is predominantly a semi-arid region located in northern Ethiopia, which is known as a main beekeeping area in the country, and for its famous white honey [40], unique colony marketing system [29] and a rapid beekeeping transformation in the last two decades. In this region, *Varroa destructor* was detected in 2010 for the first time in Ethiopia [16]. From November 2020 to November 2022, the region was heavily devastated by war that has caused significant damage also to the beekeeping sector [39]. On the other hand, Oromia is the largest region in Ethiopia, connects with almost all regions and is largely located in the central, southern and western parts of the country. Likewise, Oromia hosts the largest number of honey bee colonies in the country [30]. Our study areas in Oromia are mainly located in the west and southwest, which are largely humid.

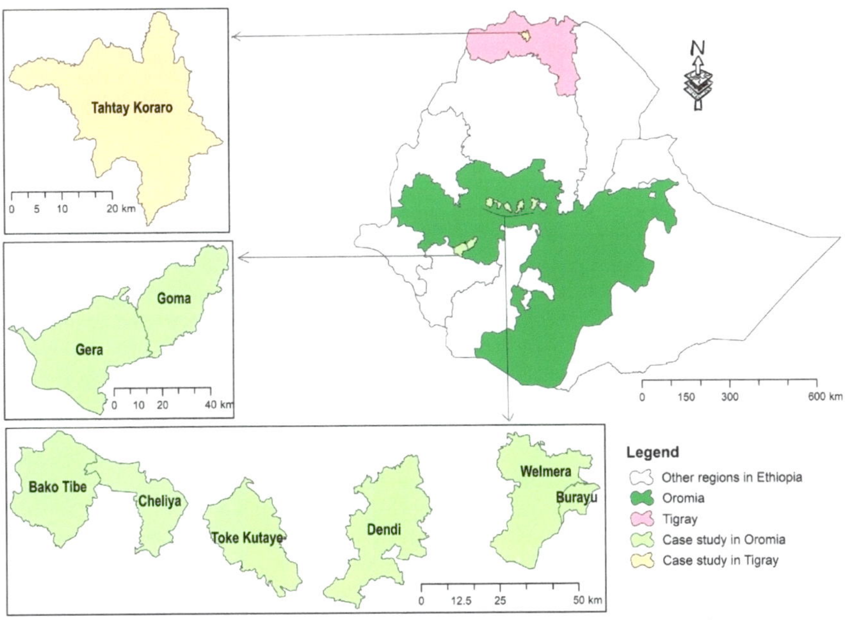

Figure 1. Location map of surveyed districts in Oromia and Tigray Regional States of Ethiopia in 2023.

Therefore, the survey was conducted in these two important beekeeping regions characterised by contrasting features of beekeeping, climate, vegetation, etc.

2.2. Survey and Data Collection

For this study, 64 beekeepers who managed a total of 1713 honey bee colonies during the survey year 2022/2023 were purposefully selected based on accessibility. The colonies were distributed over 68 apiaries and 14 villages (locally called *Kebelles*) of the two regions (Table 1). Data collection was conducted by interviewing the beekeepers face-to-face using the standardised questionnaire of the COLOSS Colony Losses Monitoring Core Project. Therefore, the results are comparable to those from other countries using the same questionnaire, in particular the loss rates. Interviews were conducted by researchers who orally translated the questionnaires from English to local languages directly when asking the beekeepers. As is established practice in Ethiopia, the aim of the survey was explained verbally to the selected households/beekeepers, who then responded to the questions voluntarily. This data collection was performed during June 2023.

Table 1. Sample regions, local areas (*Kebelles*), and numbers of apiaries, beekeepers and colonies surveyed in Ethiopia using the COLOSS questionnaire tool, version 2023.

Description	Regional State		Total
	Oromia	Tigray	
No. of Kebelles	10	4	14
No. of Beekeepers	31	33	64
No. of Apiaries	35	33	68
No. of Honey bee colonies	1305	408	1713
Average no. of Colonies per apiary	42.10	12.36	26.77

2.3. Statistical Analysis

Variables analysed and included in this paper are comprised of: (1) number of honey bee colonies managed (as a measure of beekeeping scale of operation), honey bee colony losses and loss rates; (2) beekeeping practices such as colony splitting, feed supplementation, queen replacement, and colony merging; (3) honey bee health, monitoring and pest management; (4) annual honey bee colony cycle and management calendar. These data were analysed and comparisons were made between the Oromia and Tigray regions. As the number of colonies before winter and the number of colonies in spring 2022 and spring 2023 have very skewed distributions, medians are reported as well as means, and Mann–Whitney tests are used to compare the numbers in the two regions. Fisher's exact test was used to compare the proportions of beekeepers giving each response to the management questions between regions. The data for several variables included responses in the categories labelled as "Don't know" or "Not applicable". These categories of responses were not removed for the purposes of analysis, but retained. For two variables (monitoring varroa and observing bees with deformed wings), the numbers of responses in this category were very low (Table 2) compared to other responses and were similar for the two regions, so would not affect the conclusions. For two other variables, concerning treating varroa and presence of *Vespa velutina*, there were more responses of "Not applicable" and "Don't know", respectively, for one region and very few for the other; this is discussed further in Results.

Colony loss rates are reported as the overall proportion of colonies lost, with a 95% confidence interval, and also as the average rate of loss per beekeeper. Colony loss rates per beekeeper can be assumed to be normally distributed for Oromia (Anderson–Darling test, $p > 0.05$), but not for Tigray ($p < 0.05$). Therefore, colony loss rates per beekeeper were compared between the regions using the Mann–Whitney test.

Furthermore, correlation analyses were performed to assess the degree of associations between the different components of colony losses (queen problems, natural disaster, dead colonies, empty hives) and the numbers of colonies managed at different seasons of the survey year.

Table 2. Beekeeping and colony management in Oromia and Tigray regions, including tests of a difference between regions.

Variables	Oromia	Tigray	Test of Difference	
			Method	*p*-Value
Colonies per beekeeper in spring 2022				
Mean	27.00	9.03		
Median	17.00	6.00	Mann–Whitney test	$p < 0.001$
Range	7–125	2–47		
Colonies per beekeeper before winter 2023				
Mean	42.10	12.36		
Median	27.00	8.00	Mann–Whitney test	$p < 0.001$
Range	10–200	2–60		
Colonies per beekeeper in spring 2023				
Mean	30.58	5.24		
Median	16.00	3.00	Mann–Whitney test	$p < 0.001$
Range	5–160	0–37		
Monitoring *Varroa destructor* (frequency)				
Yes	5	21		
No	26	10	Fisher's test	$p < 0.0001$
Not applicable	0	2		
Treating against *V. destructor* (frequency)				
Yes	5	0		
No	3	32	Fisher's test	$p < 0.0001$
Not applicable	23	1		
Bees with deformed wings (frequency)				
None	0	7		
Some	31	15	Fisher's test	$p < 0.0001$
Many	0	10		
Don't know	0	1		
Use of natural comb (frequency)				
Yes	31	14	Fisher's test	$p < 0.0001$
No	0	19		
Purchase of external wax (frequency)				
Yes	18	30	Fisher's test	$p < 0.01$
No	13	3		
Presence of *Vespa velutina* (frequency)				
Yes	8	29		
No	13	4	Fisher's test	$p < 0.0001$
Don't know	10	0		
Colony merging (frequency)				
Yes	21	6	Fisher's test	$p < 0.001$
No	10	27		
Colony splitting (frequency)				
Yes	26	5	Fisher's test	$p < 0.0001$
No	5	28		
Feed supplementation (frequency)				
Yes	31	17	Fisher's test	$p < 0.0001$
No	0	16		
Queen replacement (frequency)				
Yes	19	0	Fisher's test	$p < 0.0001$
No	12	33		
Proportional Colony loss rate (%)	24.1	66.4	Chi-squared test	$p < 0.0001$

Intercept-only quasibinomial generalised linear models (GLMs) were used to obtain confidence intervals (CI) for the proportional loss rates, and univariate and multivariate quasibinomial GLMs were used to assess the significance of risk factors for colony loss for both regions combined and also separately. In particular, an F-test from a univariate quasibinomial GLM, or a Chi-squared test of proportions, allowed comparison of the proportional loss rates between regions. The R software version 3.6.3 [41] and JMP Pro 17 [42] were used for the data analyses.

3. Results

3.1. Colony Management in Oromia and Tigray Regions

Table 2 shows a summary of descriptive results for the two studied regions and hypothesis tests comparing the regions. Beekeepers included in this study managed an average number of colonies of 27.00 and 9.03 in the Oromia and Tigray regions, respectively, in spring 2022. These increased to about 42.10 and 12.36 colonies on average before winter 2023 and again declined to 30.58 and 5.24 in spring 2023, respectively. Correlation analyses indicated that the number of colonies in spring 2022 and spring 2023 were strongly associated (Spearman correlation: $r = 0.92$, $p < 0.0001$). Mann–Whitney tests showed that there were significantly higher numbers of honey bee colonies in the Oromia region than in Tigray region during both 2022 and 2023 years ($p < 0.001$). A significantly higher proportion of beekeepers in Tigray monitored the status of varroa in their colonies (Figure 2). On the other hand, none of the beekeepers in Tigray applied any kind of colony treatment against varroa, whereas some beekeepers (16.1%) in Oromia applied the biotechnical method, trapping varroa mites mainly in drone brood combs and discarding them to reduce infestation without chemical treatment. While the percentage treating in Oromia is low, most of the beekeepers there (74.2%) responded that treating was not applicable.

Figure 2. Varroa monitoring (**A**, left) and treatment (**B**, right) in Oromia and Tigray regions of Ethiopia during 2022/2023 Colony Losses Monitoring Core Project survey year of COLOSS. The columns for each region in plots A and B are split between the response categories "no" (shown in mid-blue), "not applicable" (light blue) and "yes" (green) of the factor shown on the vertical axis; the relative split of the columns shows the proportions of the respondents in each category and allows comparison of the responses between the regions.

Moreover, a significantly higher number of beekeepers in Oromia use natural combs ($p < 0.0001$) and are less dependent on externally purchased beeswax ($p < 0.01$) compared to the situation in Tigray. Furthermore, there were significantly more frequent applications of improved husbandry practices such as merging of weak colonies ($p < 0.001$), queen replacement ($p < 0.0001$), feed supplementation ($p < 0.0001$) and colony splitting ($p < 0.0001$) in Oromia. However, all beekeepers in Oromia reported the detection of some bees with deformed wings, compared to 45.5% reporting some bees with deformed wings and 30.3% reporting many bees with deformed wings in Tigray. These percentages differ significantly between the regions ($p < 0.0001$). In contrast, most beekeepers in Tigray (87.9%) observed the presence of *Vespa velutina* in their apiaries, which differs from the situation in Oromia (25.8%), although in Oromia 32.3% responded "Don't know" to whether or not *V. velutina* were present (Table 2). It should be noted that these data are based on beekeepers' description and require follow-up, particularly as *V. velutina* was not previously reported from Africa.

Furthermore, correlation analysis showed that the different components of colony losses (number of colonies lost due to queen problems, natural disaster, dead colonies or empty hives) are positively related to each of the numbers of colonies managed at different seasons of the survey year (Table 3).

Table 3. Spearman correlations between numbers of colonies managed at different times, loss rates and loss components in Ethiopia, showing the degree of association between each pair of variables. Significant associations ($p < 0.05$) are marked with an asterisk (*).

	#Colonies Before Winter	#Colonies Lost Due to Queen Problems	#Colonies Lost Due to Natural Disaster	#Empty Hives or Dead Colonies	#Colonies in Spring 2022
#Colonies lost due to queen problems	0.14				
#Colonies lost due to natural disaster	0.28 *	0.41 *			
#Empty hives or dead colonies	0.68 *	0.13	0.06		
#Colonies in spring 2022	0.92 *	0.04	0.28 *	0.52 *	
#Colonies in spring 2023	0.93 *	0.01	0.11	0.58 *	0.92 *

Note: The symbol # in the table represents the words "number of".

3.2. Colony Loss Rates, Components of Loss and Risk Factors

The average annual colony loss rate in 2022/2023 in Tigray (68.3%) was far higher than that of Oromia (25.9%), and the levels of loss per beekeeper are significantly different between these two regions in northern and central Ethiopia (Mann–Whitney test, $p < 0.0001$). The loss rate varied from 0% to a maximum value of 56.7% in Oromia, whereas in Tigray it ranged from 0% to 100% (Figure 3). Among the colonies recorded as lost in Oromia were losses due to unsolvable queen problems (8% of lost colonies), natural disaster (32%), and those found as empty hives or dead colonies (60%). The same factors but with contrasting rates of loss were reported from Tigray as queen problems (10% of lost colonies), natural disaster (82%), and empty hives or dead colonies (8%). These differences could relate to the different factors influencing colony losses in the two regions of Ethiopia as presented in the following paragraphs. Furthermore, colony absconding during the survey year was significantly higher (Fisher test, $p < 0.01$) in Tigray (97.0% reported at least one absconding event in their apiary) compared to Oromia (67.7%).

The overall proportional loss rate for all beekeepers was 34.2% across both regions, which is calculated as the total number of colonies lost by all the beekeepers divided by the total number of colonies managed by all the beekeepers and expressed as a percentage (95% confidence interval (28.5%, 40.3%)). The overall proportions lost per region were 24.1% (95% CI 19.8%, 28.9%) for Oromia and 66.4% (95% CI 58.3%, 73.7%) for Tigray. These proportional loss rates were significantly different between the two regions, $p < 0.0001$ (Chi-squared test of proportions or F-test in a univariate quasibinomial model for the risk of colony loss).

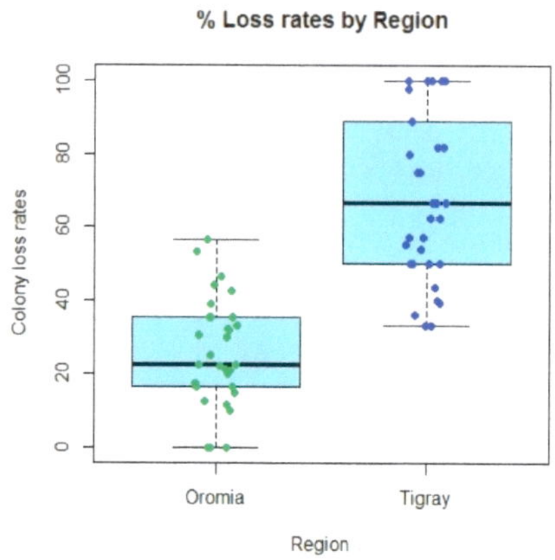

Figure 3. Colony loss rates (%) per beekeeper in Oromia and Tigray regions of Ethiopia, showing a markedly higher level of loss of colonies in Tigray compared to Oromia (Mann–Whitney test, $p < 0.0001$) based on the 2023 survey following the 2020–2022 war in northern Ethiopia. The individual loss rates are shown superimposed as points on the boxplots.

To test the effect of various factors on the risk of colony loss, quasibinomial models were fitted. In the models for both regions combined, region, varroa treatment, presence of bees with deformed wings, use of foundationless combs, queen replacement, colony splitting, merging colonies, feed supplementation, and amount of sugar supplementation were individually highly significant predictors of the risk of colony loss. In a multivariable model, region and varroa treatment were still significant, but other variables were no longer significantly associated with colony loss. Therefore, the most important predictors of colony loss were region and varroa treatment ($p < 0.0001$). In addition, models were fitted for the regions separately where there was more than one category of a factor observed in the data. For Oromia, presence of *V. velutina* and sugar supplementation were the only significant variables for predicting colony loss. Of these, only presence of *V. velutina* significantly affected the loss rate when a multivariable model was considered, and is, therefore, the most notable predictor of colony loss in Oromia ($p < 0.01$). Table S1 shows details of the best predicting models. For Tigray, none of the individual factors were significant in the fitted models, but use of foundationless combs and merging colonies were closest to being significant predictors of colony loss. These results as a whole reflect differences between the regions. Consequently, colony loss rates were significantly different between the two regions (Figure 3) due to differences in natural disasters, environmental factors and beekeeping practices. These factors include varroa monitoring, varroa treatment, use of natural combs, dependence on externally supplied beeswax, colony merging, queen replacement, feed supplementation, colony splitting and *Vespa velutina* (Table 2). In particular, the presence of *V. velutina* was the most influential factor on colony losses in Oromia. Additionally, impact of the Tigray war is a major difference between the regions.

3.3. *Annual Colony Development and Management Calendar*

To obtain more insights into the annual honey bee colony development and beekeeping activity calendar, we collected data on the major events and practices that occurred and were performed by the beekeepers during the survey year. The variables include adding supers to hives, merging weak colonies, colony absconding, queen replacement, honey harvesting, feed supplementation, colony splitting and swarming. The results are summarised in Figure 4A,B for Oromia and Tigray regions, respectively.

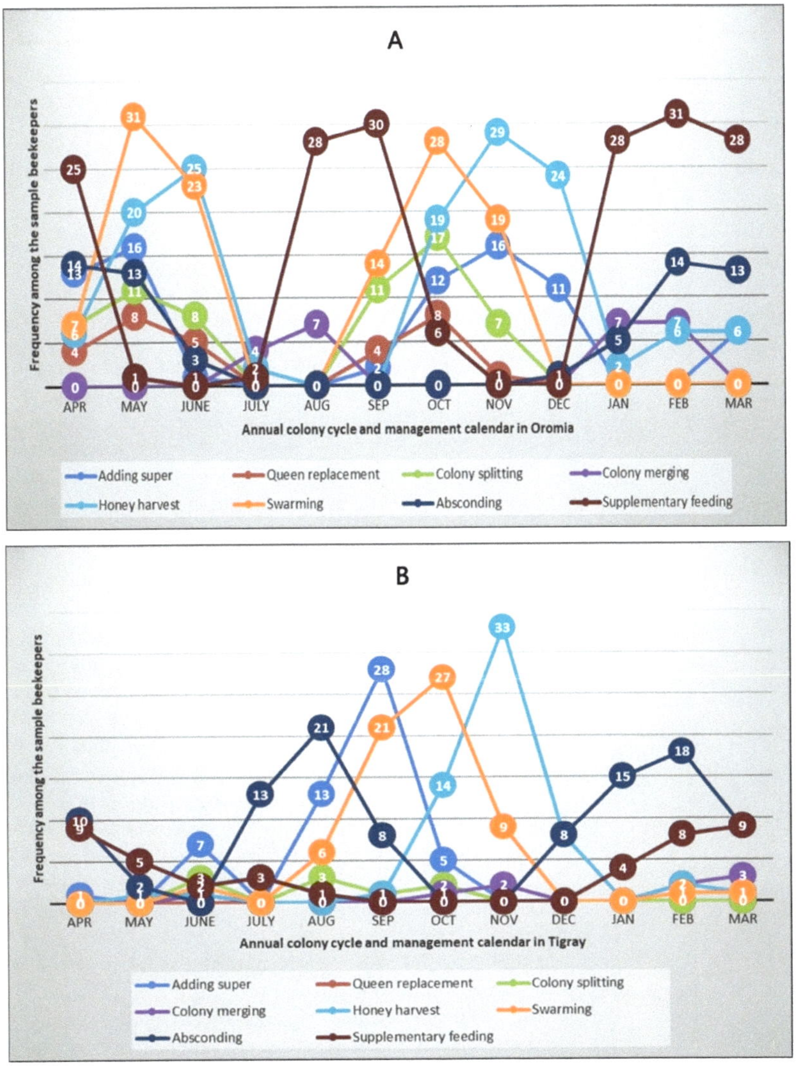

Figure 4. Annual honey bee colony cycle and beekeeping calendar in Oromia (**A**) and Tigray (**B**) regions of Ethiopia, northeast Africa. The plots show bimodal and monomodal patterns of colony development and the beekeeping calendar in Oromia and Tigray regions, respectively, which reflect two beekeeping seasons in Oromia and one season in Tigray annually. The numbers shown in the circles on the graphs indicate the frequency of beekeepers reporting that the specified events/activities occurred/were performed in the regions during each month of the survey year 2022/2023 of the Colony Losses Monitoring Core Project.

In the Tahtay Koraro district of the Tigray region, where the survey was conducted, colonies start to develop in July and peak in September. This is followed by peak reproductive swarming and honey harvesting in October and November, respectively. After honey harvesting, about half of the beekeepers (51.5%) provide supplementary feed to their colonies in the dry months from January to May, whereas most of the beekeepers (97.0%) in this region faced at least one colony absconding event (Figure 4B, Table S2). The pattern of colony development and the beekeeping calendar in the surveyed humid areas of Oromia region was different from the situation in the semi-arid or savanna areas of Tigray region [43]. The survey results showed that there is a two-cycle (double-peak) pattern of colony development, swarming, honey harvesting, and feeding in Oromia region (Figure 4A, Table S2). Another differentiating aspect is that all sampled beekeepers practised feed supplementation during the dearth periods in Oromia region, whereas 67.7% of them faced at least one absconding event in the year compared to the 97.0% reported above in Tigray region.

4. Discussion

Here, we discuss the results of our survey which was conducted in Oromia and Tigray regions of Ethiopia for the first time based on the COLOSS standardised questionnaire 2023, aiming to assess honey bee colony loss rates and driving factors with reference to regional and global reports. The discussion provides insights into the factors influencing colony losses, colony management, pests, and the annual beekeeping calendar, which are shaped by anthropogenic, climatic and environmental factors.

We found higher numbers of colonies per beekeeper (apiary) in spring 2023 in Oromia (mean = 30.58) and Tigray (mean = 5.24) compared to previously reported average estimates of 6 colonies per beekeeper for Ethiopia [44] and 2 colonies per beekeeper for Tigray [26]. It should be noted that the sampling in this survey was not random, but purposefully included well-known beekeepers and accessible apiaries such as those owned by the Oromia Apicultural Research Institute, due to limitations of logistics, time and funding. Some of the apiaries were found to implement improved beekeeping husbandry practices that are not common among the Ethiopian smallholder beekeepers. These include varroa monitoring (7.8% in Oromia; 32.8% in Tigray), varroa treatment (7.8%; 0.0%), queen replacement (29.7%; 0.0%), feed supplementation (48.4%; 26.6%), merging of weak colonies (32.8%; 9.4%) and colony splitting (40.6%; 7.8%); see Table 2 and Figure 2.

Apiaries managed in Oromia that are included in this survey applied more varroa control (biotechnical method), feed supplementation, merging and splitting ($p < 0.001$) compared to those studied from Tigray. Moreover, a significantly higher number of the surveyed beekeepers in Oromia used natural combs ($p < 0.0001$) and depended less on beeswax purchased from external suppliers ($p < 0.01$) compared to the situation in Tigray (Table 2). On the other hand, most frame-hive beekeepers in Tigray depend on externally purchased beeswax supplied from other parts of Ethiopia where traditional beekeeping is still dominant and extraction of beeswax is practised. Exchanging honey bee products and colony transportation could negatively affect honey bee population structures and health [28,45,46]. Pests, pathogens and other stressors cause colony losses, as observed by survey studies in different parts of the world [3–5,47–49]. On the other hand, application of improved beekeeping husbandry such as merging and supplementing weak colonies, replacing undesirable queens, monitoring and controlling of varroa enhances the performance, health and survival of colonies [50–53], and could reduce colony losses. Altogether, colony losses can be influenced by health, nutrition, and climatic stresses as well as anthropogenic activities.

In this survey, the average loss rate of colonies in Tigray was markedly higher (68.3%) than in Oromia (25.9%). The proportional loss rate was also significantly higher in Tigray (66.4%) than in Oromia (24.1%). A previous study [26] showed that the colony loss rate in Tigray was lower (at 15.7% overall, and ranging from 9% to 19.5% between local areas) than the results in this survey and was similar when compared to international (overall proportional) loss rate figures of COLOSS reported as 16.4% to 20.9% [3–5]. This indicates that most of the colony losses in Tigray reported in the present study can be attributed to the disaster of the two-year war in the region, which agrees with recent reports that the Tigray war caused 70% honey bee colony losses [39] and reversed decades of ecosystem restoration [37,54], agricultural development [55,56] and livelihood improvement [34] efforts. These effects could lead to habitat degradation, shortage of bee forage and increased rates of colony absconding and death. Overall, human activities such as war, fire, vandalism and honey hunting can cause detrimental impact on the survival of honey bees and the livelihood of beekeepers.

Among the sample beekeepers, 97.0% of them in Tigray responded that they had experienced at least one colony absconding event in the surveyed year compared to 67.7% in Oromia (Figure 4, Table S2). Honey bee colony absconding can be defined as seasonal migration of the colony seeking more favourable nest and/or habitat conditions. In this behavioural adaptation, the colony abandons its old nest when there is shortage of forage, extreme weather, pest infestation and/or other disturbances [57–59]. Instead, the colony searches and re-establishes itself in a suitable environment. Therefore, absconding is one form of colony loss to beekeepers, although the colony may not be lost from the overall ecosystem. This is because the colony strives in another location and continues to provide the necessary ecosystem services no matter whether it may return to its original nest when the situation is better or may be trapped by another beekeeper's bait hive. High rates of migratory and swarming behaviours in honey bees have been reported as the main coping mechanisms against pests such as varroa [60]. African honey bees are known for a high degree of migratory behaviour when the availability of floral resources is reduced [57], which could also be influenced by the local weather conditions and type of beehives [59] as well as management practices such as feed supplementation and merging of weak colonies. Therefore, colony loss rates vary with management intensity implemented by different scales of beekeeping operations, local weather, geographic location and season [3–6,20,21]. For example, in the COLOSS survey 2019/20, loss rates ranged from 7.4% in Norway to 36% in Spain, while large-scale beekeeping operations are less prone to the risk of colony losses compared to small-scale beekeepers managing fewer than 50 colonies [6]. However, this was not observed in a survey conducted in Mexico [49], which indicates the role of local/regional conditions.

Looking for annual variations in beekeeping across the regions, we observed a bimodal pattern of colony development and beekeeping calendar in the Oromia region of central and southern Ethiopia compared to a single peak of the development cycle and beekeeping activities in the Tigray region of northern Ethiopia (Figure 4, Table S2). In the semi-arid areas of northern Ethiopia, colony development, honey production and swarming occur from August to November, while the dearth period characterised by high rates of colony absconding extends from December to July despite the availability of a few trees that bloom during the dry period. These are consistent with previous studies reported from Oromia [61,62] and Tigray [63]. It is well-known that honey bees mainly depend on the nectar and pollen produced by flowering plants that are shaped by the local climate [24]. Therefore, the floral calendar and nectar flow of an area govern the development of honey bees and the beekeeping activities. Thus, honey harvesting is possible multiple times in a year in the evergreen areas of southwest Ethiopia [61,62], while it is mainly limited to September to November in the arid and semi-arid areas of northern Ethiopia [63].

In conclusion, colony loss rates in this survey showed a significant variation between the two Ethiopian regions, where beekeepers in Tigray sustained a markedly higher loss rate, which we attribute as being due to the war and other factors, compared to the loss rate in Oromia region and also compared to the global figures reviewed. In addition, the colony loss rate was significantly affected by beekeeping practices and honey bee health management. Moreover, the number of colony losses recorded as resulting from queen problems, natural disaster, dead colonies or empty hives is positively correlated with the number of colonies managed. Furthermore, annual colony development and management calendars in the two regions showed different patterns, where the beekeeping seasons were bimodal with short dearth periods and monomodal with a long dearth period in Oromia and Tigray regions, respectively.

Supplementary Materials: The following supporting information can be downloaded at: https://www.mdpi.com/article/10.3390/insects15060376/s1: Table S1: Summary of best predicting quasibinomial generalised linear models for the risk of colony loss in Oromia and Tigray regions of Ethiopia, northeast Africa; Table S2: Annual honey bee colony cycle and beekeeping calendar in Oromia and Tigray regions of Ethiopia, northeast Africa.

Author Contributions: Conceptualization, T.G.H. and A.G.; Data curation, T.G.H., A.T.A. and K.W.; Formal analysis, T.G.H. and A.G.; Investigation, A.T.A. and K.W.; Methodology, T.G.H. and A.G.; Project administration, T.G.H.; Supervision, T.G.H.; Visualization, T.G.H. and A.G.; Writing—original draft, T.G.H. and A.G.; Writing—review and editing, T.G.H., A.T.A., K.W. and A.G. All authors have read and agreed to the published version of the manuscript.

Funding: This research received no external funding.

Informed Consent Statement: Informed consent was obtained from all participants involved in the study.

Data Availability Statement: Additional data can be requested from the corresponding author.

Acknowledgments: The authors would like to thank all the beekeepers who responded to the questionnaires and are very grateful to Niguse Hagos for the study site map production using GIS tools. The authors also extend their appreciation to COLOSS as a whole and the Colony Losses Monitoring Core Project in particular for the survey tools and for providing a platform and opportunities for networking, knowledge exchange and collaboration.

Conflicts of Interest: The authors declare no conflicts of interest.

References

1. van Engelsdorp, D.; Evans, J.D.; Saegerman, C.; Mullin, C.; Haubruge, E.; Nguyen, B.K.; Frazier, M.; Frazier, J.; Cox-Foster, D.; Chen, Y.; et al. Colony Collapse Disorder: A Descriptive Study. *PLoS ONE* **2009**, *4*, e6481. [CrossRef]
2. Ellis, J.D.; Evans, J.D.; Pettis, J. Colony Losses, Managed Colony Population Decline, and Colony Collapse Disorder in the United States. *J. Apic. Res.* **2010**, *49*, 134–136. [CrossRef]
3. Brodschneider, R.; Gray, A.; Adjlane, N.; Ballis, A.; Brusbardis, V.; Charrière, J.D.; Chlebo, R.; Coffey, M.F.; Dahle, B.; de Graaf, D.C.; et al. Multi-Country Loss Rates of Honey Bee Colonies during Winter 2016/2017 from the COLOSS Survey. *J. Apic. Res.* **2018**, *57*, 452–457. [CrossRef]
4. Gray, A.; Brodschneider, R.; Adjlane, N.; Ballis, A.; Brusbardis, V.; Charrière, J.D.; Chlebo, R.F.; Coffey, M.; Cornelissen, B.; Amaro da Costa, C.; et al. Loss Rates of Honey Bee Colonies during Winter 2017/18 in 36 Countries Participating in the COLOSS Survey, Including Effects of Forage Sources. *J. Apic. Res.* **2019**, *58*, 479–485. [CrossRef]
5. Gray, A.; Adjlane, N.; Arab, A.; Ballis, A.; Brusbardis, V.; Charrière, J.D.; Chlebo, R.; Coffey, M.F.; Cornelissen, B.; Amaro da Costa, C.; et al. Honey Bee Colony Winter Loss Rates for 35 Countries Participating in the COLOSS Survey for Winter 2018–2019, and the Effects of a New Queen on the Risk of Colony Winter Loss. *J. Apic. Res.* **2020**, *59*, 744–751. [CrossRef]
6. Gray, A.; Adjlane, N.; Arab, A.; Ballis, A.; Brusbardis, V.; Bugeja Douglas, A.; Cadahía, L.; Charrière, J.D.; Chlebo, R.; Coffey, M.F.; et al. Honey Bee Colony Loss Rates in 37 Countries Using the COLOSS Survey for Winter 2019–2020: The Combined Effects of Operation Size, Migration and Queen Replacement. *J. Apic. Res.* **2023**, *62*, 204–210. [CrossRef]
7. Aurell, D.; Bruckner, S.; Wilson, M.; Steinhauer, N.; Williams, G.R. A National Survey of Managed Honey Bee Colony Losses in the USA: Results from the Bee Informed Partnership for 2020–21 and 2021–22. *J. Apic. Res.* **2024**, *63*, 1–14. [CrossRef]
8. Rosenkranz, P.; Aumeier, P.; Ziegelmann, B. Biology and Control of *Varroa destructor*. *J. Invertebr. Pathol.* **2010**, *103*, S96–S119. [CrossRef]
9. Guzmán-Novoa, E.; Eccles, L.; Calvete, Y.; McGowan, J.; Kelly, P.G.; Correa-Benítez, A. *Varroa destructor* Is the Main Culprit for the Death and Reduced Populations of Overwintered Honey Bee (*Apis mellifera*) Colonies in Ontario, Canada. *Apidologie* **2010**, *41*, 443–450. [CrossRef]
10. Noël, A.; Le Conte, Y.; Mondet, F. *Varroa destructor*: How Does It Harm *Apis mellifera* Honey Bees and What Can Be Done about It? *Emerg. Top Life Sci.* **2020**, *4*, 45–57. [CrossRef]
11. Brodschneider, R.; Schlagbauer, J.; Arakelyan, I.; Ballis, A.; Brus, J.; Brusbardis, V.; Cadahía, L.; Charrière, J.D.; Chlebo, R.; Coffey, M.F.; et al. Spatial Clusters of *Varroa destructor* Control Strategies in Europe. *J. Pest Sci. (2004)* **2023**, *96*, 759–783. [CrossRef]
12. Blacquière, T.; Boot, W.; Calis, J.; Moro, A.; Neumann, P.; Panziera, D. Darwinian Black Box Selection for Resistance to Settled Invasive *Varroa destructor* Parasites in Honey Bees. *Biol. Invasions* **2019**, *21*, 2519–2528. [CrossRef]
13. Martin, S.J.; Hawkins, G.P.; Brettell, L.E.; Reece, N.; Correia-Oliveira, M.E.; Allsopp, M.H. *Varroa destructor* Reproduction and Cell Re-Capping in Mite-Resistant *Apis mellifera* Populations. *Apidologie* **2020**, *51*, 369–381. [CrossRef]

14. Mondet, F.; Beaurepaire, A.; McAfee, A.; Locke, B.; Alaux, C.; Blanchard, S.; Danka, B.; Le Conte, Y. Honey Bee Survival Mechanisms against the Parasite *Varroa destructor*: A Systematic Review of Phenotypic and Genomic Research Efforts. *Int. J. Parasitol.* **2020**, *50*, 433–447. [CrossRef] [PubMed]
15. Luis, A.R.; Grindrod, I.; Webb, G.; Piñeiro, A.P.; Martin, S.J. Recapping and Mite Removal Behaviour in Cuba: Home to the World's Largest Population of Varroa-Resistant European Honeybees. *Sci. Rep.* **2022**, *12*, 15597. [CrossRef] [PubMed]
16. Begna, D. Occurrences and Distributions of Honeybee (*Apis mellifera*) Varroa Mite (*Varroa destructor*) in Tigray Region, Ethiopia. *J. Fish. Livest. Prod.* **2014**, *2*, 10–4172. [CrossRef]
17. Shegaw, T.; Arke, A.; Belay, N.; Habte Giorgis, D. Diagnostic Survey on Varroa Mite (*Varroa destructor*) Prevalence in South-Western Ethiopia. *Cogent Food Agric.* **2022**, *8*, 2143610. [CrossRef]
18. Gela, A.; Atickem, A.; Bezabeh, A.; Woldehawariat, Y.; Gebresilassie, A. Insights into Varroa Mite (*Varroa destructor*) Infestation Levels in Local Honeybee (*Apis mellifera*) Colonies of Ethiopia. *J. Appl. Entomol.* **2023**, *147*, 798–808. [CrossRef]
19. Robi, D.T.; Temteme, S.; Aleme, M.; Bogale, A.; Getachew, A.; Mendesil, E. Epidemiology, Factors Influencing Prevalence and Level of Varroosis Infestation (*Varroa destructor*) in Honeybee (*Apis mellifera*) Colonies in Different Agroecologies of Southwest Ethiopia. *Parasite Epidemiol. Control* **2023**, *23*, e00325. [CrossRef]
20. Overturf, K.A.; Steinhauer, N.; Molinari, R.; Wilson, M.E.; Watt, A.C.; Cross, R.M.; Williams, G.R.; Rogers, R. Winter Weather Predicts Honey Bee Colony Loss at the National Scale. *Ecol. Indic.* **2022**, *145*, 109709. [CrossRef]
21. Insolia, L.; Molinari, R.; Rogers, S.R.; Williams, G.R.; Chiaromonte, F.; Calovi, M. Honey Bee Colony Loss Linked to Parasites, Pesticides and Extreme Weather across the United States. *Sci. Rep.* **2022**, *12*, 20787. [CrossRef] [PubMed]
22. Landaverde, R.; Rodriguez, M.T.; Parrella, J.A. Honey Production and Climate Change: Beekeepers' Perceptions, Farm Adaptation Strategies, and Information Needs. *Insects* **2023**, *14*, 493. [CrossRef] [PubMed]
23. Neumann, P.; Straub, L. Beekeeping under Climate Change. *J. Apic. Res.* **2023**, *62*, 963–968. [CrossRef]
24. Rafferty, N.E.; Diez, J.M.; Bertelsen, C.D. Changing Climate Drives Divergent and Nonlinear Shifts in Flowering Phenology across Elevations. *Curr. Biol.* **2020**, *30*, 432–441.e3. [CrossRef] [PubMed]
25. McNally, L.C.; Schneider, S.S. Seasonal Cycles of Growth, Development and Movement of the African Honey Bee, *Apis mellifera scutellata*, in Africa. *Ins. Soc.* **1992**, *39*, 167–179. [CrossRef]
26. Hailu, T.G.; Wakjira, K.; Gray, A. Honey Bee Colony Population Annual Dynamics in Northern Ethiopia's Semi-Arid Region, Tigray. *J. Apic. Res.* **2024**, 1–10. [CrossRef]
27. Gebretinsae, T.; Tesfay, Y. Honeybee Colony Marketing and Its Implications for Queen Rearing and Beekeeping Development in Tigray, Ethiopia. *Int. J. Livest. Prod.* **2014**, *5*, 117–128. [CrossRef]
28. Jara, L.; Ruiz, C.; Martín-Hernández, R.; Muñoz, I.; Higes, M.; Serrano, J.; De la Rúa, P. The Effect of Migratory Beekeeping on the Infestation Rate of Parasites in Honey Bee (*Apis mellifera*) Colonies and on Their Genetic Variability. *Microorganisms* **2021**, *9*, 22. [CrossRef]
29. Adgaba, N. Selling Honeybee Colonies as a Source of Income for Subsistence Beekeepers. *Bees Dev.* **2002**, *64*, 2–3.
30. Central Statistical Agency (CSA). *Agricultural Sample Survey 2017/18 [2010 E.C.]: Report on Livestock and Livestock Characteristics*; CSA: Addis Ababa, Ethiopia, 2018; Volume II.
31. Hailu, T.G.; Chagunda, M.; Rosenkranz, P. Sustainable Development Outlooks to Subsistent Apiculture in a Transition: The Case of Ethiopia. *J. Apic. Res.* **2023**, *62*, 730–740. [CrossRef]
32. Bahadur Poudel, P.; Ram Poudel, M.; Gautam, A.; Phuyal, S.; Krishna Tiwari, C.; Bashyal, N.; Bashyal, S. COVID-19 and Its Global Impact on Food and Agriculture. *J. Biol. Todays World* **2020**, *9*, 221.
33. Kassegn, A.; Endris, E. Review on Socio-Economic Impacts of 'Triple Threats' of COVID-19, Desert Locusts, and Floods in East Africa: Evidence from Ethiopia. *Cogent Soc. Sci.* **2021**, *7*, 1885122. [CrossRef]
34. World Peace Foundation Report 2021. In *Starving Tigray*; World Peace Foundation: Somerville, MA, USA, 2021.
35. Demissie, B.; Nyssen, J.; Annys, S.; Negash, E.; Gebrehiwet, T.; Abay, F.; Wolff, E. Geospatial Solutions for Evaluating the Impact of the Tigray Conflict on Farming. *Acta Geophys.* **2022**, *70*, 1285–1299. [CrossRef]
36. Gesesew, H.; Berhane, K.; Siraj, E.S.; Siraj, D.; Gebregziabher, M.; Gebre, Y.G.; Gebreslassie, S.A.; Amdes, F.; Tesema, A.G.; Siraj, A.; et al. The Impact of War on the Health System of the Tigray Region in Ethiopia: An Assessment. *BMJ Glob. Health* **2021**, *6*, e007328. [CrossRef] [PubMed]
37. Gebrekirstos, A.; Birhane, E. The War on Tigray Wiped out Decades of Environmental Progress: How to Start Again 2023. The Conversation. Available online: https://theconversation.com/the-war-on-tigray-wiped-out-decades-of-environmental-progress-how-to-start-again-201062 (accessed on 5 April 2023).
38. Tekulu, F.B.; Asgedom, D.B.; Gebre, H.T.; Desta, G.M.; Shifare, T.H.; Hadigu, T.B. The Effect of War on Educational Institutions of Eastern Tigray Zone, Tigray State, Ethiopia. *Res. Sq.* **2023**, preprint. [CrossRef]
39. Ngussie, H.G.; Hailu, T.G. Tigray War Put Honey Bees in Peril. *Bee World* **2023**, *100*, 51–55. [CrossRef]
40. Lewoyehu, M.; Amare, M. Comparative Assessment on Selected Physicochemical Parameters and Antioxidant and Antimicrobial Activities of Honey Samples from Selected Districts of the Amhara and Tigray Regions, Ethiopia. *Int. J. Food Sci.* **2019**, *2019*, 4101695. [CrossRef] [PubMed]
41. R Core Team 2020. R: A Language and Environment for Statistical Computing. R Foundation for Statistical Computing, Vienna, Austria. Available online: https://www.R-project.org/ (accessed on 19 May 2024).
42. JMP Statistical Discovery LLC. *Discovering JMP®2022–2023*, Version 17; JMP Statistical Discovery LLC: Cary, NC, USA, 2022.

43. Food and Agriculture Organization of the United Nations (FAO); International Institute for Applied Systems Analysis (IIASA). Global Agro-Ecological Zones Data Portal. Global Agro-Ecological Zones. Available online: https://gaez.fao.org/ (accessed on 9 May 2024).
44. Gratzer, K.; Wakjira, K.; Fiedler, S.; Brodschneider, R. Challenges and Perspectives for Beekeeping in Ethiopia. A Review. *Agron. Sustain. Dev.* **2021**, *41*, 46. [CrossRef]
45. Kükrer, M.; Kence, M.; Kence, A. Honey Bee Diversity Is Swayed by Migratory Beekeeping and Trade Despite Conservation Practices: Genetic Evidence for the Impact of Anthropogenic Factors on Population Structure. *Front. Ecol. Evol.* **2021**, *9*, 556816. [CrossRef]
46. Cánovas, F.; De La Rúa, P.; Serrano, J.; Galián, J. Microsatellite Variability Reveals Beekeeping Influences on Iberian Honeybee Populations. *Apidologie* **2011**, *42*, 235–251. [CrossRef]
47. Melicher, D.; Wilson, E.S.; Bowsher, J.H.; Peterson, S.S.; Yocum, G.D.; Rinehart, J.P. Long-Distance Transportation Causes Temperature Stress in the Honey Bee, *Apis mellifera* (Hymenoptera: Apidae). *Environ. Entomol.* **2019**, *48*, 691–701. [CrossRef]
48. Simone-Finstrom, M.; Strand, M.K.; Tarpy, D.R.; Rueppell, O. Impact of Honey Bee Migratory Management on Pathogen Loads and Immune Gene Expression Is Affected by Complex Interactions with Environment, Worker Life History, and Season. *J. Insect Sci.* **2022**, *22*, 17. [CrossRef] [PubMed]
49. Medina-Flores, C.A.; López-Carlos, M.; Carrillo-Muro, O.; Gray, A. Honey Bee Colony Losses in Mexico's Semi-Arid High Plateau for the Winters 2016–2017 to 2021–2022. *Insects* **2023**, *14*, 453. [CrossRef] [PubMed]
50. Akyol, E.; Yeninar, H.; Korkmaz, A.; Çakmak, I. An Observation Study on the Effects of Queen Age on Some Characteristics of Honey Bee Colonies. *Ital. J. Anim. Sci.* **2008**, *7*, 19–25. [CrossRef]
51. Lodesani, M.; Costa, C.; Besana, A.; Dall'Olio, R.; Franceschetti, S.; Tesoriero, D.; Vaccari, G. Impact of Control Strategies for *Varroa destructor* on Colony Survival and Health in Northern and Central Regions of Italy. *J. Apic. Res.* **2014**, *53*, 155–164. [CrossRef]
52. Mortensen, A.N.; Jack, C.J.; Bustamante, T.A.; Schmehl, D.R.; Ellis, J.D. Effects of Supplemental Pollen Feeding on Honey Bee (Hymenoptera: Apidae) Colony Strength and *Nosema Spp.* Infection. *J. Econ. Entomol.* **2019**, *112*, 60–66. [CrossRef] [PubMed]
53. Gregorc, A. Monitoring of Honey Bee Colony Losses: A Special Issue. *Diversity* **2020**, *12*, 403. [CrossRef]
54. Negash, E.; Birhane, E.; Gebrekirstos, A.; Gebremedhin, M.A.; Annys, S.; Rannestad, M.M.; Berhe, D.H.; Sisay, A.; Alemayehu, T.; Berhane, T.; et al. Remote Sensing Reveals How Armed Conflict Regressed Woody Vegetation Cover and Ecosystem Restoration Efforts in Tigray, Ethiopia. *Sci. Remote Sens.* **2023**, *8*, 100108. [CrossRef]
55. Manaye, A.; Afewerk, A.; Manjur, B.; Solomon, N. The Effect of the War on Smallholder Agriculture in Tigray, Northern Ethiopia. *Cogent Food Agric.* **2023**, *9*, 2247696. [CrossRef]
56. Nyssen, J.; Negash, E.; Van Schaeybroeck, B.; Haegeman, K.; Annys, S. Crop Cultivation at Wartime–Plight and Resilience of Tigray's Agrarian Society (North Ethiopia). *Def. Peace Econ.* **2023**, *34*, 618–645. [CrossRef]
57. Schneider, S.S.; McNally, L.C. Factors Influencing Seasonal Absconding in Colonies of the African Honey Bee, *Apis mellifera scutellata*. *Insectes Soc.* **1992**, *39*, 403–423. [CrossRef]
58. Spiewok, S.; Neumann, P.; Hepburn, H.R. Preparation for Disturbance-Induced Absconding of Cape Honeybee Colonies (*Apis mellifera capensis* Esch.). *Insectes Soc.* **2006**, *53*, 27–31. [CrossRef]
59. McMenamin, A.; Mumoki, F.; Frazier, M.; Kilonzo, J.; Mweu, B.; Baumgarten, T.; Patch, H.; Torto, B.; Masiga, D.; Tumlinson, J.; et al. The Impact of Hive Type on the Behavior and Health of Honey Bee Colonies (*Apis mellifera*) in Kenya. *Apidologie* **2017**, *48*, 703–715. [CrossRef]
60. Loftus, J.C.; Smith, M.L.; Seeley, T.D. How Honey Bee Colonies Survive in the Wild: Testing the Importance of Small Nests and Frequent Swarming. *PLoS ONE* **2016**, *11*, e0150362. [CrossRef] [PubMed]
61. Tesfaye, O.; Mekonnen, E. Floral Calendar of Honeybee Plants in Kellem and West Wollega Zone, Western Ethiopia. *Int. J. For. Res.* **2023**, *2023*, 2797159. [CrossRef]
62. Bareke, T.; Roba, K.; Addi, A. Identification and Establishing Floral Calendar in Jimma Zone of Oromia Ethiopia. *J. Glob. Ecol. Environ.* **2023**, *19*, 48–61. [CrossRef]
63. Gebremedhn, H.; Tesfay, Z.; Murutse, G.; Estifanos, A. Seasonal Honeybee Forage Availability, Swarming, Absconding and Honey Harvesting in Debrekidan and Begasheka Watersheds of Tigray, Northern Ethiopia. *Livest. Res. Rural. Dev.* **2013**, *25*, 61.

insects

Article

Development of Tools to Understand the Relationship between Good Management Practices and Nest Losses in Meliponiculture: A Pilot Study in Latin American Countries

Joseline Sofía Ocaña-Cabrera [1], Sarah Martin-Solano [2] and Claude Saegerman [1,*]

[1] Research Unit of Epidemiology and Risk Analysis Applied to Veterinary Sciences (UREAR-ULiège), Fundamental and Applied Research for Animal and Health (FARAH) Center, Department of Infections and Parasitic Diseases, Faculty of Veterinary Medicine, University of Liege, 4000 Liège, Belgium; jocana@doct.uliege.be

[2] Grupo de Investigación en Sanidad Animal y Humana (GISAH), Carrera de Ingeniería en Biotecnología, Departamento de Ciencias de la Vida y de la Agricultura, Universidad de las Fuerzas Armadas ESPE, P.O. Box 171-5-231, Sangolquí 171103, Ecuador; ssmartin@espe.edu.ec

* Correspondence: claude.saegerman@uliege.be; Tel.: +32-4-366-45-79

Simple Summary: The overall decline of bees may be exacerbated by the simultaneous presence and interaction of multiple causal factors. To elucidate how these factors interact and their collective impact, it is of the utmost importance to develop effective analytical tools. We collected data through an online questionnaire. We started estimating the annual mortality of stingless bee nests at 15%. Four risks to stingless bee survival were identified: invasive species (73%), the proximity of nests to sources of environmental pollution (61%), the presence of honey bees as potential transmitters of diseases (57%), and unusual behavior reports (44%). The biosecurity practices with the highest compliance rates were hand washing (79%), sterilization (75%), storage conditions for product quality (66%), and the use of protective equipment (40%). The spider web and barometer tools facilitate a unified observation of the status of implementation or non-implementation of biosecurity measures, actions to care for the environment in which stingless bees live, the quality and efficiency of nest management techniques, and the monitoring of the health status of stingless bees. The comprehensive evaluation of these factors within best management practices (BMPs) facilitates immediate decision-making and the implementation of enhancements, as well as individual and collective feedback.

Abstract: Insect pollination services amount to USD 235–577 billion. Seventy five percent of agricultural production for human consumption depends on pollination, mainly by bees. A decline in pollinators, including Meliponini tribe bees, will impact the economy, food security, human health, and ecosystem stability, especially in tropical forests where stingless bees are the main pollinators. The objective of this survey was to understand the relationship between good management practices and nest losses in meliponiculture, encompassing biosecurity and conservation criteria. A 36-question survey was organized and spread. We received 92 responses, representing 4548 managed nests. The primary motivation for engaging in meliponiculture was biodiversity conservation (92%). More than 50% of the questions on biosecurity were answered as "applied". Hand washing before any activity with bees was the main rule, followed by material sterilization and personal protective equipment use. The annual mortality rate of stingless bee nests was estimated at 15%. Nest invaders (72%) and nearby sources of pollution (60%) were identified as the main potential causes of nest losses. From a general perspective, meliponiculture practices continue to expand remarkably. The implementation of effective nest management strategies is associated with a reduction in nest losses. It is important to consider One Health's perspective to ensure optimal management practices.

Keywords: stingless bees; management; practices; biosecurity; nest loss; Latin America; evaluation tools

Citation: Ocaña-Cabrera, J.S.; Martin-Solano, S.; Saegerman, C. Development of Tools to Understand the Relationship between Good Management Practices and Nest Losses in Meliponiculture: A Pilot Study in Latin American Countries. *Insects* **2024**, *15*, 715. https://doi.org/10.3390/insects15090715

Academic Editors: Ivana Tlak Gajger and Franco Mutinelli

Received: 8 August 2024
Revised: 7 September 2024
Accepted: 12 September 2024
Published: 19 September 2024

1. Introduction

The global economic value of pollination services amounted to USD 235–577 billion, representing 10% of the total value of agricultural production for human consumption in 2021. Around 75% [1] of this agricultural production depends on pollinators, especially bees [2]. It is evident that the decline of main pollinators, including stingless bee species [3], will have a great economic impact on food security, human health, and ecosystem stability.

The available data indicate that the Neotropics are home to more than 15,150 species of bees [4], and it is only a third of the total animal species richness in that region. Worldwide, the number of stingless bee species exceeds 500 [5]. There is the possibility of finding subspecies or cryptic species due to the complexity of certain genera such as *Melipona beecheii* [6] or the taxonomic updating of stingless bees [7]. In Ecuador, a great contribution showed the presence of >200 [8,9], consolidating the megadiverse label despite the small size of the country with other neighbors.

There are multiple approaches to practicing meliponiculture, and they are contingent upon the motivations, needs, and objectives of the practitioner [10]. Meliponiculture represents a fusion of ecological (from the academy) and cultural (empirical local) knowledge, and both, along with stingless bees, serve as interesting fusion that facilitates the transition to sustainable practices within complex farming systems [11].

The five major threats for native tropical bees are deforestation, agriculture intensification, the spread of exotic species [12], climate change, and resource–habitat loss [13]. The introduction of non-native pollinators modifies socioecological interactions between insects and environmental health, i.e., by competing with native insects for floral resources or due to the spread of new diseases [14] for which the native insects have no immune defense [15]. The effects of deforestation include habitat loss and fragmentation [16], which are mainly caused by the expansion of crops such as potatoes in the Colombian and Ecuadorian Andes [17], soybeans in the Brazilian Amazon rainforest [18], or the expansion of areas focused on cattle breeding [12].

Meliponiculture practices that include harvesting honey and pollen, dividing nests, and selling nest products have faced several other menaces, such as the loss of numerous daughter colonies from a single mother, inbreeding, and queen succession problems in *Scaptotrigona* and *Cephalotrigona* species [19]. There are mainly two stingless bee nest invasive insect problems. The first, *Lestrimelitta* sp., is a kleptobiotic stingless bee, considered a resource thief that uses a chemical trickery mechanism based on its cuticular characteristics [20]. The other major invasive insect problem is Phoridae flies (*Pseudohypocera kerteszi*), which, avoiding the guardians at the nest entrance, lay eggs in pollen pots, containers, and near the brood, which will develop into white larvae that feed on the bee bread [21].

A study of the population dynamics of stingless bees in seasonal dry lowlands in Costa Rica reveals that they invest more efforts in colony survival rather than in increasing their reproductive rates, which means that, under better life conditions, these stingless bees can survive around 23 years [22], but the most recent study of colony loss in Latin America indicated a 39.6% loss of stingless bee colonies per year across the region. Furthermore, the study found that losses were highest in summer and increased with farm size [23]. These findings suggest that maintaining the overall health of bee colonies is challenging, which could have significant implications for the economic survival of stingless bee keepers. The role of stingless bee keepers is an option to care for intangible heritage and the conservation of natural resources [9], as well as their training and adoption of best practices to preserve the life of stingless bees and thus the environment.

The FAO, the WHO, and the European Commission have recognized good farming practices in beekeeping and describe their advantages, such as improved bee colony health, decreased medicinal costs, increased hive production, and the yield of healthier and higher quality honey [24]. In this sense, stingless beekeeping also needs the application of good management practices, since it has been recognized as an informal activity with poor management [25] which continues to grow and expand, especially in Latin America, at an accelerated rate [26–29]. Good practices in the management of stingless bees are a means to reduce risks associated with human error that impact human public health due to the consumption of nest products, such as honey, contaminated with agrochemicals [30]. In addition, the same risks can affect bee health, as pesticide residues can bioaccumulate in bees' bodies, in their food, and in nest structures, affecting their health, condition, and ability to survive.

Ecuadorian meliponiculture has developed depending on the climatic region. The southern highlands region, especially the province of Loja, has the highest development at the national level in dry tropical forest meliponiculture, followed by Amazon rainforest meliponiculture, urban tourist-productive meliponiculture in the coastal region, and conservations projects in protected areas.

In terms of regulations on stingless bee products, the Ecuadorian Service of Normalization (INEN) does not contemplate quality standards for pot honey or pot pollen [31]. Regarding good management practices, the Agency for Regulation and Phytosanitary—Zoo Sanitary Control (AGROCALIDAD) has only issued beekeeping guidelines [32]. In terms of bee health, the capital of the country, Quito, recently issued an ordinance banning some herbicides and pesticides [33].

A more comprehensive approach to the assessment of the impact of stingless bee breeding and management is required, encompassing social, ecological, and cultural dimensions. This approach will facilitate the development of more effective pollinator-friendly strategies and diversified agricultural systems [34].

Thus, in response to the need to develop tools to improve decision-making and provide guidance for practical actions to reduce and prevent pollinator decline, this survey aims to (i) collect stingless bee keepers' knowledge about the management of stingless bee nests (from the origin of the nest to the harvesting of products); (ii) estimate the nest death rate; (iii) identify specific health risk factors for stingless bee nests; and (iv) develop tools to correlate the application of good management practices with nest losses.

2. Materials and Methods

2.1. Online Survey Development

The free software KoboToolbox (v2022 1.2.) was used to prepare an online questionnaire with 36 questions (Table S1). All questions were configured as mandatory to ensure that all were answered. The anonymity of respondents was maintained. The survey was organized into 4 sections: (i) socio-demographic variables, (ii) biosecurity and product management, (iii) nest management and infrastructure of the farm, and (iv) sanitary and environmental aspects. The questions used for nest death rate estimation were not included in any of the previous groups since the data obtained were directly processed with the formula in Section 2.3 (namely, "Statistical Analysis"). The types of questions included in the questionnaire were single-choice, multiple-choice, and open-ended questions. The survey was available from 23 March 2022 to 31 December 2022, in two languages: Spanish and Portuguese. The target audience was meliponicultors (stingless bee keepers) with experience in managing at least one (1) nest of any stingless bee species in any country of Latin America.

Before the public launch, the questionnaire was reviewed by three experienced stingless bee keepers. They gave points for improvement and suggestions for the survey, for a better understanding of the target audience. After adding these modifications, the survey was officially launched online. The survey link (https://ee.kobotoolbox.org/x/HVbthWiD, accessed on 31 July 2023) was disseminated through social networks (meliponicultors' groups on WhatsApp and Facebook) as well as through e-mails sent to local meliponiculture organizations (when available) and to the authors of scientific articles related to stingless bees. The rationale behind selecting this particular methodology for the survey spread is twofold. Firstly, this is a pilot study designed to test the operationality of a data relation–visualization tool. Secondly, according to the Ecuadorian Observatory of Information and Communication Technologies (TIC), 82.88% of citizens in rural areas with access to a phone use social networks as their primary source of information. Together with

Brazil, Colombia, Costa Rica, and México are included in the medium- and high-Significant Rural Connectivity Index countries [35]. Third, without a national official registry of meliponicultors, we used social networks as a census tool.

2.2. Scoring System Development

The questions in section (i), socio-demographic variables, and other open-ended questions of the inquiry were not included in the subsequent phase of the study.

All answer options, from single-choice and multiple-choice questions, were numerically scored by the authors. The lowest score represented the "worst situation" and the highest score represented the "best situation". The criteria for this scoring considered those answers that were based on scientific evidence and focused on the conservation and guarantee of the best living conditions for stingless bees as a priority and of greater weight. In addition, a consensus was reached among a panel of four experts in biology, epidemiology, meliponiculture, and biosecurity. The panel agreed on the options for each question, from "worst situation" to "best situation".

Each question had different maximum scores. Each section—(ii) biosecurity and product management, (iii) nest management and infrastructure of the farm, and (iv) sanitary and environmental aspects—had a different number of questions. To ensure the fairness, consistency, and accuracy of the weighting of each section on the results, the maximum score was normalized and the minimum difference in the number of questions within each section was targeted.

2.3. Statistical Analysis

Questions were classified into five groups, one including socio-demographic information (INF) and four explaining the application of good management practices (GMP) in meliponiculture: (i) environment and conservation (ENV PROTEC), (ii) producer training and modern techniques (TECHN), (iii) the use of personal protective equipment and biosecurity measures (BIOSEC), and (iv) health care (HEALTH).

The scoring of the questions was applied to those from which quantitative information could be obtained. The maximum was calculated for each question based on the response options and we categorized these options as "best" if they adhered to conservation criteria and "worst" if they were far from it (called "theoretical best score"). To verify the analyses, the same procedure was performed, except that the maximum this time was taken according to the "best" answer given by the respondents (called "best meliponicultor score").

An overall score for each respondent was calculated using the sum of scores obtained for all their responses and the sum of the "best" scores for each question (for explanation, see Table S1).

The calculation of the nest death rate (NDR) of stingless bees was calculated as follows according to the formula modified from [36]:

$$\text{Nest death rate (NDR)} = \frac{\#\text{nest dead}}{\#\text{nest until 2021} + \#\text{nest IN} + \#\text{nest OUT}} \quad (1)$$

The terms inside the numerator and denominator are explained as follows:

#nests dead—the number of nests of stingless bees that died the last year (question (Q) 28);

#nests until 2021—the number of nests of stingless bees that existed until 2021 (Q 27);

#nests IN—the number of nests of stingless bees that were added during the last year (Q 20);

#nests OUT—the number of nests of stingless bees that were sold, donated, or given away during the last year (Q 21).

To determine any relation between the NDR (independent variable) and the overall score (dependent variable), we made a linear correlation test to obtain the Pearson's coefficient. To check the normality of the data (both overall score and NDR), a Kernel density estimation and a Shapiro–Wilk test were performed. A two-sample Wilcoxon rank sum test (Mann–Whitney) was used to test whether melipolicultors who had an NDR of less than 15% and an NDR equal or above 15% belonged to the same population or not.

2.4. Spider Web and Barometer Tools

For a general visualization of the status of meliponiculture, as an activity that must include minimum standards of compliance with GMPs in each group of questions, two tools were developed. The first one, the spider web tool, contrasts the status of each area: information sources, the application of basic biosecurity standards and the use of personal protective equipment, monitoring in health care, and conservation actions. For this purpose, we used the total score obtained per respondent and an average obtained per question group (see Table S2). The result (percentage) given for each group of questions indicates how closely the practices are aligned with what is expected according to scientifically based theoretical criteria. The closer the result is to 100%, the better the practices are considered, and the closer the result is to 0%, the more there is an opportunity for intervention and improvement in that area.

The second one, the barometer tool, ranks the overall status using the average of the above values. It means that from a global perspective, meliponiculture is evaluated and qualified. To determine the status, we divided the barometer bar into three zones, using quartiles (Q1 and Q3) of the overall score. Each zone has an action proposal, i.e., red zone: to write an action plan, implement it, and audit again within a month; orange zone: to take corrective actions and check their implementation; green zone: the management and practices are the best.

3. Results

We collected a total of 94 surveys, of which only 92 were used because two were eliminated during data cleaning and validation. Surveys were collected from 14 Latin American countries (Figure 1). In terms of academic level, a university degree was obtained by the largest percentage of respondents (38%). The mean age of the respondents was 43 years. Experience as a stingless bee keeper ranged from 5 months to 52 years. An average of 48 nests per meliponicultor was calculated. The total number of nests among all respondents amounted to 4548 (by nest, the median = 17, min = 1, and max = 700). Most respondents spent part of their time (about 8 h per week) on the care and management of stingless bees. The individual product with the highest percentage of harvest was honey (16%), followed by a combined harvest that included honey, cerumen, pollen, and geopropolis (63%) (Table 1).

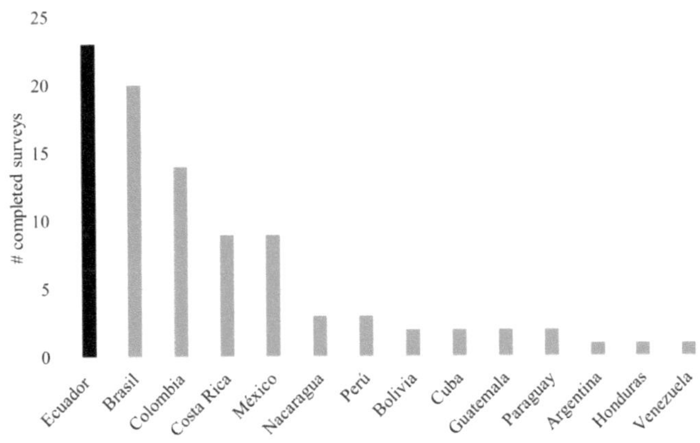

Figure 1. Survey participation by country. The bars represent the number of completed surveys (y-axis) per country (x-axis).

Table 1. Summary of the main socio-demographic variables.

Variable	Range		Percentage
Age (years)	Young	≤28	25
	Adult	>28 and ≤60	63
	Old adult	>60	12
Stingless beekeeping experience (years)	Beginner	≤5	54
	Upper beginner	>5 and ≤10	18
	Intermediate expert	>10 and ≤20	16
	Expert	>20	11
Full academic level	Elementary		1
	High School		29
	Technology		12
	University		38
	Post grade		20
Spending time	Full Time (≥8 h/day)		9
	Part-time (<8 h/day)		23
	Hobby (~8 h/week)		68
Amount of nests (quantity)	≤10		34
	>10 and ≤50		47
	>50 and ≤100		8
	>100		12
Main product harvested from nests	Honey		16
	Geopropolis		4
	Cerumen		3
	Honey, cerumen, geopropolis, pollen		63
	Other reason for nest keeping *		13

* Among other reasons for keeping nests of stingless bees were (i) nest multiplication for sale, (ii) stingless bee conservation, and (iii) protection.

3.1. Environment and Conservation (GMP–CONSERV)

A total of 61% of stingless bee keepers consider that there are one or more sources of pollution around their nests. From the highest to lowest number of reports, there were plantations using agrochemical products for pest control, companies extracting oil and oil derivates (plastics), mining, city pollution (urban meliponiculture), and polluted rivers. In addition, 96% of respondents consider that climate change affects or will affect the life of bees. The same percentage of respondents take climate-friendly actions such as recycling, saving energy, not using agrochemicals for pest control, and planting more plants, and a small percentage of producers (n = 4/92, 4%) mention "agroecology" as a new climate-friendly practice.

The main reason for keeping stingless bees was the conservation of land (93%), pollinators, or biodiversity in general and the conservation of ancestral agricultural heritage in particular. Respondents (n = 21/92, representing 22%) purchased whole nests or brood disks to obtain more stingless bee nests. In general, those who buy nests try to get them from nearby areas (n = 10/21, representing 48%), same region (n = 6/21, representing 28%), or same country (n = 2/21, representing 9%), except in one case (n = 1/21, representing 4%) (international purchase).

A total of 60% of stingless bee keepers feed their managed stingless bees with water, *Apis mellifera* honey, honey from other stingless bee species, commercial food, and processed substances such as sugar, flour, or lemon juice. They do it according to stingless bees' necessity, i.e., breeding seasons, winter/non-flowering, new splits, weak nests/no reserves, and also for the maintenance and stimulation of nests.

3.2. Producer Training and Modern Techniques (GMP–TECHN)

To obtain their first nest, 76% of respondents practiced trapping in the wild. It is important to notice that some other meliponicultors (8%) obtained their first nest by rescuing stingless bee nests that were in significant danger. Respondents (37%) mentioned that they received expert support or some previous training for the transfer of natural nests to wooden boxes for technical nest management. However, a percentage of respondents (n = 12/92, representing 13%) keep nests in natural structures (i.e., hollowed tree trunks).

During nest division, stingless bee keepers confirmed that they ensure the following conditions: the existence of a viable virgin queen and old virgin, the health of and sufficient food for the old nest and the new nest, the seasonal flowering of plants (summer), positioning the new nest and scheduling the time of bees' work that avoids damages or loss of workers, the existence of mature–viable brood discs, an abundant population, and a strong and disease-free nest of origin. Excluding urban meliponiculture, 92% of the producers maintain their nests in open spaces with plants.

The organization of stingless bee nests (meliponaries) was attributed to being specific to the species managed, the size of the bees, their behavior, and the ease with which the nests can be harvested. The most reported conditions are described as follows: at least 1 m above the ground, one nest next to the other, minimum separation between nests of 0.40 to 3 m, nests stacked one on top of the other (condominium or tower blocks), and nests directly on the ground. This survey did not ask species-specific questions about nest organization in a meliponary; thus, the conditions detailed above are a general guide.

Among the places where respondents located their meliponaries were their own land (n = 61/92, representing 67%), common land (n = 21/92, representing 23%), association land (n = 4/92, representing 4%), natural tourist spaces (n = 4/92, representing 4%), and land belonging to academic institutions (n = 2/92, representing 2%).

Academia is the main source of producer training or teaching (n = 57/92, representing 62%). Knowledge sharing among producers is strong (around 28%), with social networks being the main channel of information transfer, where experienced meliponicultors share their knowledge with those who are new to the activity.

3.3. *Use of Personal Protective Equipment and Biosecurity Practices (GMP–BIOSEC)*

One person manages the meliponary in 73% of the cases, while 27% of respondents stated that they do not carry out meliponiculture alone. The accompaniment for activities in the meliponary ranged from 2 to associations of 25 people (Ecuadorian example).

The application of biosecurity practices and the use of appropriate materials are summarized in Figure 2. It is important to mention that hand washing and the use of personal protective equipment (PPE) during regular nest checks had the same behavior in both management cases (one person or more than one person). The main PPE and instruments used for different activities at the surveyed meliponaries are summarized in Table 2. The use of a sterilized material for product storage (n = 75/92, representing 82%) as a biosafety measure was also emphasized in the survey. The main storage conditions for products were as follows: refrigeration (4 °C) (n = 34/75, representing 45%), protection from humidity (n = 25/75, representing 33%), protection from light (n = 12/75, representing 16%), and environmental temperature and freezing (−20 °C) (n = 4/75, representing 5%).

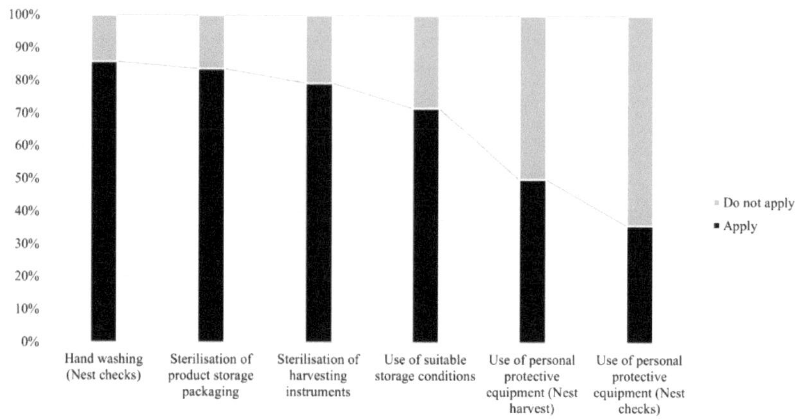

Figure 2. Application of basic biosecurity standards in stingless bee nests. Percentage of compliance (y-axis) with specific biosecurity standards in stingless bee nests (x-axis). Ordered from highest to lowest and differentiated by stage during nest management.

Table 2. Summary of the main biosecurity measures complied with in the key stages of meliponiculture (regular check, harvesting, product storage).

Item	**Activity in the Nest Set (Meliponary)**	
	Regularly Check (n = 33)	Harvesting (n = 45)
(a) Personal Protective Equipment		
Head coverings	30	18
Sterile gloves	15	28
Face mask	9	31
Clean boots	9	
Clothing cover	10	
Protective glasses	5	
Tent for creating a sterile environment		5
	Harvesting (n = 71)	Product storage (n = 75)
(b) Instruments		
Food-grade containers	45	
Spoons or paddles	31	
Syringes	51	
Filters	41	
Palette, knife, scrapers	7	
Vacuum pumps	3	
Glass bottles with lids		66
Plastic bottles with lids		24
Plastic bags with hermetic seals		8
Glass bottles with gas release		1

3.4. Health Care (GMP–HEALTH)

Meliponicultors (n = 62/92, representing 57%) kept a record of activities carried out in their meliponaries. In these records, they have been able to observe aspects such as insects/organisms invading stingless bee nests (73%) and unusual behavior (44%), detailed from the highest to lowest rates of sighting in Figures 3 and 4.

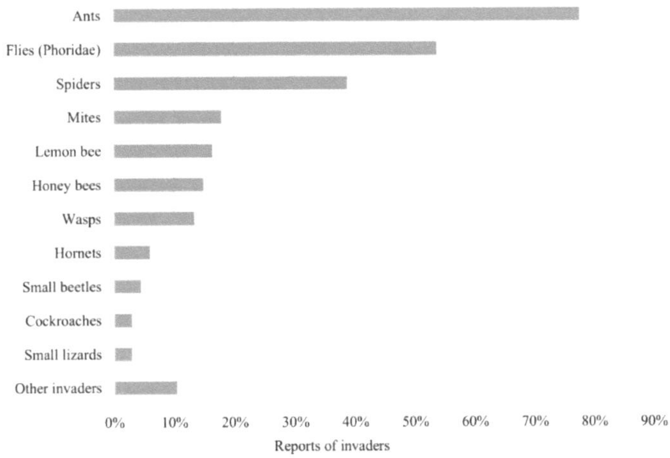

Figure 3. Presence of invaders in stingless bee nests. List (y-axis) and percentage of stingless bee nest invaders reported (x-axis). Sorted from highest to lowest number of reports. Other invaders include just one report of Euglossini and Bombini bees, crickets, mammals, blank soldier fly (*Hermetia illucens*), termites, and arapuá bee (*Trigona spinipes*).

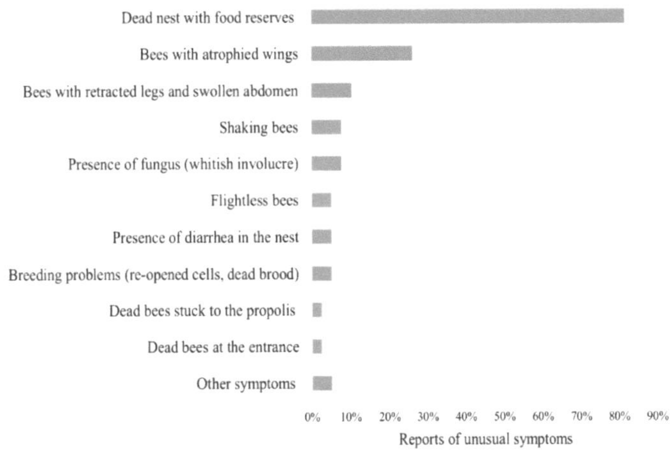

Figure 4. Presence of unusual symptoms in stingless bees. List (y-axis) and percentage of reported unusual clinical symptoms in stingless bees (x-axis). Sorted from most to least severe. Other symptoms include just one report of death by pesticides and invasion by the same species.

The first place in terms of the most commonly reported invaders of stingless bee nests is occupied by ants, followed by Phoridae flies and spiders. The two best-known problem insects for meliponiculture are the phorid fly and the lemon bee (ranked fifth in this study as an invader).

Respondents (n = 28/92, representing 30%) know about nosemosis (no statistically significant effect on NDR, Mann–Whitney test, p-value = 0.262). More than half of the total respondents (n = 52/92, representing 57%) confirmed the existence of apiaries near their meliponaries (no statistically significant effect on NDR, Mann–Whitney test, p-value = 0.733).

Knowledge of nosemosis was not associated with the existence of honey bees near stingless bee nests (no statistically significant correlation between the variables in question, Pearson product–moment correlation test, p-value = 0.219).

Only one meliponicultor replied that he treated his bees with veterinary medicine and did not store this medicine after it was opened (this survey did not collect data regarding the specific type of medicine employed by stingless bee keepers for the treatment of their bees). Among the sources of reference to face and solve unusual health concerns in nests, the meliponicultors answered that 69% prefer to ask other stingless bee keepers, 13% consult an expert (veterinarian), another 13% prefer to experiment by themselves, 9% treat the bees by themselves since they have previous knowledge, and a small 1% go to academic bibliographic sources or theses.

3.5. Relationship between the NDR and the Application of Good Practices in Meliponiculture

Normality was verified for the overall score (dependent variable) (Shapiro–Wilk test, p-value =0.614 for theoretical best score, p-value =0.617 for best meliponicultor score) but not for the NDR (independent variable), giving us a cut-off point = 0.15 (i.e., 15%), which divides the population into two groups based on nest losses (Figure 5).

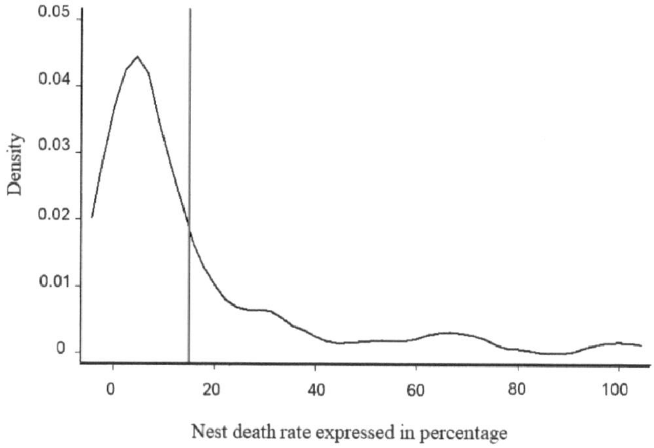

Figure 5. The kernel density estimate of the nest death rate. X-axis: probability density. Y-axis: nest death rate calculated and expressed as a percentage (scale between 0 and 100%). The red vertical line at 15% represents the observed cut-off point to separate the population into two parts.

An inverse relationship was observed between compliance with GMPs and NDR (Figure 6). The linear correlation between variables explained 8% of the NDR concerning the overall score (p-value = 0.005).

The overall scores are significantly different in the two sub-groups of meliponicultors depending on the NDR and considering the cut-off point of 15% (Mann–Whitney test, p-value = 0.001) (Figure 7). The last three calculations were verified by both methods using the best theoretical and best meliponicultor scores.

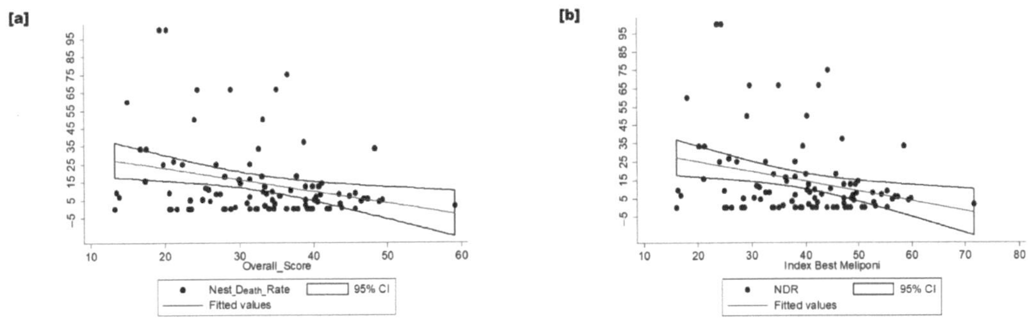

Figure 6. The relation between dependent and independent variables. (**a**) The inverse relation between the overall score and nest death rate. (**b**) The inverse relation between the index of the best meliponicultor and the nest death rate. Legend: NDR—nest death rate.

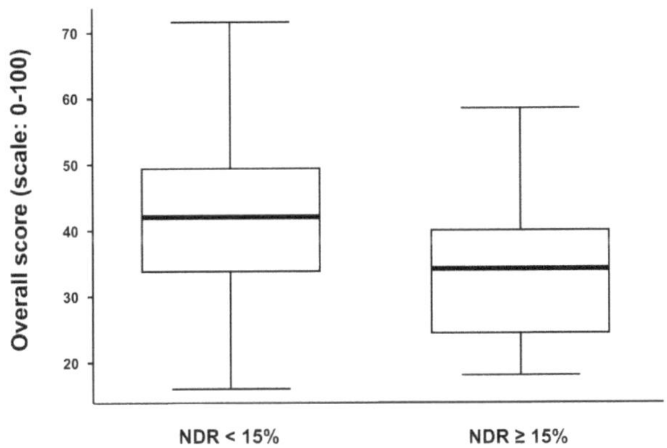

Figure 7. A boxplot of the overall score (y-axis) vs. the estimated nest death rate (x-axis). Population division is visualized considering the estimated mortality rate. NDR: nest death rate. Legend: The horizontal bold line in the rectangle represents the median of the overall score; the solid lines at the top and bottom of each rectangle represent, respectively, the first and third quartiles; adjacent lines to the whiskers represent the limits of the 95% confidence interval.

3.6. Spider Web and Barometer Tools

The spider web tool showed a great socio-demographic status (65.4% of compliance). Items better aligned with scientific theoretical criteria, from the highest to lowest percentage of compliance were as follows: GMPs applied to training and modern techniques, GMPs in healthy controls, GMPs in biosecurity practices, and environmental protection actions (Figure 8a). However, when it is differentiated by the best meliponicultor score, GMP—HEALTH comes in second place, followed by GMP—TECHN, GMP—BIOSEC, and GMP- ENV PROTEC (Figure 8b).

The barometer tool gave a result of 32.6% for the theoretical best score (Figure 9a) and 39.5% for the best meliponicultor (Figure 9b), both right in the middle of the orange zone, which asks respondents to take corrective actions and check their implementation.

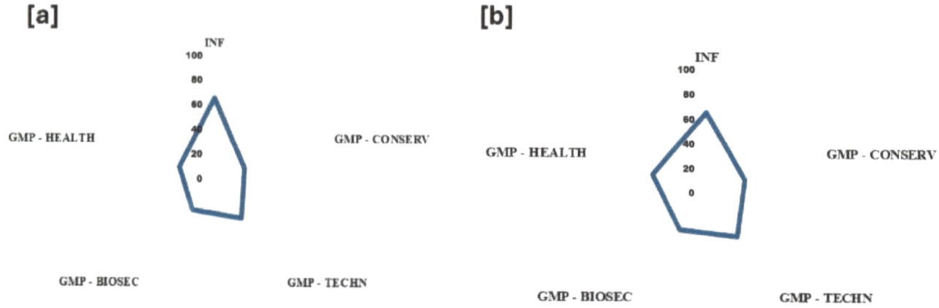

Figure 8. Spider tool. Overview of compliance in each analyzed area: social aspects, modern techniques, health, biosafety, and conservation. (**a**) The percentage of compliance based on the theoretical best score. (**b**) The percentage of compliance based on the score obtained by the best meliponicultor. INF: socio-demographic information. GMP: good management practices. CONSERV: environment and conservation. BIOSEC: biosecurity measures. TECHN: producer training and modern techniques.

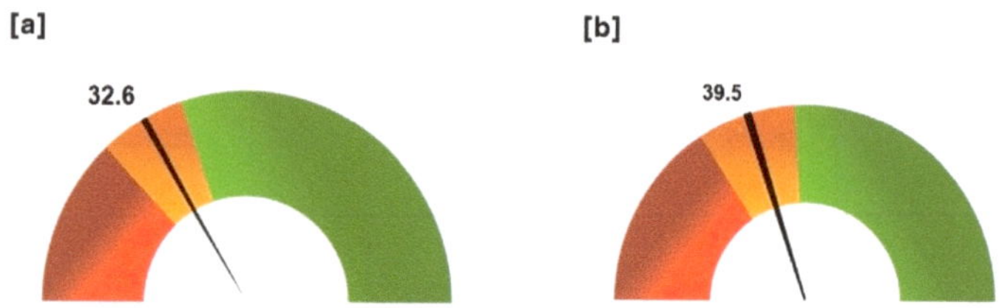

Figure 9. Barometer tool. Summary of the general status of the stingless bee keeper respondent population in terms of good management practice compliance. (**a**) Status based on the theoretical best score (Q1 = 25.8, Q3 = 39.6). (**b**) Status based on the score obtained by the best meliponicultor (Q1 = 31.3, Q3 = 48.03). Legend: The barometer was divided in three zones, using quartiles (Q1 and Q3) of the overall score. Each zone has an action proposal, i.e., red zone: to write an action plan, implement it, and audit again within a month; orange zone: to take corrective actions and check their implementation; green zone: the management and practices are the best.

4. Discussion

This pilot study mainly reached a "sector" of the stingless bee keepers population with access to the internet, a cell phone, or a computer, as well as to studies, which is reflected in the highest percentages of respondents with university and high school education, which may be surprising given the rural reality of the world. In Ecuador, a 2019 study showed a shift in university enrollment among rural youth in a coastal province, largely due to the confidence parents now have in university education [37]. The rise in student demand for distance education has reached 10% per semester, an alternative modality to solve the problem of remote locations, through a system of grants for the implementation of technology at home.

It is important to note that the current statistics about education enrollment do not reflect the reality of the entire rural youth population of Ecuador, let alone Latin America. However, they do provide an approximation of meliponiculture and the potential loss of its 'rurality' in the context of a globalized world. This could potentially result in the loss of ancestral knowledge on meliponiculture, which has been practiced for a considerable length of time [38,39], more than 2000 years [11].

Furthermore, the utilization of technologies, such as these online surveys, facilitated the gathering of data and insights into the contemporary practices and management of meliponiculture. A significant approach was to gain an understanding of the processes involved in the care of stingless bee nests, which is predominantly a collective endeavor involving family members or associations. Thus, knowledge is still inherited, and teamwork [25] helps to reduce errors, since each person assumes a single task.

The survey also shows the participation of the academy with the provision of institutional lands as a strategy for mutual benefit between producers and research. This community work extends knowledge among stingless bee keepers [40]. The hybridization between traditional knowledge and modern stingless beekeeping improves local practices, thus increasing production. If this were the case, above all, it would reduce the chance of colony losses [41].

This study highlights the role of more experienced meliponicultors, since they become sources of new knowledge and promoters of stingless beekeeping. While these examples of collaborative behavior and knowledge transfer are commendable, there is a need to recognize the continued risk associated with the perpetuation of less ethical practices in this field, especially risks associated with the introduction of animal or plant species (nectiferous) that may facilitate the spread of diseases or new predators/competitors. This is exemplified by the case of African tulips [42].

The mean age of stingless bee keepers as well as the variability in years of experience in this study compares with another Ecuadorian study [43], with ages from 22 to 72 years old, and with the average age of Brazilian meliponicultors being 44.1 ± 2.14 for women and 43.4 ± 0.78 for men, including 5.9 ± 0.5 years of experience in stingless beekeeping [44].

As a field activity, stingless beekeeping is a side job in families that practice it, even though the marketing value of honey is around USD 133–200/Kg [45]. As it is a secondary activity, people invest 8 h per week on average. Taking time between revisions helps to keep nests free of pests. Even in critical periods, such as the time after the split, experts recommend checking the new nest every three days for three weeks, and then once a week [46], but above all, meliponicultors should not over-manipulate the brood comb [47].

4.1. Environment and Conservation (GMP–CONSERV)

Regarding stingless bee conservation aspects, a low percentage of respondents purchase nests from outlying areas from meliponaries. However, interregional and one international sale were reported in this survey, making it imperative to create awareness programs on the impact of colony displacement. The consequences of anthropogenic nest displacement have been widely reported [28,48,49].

Feeding stingless bees is appropriate at specific times, i.e., after honey harvest (low nutritional reserves) [50], during non-flowering seasons or harsh winters [51], to strengthen colonies after a split [52], and under pollination greenhouses [53], as well as the cases of urban meliponiculture found in this study. Feeding may include nectar (energy source) or pollen (protein source) replacement, such as the protein substitute in the diet of *Melipona flavolineata* that was tested and accepted under laboratory conditions [54].

It is our contention that the utilization of flour as a pollen substitute in stingless bees is a matter of concern. A study was conducted to evaluate the acceptance of four types of flours in a mixture of honey and water by honey bees. The results demonstrated that all mixtures were accepted, with soybean meal being the most accepted [55]. The quality of nutrition is associated with alterations in the gut microbiota of honey bees, which in turn impact their immune system and susceptibility to pathogens [56]. The impact of flour as a protein substitute in stingless bees remains largely unstudied.

4.2. Producer Training and Modern Techniques (GMP–TECHN)

Producers who followed training courses in meliponiculture were able to make nest divisions and provide adequate supplementary feeding according to the nests' needs [52]. Good nest management depends mainly on the practice and continuity with which it is practiced and the support that can be provided by the academy [57] or field technicians.

A disadvantage of maintaining nests in their natural structures, i.e., tree logs, is difficulty during honey harvesting and the possibility of contamination, as it passes through waste areas [50]. In addition, shaking and turning the nest upside down to let the honey fall by gravity induces the loss of eggs that sink in the larval food, causing nest collapse [57]. Thus, the management suggestion is the use of technical boxes with vertical divisions and separate cavities for the brood chamber as well as for honey and pollen pots so that during the honey harvest, only the storage modules are removed and it would be possible to continue using the gravity honey harvesting technique.

In the case of Mexican "jobones", whose structures are horizontal, single-story structures for brood chambers and food storage, the technique of gravity honey harvesting has been used since the pre-Hispanic Mayas [38] with no major reports of brood collapse. It is therefore possible to attribute this to the density of larval food and suggest that the bee larvae do not 'drown' but remain afloat for a certain time during the gyrus downwards from the nest for harvesting. This last topic merits further in-depth study, as well as the application of vacuum pumps or automated suction devices for honey extraction reported in this study.

The artificial division of colonies is recommended once a year [52]. Among the precautions to be taken during the division of nests are that the nest of origin must have abundant brood discs, a large population, and reserves of honey and pollen [46]. It should be performed at night or in an enclosed space with a mosquito mesh to avoid fly (Phoridae) infestation [58].

A 50/50 method for nest multiplication is being practiced [59]. Thanks to this study, it is possible to add the following suggestions: First, 4–6 brood disks should be transferred to the new nest. In species that build a queen cell, it is recommended that this queen cell should be included in one of the brood disks. It is preferred to feed the new nest 24 h after being transferred and to check it at least twice a week. It is not recommended to transfer pots of honey or pollen in poor conditions [60,61]. All these considerations contribute to making the propagation techniques sustainable and self-sufficient because they will always have new queens available [62].

Trap nests are considered a viable tool to study stingless bee colonies for meliponicultors, researchers, and conservationists [54]. Traps are used to identify species and differentiate their distribution in primary and degraded forests [63]. The use of traps should not be for the over-exploitation of natural resources, as this may generate a disturbance in the ecological balance [64].

The primary motivation for engaging in meliponiculture was conservation, while the primary source of meliponicultors' initial nests was through trapping. This does not necessarily indicate a contradiction but rather a potential deficiency in understanding the true nature of conservation. Trapping may potentially contribute to the unnecessary extraction of stingless bee nests from the wild. The removal of nests from their natural habitat should only occur when stingless bees are at risk, e.g., due to deforestation.

4.3. Use of Personal Protective Equipment and Biosecurity Practices (GMP–BIOSEC)

The implementation of biosecurity measures on a farm prevents the introduction and spread of infectious agents and diseases [65]. For example, the use of personal protection equipment and hygiene were considered protective factors against colony loss in Belgian beekeeping [66]. The use of personal protective equipment as well as sterilized instruments are keys to improving nest management because stingless bee keepers can focus their attention on an activity free of bites or any discomfort that these species can cause [67].

The maintenance of colony hygiene is directly correlated with the safeguarding of bee health and the protection of bee products. Disinfection represents a hygienic measure that is designed to prevent and eliminate agents that are capable of causing infectious diseases in bees. Furthermore, it serves to avoid the contamination of honey and other bee products with harmful microorganisms [68]. Given the toxicity and other negative effects of chemical disinfectants, it is recommended that physical methods of disinfection be employed wherever feasible.

In regard to physical methods of disinfection, the following is recommended for implementation in the field: boiling the instruments in water at normal (atmospheric) pressure for a period of 30 min. It is recommended that instruments be washed with hot water at a temperature of 90 °C or use hot air (110 °C and 150 °C) [69].

Stingless bee honey is characterized by having high moisture in comparison with *A. mellifera* honey, causing a natural fermentation process [70]. This fermentation process made by symbiont microorganisms contributes to the preservation of honey and the transformation of pollen into bee bread [71]. The findings of this study allow us to propose storage conditions for honey: refrigeration (4 °C) and containers that protect from humidity and light.

4.4. Health Care (GMP–HEALTH)

Local experts in Mexico reported attacks on stingless bee nets by different predators [72]. Indeed, the list of predators includes skunks (*Mephitis* sp), *Canis latrans*, *Dasypus novemcinctus*, ants, wasp rams, kleptobiotic stingless bees (*Lestrimelitta chamelensis*), and *A. mellifera*. Some of those predators were reported in this study in Brazil, Costa Rica, Colombia, Ecuador, and Perú.

Both Phoridae flies (*Pseudohypocera kerteszi*) and *Lestrimelitta* sp. can cause the complete loss of stingless bee nests, but the Phoridae fly is considered the most representative risk as far as stingless bee plagues are concerned. At least, *Lestrimelitta* sp is considered a biological population controller of stingless bees. Therefore, the recommendation that is under the control of the stingless bee keepers is the maintenance of hygiene in the nests, especially at the beginning of a transfer from a natural nest to a technical one.

For Phoridae flies, a useful recommendation is to collect all honey and pollen from the nest pots to prevent fly eggs from hatching and to constantly check these three areas of the nest which are the favorite places to start an invasion, and the use of white or red vinegar traps inside the nests [73].

Unusual signs in stingless bees such as extended proboscis, expanded or unhooked wings, wrinkled bodies, and defecation on cage covers are visible signs of poisoning with some agrochemicals (e.g.,): fipronil, cypermethrin, dimethoate, imidacloprid, and indoxacarb [74]. Crippled wings and a contracted abdomen are visible indicators of a possible infection with deformed wing virus (DWV), Israeli acute paralysis virus, and Kashmir bee virus (KBV) [75]. Trembling movements in bees and the inability to fly are reported as signs of acute bee paralysis virus (ABPV) infestation [76]. These unusual behaviors raise alarm bells regarding the health of stingless bees since their signs are similar to those described in *A. mellifera*. However, there are no reports in native bees, except for *Vairimorpha ceranae (Nosema ceranae)* [77].

In this study, the reported proximity of *Apis mellifera* to meliponid sets may present a risk to the health of stingless bees, given the potential for their interaction in the same floral resource during foraging [14]. It has been demonstrated that honeybee pollen loads frequently contain pathogenic protozoa and microsporidia [78]. The utilization of this pollen as a food source for stingless bee nests suggests a heightened probability of the transmission of infectious agents. Nevertheless, research has demonstrated that propolis derived from stingless bees can effectively mitigate the progression of *Nosema* infections in honey bees [79]. It is possible that propolis, a resinous substance used by stingless bees in the construction of their nests, may offer protection against *Nosema* infection.

It is therefore recommended that the use of honey bee products in stingless bee nests be avoided. In cases where the use of such products is unavoidable and within the reach of stingless bee keepers, it is advised that they verify that the products do not contain any agents or substances that could prove harmful to the stingless bees.

The natural ecology of stingless bees includes natural biological controllers such as lemon bees and phorids [80], as well as their natural competitive relationships, such as fights with solitary bees for resources [81]. These examples also cause morphological damage and even death to stingless bees. It is recommended to examine this symptomatology in depth and make accurate diagnoses of possible viruses or bacteria that are pathogenic to native bees.

Registering activities such as unusual behaviors, invasions, death, and other aspects in the meliponary [25] can be used as a basis for creating or providing records that can be submitted to or socialized with legal entities for regularization and health surveillance purposes.

4.5. Developed Tools

The annual calculation of the death rate under technical management conditions and without considering the difference in calculation between forage and non-forage stingless bees compares with the natural nest death rate reported at 13% for stingless bees [82] and 10% for honey bees [83]. Therefore, this value of the death rate in stingless bees should be considered an acceptable level of colony loss rates under domestic management. It was also verified that the better the compliance with good management practices, the lower the loss or mortality (inverse relationship).

The three main groups of causes associated with an increase in nest loss, namely GMP-CONSERV, can be attributed to two key factors: the high prevalence of polluting sources in close proximity to the meliponaries and the growing consensus regarding the adverse impact of climate change. Additionally, the GMP-BIOSEC group is included due to the dearth of adherence to fundamental biosecurity standards during nest inspection and product harvesting. This is a significant concern for the preservation of nest health and the quality of the products obtained. Finally, the GMP-HEALTH group is of note for the high number of reports of nest-invading insects causing nest collapse, as well as the observation of unusual behaviors in bees. These observations are comparable to those made in honey bees, but it is unclear whether the same causal and effect relationships can be applied to stingless bees.

The spider web and barometer tools are pedagogic instruments to interact with meliponicultors and identify margins of improvement. The interpretation of the spider tool means that the sources of information, experience, and management practices of meliponicultors are alienated to extend stingless bees' life, as well as environmental protection, according to scientific theoretical criteria. At the same time, the barometer tool confirms the widely discussed need for the implementation of good management practices.

The benchmarking made for score assignment showed that meliponiculture should have its guidelines, and even within meliponiculture, management should be separated according to the stingless bee species being managed, according to the region where the activity is developed, and according to the scientific information that each country generates.

The limitations of the present pilot study can be attributed to the continuous growth of meliponiculture and therefore research, since we only have three examples of developing tools for the evaluation of stingless beekeeping, in Mexico, Brazil, and Costa Rica. People dedicated to this activity are located mainly in rural zones, and the lack of access to internet sources (the main medium of dispersion of this pilot survey), is a limitation. The reliability that researchers can create with producers must be considered.

Despite evidence of the positive influence of the training and education of stingless bee keepers, more programs of this kind should be created or research results should be disseminated in the language of stingless bee keepers and on freely accessible platforms, as a large percentage base their management practices on the advice of others meliponicultors. Improved management and risk control in meliponiculture should be addressed using this economic activity as a tool inside agroecological systems. A loss/death rate calculation will improve long-term nest management conditions. Finally, we recommend the application and socialization of spider and barometer tools with meliponicultors in the field through an app.

5. Conclusions

Stingless beekeeping in Latin America, especially in Ecuador, is growing rapidly. Fortunately, guidelines related to biosecurity show acceptable nest management. However, some items need to be addressed to ensure better health: global compliance with biosecurity measures, actions for the care of the environment in which stingless bees live, the quality and efficiency of technology in the handling and management of nests, and the diagnosis/monitoring of the health status of stingless bees.

Hand washing and sterilization are applied during management and constitute a very good basis for turning meliponiculture into a sustainable practice.

Risk factors for the conservation of stingless bees include the effect of the introduction of species such as the European honey bee as a potential disease disperser, the use of agrochemicals, the pollution that bees face, and the effect of anthropogenic activities such as colony movement that are not aligned with good management practices.

Honey, as the main product harvested, must have an adequately good management procedure from harvesting to storage, due to its unique physical and chemical characteristics. However, it can become complex as the number of nests increases.

The nest death rate calculated here does not exceed the naturally calculated rate by far. It is a good indicator that the human practice is performed in a good way. However, the application of practices that were found to be missing in this study could reduce this percentage to a more acceptable number.

Graphic tools such as the spider and the barometer are instruments for the empowerment of each meliponicultor, as they help in the field and instantly help detect shortcomings to be corrected after entering some parameters.

Supplementary Materials: The following supporting information can be downloaded at: https://www.mdpi.com/article/10.3390/insects15090715/s1, Table S1: Dispersed online questionnaire 'Good management practices for stingless bees'. Table S2: Example of calculation of the score by group of questions and the overall score.

Author Contributions: Conceptualization, J.S.O.-C., S.M.-S. and C.S.; Methodology, J.S.O.-C. and C.S.; Software, J.S.O.-C. and C.S.; Validation, S.M.-S. and C.S.; Formal analysis, J.S.O.-C.; Investigation, J.S.O.-C.; Resources, C.S.; Data curation, J.S.O.-C. and S.M.-S.; Writing—original draft preparation, J.S.O.-C.; Writing—review and editing, S.M.-S. and C.S.; Visualization, S.M.-S. and C.S.; Supervision, S.M.-S. and C.S.; Project administration, C.S.; Funding acquisition, C.S. All authors have read and agreed to the published version of the manuscript.

Funding: This research received no external funding. The first author of this article has a doctoral mobility grant funded by ARES (Academy for Research and Higher Education).

Informed Consent Statement: Informed consent was obtained from all subjects involved in the study.

Data Availability Statement: The data that support the findings of this study are available from the corresponding authors upon reasonable request.

Conflicts of Interest: The authors declare no conflicts of interest. The public funders had no role in the design of the study; in the collection, analysis, or interpretation of data; in the writing of the manuscript; or in the decision to publish the results.

References

1. Ritchie, H. How Much of the World's Food Production Is Dependent on Pollinators? Our World Data [Internet]. 2021. Available online: https://ourworldindata.org/pollinator-dependence (accessed on 28 August 2024).
2. Klein, A.M.; Vaissière, B.E.; Cane, J.H.; Steffan-Dewenter, I.; Cunningham, S.A.; Kremen, C.; Tscharntke, T. Importance of pollinators in changing landscapes for world crops. *Proc. R. Soc. B Biol. Sci.* **2007**, *274*, 303–313. [CrossRef] [PubMed]
3. Villanueva, R.; Roubik, D.W.; Colli-Ucán, W. Extinction of *Melipona beecheii* and traditional beekeeping in the Yucatán peninsula. *Bee World* **2005**, *86*, 35–41. [CrossRef]
4. Moure, J.S.; Urban, D.; Melo, G. *Catalogue of bees (Hymenoptera, Apoidea) in the Neotropical Region [Internet]*; Sociedade Brasileira de Entomologia: Curitiba, Brazil, 2013; 1058p. Available online: http://www.moure.cria.org.br/catalogue (accessed on 11 November 2023).
5. Michener, C.D. The Meliponini. In *Pot-Honey: A Legacy of Stingless Bees [Internet]*; Vit, P., Pedro, S.R.M., Roubik, D., Eds.; Springer: New York, NY, USA, 2013; pp. 3–17. [CrossRef]
6. May-Itzá, W.d.J.; Peña, W.L.; De la Rúa, P.; Quezada-Eúan, J.J.G. A genetic and morphological survey to trace the origin of *Melipona beecheii* (Apidae: Meliponini) from Cuba. *Apidologie* **2019**, *50*, 859–870. [CrossRef]
7. Engel, M.S.; Rasmussen, C.; Ayala, R.; de Oliveira, F.F. Stingless bee classification and biology (Hymenoptera, Apidae): A review, with an updated key to genera and subgenera. *ZooKeys* **2023**, *1172*, 239–312. [CrossRef] [PubMed]

8. Roubik, D.W. 100 Species of Meliponines (Apidae: Meliponini) in a Parcel of Western Amazonian Forest at Yasuní Biosphere Reserve, Ecuador. In *Pot-Pollen in Stingless Bee Melittology [Internet]*; Vit, P., Pedro, S.R.M., Roubik, D.W., Eds.; Springer International Publishing: Cham, Switzerland, 2018; pp. 189–206. [CrossRef]
9. Vit, P.; Pedro, S.R.M.; Maza, F.; Ramírez, V.M.; Frisone, V. Diversity of Stingless Bees in Ecuador, Pot-Pollen Standards, and Meliponiculture Fostering a Living Museum Meliponini of the World. In *Pot-Pollen in Stingless Bee Melittology [Internet]*; Vit, P., Pedro, S.R.M., Roubik, D.W., Eds.; Springer International Publishing: Cham, Switzerland, 2018; pp. 207–227. Available online: http://link.springer.com/10.1007/978-3-319-61839-5_15 (accessed on 26 July 2021).
10. Chan Mutul, G.A.; Vera Cortés, G.; Aldasoro Maya, E.M.; Sotelo Santos, L.E. Retomando saberes contemporáneos. Un análisis del panorama actual de la meliponicultura en Tabasco. *Estud. Cult. Maya* **2019**, *LIII*, 289–326. [CrossRef]
11. Aldasoro Maya, E.M.; Rodríguez Robles, U.; Martínez Gutiérrez, M.L.; Chan Mutul, G.A.; Avilez López, T.; Morales, H.; Ferguson, B.G.; Mérida Rivas, J.A. Stingless bee keeping: Biocultural conservation and agroecological education. *Front. Sustain. Food Syst.* **2023**, *6*. Available online: https://www.frontiersin.org/articles/10.3389/fsufs.2022.1081400 (accessed on 27 April 2023). [CrossRef]
12. Freitas, B.M.; Imperatriz-Fonseca, V.L.; Medina, L.M.; Kleinert, A.d.M.P.; Galetto, L.; Nates-Parra, G.; Quezada-Euán, J.J.G. Diversity, threats and conservation of native bees in the Neotropics. *Apidologie* **2009**, *40*, 332–346. [CrossRef]
13. Decourtye, A.; Alaux, C.; Le Conte, Y.; Henry, M. Toward the protection of bees and pollination under global change: Present and future perspectives in a challenging applied science. *Curr. Opin. Insect Sci.* **2019**, *35*, 123–131. [CrossRef]
14. Singh, R.; Levitt, A.L.; Rajotte, E.G.; Holmes, E.C.; Ostiguy, N.; vanEngelsdorp, D.; Lipkin, W.I.; dePamphilis, C.W.; Toth, A.L.; Cox-Foster, D.L. RNA Viruses in Hymenopteran Pollinators: Evidence of Inter-Taxa Virus Transmission via Pollen and Potential Impact on Non-Apis Hymenopteran Species. *PLoS ONE* **2010**, *5*, e14357. [CrossRef]
15. Macías-Macías, J.O.; Tapia-Rivera, J.C.; De la Mora, A.; Tapia-González, J.M.; Contreras-Escareño, F.; Petukhova, T.; Morfin, N.; Guzman-Novoa, E. *Nosema ceranae* causes cellular immunosuppression and interacts with thiamethoxam to increase mortality in the stingless bee *Melipona colimana*. *Sci. Rep.* **2020**, *10*, 17021. [CrossRef]
16. Nates-Parra, G. Las Abejas sin aguijón (Hymenoptera: Apidae: Meliponini) de Colombia. *Biota Colomb [Internet]* **2001**, *2*. Available online: http://revistas.humboldt.org.co/index.php/biota/article/view/101 (accessed on 13 February 2023).
17. Ross, C.; Fildes, S.; Millington, A. Land-Use and Land-Cover Change in the Páramo of South-Central Ecuador, 1979–2014. *Land* **2017**, *6*, 46. [CrossRef]
18. Fearnside, P. The Roles and Movements of Actors in the Deforestation of Brazilian Amazonia. *Ecol. Soc.* **2008**, *13*. Available online: https://www.ecologyandsociety.org/vol13/iss1/art23/ (accessed on 13 February 2023). [CrossRef]
19. Stierlin, E.; Szabo, H. *Manual de manejo de abejas nativas: Suro y obobosi (Scaptotrigona spp.)*, 1st ed.; Aguarague: Abanay, Bolivia, 2004.
20. von Zuben, L.G.; Nunes, T.M. A scientific note on the presence of functional tibia for pollen transportation in the robber bee *Lestrimelitta limao* Smith (Hymenoptera: Apidae: Meliponini). *Sociobiology* **2014**, *61*, 570–572. [CrossRef]
21. Vázquez, M.; Muñoz, D.; Medina, R.; Paxton, R.J.; de Oliveira, F.F.; Quezada-Euán, J.J.G. Sympatric cleptobiotic stingless bees have species-specific cuticular profiles that resemble their hosts. *Sci. Rep.* **2022**, *12*, 2621. [CrossRef]
22. Slaa, E.J. Population dynamics of a stingless bee community in the seasonal dry lowlands of Costa Rica. *Insectes Sociaux* **2006**, *53*, 70–79. [CrossRef]
23. Requier, F.; Leyton, M.S.; Morales, C.L.; Garibaldi, L.A.; Giacobino, A.; Porrini, M.P.; Rosso-Londoño, J.M.; Velarde, R.A.; Aignasse, A.; Aldea-Sánchez, P.; et al. First large-scale study reveals important losses of managed honey bee and stingless bee colonies in Latin America. *Sci. Rep.* **2024**, *14*, 10079. [CrossRef]
24. Formato, G.; Smulders, F.J.M. Risk management in primary apicultural production. Part 1: Bee health and disease prevention and associated best practices. *Vet. Q.* **2011**, *31*, 29–47. [CrossRef]
25. Cortopassi-Laurino, M.; Imperatriz-Fonseca, V.L.; Roubik, D.W.; Dollin, A.; Heard, T.; Aguilar, I.; Venturieri, G.C.; Eardley, C.; Nogueira-Neto, P. Global meliponiculture: Challenges and opportunities. *Apidologie* **2006**, *37*, 275–292. [CrossRef]
26. Ramírez, J. Producción y Comercialización de miel de abejas Meliponas en la ciudad de Quito [Internet]. Tesis de pregrado, Universidad de las Américas, Quito, Ecuador, 2016. Available online: https://dspace.udla.edu.ec/bitstream/33000/5958/1/UDLA-EC-TINI-2016-134.pdf (accessed on 27 November 2023).
27. Secretaria de Medio Ambiente y Recursos Naturales. La meliponicultura en México: Un acercamiento a las prácticas tradicionales y a las perspectivas de su manejo contemporáneo. [Internet]. México. 2023; 144p. Available online: https://www.gob.mx/semarnat/polinizadores (accessed on 27 November 2023).
28. Quezada-Euán, J.J.G.; May-Itzá, W.J.; de la Rúa, P.; Roubik, D.W. From neglect to stardom: How the rising popularity of stingless bees threatens diversity and meliponiculture in Mexico. *Apidologie* **2022**, *53*, 70. [CrossRef]
29. Rosso, J.M.; Imperatriz-Fonseca, V.L.; Cortopassi-Laurino, M. Meliponicultura en Brasil. In *I: Situacion em 2001 y Perspectivas*; El Colegio de la Frontera Sur: Mérida, Mexico, 2001; pp. 28–35.
30. Al Naggar, Y.; Estrella-Maldonado, H.; Paxton, R.J.; Solís, T.; Quezada-Euán, J.J.G. The Insecticide Imidacloprid Decreases Nannotrigona Stingless Bee Survival and Food Consumption and Modulates the Expression of Detoxification and Immune-Related. *Genes Insects* **2022**, *13*, 972. [CrossRef] [PubMed]
31. INEN. *Norma Técnica Ecuatoriana Obligatoria. Miel de Abejas, Requisitos*; NTE INEN 1572:2016; INEN: Quito, Ecuador, 2016.
32. Vizcaíno, D.; Betancourt, R. *Buenas Prácticas Apícolas [Internet]*; AGROCALIDAD: Quito, Ecuador, 2015; 72p. Available online: https://www.agrocalidad.gob.ec/guia-de-buenas-practicas-apicolas/ (accessed on 28 November 2023).

33. Concejo Metropolitano de Quito. *Ordenanza Metropolitana No. 041-2022 [Internet]*; Concejo Metropolitano de Quito: Quito, Ecuador, 2022; p. 38. Available online: https://www7.quito.gob.ec/mdmq_ordenanzas/Administraci%C3%B3n%202019-2023/Ordenanzas/2022/ORD-041-2022-MET-Arbolado-Urbano.pdf (accessed on 28 November 2023).
34. Martínez-Fortún, S.; Ruiz, C.; Quijano, N.A.; Vit, P. Rural-Urban Meliponiculture and Ecosystems in Neotropical Areas. *Scaptotrigona*, a Resilient Stingless Bee? In *Pot-Pollen in Stingless Bee Melittology [Internet]*; Vit, P., Pedro, S.R.M., Roubik, D.W., Eds.; Springer International Publishing: Cham, Switzerland, 2018; pp. 421–434. Available online: http://link.springer.com/10.1007/978-3-319-61839-5_30 (accessed on 26 July 2021).
35. Ziegler, S.; Arias Segura, J.; Agroalimentaria, P.D.A. *Conectividad rural en América Latina y el Caribe. Estado de situación y acciones para la digitalización y desarrollo sostenible [Internet]*; IICA: San José, Costa Rica, 2022. Available online: https://repositorio.iica.int/handle/11324/21350 (accessed on 7 August 2024).
36. van Engelsdorp, D.; Lengerich, E.; Spleen, A.; Dainat, B.; Cresswell, J.; Baylis, K.; Nguyen, B.K.; Soroker, V.; Underwood, R.; Human, H.; et al. Standard epidemiological methods to understand and improve *Apis mellifera* health. *J. Apic. Res.* **2013**, *52*, 1–16. [CrossRef]
37. Véliz Briones, V.F.; Zambrano Cedeño, E.R. Zona rural y su nueva visión de la Educación Superior en Ecuador. *Rev. Espac.* **2019**, *40*, 10.
38. Martínez-Puc, J.F.; Cetzal-Ix, W.; Magaña-Magaña, M.A.; López-Castilla, H.M.; Noguera-Savelli, E. Ecological and socioeconomic aspects of meliponiculture in the Yucatan Peninsula, Mexico. Agro Product [Internet]. 28 February 2022. Available online: https://revista-agroproductividad.org/index.php/agroproductividad/article/view/2108 (accessed on 27 November 2023).
39. Kidane, A.A.; Tegegne, F.M.; Tack, A.J.M. Indigenous knowledge of ground-nesting stingless bees in southwestern Ethiopia. *Int. J. Trop Insect Sci.* **2021**, *41*, 2617–2626. [CrossRef]
40. Jaffé, R.; Pope, N.; Carvalho, A.T.; Maia, U.M.; Blochtein, B.; de Carvalho, C.A.L.; Carvalho-Zilse, G.A.; Freitas, B.M.; Menezes, C.; Ribeiro, M.d.F.; et al. Bees for Development: Brazilian Survey Reveals How to Optimize Stingless Beekeeping. *PLoS ONE* **2015**, *10*, e0121157. [CrossRef]
41. May-Itzá, W.d.J.; Martínez-Fortún, S.; Zaragoza-Trello, C.; Ruiz, C. Stingless bees in tropical dry forests: Global context and challenges of an integrated conservation management. *J. Apic. Res.* **2022**, *61*, 642–653. [CrossRef]
42. Queiroz, A.C.M.; Contrera, F.A.L.; Venturieri, G.C. The effect of toxic nectar and pollen from *Spathodea campanulata* on the worker survival of *Melipona fasciculata* Smith and *Melipona seminigra* Friese, two Amazonian stingless bees (Hymenoptera: Apidae: Meliponini). *Sociobiology* **2014**, *61*, 536–540. [CrossRef]
43. Martínez-Fortún Martínez, M.d.l.S. Desarrollo sostenible y conservación etnoecológica a través de la meliponicultura, en el sur de Ecuador [Internet]. Master's Thesis, Universidad Internacional de Andalucía, Sevilla, Spain, 2015. Available online: https://dspace.unia.es/handle/10334/3519 (accessed on 3 September 2024).
44. Dias de Freitas, C.; Oki, Y.; Resende, F.M.; Zamudio, F.; Simone de Freitas, G.; Moreira de Rezende, K.; Amaro de Souza, F.; De Jong, D.; Quesada, M.; Siqueira Carvalho, A.; et al. Impacts of pests and diseases on the decline of managed bees in Brazil: A beekeeper perspective. *J. Apic. Res.* **2023**, *62*, 969–982. [CrossRef]
45. Zawawi, N.; Zhang, J.; Hungerford, N.L.; Yates, H.S.A.; Webber, D.C.; Farrell, M.; Tinggi, U.; Bhandari, B.; Fletcher, M.T. Unique physicochemical properties and rare reducing sugar trehalulose mandate new international regulation for stingless bee honey. *Food Chem.* **2022**, *373*, 131566. [CrossRef]
46. Guzmán Díaz, M.Á.; Balboa Aguilar, C.C.; Vandame, R.; Albores González, M.L.; González Acereto, J.Á. (Eds.) *Manejo de las Abejas Nativas sin Aguijón en México: Melipona Beecheii y Scaptotrigona Mexican*; San Cristóbal de Las Casas; El Colegio de la Frontera Sur: Chiapas, Mexico, 2011; 59p.
47. Sommeijer, M.J. Beekeeping with stingless bees: A new type of hive. *Bee World* **1999**, *80*, 70–79. [CrossRef]
48. Byatt, M.A.; Chapman, N.C.; Latty, T.; Oldroyd, B.P. The genetic consequences of the anthropogenic movement of social bees. *Insectes Sociaux* **2016**, *63*, 15–24. [CrossRef]
49. Chapman, N.C.; Byatt, M.; Cocenza, R.D.S.; Nguyen, L.M.; Heard, T.A.; Latty, T.; Oldroyd, B.P. Anthropogenic hive movements are changing the genetic structure of a stingless bee (*Tetragonula carbonaria*) population along the east coast of Australia. *Conserv. Genet.* **2018**, *19*, 619–627. [CrossRef]
50. León Contrera, F.A.; Menezes, C.; Venturieri, G.C. New horizons on stingless beekeeping (Apidae, Meliponini). 2011. Available online: https://www.alice.cnptia.embrapa.br/alice/handle/doc/906787 (accessed on 28 April 2023).
51. Vazhacahrickal, P.J.; Jose, S. *Stingless Bees Culture (Meliponiculture) in Kerala: Hand Book for Farmers*; Amazon Publishers: Seattle, WA, USA, 2018; 72p.
52. Nates-Parra, G.; Rosso-Londoño, J.M. Diversity of *Stingless Bees* (Hymenoptera: Meliponini) Used in Meliponiculture in Colombia. *Acta Biológica Colomb.* **2013**, *18*, 415–426.
53. Venturieri, G. Uso de Melipona (Apidae, Meliponini) na polinização de solanáceas em casa de vegetação. In *Encontro sobre Abelhas*; Holos, Editora: Ribeirão Preto, Brazil, 2010; pp. 220–224.

54. Costa, L.; Venturieri, G.C. Diet impacts on *Melipona flavolineata* workers (Apidae, Meliponini). *J. Apic. Res.* **2009**, *48*, 38–45. [CrossRef]
55. Usha, U.; Srivastava, P.; Goswami, V.; Khan, M.S. Exploration of various flours as pollen substitutes for *Apis mellifera*, L. during Dearth period at Tarai region of Uttarakhand, India. *J. Appl. Nat. Sci.* **2014**, *6*, 812–815. [CrossRef]
56. Ricigliano, V.A.; Williams, S.T.; Oliver, R. Effects of different artificial diets on commercial honey bee colony performance, health biomarkers, and gut microbiota. *BMC Vet. Res.* **2022**, *18*, 52. [CrossRef]
57. Jungnickel, H.; Velthuis, H.H.W.; Imperatriz Fonseca, V.L.; Morgan, E.D. Chemical properties allow stingless bees to place their eggs upright on liquid larval food. *Physiol. Entomol.* **2001**, *26*, 300–305. [CrossRef]
58. Parra Argüello, F.Y.; Martin Calderon, E.V.; Navarrete Cante, R.A. La Meliponicultura una práctica tradicional para el Desarrollo Regional de la Comunidad de Maní, Yucatán. In *México: Universidad Nacional Autónoma de México y Asociación Mexicana de Ciencias para el Desarrollo Regional*. 2018; pp. 720–738. Available online: http://ru.iiec.unam.mx/3854/ (accessed on 14 June 2023).
59. Kerr, W.E. Criando as abelhas indígenas. *Chác E Quintais* **1945**, *72*, 472–473.
60. González-Acereto, J.; de Araujo Freitas, J. *Manual de Meliponicultura Mexicana*; Fundación Produce Guerrero, Universidad Autónoma de Yucatán: México, Mexico, 2005.
61. Nogueira-Neto, P. *Vida e Criação de Abelhas indígenas sem ferrão*; Editora Nogueirapis: São Paulo, Brazil, 1997; 445p.
62. FAO. Commission on Genetic Resources for Food and Agriculture. 2021. Sustainable Use and Conservation of Invertebrate Pollinators, Including Honeybees. Available online: https://www.fao.org/3/ng879en/ng879en.pdf (accessed on 7 July 2024).
63. Samejima, H.; Marzuki, M.; Nagamitsu, T.; Nakasizuka, T. The effects of human disturbance on a stingless bee community in a tropical rainforest. *Biol. Conserv.* **2004**, *120*, 577–587. [CrossRef]
64. Alonso-Fernández, P.; Regueiro-Ferreira, R.M. Extractivism, ecologically unequal exchange and environmental impact in South America: A study using Material Flow Analysis (1990–2017). *Ecol. Econ.* **2022**, *194*, 107351. [CrossRef]
65. Arzey, G. *NSW Biosecurity Guidelines for Free Range Poultry Farms*; NSW Department of Primary Industries: Camden, NSW, Australia, 2007; 15p.
66. El Agrebi, N.; Steinhauer, N.; Tosi, S.; Leinartz, L.; de Graaf, D.C.; Saegerman, C. Risk and protective indicators of beekeeping management practices. *Sci. Total Environ.* **2021**, *799*, 149381. [CrossRef] [PubMed]
67. Ingrao, A.J. Equipment and Safety. In *Honey Bee Medicine for the Veterinary Practitioner [Internet]*; John Wiley & Sons, Ltd.: Hoboken, NJ, USA, 2021; pp. 149–166. Available online: https://onlinelibrary.wiley.com/doi/abs/10.1002/9781119583417.ch11 (accessed on 23 May 2023).
68. Bojanic Rasovic, M. The most important methods of disinfection in Beekeeping. *Agric. For.* **2021**, *67*. Available online: http://www.agricultforest.ac.me/paper.php?id=3071 (accessed on 4 September 2024). [CrossRef]
69. Bedná, M.; Dolínek, J.; Haklová, M.; Jaša, T.; Kamler, F.; Tit, D.; Veselý, V. *Hygiene in the Apiary (A Manual for Hygienic Beekeeping)*; EU FP6 research project BeeShop; Bee Research Institute: Maharashtra, India, 2009; 30p.
70. Sodré, G.d.S.; Carvalho, C.A.L.d.; Oliveira Fonseca, A.A.; Alves, R.M.d.O.; Souza, B.d.A. Perfil sensorial e aceitabilidade de méis de abelhas sem ferrão submetidos a processos de conservação. *Food Sci. Technol.* **2008**, *28*, 72–77. [CrossRef]
71. Toninato de Paula, G.; Menezes, C.; Pupo, M.T.; Rosa, C.A. Stingless bees and microbial interactions. *Curr. Opin. Insect Sci.* **2021**, *44*, 41–47. [CrossRef] [PubMed]
72. Reyes-González, A.; Camou-Guerrero, A.; Reyes-Salas, O.; Argueta, A.; Casas, A. Diversity, local knowledge and use of stingless bees (Apidae: Meliponini) in the municipality of Nocupétaro, Michoacan, Mexico. *J. Ethnobiol. Ethnomedicine* **2014**, *10*, 47. [CrossRef] [PubMed]
73. Martins de Oliveria, A.; Venturieri, G.; León, F. Studies on the Control of Phorid flies (Diptera, Phoridae) parasites of Stingless bees (Apidae, Meliponini). In *Anais do X Encontro sobre Abelhas*; Ribeirão Preto: Câmara Brasileira do Livro, SP, Brazil, 2012; p. 533.
74. Chapuisat, M.; Oppliger, A.; Magliano, P.; Christe, P. Wood ants use resin to protect themselves against pathogens. *Proc. Biol. Sci.* **2007**, *274*, 2013–2017. [CrossRef]
75. de Miranda, J.R.; Genersch, E. Deformed wing virus. *J. Invertebr. Pathol.* **2010**, *103*, S48–S61. [CrossRef]
76. de Miranda, J.R.; Cordoni, G.; Budge, G. The Acute bee paralysis virus–Kashmir bee virus–Israeli acute paralysis virus complex. *J. Invertebr. Pathol.* **2010**, *103*, S30–S47. [CrossRef]
77. Porrini, M.P.; Porrini, L.P.; Garrido, P.M.; de Melo e Silva Neto, C.; Porrini, D.P.; Muller, F.; Nuñez, L.A.; Alvarez, L.; Iriarte, P.F.; Eguaras, M.J. *Nosema ceranae* in South American Native Stingless Bees and Social Wasp. *Microb Ecol.* **2017**, *74*, 761–764. [CrossRef]
78. Graystock, P.; Jones, J.C.; Pamminger, T.; Parkinson, J.F.; Norman, V.; Blane, E.J.; Rothstein, L.; Wäckers, F.; Goulson, D.; Hughes, W.O.H. Hygienic food to reduce pathogen risk to bumblebees. *J. Invertebr. Pathol.* **2016**, *136*, 68–73. [CrossRef]
79. Naree, S.; Benbow, M.E.; Suwannapong, G.; Ellis, J.D. Mitigating *Nosema ceranae* infection in western honey bee (*Apis mellifera*) workers using propolis collected from honey bee and stingless bee (*Tetrigona apicalis*) hives. *J. Invertebr. Pathol.* **2021**, *185*, 107666. [CrossRef] [PubMed]
80. Disney, H. *Scuttle Flies: The Phoridae*; Springer Science & Business Media: Berlin/Heidelberg, Germany, 2012; 479p.
81. Wcislo, W.T.; Cane, J.H. Floral Resource Utilization by Solitary Bees (Hymenoptera: Apoidea) and Exploitation of Their Stored Foods by Natural Enemies. *Annu. Rev. Entomol.* **1996**, *41*, 257–286. [CrossRef] [PubMed]

82. Roubik, D.W. Stingless bee nesting biology. *Apidologie* **2006**, *37*, 124–143. [CrossRef]
83. El Agrebi, N.; Steinhauer, N.; Renault, V.; de Graaf, D.C.; Saegerman, C. Beekeepers perception of risks affecting colony loss: A pilot survey. *Transbound. Emerg. Dis.* **2022**, *69*, 579–590. [CrossRef]

insects

MDPI

Article

Japanese Honeybees (*Apis cerana japonica* Radoszkowski, 1877) May Be Resilient to Land Use Change

Philip Donkersley [1,*], Lucy Covell [1] and Takahiro Ota [2]

1 Lancaster Environment Centre, Lancaster University, Lancaster LA1 4YQ, UK; lucy.covell@gmail.com
2 Graduate School of Fisheries and Environmental Science, Nagasaki University, Nagasaki 852-8521, Japan; takahiro@nagasaki-u.ac.jp
* Correspondence: donkersleyp@gmail.com

Simple Summary: Pollinators are threatened globally by growing urban sprawl and agriculture. The Western Honeybee (*Apis mellifera*) readily adapts to whatever food is available, so people have made it the most widely distributed pollinator across the world. Previous research has suggested that the Western Honeybee may be less resilient to land use change outside of its natural range. This study examines a different honeybee species—the Japanese Honeybee (*Apis cerana japonica*). Unlike the Western Honeybee, this species is found almost exclusively in its natural range in Japan. Consequently, it may be better adapted to its local food sources and therefore more resilient. Working in southern Japan, in the Nagasaki and Saga prefectures, we looked at the nectar and pollen that the Japanese Honeybee feeds on. Their food intake was then examined in relation to local land use composition. We found minimal impact of increasing urban sprawl on the forage of the Japanese Honeybee. This goes against previous studies on the Western Honeybee elsewhere in the world. Though in need of a direct comparison with Western Honeybee, these preliminary results could be due to differences in urban green infrastructure in Japan, or due to an adaptation by the Japanese honeybee to its surroundings.

Abstract: Pollinators are being threatened globally by urbanisation and agricultural intensification, driven by a growing human population. Understanding these impacts on landscapes and pollinators is critical to ensuring a robust pollination system. Remote sensing data on land use attributes have previously linked honeybee nutrition to land use in the Western Honeybee (*Apis mellifera* L.). Here, we instead focus on the less commonly studied *Apis cerana japonica*—the Japanese Honeybee. Our study presents preliminary data comparing forage (honey and pollen) with land use across a rural-urban gradient from 22 sites in Kyushu, southern Japan. Honey samples were collected from hives between June 2018 and August 2019. Pollen were collected and biotyped from hives in urban and rural locations (n = 4). Previous studies of honey show substantial variation in monosaccharide content. Our analysis of *A. cerana japonica* honey found very little variation in glucose and fructose (which accounted for 97% of monosaccharides), despite substantial differences in surrounding forage composition. As expected, we observed temporal variation in pollen foraged by *A. cerana japonica*, likely dependent on flowering phenology. These preliminary results suggest that the forage and nutrition of *A. cerana japonica* may not be negatively affected by urban land use. This highlights the need for further comparative studies between *A. cerana japonica* and *A. mellifera* as it could suggest a resilience in pollinators foraging in their native range.

Keywords: pollinator; landscape; land use; urban rural gradient; Japanese honeybee; honey; pollen; nutrition

Citation: Donkersley, P.; Covell, L.; Ota, T. Japanese Honeybees (*Apis cerana japonica* Radoszkowski, 1877) May Be Resilient to Land Use Change. *Insects* **2021**, *12*, 685. https://doi.org/10.3390/insects12080685

Academic Editors: Ivana Tlak Gajger, Franco Mutinelli and Cristina Botías

Received: 15 July 2021
Accepted: 29 July 2021
Published: 30 July 2021

1. Introduction

Globally, land use change is driving biodiversity loss. Primarily driven by anthropogenic uses, such as agricultural intensification and urban development, these losses threaten ecosystem services, and impact the growing human population [1,2]. One such ecosystem service is provided by insect pollinators. Pollinator declines are linked to a reduction in both nesting sites (for wild bees) and available forage (for all pollinators) [3]. Land use change has damaged the availability and variety of pollen [4,5]. This is particularly evident in specialist pollinators, which can only utilise certain suitable flora [6,7]. Consequently, this is particularly important for specialist pollinator species that are under selection pressure to make a transition to newly available sources of food [5,8].

In a landscape context, pollinator activity shifts depending on the pollinator species' forage preferences, and the nutritional value of the pollen [9–13]. Pollinator health and ecosystem service provision may effectively be determined by land use change, in combination with other significant factors [14–16]. These effects have only recently begun to be incorporated into the conservation literature, where papers highlight the importance of landscape heterogeneity, trees, matrix effects, habitat loss and fragmentation on pollinator diversity and success [17–19].

Amongst all animal pollinators, bees in particular have been affected on a global scale by land use change [19–22]. Western Honeybees (*Apis mellifera L.*) remain a model species of understanding pollinator health, nutrition, behaviour [23–25] and are key organisms in the development of mathematical algorithms and behavioural models for pollinators [26–28], and as such are still used an important model species for studying insect pollinators globally. Yet, this simplicity may obscure the importance of considering all pollinator species and the impacts of these factors on wild pollinators. The behavioural, ecological and evolutionary differences between *A. mellifera* and other genera of insect pollinators, for example: pseudo-social bees (e.g., *Bombus terrestris*) or solitary bees (e.g., *Osmia bicornis*) are vast. In this study, we examine substantial differences within the genus *Apis*, further highlighting the inadequacy of *A. mellifera* as a "catch-all" species for pollinator decline.

In Japan, both *A. mellifera* (the western honeybee) and *Apis cerana japonica* (the Japanese honeybee) are managed for their pollination, honey production and cultural/heritage values. Arguably, elsewhere in the world the Western honeybee has become one of several factors negatively impacting the health of native, wild pollinators. Tatsuno and Osawa [29] found that the native Japanese honeybee pollinate more native species and more species overall, making them potentially one of the more important pollinators in the country. Despite this, Western honeybee are dominant in the Japanese beekeeping industry [30], their widespread use, equivocal with "livestock" management, highlights the importance of examining how the Western honeybee may be impacting other native pollinators, like the Japanese honeybee.

1.1. Japanese Beekeeping

Western honeybees were originally introduced to Japan as they have a higher honey production and lower swarming rate than Japanese honeybees. There was a significant decline in the prevalence of native beekeeping in Japan until 2005, determined by an increasing reliance on imported honey (more than ten times greater than domestic supplies) [30]. This was not only due to a decline in beekeeping as a profession, but also a decline in nectar sources, especially orange trees [30]. This reduction in pollinators has reduced the pollination services provided to cultivated crops, thus negatively affecting yield and quality of produce [30].

This decline has now stabilised, due to increases in urban and small-scale beekeeping [31]. The Japanese Government amended the Apiculture Promotion Act in 2012, which required hobbyist beekeepers to report their number of hives. The most recent statistics available estimates 10,000 active beekeepers in Japan, growing from approximately 2000 when the Apiculture Promotion Act was introduced in 1995 [32]. For context, the National Bee Unit in England estimates the number of beekeepers at 44,000, a number that has remained relatively stable over the past 5 years [33].

Local expert knowledge from beekeepers is key to understanding the behaviour and adaptations of their bees. Beekeepers in the Nagasaki prefecture have said that the Western honeybee has adapted to forage more on "mass-flowering plants" than the Japanese honeybee. Beekeepers and academics have observed that the Japanese honeybee instead forages on more diverse flower sources, collecting nectar and pollen from anything available, rather than through the "flower constancy" behaviour observed in Western honeybee [34,35]. The beekeepers believe that "if nectar is very scarce (like in urban area), *cerana* can more successfully find small patches on which to survive (maybe in a small garden)" [36]. Observations by this group comparing Western honeybee and Japanese honeybee when kept in the same apiary have led the beekeepers to believe that "*mellifera* is not good at collecting nectar in summer compared with *cerana*", especially in urban environments with less spatially extensive flower patches [36]. It has recently been identified that Western honeybee are more susceptible than Japanese honeybee to predation from various native hornets in Japan [29,30].

Beekeeping is growing in Japan, as more people take it up as a hobby or a business interest. Investigating the threats to pollinators and their ecosystem services are equally increasing in importance. Identifying land uses and plant species that support *A. cerana japonica* may be key in maintaining the success of both urban and rural commercial and hobbyist beekeepers, as their numbers continue to grow.

1.2. Aims and Scope

The sugar composition of honey is directly linked to the flowers that have been foraged by bees to make it. Phytochemicals derived from the flowers convey unique biochemical properties to honey, which has led to marketization of particular honeys for their medicinal properties (such as "Manuka") [37–39]. The glucose/fructose ratio of honey can impact its palatability for humans [40,41], and higher glucose levels contribute to the production of hydrogen peroxide, an important antimicrobial compound found in honey [42,43].

This study aims to investigate pollen and nectar foraging by *A. cerana japonica*, studying urban and rural populations maintained by hobbyists in Nagasaki and Saga prefectures, on Japan's southern island of Kyushu. Here, beekeeping is practiced by a small beekeeping community, largely producing honey part-time for personal use and sale, with a few full-time commercial producers. *A. cerana japonica* are almost exclusively chosen by hobbyist beekeepers in Nagasaki (though *A. mellifera* continue to be favoured by commercial beekeepers). The factors affecting pollinator health and honey production in Japan are not well studied, so this study focuses on answering the two following key questions:

1. How does land use in Japan affect the honey produced by *A. cerana japonica*?
2. How does the time of year and location affect the pollen collected and honey produced by *A. cerana japonica*?

2. Materials and Methods

2.1. Honey Sampling

Honey was collected from 22 hives across Nagasaki and Saga (Figure 1) between June 2018 and August 2019. A questionnaire was given to the owner of each hive to determine information such as the species of bee, the location of their hive and the environment surrounding the hive (Supplementary Materials 1: Questionnaire).

Honey samples were analysed using high-performance liquid chromatography (HPLC) to determine sugar composition (fructose, glucose and maltose), following methods for honey analysis used in Ouchemoukh et al. [44]. These sugars were selected due to previously observed geographic variation in composition [45].

Crystallised honey samples were heated in a 90 °C water bath until clear and left to cool to room temperature. All samples were then diluted using High Performance Liquid Chromatography (HPLC)-grade water to a ratio of 1 μL honey per 100 mL water. Two repeats of 10 μL solution were run through the HPLC, testing for fructose, glucose and maltose levels. Three repeats per sample were conducted and the means taken.

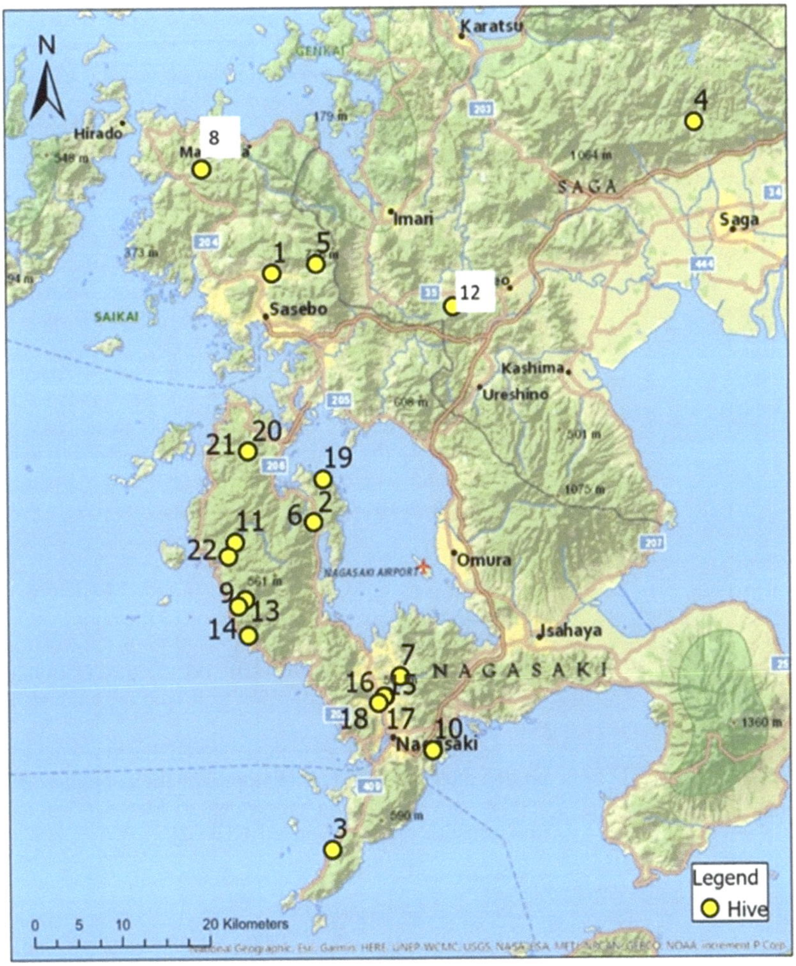

Figure 1. Locations of 22 hives from which honey samples were collected, clustered hive locations (Hives 7, 15–18) are also available in Supplementary Figure S1.

Aliquots were analysed on an Agilent Analytical 1200LC HPLC machine (Agilent Systems, UK) using a Thermo Dionex CarboPac PA20 Analytical column, 3 × 150 mm (Thermofisher, UK). A Pulsed Amperometric Detection (PAD) detector was used, and samples were transported in HPLC grade water/200 nM NaOH. Detection peaks were quantified against a dilution series of standards for fructose, glucose and maltose.

2.2. Pollen Sampling

Pollen samples were collected at least once every three weeks for nine weeks from 13.6.2019 to 8.8.2019 at two urban (Hives 15, 16) and two rural hives (Hives 11, 19) in Nagasaki-ken. Pollen was sampled via pollen trapping, direct pollen basket collection and brood chamber sampling, as these methods have previously been used to sample pollen successfully [46,47]. Samples at all hives were collected from the brood chamber, as this was found to be the quickest and most effective method, as well as arguably the least disruptive. Typhoons during weeks four and seven limited sampling during these periods; one hive was abandoned by the colony in week eight (detailed information available in Table S1).

Biotyping of pollen grains was used to provide a descriptive measure of the diversity of foraged pollen in samples [48]. Pollen samples were purified via acetolysis using methods based on Jones [49]; each sample was imaged five times with a digital microscope. Pollen grains were then counted and identified into biotypes based on physical properties (Supplementary Appendix 1: Pollen Identification), due to low sample site replication, no further analysis was performed on the pollen data.

2.3. Land Use Composition

To analyse the correlation between land cover and Japanese honeybee nutrition, data were sourced from the Japanese Ministry of the Environment's Biodiversity Centre (Table S2). The composition and configuration of different land uses in the surrounding 1, 3 and 5 km radii of each hive were measured using the *radius* tool in ArcGIS 10.8.1 (ESRI, US).

Radii of 1, 3 and 5 km were chosen as these cover the range of foraging distances travelled by bees from their hive [47]. Land cover classes that accounted for <0.5% of total cover within a buffer zone were excluded from analysis. The dominant land use (the land use contributing the greatest percentage of land cover) and the ratio of urban-to-rural land uses were then calculated using these data. The distance from each hive to the nearest urban area was also calculated based on methods established in Clermont et al. [50].

2.4. Statistical Analysis

Honey composition data were analysed to determine inter-hive variance in sugar content. Monosaccharide composition of the honey samples was analysed by Pearson's correlation with three land use factors: distance to urban areas, dominant surrounding land use and ratio of rural-to-urban land use. Critical P-scaling was performed to control false discovery rate (FDR) on these multiple land use analyses [51], for these analyses critical P was set at $\alpha = 0.0115$.

Pollen count data were not analysed due to low hive level replication. The number of observed biotypes (as a rough approximation of pollen species richness), and sample date and hive are presented in Supplementary Materials. All other analyses were all carried out in *R* statistical software version 4.0.5 [52].

3. Results

3.1. Honey Sugar Composition

Honey samples from 22 different hives across Japan, were analysed for fructose, glucose and maltose content. Glucose accounted for the majority of mono-saccharides within the honey samples, accounting for 59.3 $\pm$ 0.4% (mean $\pm$ S.E.); followed by fructose at 38.1 $\pm$ 0.5%, and maltose sugars being present in trace amounts at 2.5 $\pm$ 0.2%. The sugar ratios of honey samples did not change significantly between hives and apiary location ($H = 21$, df = 21, $p = 0.459$; Figure 2).

Although fructose and glucose content remained consistent between hives, significant inter-hive variance was observed in maltose content ($F = 0.465$, df = 21, $p = 0.029$). The distance to the nearest urban area had no effect on the proportions of sugars in honey samples (*fructose*: $r_s = -0.099$, $n = 22$, $p = 0.660$; *glucose*: $r_s = -0.284$, $n = 22$, $p = 0.200$; *maltose*: $r_s = -0.055$, $n = 22$ $p = 0.807$).

The proportions of sugars found in the honey samples did not change depending on the date the sample was collected (*fructose*: $H = 19.791$, $n = 22$, $p = 0.285$; *glucose*: $H = 20.146$, $n = 22$, $p = 0.267$; *maltose*: $H = 19.708$, $n = 22$, $p = 0.289$).

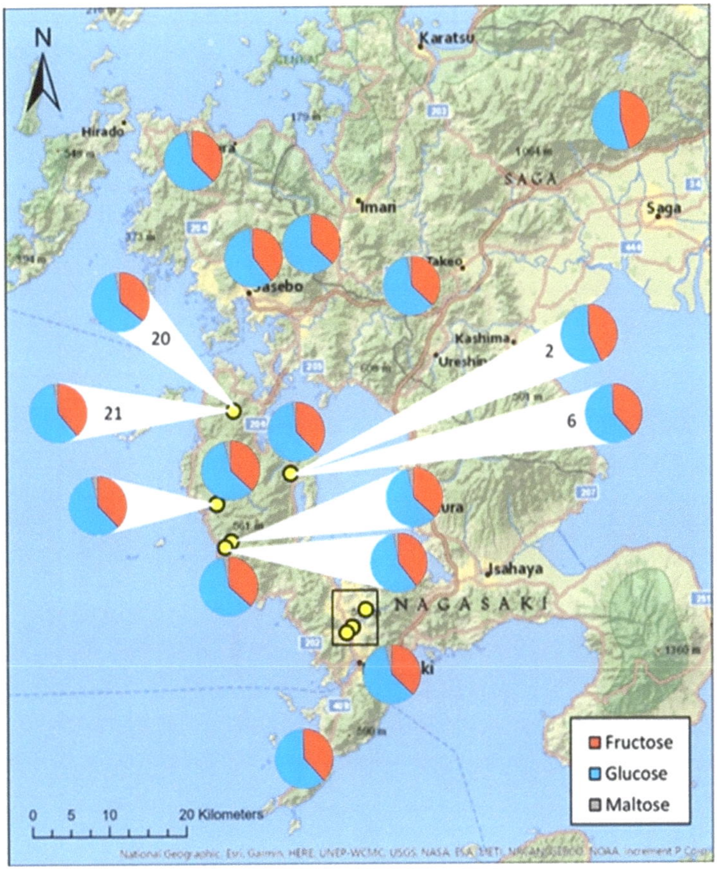

Figure 2. Proportion of fructose, maltose and glucose in honey samples collected from 22 *Apis* species hives across Kyushu, Japan. Yellow dots represent hives where pie chart could not be placed directly in the correct location. Where multiple hives had the same location, pie charts are labelled with hive number. Sugar proportions were calculated using HPLC. Clustered hives located in Nagasaki, within the black rectangle, can be found in Figure S2.

3.2. Pollen Biotype Composition

In the process of biotyping, 262 microscope images were counted across 131 samples from four *A. cerana japonica* hives collected between June and September 2019. Across these samples, 50 biotypes were identified (Supplementary Materials: Pollen). Biotypes 1, 3, 10, 14, 17 and 20 had a total abundance greater than 1000 grains across all sample images, and thus were deemed 'dominant' biotypes. A description of these biotypes can be found in Table 1. Statistical analyses of pollen biotypes use these groups throughout, as well as 'others' (combined totals from biotypes 2, 4–9, 11–13, 15, 16, 18, 19 and 20–52).

Based on acetolysis, the most common species of pollen found within the study area were provisionally identified as Japanese angelica tree (*Aralia elata*), wutong (*Firmiana simplex*), Japanese Oak (*Lithocarpus edulis*), East Asian mallotus (*Mallotus japonicus*), kumquat (*Citrus/Fortunella crassifolia*), Tree Ivy (*Dendropanax trifidus*) and Japanese spindle (*Euonymus japonicus*).

3.3. Land Use Composition

Within a 1 km radius of the 22 hives, eight land types were dominant (Table S2). Within a 3 km radius, this was six land types, and in a 5 km radius, this was five land types. On average, within a 1 km radius around each hive, rural land use was dominant, covering an average of 80.0 ± 6.7% of all land. This land cover varied between hives (Figure 3 and Figure S2), but this variation was not reflected in fructose or glucose content of honey. The maltose content of honey increased as rural land cover within 1 km radius of the hive increased, but not significantly under P-scaling to control FDR (Table 2).

Table 1. Dominant biotype classifications and descriptions used to identify pollen grains found in samples collected from five *Apis cerana japonica* hives in Japan, between June and September 2019. Potential family and species names are also given where possible, based on pollen samples collected directly from flowering plants and identified using a palynomorph guide of Japanese flora by Shimakura [53]. Reference images are approximately relatively sized, as indicated by scale bar (50 μm).

Biotype	Possible Species and Family Name (s)	Description	Reference Image
1	*Aralia elata* (Araliaceae)	Small, circular/semi-circular	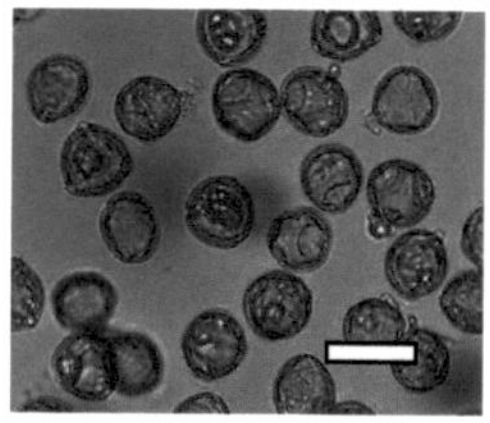
3	*Firmiana simplex* (Malvaceae)	Medium, dark, 3-way symmetry, rounded	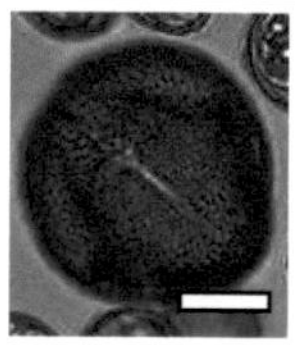
10		Very small, light, circular/semi-circular.	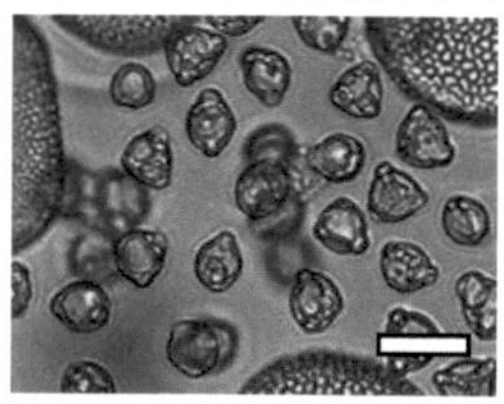
14		Medium, oblong, slightly pointed at ends, arcing lines through	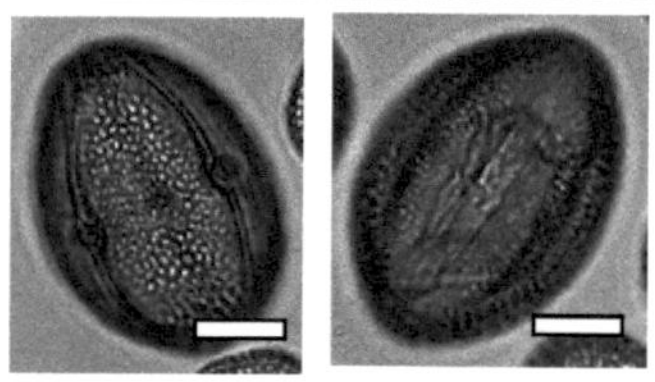
17	*Lithocarpus edulis* (Fagaceae)	Small, oval, lines arcing through.	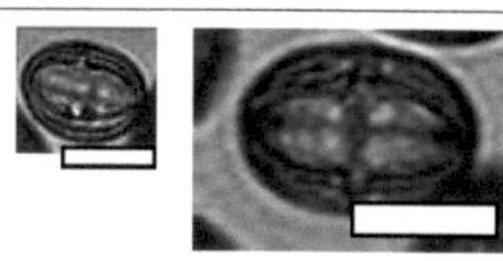

Table 1. *Cont.*

Biotype	Possible Species and Family Name (s)	Description	Reference Image
20	*Mallotus japonicus* (Euphorbiaceae) *Citrus/Fortunella crassifolia* (Rutaceae) *Dendropanax trifidus* (Arialaceae) *Euonymus japonicus* (Celastraceae)	Similar to 16, but more indents, elongated. From end on: small-medium, three rounded sides, triangle inside, with points between indents.	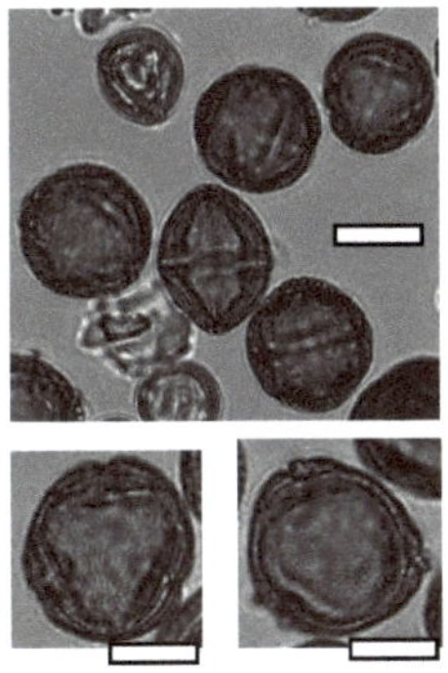

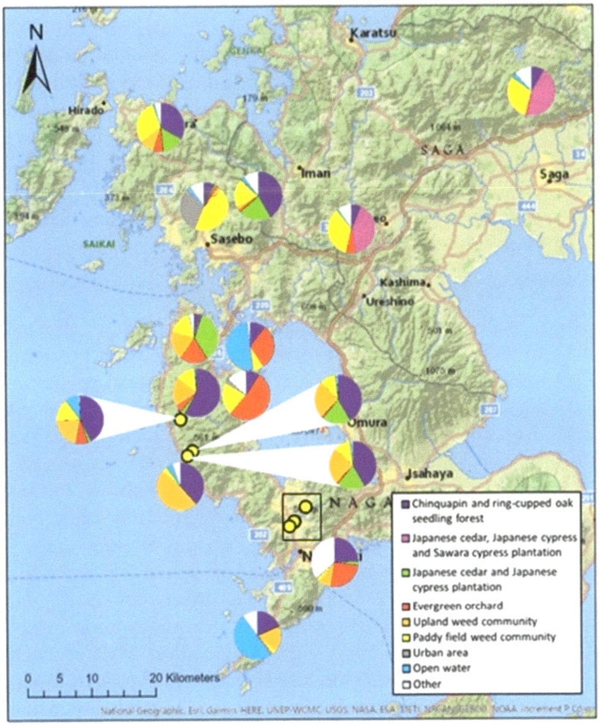

Figure 3. Proportion of land uses in a 1 km radius surrounding 17 hives across Kyushu, Japan. Land uses which were dominant for at least one hive are shown, along with all other land types, grouped into 'other'. Yellow dots represent hives where pie chart could not be placed directly in the correct location. Hives 2 and 6 as well as 20 and 21 were located in the same apiaries, so are shown using one pie chart per location. Clustered hives in Nagasaki, within the black rectangle, can be found in Supplementary materials. Land use was calculated using vegetation maps freely available from the Japanese Ministry of the Environment's Biodiversity Centre.

Table 2. Statistical analysis of the effect of rural-to-urban ratio of land use on composition of sugars found in honey collected by 22 colonies of Japanese honeybee in various locations within Kyushu, Japan, between June 2018 and August 2019. Statistical values for rural-to-urban ratios within radii of 1, 3 and 5 km were calculated. N for all statistics was 22. All maltose values were calculated using Pearson's product-moment correlation (r) and all fructose and glucose values were calculated using Spearman's rank correlation (r_s).

Sugar	1 km		3 km		5 km	
	r_s/r Value	*p*-Value	r_s/r Value	*p*-Value	r_s/r Value	*p*-Value
Fructose	-0.230	0.303	0.079	0.726	0.174	0.439
Glucose	−0.095	0.675	−0.051	0.822	−0.297	0.179
Maltose	0.465	0.029	0.189	0.399	0.043	0.848

4. Discussion

In this study, fructose and glucose sugars made up the majority (an average of 97.5 ± 0.9%) of the three monosaccharide sugars analysed from the honey of the Japanese honeybee. Our study was based on sugar analysis from sites across south Japan (n = 22), as well as pollen collections from a smaller number of sites (n = 4), sampled repeatedly (n = 13 per hive). Our results show minimal variance in the sugar composition across a wide geographical area, and a broad diversity of pollen types associated with these bees.

Despite the small replication size of pollen trapping, the sampling regime we attempted was able to show temporal variation in pollen forage composition. Flowering date is a significant factor impacting pollinator foraging [9,54,55], and our study was no exception. As a result, it is important to focus on ensuring a consistent provision of pollen sources throughout the year, by maintaining heterogeneity of flowering times within vegetation cover [12,48].

4.1. The Importance of Honey Sugar Composition

The biochemical properties (i.e., antibacterial activity) of honey varies according to its sources, as do its physical and chemical properties [38,39,56]. These properties are as important to humans for their medical properties as they are to the honeybees for their potential to act as a "bee pharmacy" [57,58]. Variation in these properties has been suggested to be part of a suite of factors negatively impacting honeybee health globally [16,59].

Consequently, flower resources (availability and botanical origin) in the urban environment and subsequent variance in honey sugar composition have the potential to substantially impact bee health. Previous research in the Western honeybee from western urban-rural gradient landscapes has shown that increasing urban land use has a negative impact on the nutrition of these eusocial bees [9–13]. Yet, in our study measures of land use composition, for example rural-to-urban ratio or distance to urban landscapes, did not significantly impact sugar composition. Maltose levels appeared to vary substantially more between hives. Maltose is often added to man-made products, as it is not commonly naturally occurring [60]. However, studies have shown that the sugar composition of honey changes over time, due to various chemical processes enabled by the heat of hives [40,61]. This can lead to the production of maltose from glucose and fructose, meaning that the maltose may have formed during storage [61].

No link between nutrition of the Japanese honeybee and the composition of their environment suggests that Japanese honeybees in Japan may not be perturbed by the negative effects of urbanisation observed in other studies of honeybees. The question remains, as to why: is it land use being less detrimental, or are Japanese honeybee being more resilient to land use change?

4.2. Could Urban Land Use Be Less Detrimental?

As observed by beekeepers involved in our study, when comparing Western honeybee with Japanese honeybee, the latter may prefer to forage on a wider array of available forage [36]. The low variance in monosaccharide composition in our study of Japanese honeybee, compared with the significant inter-hive variance in sugar composition from studies of Western honeybee [45], suggests Japanese honeybee may be better at regulating the balance of sugars in their nectar forage intake.

One could suggest this is due to differences in the landscape composition impacting the bees. Within the context of our study (Nagasaki and Saga), the rural-urban matrix of landscapes was significantly variable between sites (Figures S1 and S2). Our findings may show that Japanese honeybee could be successfully exploiting the diverse nectar sources available to them in these environments. This may suggest that Japanese urban landscapes are not as detrimental to Japanese honeybee as western urban environments are to Western honeybee, as found in existing studies that have explored the impact of urban environments on pollen forage and pollinator health [50,62,63]. Few studies have explicitly studied the link between landscape "quality" and nectar foraging as has been attempted in our study. Homogenous landscapes (i.e., agricultural monocultures) have previously been shown to negatively impact the nectar foraging of Western honeybees [64]. The effect of urban environments on nectar foraging remains largely unexplored for Western honeybees however. Furthermore, there is a growing literature on urban green infrastructure that suggests cultural and demographic factors influence the distribution and composition of urban green spaces [65,66].

Elsewhere in the world, where urban land uses dominate, the managed floral composition and overall lack of availability of nectar has been shown to have detrimental effects on honeybee nutrition [50,63]. Previous studies directly comparing the effects of land use composition on honeybee nutrition found significant effects on pollinator foraging and nutrition [9,67,68]. The lack of impact from landscape composition on nutrition of the Japanese honeybee here suggests urban flora may not be perturbing foraging by Japanese honeybee. Cultural, historic and heritage factors that impact the composition of natural elements in urban settings may explain these differences. If urban green spaces are providing a stable and sufficient nectar source for Japanese honeybee [65,69], prefectures like Nagasaki and Saga may become important case studies in urban landscapes that support local bee populations.

The social construct of what constitutes "urban" in Japan is observably distinct to other sites in which pollinator ecology has been researched, in terms of extent and age of urban green spaces, the social expectation for preserving and cultivating these spaces, as well as social engagement with natural history. [65,70–72]. In previous studies, urban environments are often dominated by invasive or alien species, supplanting native flora [73,74]. Cultural factors leading to an emphasis on native flora [65,71], as well as strict plant quarantine and biosecurity measures [75], may have contributed to urban environments in Japan being more "native" than western urbanised landscapes. Emergent analyses of urban green infrastructure suggest remote sensing data may bear out this comparison [65,66,76], but a direct comparative analysis with sufficiently controlled parameters is currently lacking in the literature.

4.3. Could Japanese Honeybees Be More Resilient?

The other driver of our observed minimal disturbance of Japanese honeybee within urban environments and agricultural rural environments is that Japanese honeybee itself may be better adapted to these environments. Due to the lack of significant correlation with land use composition on honeybee nutrition, we suggest that in Nagasaki and Saga, Japanese honeybee may in fact be "coping" with the presence and expansion of urbanisation. Previous studies by Garbuzov et al. [69] and Lowenstein et al. [8] suggest that pollinators can adapt successfully to urban living. However, notably, due to a lack of direct comparison with Western honeybee or experiments relocating Japanese honeybee to a non-localised environment, determining this link remains speculative.

Some studies suggested that honeybees may prefer key plant species because of their ability to perceive higher nutritional value in these plants [77,78]. We suggest that it is possible the Japanese honeybee may have co-evolved with their native flora to have these advantageous preferences. Whereas the Western honeybee, being both a global generalist and alien species to this location, is not connected with the local flora via eco-evolutionary interactions and therefore is lacking this foraging and nutritional advantage.

When comparing the biology of Western honeybee with Japanese honeybee, studies of the "waggle dance" properties suggest that the flight capacity of Western honeybee may be up to two times larger than Japanese honeybee [35,79]. The suggestion, that different species of honeybee might possess distinct 'dialects' of the waggle dance, remains controversial. Direct comparisons of waggle dance properties between these two species suggest that their "dialects" (i.e., the forage distance-to-dance duration) scale differently [29,80]. Foraging over larger distances is typically advantageous to generalist foragers like the Western honeybee, though it suggests that Japanese honeybee may be better adapted to smaller areas with forage sources that are more difficult to locate [80].

Denser, more homogeneous and/or larger urban areas may have a significant effect on Japanese honeybee [54]. The area studied in this paper did not have many high population density (Nagasaki: 335 people/km^2, Saga: 342 people/km^2) or large urban areas, so this may be a relevant factor to incorporate in future studies. However, based on the results of this study alone, we found that the proximity of hives to urban areas is not an issue for regulation of honey composition, which means that the Japanese honeybee may be resilient in increasingly urbanised areas. This contradicts research on Western honeybee, which has highlighted the threats of urban and agricultural expansion. Here, we may be observing a difference between Western honeybee and Japanese honeybee, a resilience to land use variance in Japanese honeybee not found in other honeybee species [12,63]. One key factor when considering local environmental impacts on pollinators is activity outside of their native range. Whereas Western honeybee has a cosmopolitan distribution, well outside of its native range [81], Japanese honeybee is almost exclusively present within its native range in Japan. Any apparent resilience of this latter species may be linked with this, highlighting it as an important factor in future studies of pollinator resilience relative to its range/distribution.

5. Conclusions

Our study found that the sugar composition of honeys collected by Japanese honeybee does not vary significantly between hives across a wide geographical area. Furthermore, no significant correlation was found with land use composition surrounding each hive. Our understanding of the importance of sugar composition for hive health, the impact of botanical origins on sugar composition, and local knowledge on the foraging behaviour of Japanese honeybee leads us to conclude that this particular bee species may be sufficiently able to adapt to local land use conditions. By this, we mean the Japanese honeybee honey forage is not negatively affected by land use, as has been seen in studies of the Western honeybee [64].

Comparing Japanese honeybee to Western honeybee would potentially identify factors allowing the Japanese honeybee to successfully exploit both urban and rural environments in Japan, when the Western honeybee tends to be less successful across the planet. It would also be beneficial to identify local plant species abundances relative to the pollen spectrum foraged by the Japanese honeybee, to maximise benefits of any vegetation conservation efforts.

To discover more of the effects of various external factors on honey composition, more complex compositional analyses, including other sugars such as melezitose and sucrose, as well as other components such as pollen grains could be relevant [44,82,83]. In summary, our study has provided a first-look at the nutrition of *A. cerana japonica* in its native habitat, which suggests these bees in particular may be resilient to the effects of land use change that have negatively affected other honeybee species elsewhere in the world.

Supplementary Materials: The following are available online at https://www.mdpi.com/article/10.3390/insects12080685/s1, Supplementary Materials 1: Questionnaire. Appendix 1: Pollen biotype identification; Table S1. Pollen collection dates for Apis cerana japonica hives in Nagasaki-ken, Japan; Table S2. The land types used to classify vegetation cover across Nagasaki and Saga provinces of Japan. Figure S1. Proportion of fructose, maltose and glucose in honey samples collected from five Apis cerana japonica hives in Nagasaki, Japan; Figure S2. Proportion of land uses in a 1 km radius surrounding five Apis cerana japonica hives in Nagasaki, Japan. Supplementary materials: Questionnaire data.

Author Contributions: All authors contributed to conceiving and writing the manuscript. P.D. performed statistical analysis, L.C. and T.O. performed honey and pollen sampling, P.D. and L.C. performed chemical analyses, species identification were performed by L.C. All authors have read and agreed to the published version of the manuscript.

Funding: P.D. is supported by an ESRC Impact Acceleration Account grant awarded by Lancaster University. Fieldwork by L.C. was supported by the Japan Student Services Organisation (JASSO) exchange programme between Lancaster University and Nagasaki University. T.O. acknowledges departmental support from Nagasaki University.

Institutional Review Board Statement: Not applicable.

Data Availability Statement: All data generated or analysed during this study are included in this published article and its supplementary information files.

Acknowledgments: The authors would like to thank Gabriella Toledo-Ortiz, without their tireless efforts in building the Lancaster-Nagasaki partnership through JASSO, this research could not have happened. We would also like to thank the peer reviewers for their efforts in vastly improving the quality of the manuscript before publication.

Conflicts of Interest: The authors declare no conflict of interest.

References

1. Sponsler, D.B.; Grozinger, C.M.; Hitaj, C.; Rundlöf, M.; Botías, C.; Code, A.; Lonsdorf, E.V.; Melathopoulos, A.P.; Smith, D.; Suryanarayanan, S.; et al. Pesticides and pollinators: A socioecological synthesis. *Sci. Total Environ.* **2019**, *662*, 1012–1027. [CrossRef]
2. Aznar-Sánchez, J.A.; Piquer-Rodríguez, M.; Velasco-Muñoz, J.F.; Manzano-Agugliaro, F. Worldwide research trends on sustainable land use in agriculture. *Land Use Policy* **2019**, *87*, 104069. [CrossRef]
3. Vanbergen, A.J.; Garratt, M.P.; Vanbergen, A.J.; Baude, M.; Biesmeijer, J.C.; Britton, N.F.; Brown, M.J.F.; Brown, M.; Bryden, J.; Budge, G.E.; et al. Threats to an ecosystem service: Pressures on pollinators. *Front. Ecol. Environ.* **2013**, *11*, 251–259. [CrossRef]
4. Hadley, A.S.; Betts, M.G. The effects of landscape fragmentation on pollination dynamics: Absence of evidence not evidence of absence. *Biol. Rev.* **2011**, *87*, 526–544. [CrossRef]
5. Betts, M.G.; Hadley, A.S.; Kormann, U. The landscape ecology of pollination. *Landsc. Ecol.* **2019**, *34*, 961–966. [CrossRef]
6. Menz, M.H.; Phillips, R.D.; Winfree, R.; Kremen, C.; Aizen, M.; Johnson, S.; Dixon, K. Reconnecting plants and pollinators: Challenges in the restoration of pollination mutualisms. *Trends Plant Sci.* **2011**, *16*, 4–12. [CrossRef]
7. Hopwood, J.; Black, S.H.; Lee-Mäder, E.; Charlap, A.; Preston, R.; Mozumder, K.; Fleury, S. Literature Review: Pollinator Habitat Enhancement and Best Management Practices in Highway Rights-of-way. *Fed. Highw. Adm.* **2015**. Prepared by: The Xerces Society for Invertebrate Conservation in collaboration with ICF International. Available online: https://xerces.org/sites/default/files/2018-05/15-055_01_pollinators_BMPs_in_highway_ROW.pdf (accessed on 29 July 2021).
8. Lowenstein, D.M.; Matteson, K.C.; Minor, E.S. Diversity of wild bees supports pollination services in an urbanized landscape. *Oecologia* **2015**, *179*, 811–821. [CrossRef] [PubMed]
9. Donkersley, P.; Rhodes, G.; Pickup, R.W.; Jones, K.C.; Power, E.F.; Wright, G.A.; Wilson, K. Nutritional composition of honeybee food stores vary with floral composition. *Oecologia* **2017**, *185*, 749–761. [CrossRef] [PubMed]
10. Sponsler, D.B.; Matcham, E.G.; Lin, C.-H.; Lanterman, J.L.; Johnson, R.M. Spatial and taxonomic patterns of honey bee foraging: A choice test between urban and agricultural landscapes. *J. Urban Ecol.* **2017**, *3*. [CrossRef]
11. Ruedenauer, F.A.; Raubenheimer, D.; Kessner-Beierlein, D.; Grund-Mueller, N.; Noack, L.; Spaethe, J.; Leonhardt, S.D. Best be(e) on low fat: Linking nutrient perception, regulation and fitness. *Ecol. Lett.* **2020**, *23*, 545–554. [CrossRef]
12. Danner, N.; Molitor, A.M.; Schiele, S.; Härtel, S.; Steffan-Dewenter, I. Season and landscape composition affect pollen foraging distances and habitat use of honey bees. *Ecol. Appl.* **2016**, *26*, 1920–1929. [CrossRef] [PubMed]

13. Osborne, J.L.; Smith, A.; Clark, S.J.; Reynolds, D.R.; Barron, M.C.; Lim, K.S.; Reynolds, A.M. The Ontogeny of Bumblebee Flight Trajectories: From Naïve Explorers to Experienced Foragers. *PLoS ONE* **2013**, *8*, e78681. [CrossRef]
14. Donkersley, P.; Elsner-Adams, E.; Maderson, S. A One-Health Model for Reversing Honeybee (*Apis mellifera* L.) Decline. *Veter. Sci.* **2020**, *7*, 119. [CrossRef] [PubMed]
15. Rader, R.; Bartomeus, I.; Garibaldi, L.A.; Garratt, M.P.D.; Howlett, B.; Winfree, R.; Cunningham, S.; Mayfield, M.; Arthur, A.D.; Andersson, G.; et al. Non-bee insects are important contributors to global crop pollination. *Proc. Natl. Acad. Sci. USA* **2015**, *113*, 146–151. [CrossRef]
16. Becher, M.A.; Osborne, J.L.; Thorbek, P.; Kennedy, P.J.; Grimm, V. REVIEW: Towards a systems approach for understanding honeybee decline: A stocktaking and synthesis of existing models. *J. Appl. Ecol.* **2013**, *50*, 868–880. [CrossRef]
17. Donkersley, P. Trees for bees. *Agric. Ecosyst. Environ.* **2019**, *270–271*, 79–83. [CrossRef]
18. Senapathi, D.; Goddard, M.; Kunin, W.E.; Baldock, K.C.R. Landscape impacts on pollinator communities in temperate systems: Evidence and knowledge gaps. *Funct. Ecol.* **2016**, *31*, 26–37. [CrossRef]
19. Powney, G.D.; Carvell, C.; Edwards, M.; Morris, R.K.A.; Roy, H.E.; Woodcock, B.A.; Isaac, N.J.B. Widespread losses of pollinating insects in Britain. *Nat. Commun.* **2019**, *10*, 1–6. [CrossRef]
20. Ollerton, J. Pollinator Diversity: Distribution, Ecological Function, and Conservation. *Annu. Rev. Ecol. Evol. Syst.* **2017**, *48*, 353–376. [CrossRef]
21. Eggleton, P. The State of the World's Insects. *Annu. Rev. Environ. Resour.* **2020**, *45*, 61–82. [CrossRef]
22. Hanley, N.; Breeze, T.; Ellis, C.; Goulson, D. Measuring the economic value of pollination services: Principles, evidence and knowledge gaps. *Ecosyst. Serv.* **2015**, *14*, 124–132. [CrossRef]
23. Wood, T.J.; Michez, D.; Paxton, R.J.; Drossart, M.; Neumann, P.; Gérard, M.; Vanderplanck, M.; Barraud, A.; Martinet, B.; Leclercq, N.; et al. Managed honey bees as a radar for wild bee decline? *Apidologie* **2020**, *51*, 1100–1116. [CrossRef]
24. López-Uribe, M.M.; Ricigliano, V.A.; Simone-Finstrom, M. Defining Pollinator Health: Assessing Bee Ecological, Genetic, and Physiological Factors at the Individual, Colony, and Population Levels. *Annu. Rev. Anim. Biosci.* **2020**, *8*, 269–294. [CrossRef] [PubMed]
25. Carreck, N.L.; Dietemann, V.; Neumann, P.; Ellis, J.D. The COLOSS BEEBOOK: Global standards in honey bee research. *J. Apic. Res.* **2020**, *59*, 1–4. [CrossRef]
26. Menzel, R. The honeybee as a model for understanding the basis of cognition. *Nat. Rev. Neurosci.* **2012**, *13*, 758–768. [CrossRef] [PubMed]
27. Rivière, J.; Alaux, C.; Le Conte, Y.; Layec, Y.; Lozac'h, A.; Rodin, V.; Singhoff, F. Toward a complete agent-based model of a Honeybee colony. In *Highlights of Practical Applications of Agents, Multi-Agent Systems, and Complexity: The PAAMS Collection*; Springer: Cham, Switzerland, 2018; pp. 493–505.
28. Bagheri, S.; Mirzaie, M. A mathematical model of honey bee colony dynamics to predict the effect of pollen on colony failure. *PLoS ONE* **2019**, *14*, e0225632. [CrossRef]
29. Tatsuno, M.; Osawa, N. Flower visitation patterns of the coexisting honey bees *Apis cerana japonica* and *Apis mellifera* (Hymenoptera: Apidae). *Èntomol. Sci.* **2016**, *19*, 255–267. [CrossRef]
30. Kohsaka, R.; Park, M.S.; Uchiyama, Y. Beekeeping and honey production in Japan and South Korea: Past and present. *J. Ethn. Foods* **2017**, *4*, 72–79. [CrossRef]
31. Mitsumori, Y. An Analysis of Impact of Urban Beekeeping Projects on Community: Ginza Bee Projects brought not only bees, but also a more sophisticated image to Ginza. In Proceedings of the 2020 IEEE 8th R10 Humanitarian Technology Conference (R10-HTC), Kuching, Malaysia, 1–3 December 2020; pp. 1–7. [CrossRef]
32. Ministry of Agriculture Forestry Fisheries. Situation of Beekeeping; 令和2年11月農林水産省 生産局 畜産部 (Reiwa Livestock Department, Production Bureau, Ministry of Agriculture, Forestry and Fisheries), Japan. 2020. Available online: https://www.maff.go.jp/j/chikusan/kikaku/lin/sonota/attach/pdf/bee-8.pdf (accessed on 29 July 2021).
33. National Bee Unit. The Hive Count; FERA, York. 2021. Available online: nationalbeeunit.com/index.cfm?pageid=362 (accessed on 29 July 2021).
34. Wells, H.; Wells, P.H. Optimal Diet, Minimal Uncertainty and Individual Constancy in the Foraging of Honey Bees, *Apis mellifera*. *J. Anim. Ecol.* **1986**, *55*, 881. [CrossRef]
35. Yoshida, T. *Nihonmitsubachi no Shiikuho to Seitai (Keeping Method and Ecology of Japanese Honey Bee)*; Tamagawa University Press: Tokyo, Japan, 2000.
36. Nagasaki University Community Beekeepers. Personal communication, 2021.
37. Adams, C.J.; Boult, C.H.; Deadman, B.; Farr, J.M.; Grainger, M.; Manley-Harris, M.; Snow, M.J. Isolation by HPLC and characterisation of the bioactive fraction of New Zealand manuka (*Leptospermum scoparium*) honey. *Carbohydr. Res.* **2008**, *343*, 651–659. [CrossRef]
38. Combarros-Fuertes, P.; Estevinho, L.M.; Teixeira-Santos, R.; Rodrigues, A.G.; Pina-Vaz, C.; Fresno, J.M.; Tornadijo, M.E. Antibacterial Action Mechanisms of Honey: Physiological Effects of Avocado, Chestnut, and Polyfloral Honey upon Staphylococcus aureus and Escherichia coli. *Molecules* **2020**, *25*, 1252. [CrossRef]
39. Bucekova, M.; Jardekova, L.; Juricova, V.; Bugarova, V.; Di Marco, G.; Gismondi, A.; Leonardi, D.; Farkasovska, J.; Godocikova, J.; Laho, M.; et al. Antibacterial Activity of Different Blossom Honeys: New Findings. *Molecules* **2019**, *24*, 1573. [CrossRef]
40. Escuredo, O.; Carmen Seijo, M. Honey: Chemical composition, stability and authenticity. *Foods* **2019**, *8*, 577. [CrossRef]

41. Anupama, D.; Bhat, K.; Sapna, V. Sensory and physico-chemical properties of commercial samples of honey. *Food Res. Int.* **2003**, *36*, 183–191. [CrossRef]
42. White, J.W., Jr.; Subers, M.H.; Schepartz, A.I. The identification of inhibine, the antibacterial factor in honey, as hydrogen peroxide and its origin in a honey glucose-oxidase system. *Biochim. Biophys. Acta Bioenerg.* **1963**, *73*, 57–70. [CrossRef]
43. Brudzynski, K. A current perspective on hydrogen peroxide production in honey. A review. *Food Chem.* **2020**, *332*. [CrossRef]
44. Ouchemoukh, S.; Schweitzer, P.; Bey, M.B.; Djoudad-Kadji, H.; Louaileche, H. HPLC sugar profiles of Algerian honeys. *Food Chem.* **2010**, *121*, 561–568. [CrossRef]
45. Wang, J.; Kliks, M.M.; Jun, S.; Jackson, M.; Li, Q.X. Rapid Analysis of Glucose, Fructose, Sucrose, and Maltose in Honeys from Different Geographic Regions using Fourier Transform Infrared Spectroscopy and Multivariate Analysis. *J. Food Sci.* **2010**, *75*, C208–C214. [CrossRef]
46. Giesecke, T.; Fontana, S.L.; Van Der Knaap, W.O.; Pardoe, H.S.; Pidek, I. From early pollen trapping experiments to the Pollen Monitoring Programme. *Veg. Hist. Archaeobotany* **2010**, *19*, 247–258. [CrossRef]
47. Donkersley, P.; Rhodes, G.; Pickup, R.W.; Jones, K.C.; Wilson, K. Honeybee nutrition is linked to landscape composition. *Ecol. Evol.* **2014**, *4*, 4195–4206. [CrossRef]
48. Lau, P.; Bryant, V.; Ellis, J.D.; Huang, Z.Y.; Sullivan, J.; Schmehl, D.R.; Cabrera, A.R.; Rangel, J. Seasonal variation of pollen collected by honey bees (*Apis mellifera*) in developed areas across four regions in the United States. *PLoS ONE* **2019**, *14*, e0217294. [CrossRef]
49. Jones, G.D. Ollen analyses for pollination research, acetolysis. *J. Pollinat. Ecol.* **2014**, *13*, 203–217. [CrossRef]
50. Clermont, A.; Eickermann, M.; Kraus, F.; Hoffmann, L.; Beyer, M. Correlations between land covers and honey bee colony losses in a country with industrialized and rural regions. *Sci. Total Environ.* **2015**, *532*, 1–13. [CrossRef]
51. Benjamini, Y.; Yekutieli, D. The control of the false discovery rate in multiple testing under dependency. *Ann. Stat.* **2001**, *29*, 1165–1188. [CrossRef]
52. R Development Core Team. *R: A Language and Environment for Statistical Computing v4.0.5*; R Foundation for Statistical Computing: Vienna, Austria, 2020.
53. Shimakura, M. Palynomorphs of Japanese plants. *Spec. Publ. Osaka Mus. Nat. Hist.* **1973**, *5*, 1–60.
54. Steffan-Dewenter, I.; Kuhn, A. Honeybee foraging in differentially structured landscapes. *Proc. R. Soc. B Boil. Sci.* **2003**, *270*, 569–575. [CrossRef]
55. Nottebrock, H.; Schmid, B.; Mayer, K.; Devaux, C.; Esler, K.J.; Böhning-Gaese, K.; Schleuning, M.; Pagel, J.; Schurr, F.M. Sugar landscapes and pollinator-mediated interactions in plant communities. *Ecography* **2016**, *40*, 1129–1138. [CrossRef]
56. Nzeako, B.C.; Hamdi, J. Antimicrobial potential of honey on some microbial isolates. *Sultan Qaboos Univ. Med. J.* **2000**, *2*, 75–79.
57. Erler, S. Pharmacy bee hive—Use of natural drugs to increase pollinator health. In Proceedings of the 2016 International Congress of Entomology, Orlando, FL, USA, 26 September 2016. [CrossRef]
58. Tauber, J.P.; Collins, W.R.; Schwarz, R.S.; Chen, Y.; Grubbs, K.; Huang, Q.; Lopez, D.; Peterson, R.; Evans, J.D. Natural Product Medicines for Honey Bees: Perspective and Protocols. *Insects* **2019**, *10*, 356. [CrossRef]
59. Erler, S.; Moritz, R.F.A. Pharmacophagy and pharmacophory: Mechanisms of self-medication and disease prevention in the honeybee colony (*Apis mellifera*). *Apidologie* **2015**, *47*, 389–411. [CrossRef]
60. Fujlta, I. Determination of Maltose in Honey. *Int. J. Food Sci. Nutr. Diet* **2012**, *1*, 1–2. [CrossRef]
61. Ball, D.W. The Chemical Composition of Honey. *J. Chem. Educ.* **2007**, *84*. [CrossRef]
62. Tangtorwongsakul, P.; Warrit, N.; Gale, G.A. Effects of landscape cover and local habitat characteristics on visiting bees in tropical orchards. *Agric. For. Èntomol.* **2017**, *20*, 28–40. [CrossRef]
63. Harrison, T.; Gibbs, J.; Winfree, R. Anthropogenic landscapes support fewer rare bee species. *Landsc. Ecol.* **2017**, *34*, 967–978. [CrossRef]
64. Nürnberger, F.; Steffan-Dewenter, I.; Härtel, S. Combined effects of waggle dance communication and landscape heterogeneity on nectar and pollen uptake in honey bee colonies. *PeerJ* **2017**, *5*. [CrossRef]
65. Rupprecht, C.D.D. Informal Urban Green Space: Residents' Perception, Use, and Management Preferences across Four Major Japanese Shrinking Cities. *Land* **2017**, *6*, 59. [CrossRef]
66. O'Neil, J.A.; Gallagher, C.E. Determining What is Important in Terms of the Quality of an Urban Green Network: A Study of Urban Planning in England and Scotland. *Plan. Pr. Res.* **2014**, *29*, 202–216. [CrossRef]
67. Sutter, L.; Jeanneret, P.; Bartual, A.M.; Bocci, G.; Albrecht, M. Enhancing plant diversity in agricultural landscapes promotes both rare bees and dominant crop-pollinating bees through complementary increase in key floral resources. *J. Appl. Ecol.* **2017**, *54*, 1856–1864. [CrossRef]
68. Hall, M.A.; Nimmo, D.G.; Cunningham, S.; Walker, K.; Bennett, A.F. The response of wild bees to tree cover and rural land use is mediated by species' traits. *Biol. Conserv.* **2019**, *231*, 1–12. [CrossRef]
69. Garbuzov, M.; Schürch, R.; Ratnieks, F.L. Eating locally: Dance decoding demonstrates that urban honey bees in Brighton, UK, forage mainly in the surrounding urban area. *Urban Ecosyst.* **2014**, *18*, 411–418. [CrossRef]
70. Rupprecht, C.; Byrne, J.; Ueda, H.; Lo, A. 'It's real, not fake like a park': Residents' perception and use of informal urban green-space in Brisbane, Australia and Sapporo, Japan. *Landsc. Urban Plan.* **2015**, *143*, 205–218. [CrossRef]
71. Todorova, A.; Asakawa, S.; Aikoh, T. Preferences for and attitudes towards street flowers and trees in Sapporo, Japan. *Landsc. Urban Plan.* **2004**, *69*, 403–416. [CrossRef]

72. Petrova, E.G.; Mironov, Y.V.; Aoki, Y.; Matsushima, H.; Ebine, S.; Furuya, K.; Petrova, A.; Takayama, N.; Ueda, H. Comparing the visual perception and aesthetic evaluation of natural landscapes in Russia and Japan: Cultural and environmental factors. *Prog. Earth Planet. Sci.* **2015**, *2*, 6. [CrossRef]
73. Penick, C.A.; Crofton, C.A.; Appler, R.H.; Frank, S.D.; Dunn, R.R.; Tarpy, D.R. The contribution of human foods to honey bee diets in a mid-sized metropolis. *J. Urban Ecol.* **2016**, *2*. [CrossRef]
74. Piroux, M.; Lambert, O.; Puyo, S.; Farrera, I.; Thorin, C.; L'Hostis, M.; Vigues, B.; Bastian, S. Correlating the pollens gathered by apis mellifera with the landscape features in western france. *Appl. Ecol. Environ. Res.* **2014**, *12*, 423–439. [CrossRef]
75. Jiao, Y.; Ding, Y.; Zha, Z.; Okuro, T. Crises of Biodiversity and Ecosystem Services in Satoyama Landscape of Japan: A Review on the Role of Management. *Sustainability* **2019**, *11*, 454. [CrossRef]
76. Haaland, C.; Bosch, C.K.V.D. Challenges and strategies for urban green-space planning in cities undergoing densification: A review. *Urban For. Urban Green.* **2015**, *14*, 760–771. [CrossRef]
77. Hendriksma, H.P.; Shafir, S. Honey bee foragers balance colony nutritional deficiencies. *Behav. Ecol. Sociobiol.* **2016**, *70*, 509–517. [CrossRef]
78. Anderson, K.E.; Carroll, M.J.; Sheehan, T.; Mott, B.M.; Maes, P.; Corby-Harris, V. Hive-stored pollen of honey bees: Many lines of evidence are consistent with pollen preservation, not nutrient conversion. *Mol. Ecol.* **2014**, *23*, 5904–5917. [CrossRef]
79. Yoshida, T.; Saito, J.; Kajigaya, N. The mating flight times of native *Apis cerana* japonica Radoszkowski and *introduced Apis mellifera* L. in sympatric conditions. *Apidologie* **1994**, *25*, 353–360. [CrossRef]
80. Su, S.; Cai, F.; Si, A.; Zhang, S.; Tautz, J.; Chen, S. East Learns from West: Asiatic Honeybees Can Understand Dance Language of European Honeybees. *PLoS ONE* **2008**, *3*, e2365. [CrossRef]
81. Wallberg, A.; Han, F.; Wellhagen, G.; Dahle, B.; Kawata, M.; Haddad, N.; Simões, Z.L.P.; Allsopp, M.H.; Kandemir, I.; De La Rúa, P.; et al. A worldwide survey of genome sequence variation provides insight into the evolutionary history of the honeybee Apis mellifera. *Nat. Genet.* **2014**, *46*, 1081–1088. [CrossRef] [PubMed]
82. Sniderman, J.M.K.; Matley, K.A.; Haberle, S.; Cantrill, D.J. Pollen analysis of Australian honey. *PLoS ONE* **2018**, *13*, e0197545. [CrossRef] [PubMed]
83. Ponnuchamy, R.; Bonhomme, V.; Prasad, S.; Das, L.; Patel, P.; Gaucherel, C.; Pragasam, A.; Anupama, K. Honey Pollen: Using Melissopalynology to Understand Foraging Preferences of Bees in Tropical South India. *PLoS ONE* **2014**, *9*, e101618. [CrossRef] [PubMed]

insects

MDPI

Article

Bumble Bee Foraged Pollen Analyses in Spring Time in Southern Estonia Shows Abundant Food Sources

Anna Bontšutšnaja [1,2,*], Reet Karise [1,*], Marika Mänd [1] and Guy Smagghe [2]

1 Institute of Agricultural and Environmental Sciences, Estonian University of Life Sciences, 51006 Tartu, Estonia; marika.mand@emu.ee
2 Laboratory of Agrozoology, Department of Plants and Crops, Ghent University, 9000 Ghent, Belgium; guy.smagghe@ugent.be
* Correspondence: anna.bontsutsnaja@emu.ee (A.B.); reet.karise@emu.ee (R.K.); Tel.: +372-731-3457 (R.K.)

Simple Summary: Pollinators make a strong contribution to ecosystem stability. However, nowadays, they also need protection and sustainable habitat to live and develop. Not all regions can provide suitable habitats due to agricultural intensification, urbanization, climate changes and corresponding impacts. Our study was conducted in the late spring in south Estonia where arable lands were surrounded by forest patches and rural areas. For better performance, we used both light microscopy and DNA metabarcoding methods for pollen identification. We found that bumble bees foraged on the diverse food sources showing preferences for several main plant families. Additionally, in our case, land-use types did not show important effects on bumble bee food choices and foraging decisions. Various landscape features can provide diverse food sources at the early development stages and support nest longevity. Here, we can say that a better understanding of pollinators' food preferences can help in the application of more suitable measures for their conservation.

Abstract: Agricultural landscapes usually provide higher quantities of single-source food, which are noticeably lacking in diversity and might thus have low nutrient value for bumble bee colony development. Here, in this study, we analysed the pollen foraging preferences over a large territory of a heterogeneous agricultural landscape: southern Estonia. We aimed to assess the botanical diversity of bumble bee food plants in the spring time there. We looked for preferences for some food plants or signs of food shortage that could be associated with any particular landscape features. For this purpose, we took *Bombus terrestris* commercial hives to the landscape, performed microscopy analyses and improved the results with the innovative DNA metabarcoding technique to determine the botanical origin of bumble bee-collected pollen. We found high variability of forage plants with no strong relationship with any particular landscape features. Based on the low number of plant species in single flights, we deduce that the availability of main forage plants is sufficient indicating rich forage availabilities. Despite specific limitations, we saw strong correlations between microscopy and DNA metabarcoding data usable for quantification analyses. As a conclusion, we saw that the spring-time vegetation in southern Estonia can support bumble bee colony development regardless of the detailed landscape structure. The absence of clearly dominating food preference by the tested generalist bumble bee species *B. terrestris* makes us suggest that other bumble bee species, at least food generalists, should also find plenty of forage in their early development phase.

Keywords: bumble bees; food source richness; pollen determination with microscopy and DNA metabarcoding

Citation: Bontšutšnaja, A.; Karise, R.; Mänd, M.; Smagghe, G. Bumble Bee Foraged Pollen Analyses in Spring Time in Southern Estonia Shows Abundant Food Sources. *Insects* **2021**, *12*, 922. https://doi.org/10.3390/insects12100922

Academic Editors: Ivana Tlak Gajger and Franco Mutinelli

Received: 2 August 2021
Accepted: 1 October 2021
Published: 9 October 2021

1. Introduction

Healthy pollinator assemblages in the vital beekeeping sector is recognized as critical to sustainable ecosystems [1]. The new realities imposed by climate changes necessitate the present-day understanding of the actual status of available resources among our landscapes to create or preserve areas really needed by pollinators. Smart land use is part of climate policy [2].

The importance of diet heterogeneity is variable by pollinator species, but improves their development and sustainability. The high-quality diet not only supports nutrition but also positively affects the bee immune system suppressing the negative impact of toxins and pesticides present in our landscapes [3]. Moreover, food source variability allows pollinators to make quality and quantity decisions according to forage preferences and behaviour [4–6]. Generalist pollinators can show food plant constancy by foraging on a limited number of certain plant families, which provide highly valuable nutrient content [7,8]. These highly attractive families often support several bee species. Knowing the food plant availabilities in specific landscapes would be an important measure for wild pollinator conservation.

Agricultural landscapes usually provide higher quantities of single-source food, which are noticeably lacking in diversity [9]. Seminatural areas thereafter might support the bees with food of higher value. Smart et al. [10] demonstrated that the lipid content of honey bee (*Apis mellifera* L.) abdomens was positively correlated with the proportion of seminatural landscapes. While seminatural and even natural landscapes are tightly related to agriculture and pesticide residues are known to affect organisms even far away from fields, high-value food can relieve the negative impacts of the chemicals on bees [11].

Several interventions have been developed to sustain pollinator-friendly agro-ecosystems [12]. The unmanaged grasslands often turn to hay without interference; the field boundaries need attention to prevent the growth of invasive weeds or hay. Marja et al. [13] described that even if the intervention does not lead to major changes, these help to preserve the situation. The crops themselves support bee species differently, while the foraging strategies and dependence on alternative plant species differ much from one bee species to another. Raimets et al. [14] even demonstrated that within the same landscape, honey bees were less attracted to spring compared to summer flowering oilseed rape. This allows suggesting that the plant species availability was better in spring time and the pressure to fly long distances was lower. Bumble bees, however, fly notably shorter distances [15,16] which limits the access to rich flower resources and increases the environmental pressure on colony development.

One way to assess the forage plant availabilities is to use bee-collected pollen. In the past, plant species identification has been performed only by light microscopy, which is very laborious, and usually, the identification stays at the family level [17]. Nowadays, more accurate methods are available. Consequently, the traditional techniques could be supplemented by innovative DNA metabarcoding techniques [18–20]. Plant species taxonomic identification has been achieved by using, for instance, the second internal transcribed spacer (ITS2), a widely used marker, and constantly updated databases of sequences. ITS2 should discriminate not only plant families but also differentiate between plant species allowing going deeper with pollen analyses. However, there is still no correct calibration of the numbers of sequence reads from DNA metabarcoding to actual proportions of pollen grains in the sample. Some studies state that the relative amounts of reads of any particular taxa are comparable to their actual proportions [21,22], but this might be organism-group or taxa-specific [23,24]. Thus, until now, the DNA metabarcoding for processing mixed samples has been used, rather, to obtain a broader overview of the pollen taxonomic origin but exclud quantity assessments [18,23]. Shortcomings also occur with microscopy technique as this does not determine every single pollen grain in a sample, which is why this technique might miss too many plant species limiting the accuracy of proportional estimations, too. Reliable determination of plant species proportions in bee-collected pollen samples is an approach that needs further development.

Here, in this study, we aimed to assess the botanical diversity of bumble bee food plants in the spring time in heterogeneous landscapes of southern Estonia. We used bumble bee hives to assess the landscape for pollinator support as described by [25]. We wanted to see whether there is a preference for some food plants or signs of food shortage that could be associated with any particular landscape features. In addition, we raised the question about the comparability of microscopy and DNA metabarcoding techniques in the determination of the botanical origin of bumble bee-collected pollen.

2. Materials and Methods

2.1. Study Environment

We selected 15 sites for our study in 2017 to understand the availability of forage resources in southern Estonia (Tartumaa, Viljandimaa, Põlvamaa and Võrumaa counties) (Figure 1). Sites were chosen according to variability, logistical limitations and local owners' agreements. The region is characterised by heterogeneous landscapes (examples are presented in Supplementary materials, Figure S1a–c), where rather extensive agriculture is practiced. The landscape is slightly hilly, with lots of lakes, rivers and forest patches. Due to high patchiness, the local weather conditions create suitable microclimates for growing fruits and berries. The human inhabitation is rather low, 10–20 residents per km^2 [26], but still evenly distributed. The density of households is scattered [27], mostly single-family houses. The houses are surrounded by yards, where fruit trees, flowerbeds and berries are common. Although a large share of Estonian territory is covered with forests, there are plenty of forest patches, clear-cuts, and brush. In spring time these landscapes usually are with plenty of flowers different trees and berries, ornamental plants in yards, small flowers on underwood and early season flowering plants on field edges and meadows.

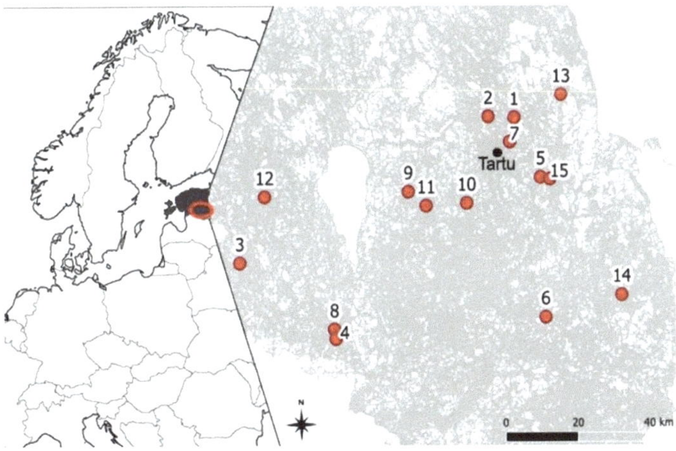

Figure 1. A schematic overview of the location of the study area on the European map (**left**) with detailed southern Estonian region (**right**) zoomed out. The 15 study locations around Tartu city are indicated with red circles and numbers.

2.2. Hive Locations

Thirty bumble beehives were located in various environments with one common feature, strawberry field, next to the hives. We selected 15 sites; in each, we located two hives next to each other with about a 1 m gap between them. We located them on the strawberry field borders so that at least 50% of the foraging territory was a non-crop area. Strawberry was chosen to guarantee at least one food source for our experimental colonies. While strawberry is not the preferred food source for *Bombus terrestris* [28,29], we expected bumble bees to forage on more suitable plants with the same flowering period. Thus, the high strawberry pollen proportion might have reflected an absence of more suitable food sources.

We searched for contrasting landscapes in 1000 m radiuses around beehives based on a share of different land-use types. The 1000 m radius was selected, because it is commonly the maximum bumble bee foraging range [15,16], although sometimes few marked bumble bees have been seen also at a distance of 1500 m [30]. The site selection excluded foraging area overlapping. Distances between study sides varied between 3.7 km and 104 km (Figure 1).

We used maps of Estonian Land Board and QGIS software (QGIS Development Team, 2018. QGIS Geographic Information System. Open Source Geospatial Foundation Project. http://qgis.osgeo.org, accessed on 2 August 2021) to calculate the proportions of land use characteristics around the bumble bee nests to understand which land-use types better explain the pollen choice of bumble bees. The characteristics used were: shares of woodland, arable land, grassland, yards and fruit gardens. While no clear patterns occurred when using these land-use parameters in statistical analyses, we additionally divided the land use parameters into the categories based on the three most important ones and named them as Forest (5 sites), Field (7 sites) and Orchards (3 sites) (Figure 2, illustrative material provided in supplementary materials). When a site had more than 50% of the area covered by woodland, it was classified to Forest category; when more than 50% was covered by arable lands, it was classified to Field category. The Orchard category was very variable while fruit gardens are mostly located in agricultural landscapes, which explains the large share of "arable land"; the share of "woodland" stayed around 20% there. Preliminary statistical analysis showed that land-use types "grasslands" and "yards" did not differ between the categories (grasslands: $p = 0.91$, yards: $p = 0.22$).

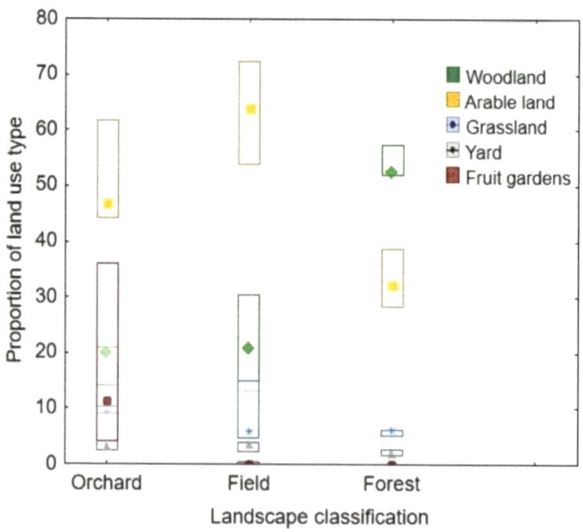

Figure 2. The share of land-use types (woodland, arable land, grassland, yards and fruit gardens) across the three land-use categories (Orchard, Field and Forest) within a radius of 1000 m around hives.

2.3. *Bumble Bees*

The bumble bee *Bombus terrestris* L. is a common short-tongued bumble bee with a broad food plant spectrum. *B. terrestris* is a species that is related to open landscapes and its abundance is negatively correlated with the share of woodlands [31]. *B. terrestris* is considered a spatial generalist because of its large foraging distances [32,33]. They might prefer open landscapes, but forest patches are not barriers [34]. This species has relatively large colonies, therefore they need rich foraging areas to cover the needs of the developing brood.

Commercially reared bumble bee *B. terrestris* hives were obtained from Biobest, Westerlo, Belgium. The hives were Standard queenright colonies with about 80 workers at the beginning of the experiment. The hives were equipped with sugar solution (Biogluc bottle, from the producer) and polystyrene outer coating (Bee-Coat, from the producer) to keep the colonies warm even in cool nights which may occur in this time of year.

Before taking the colonies to the field, they had no previous foraging experiences in these landscapes nor with these local plant availabilities. If the bumble bees had developed some food preference based on the food the hives were provided during the early development of the colonies [35], we suppose all colonies had similar experiences. The hive entrances were opened two days before sampling of pollen, to allow foragers to learn the foraging territory and food plants available.

All the colonies used were in their developing phase and while all the food provided by the producer was eliminated, we expected them to forage actively. All hives were weighed before and after the experiment and the weight increase was obvious 119.6 ± 17.1 g. All hives were in the middle of the development stage (brood cells were present, no gynes or males detected).

2.4. Collection of Pollen

Pollen sampling was timed to the mid flowering of strawberry fields to guarantee the highest number of strawberry flowers available. Pollen was collected twice from each hive and location with an in-between period of 5–7 days, covering the whole strawberry flowering period. Moreover, at the same time orchards are in their late flowering state, winter oilseed rape is in the middle or final stage of flowering depending on the cultivar. Additionally, natural plants provide high flower density. The foraging activity of bumble bees from different hives was variable, therefore the number of pollen samples varied, being 4–10 pollen loads from individual bumble bees. Each sampling lasted roughly one hour per site between 9:00 to 18:00. Each hive was monitored once in the morning and once in the afternoon and thus we suppose the time of day did not affect the outcome. Both pollen loads were collected from the homing foragers. The hive entrances were closed with plastic jars and the bee was cached into it. Each jar was thoroughly cleaned after every usage to avoid cross-contamination of the samples. Bumblebees were released after pollen loads were taken. Individual forager pollen loads were packed into paper bags and labelled with the date, time, hive number and study site. In total, 427 samples were collected during the study period with a mean of 8.9 ± 0.13 individuals from each hive each date. Bags were air-dried at room temperature, then the pollen loads consisting of two pollen pellets were weighed and separated into two Eppendorf tubes one for light microscopy and the other for DNA metabarcoding.

2.5. The Plant Species Determination of Pollen

2.5.1. Light Microscopy

For light microscopy, pollen grain structure needs to be recognisable. We followed the protocol described by [36] to purify the pollen pellets from excessive materials. For that 1 mL of 99.6% acetic acid was added for 48 h to separate the pollen kit and to clarify pollen grain external structure. After that, the solution was homogenized by a glass stirring rod and centrifuged for 5 min at 13,400 rpm. For the next step, acetic acid was discarded and replaced by 1 mL of the demineralised water. Afterwards, samples are centrifuged for 5 min at 13,400 rpm, demineralised water was changed and samples were stored at room temperature. Then, the sample was transferred to a slide by the glass stirring rod and covered by the cover glass without adding any dye. From each sample, 200 pollen grains were determined at 400× *g* magnification (Olympus CX 31 RBSF) to genus level, and where possible to species level, using 1000× *g* magnification (Nikon H550L) in cases, where the structure was not seen or ambiguous. As references were used flowering plants (about 70 plant species) pollen collected from the study sites during the field-work period and prepared by the same protocol. An example of the strawberry pollen grain seen through light microscopy is presented in Supplementary materials Figure S2. Reference publications were also used [36,37].

2.5.2. DNA Metabarcoding

For the first step of DNA metabarcoding, dried pollen was pooled from the same hive and day to obtain a larger sample size. After that mixed pollen was homogenised by mortar. In total, 54 samples were used for DNA extraction. An amount of 50 mg of powdered pollen was placed into a 2 mL Eppendorf tube, adding 800 µL of RTL buffer and 8µl of beta-mercaptoethanol, after 5 s vortex, 0.3 g of silica beads were added. Samples were lysed (2 min at 30 Hz) and homogenised (2 min 20 Hz) using the TissueLyser II (Qiagen, Venlo, the Netherlands). Subsequently, samples were centrifuged for 2 min at 20,000× *g* and 200 µl supernatant was used for DNA extraction. DNA was extracted using the Invisorb Spin Tissue Mini Kit (Stratec molecular GMBH, Berlin, Germany) according to the manufacturer's protocol No. 5. In between work, samples were stored at −20 °C. After DNA isolation, the ITS2 region was amplified by PCR (polymerase chain reaction) using dual-barcoding as described by [20]. Each sample was amplified in triplicate. Samples were initially denatured at 95 °C for 4 min, then amplified within 36 cycles at 95 °C for the 40 s, 49 °C for 40 s and 72 °C for 40 s with a 5 min final extension (72 °C). After PCR amplification, replicates were pooled and quantified using the Quan-IT™ PicoGreen™ dsDNA kit (Thermo Fisher Scientific, MA, USA). The quantified samples were normalized and subsequently purified and concentrated using E.Z.N.A.® Cycle Pure Kit (Omega Bio-Tek, Norcross, GA, USA)and AmiconUltra-0.5 columns (Millipore Cooperation, Billercia, MA, USA), respectively. Samples were sent for sequencing using Illumina Miseq PE250 (NXTGNT, Ghent University, Ghent, Belgium).

Raw reads were combined and the quality was checked using MOTHUR 1.34.4 [38]. Combined reads were filtered on zero ambiguity and amplicon length between 100 bp and 360 bp. Unique sequences were blasted against a custom database that contained the ITS2 region of plants (compiled from [39]). Clustering was performed on 97% identity. The most abundant sequence of each cluster was used as a representative sequence which was identified to the genus level using the Blast function in NCBI. We looked at the results to filter out plants not occurring in Estonia.

2.6. Statistical Analyses

We assessed five land-use parameters in the 400 m and 1000 m radius of the hives: woodland, arable land, grassland, yards and fruit gardens. Exact proportions of different land-use types were calculated using the QGIS (version 3.4.2), Microsoft Excel 2016 (Microsoft, Washington, DC, USA) software and Estonian Basic Map 1:10,000 (Estonian Land Board 2018).

One way ANOVA was used in analyses of variability between different sites or landscape categories at levels of plant families and genera, while Wald test was used in analyses at the plant species level. Kruskal–Wallis median test was used to estimate differences in strawberry pollen abundance between landscape categories and confirm land use parameter differences between selected landscape categories, *Fragaria* reads or Strawberry pollen counts in landscape categories. Pearson's correlation coefficient r was used to measure the strength of relationships between the proportions of strawberry pollen and diversity of plant families foraged, between proportions of strawberry pollen and *Fragaria* reads; between proportions of plant families and land use parameters (see Supplementary materials Table S1 for all the r-value color codes of all the individual correlations). All the analyses were made using statistical software Statistics version 13 (© 2021 StatSoft Europe, Hamburg, Germany).

3. Results

3.1. Taxonomic Variability

We compared the study sites and thereafter three landscape categories at different plant taxonomic levels: family, genera, and species. The microscopy technique helped us to define on average 4.3 ± 0.2 SE plant families from each hive and date, a result, which is low compared to the 9.7 ± 0.2 SE determined with DNA metabarcoding. The innovative technology allows us to seek deeper into the dataset. Using DNA data, we defined pollen from 18.9 ± 0.4 SE plant genera. There was no variation between sites nor landscape categories (Figure 3, Table 1). We saw that despite being also geographically distinct, the numbers of plant families and genera foraged in each site or landscape category did not differ (except at the site level according to the DNA barcoding results).

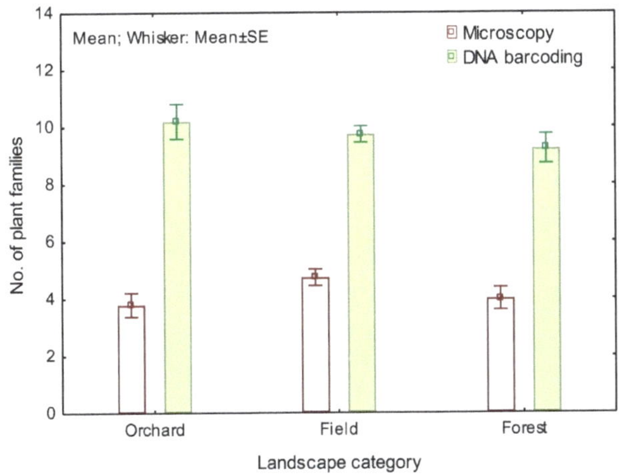

Figure 3. A number of plant families were determined with microscopy and DNA metabarcoding in bumble bee pollen forage from different landscape categories.

Table 1. Statistical analysis of variance of different pollens at a different level of plant taxa determination. The analyses at the Family and Genera level are made using ANOVA, the analyses at Species level are made using Wald statistics from LOG Generalized linear model.

		Microscopy			DNA Metabarcoding		
Taxonomic Level	**Comparison Level**	*F*	*df*	*p*	*F*	*df*	*p*
Family	Site	1.57	14; 39	0.13	2.66	14; 39	0.008
	Landscape category	2.05	2; 51	0.14	0.96	2; 51	0.39
Genera	Site	–	–	–	1.04	2; 51	0.36
	Landscape category						
		Wald. Stat.	*df*	*p*			
Species	Site	0.19	14; 427	0.65	–	–	–
	Landscape category	0.06	14; 427	0.80	–	–	–

The data pooled over hive, site and date showed that although the maximum proportion of strawberry pollen was 64.8%, the median was as low as 5.6 % ± 20.5 SD. There was no correlation between the proportion of strawberry (determined by microscopy) foraged and diversity of plant families foraged (microscopy: $r = 0.06$, $p = 0.68$, $r^2 = 0.003$; DNA barcoding: $r = 0.02$; $p = 0.91$, $r^2 < 0.001$) and no difference between landscape categories (KW-H(2; 54) = 2.12, $p = 0.35$). There were significant correlations between some land-use types and strawberry foraged: the abundance of yards tended to decrease and the abundance of fruit gardens increased strawberry pollen forage (Supplementary materials Table S1).

The median number of plant species per single foraging flight (analysed by microscopy technique) was 1.65 ± 1.09 SD (Figure 4) with no significant difference neither at site nor landscape category level (Table 1). Although the strawberry was the closest pollen source available in each site, it was not preferred by bumble bees. Considering samples, where one species formed more than 70% of all pollens determined, it was strawberry only in

15% of samples, while being Rosaceae in 24% and Fabaceae in 35% of samples. Both main plant families were accompanied by several plant Families. However, the Rosaceae was accompanied by Fabaceae in 54% of cases, but Fabaceae was accompanied with Rosaceae only in 23% of cases.

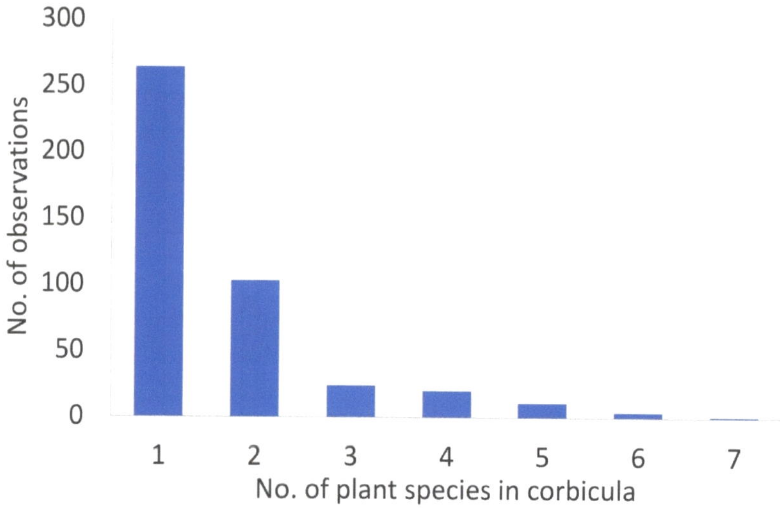

Figure 4. The number of bumble bee corbicular pollen loads containing a different number of plant species.

3.2. *Proportional Variability of Pollens between Landscape Categories (Plant Family Level)*

Because both of the techniques have limitations in their plant species determination ability, we made the quantifying analysis at the level of plant families. We saw three major plant families represented in pollens foraged by bumble bees. These were Fabaceae, Rosaceae and Brassicaceae. Although there were some differences in quantification reliability between the two pollen identification techniques (detailed analyses are presented in Supplementary materials Figure S3), we saw many variations between sites, but much less between landscape categories (Statistical details presented in Supplementary materials Table S2). Both techniques indicated the most abundant plant family in bumble bee forage being Fabaceae, which was prevailing in Orchards and least common in Field. Brassicaceae was prevailing in landscape category Field, less, but still abundant in category Forest, but almost absent in Orchards, surprisingly (Figure 5). Although the detection rate of Rosaceae pollens was significantly lower with DNA metabarcoding. There were no differences between landscape categories. The DNA metabarcoding detected four less represented families in addition to those identified with microscopy. The Lamiaceae pollen, which comprised even more than 40% of total pollen counts in some hives and dates by microscopy, was almost absent based on DNA metabarcoding. The same with Ericaceae pollen, which was detected in extremely low levels with DNA metabarcoding, although the microscopy technique showed high prevalence in one forested site.

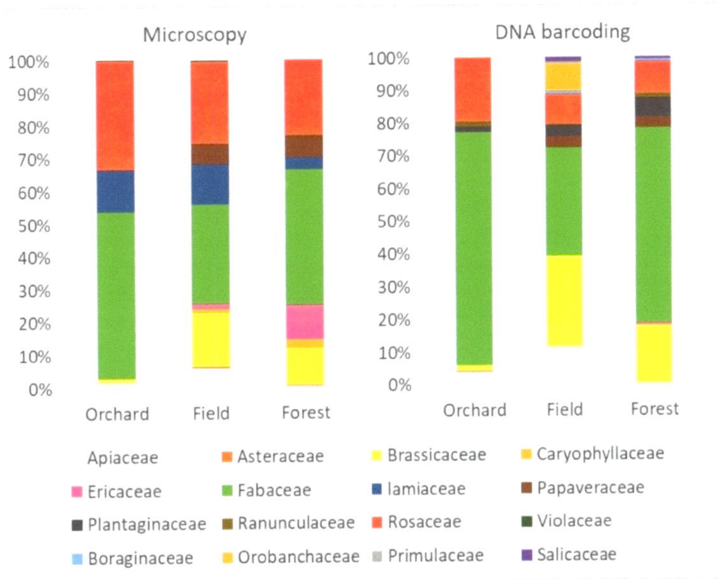

Figure 5. The percentage of bumble bee corbicular pollen loads containing the different number of plant species.

3.3. Correlations of Pollen Proportions and Share of Land Use Parameters (Plant Family Level)

Correlation analysis with proportions of pollens collected and share of land-use types (Supplementary materials, Table S1) in each site at radiuses of 400 m and 1000 m showed few significant results: positive correlation occurred between forests and Ericaceae, grasslands and Brassicaceae, yards and Fabaceae at 400 m radius. Lengthening the radius added positive correlations between arable land and Apiaceae and also Lamiaceae. Negative correlations occurred between orchards and Brassicaceae, arable land and Ericaceae, yards and Rosaceae at 400 m radius. At 1000 m radius, also between yards and Ericaceae, forests and Lamiaceae, grasslands and Rosaceae. Based on microscopy data, the number of families and share of grasslands within 400 m radius showed a significant positive correlation, however, this significance got lost when analysed with DNA metabarcoding or when the distance estimation was lengthened. The pollens of less preferred plant families Orobanchaceae and Primulaceae showed also a significant positive correlation with grasslands and yards in a shorter distance.

3.4. Reliability of DNA Metabarcoding Quantification

We saw good correlations between microscopy and metabarcoding data (look at Supplementary materials Figure S2). Some plant families tended to show positive correlations between the results from microscopy pollen grain counting and relative proportions of reads from DNA metabarcoding. Best overlaps between these two datasets were revealed with these families, which are the most probable forage plants for bumble bees and were also most commonly detected in this study. These families are Brassicaceae (b = 1.02, $r = 0.48$), Fabaceae (b = 0.99, $r = 0.70$) and Rosaceae (b = 0.56, $r = 0.65$). Good correlations were detected also in Apiaceae (b = 1.13, $r = 0.50$) and Papaveraceae (b = 0.58, $r = 0.87$), but these were of minor importance to bumble bees. In comparison with the microscopy technique, the DNA metabarcoding showed a proportional detection rate only for Brassicaceae and Fabaceae, while overestimating Apiaceae, Plantaginaceae and Ranunculaceae pollens. Rosaceae, Papaveraceae. Asteraceae, Caryophyllaceae, Ericaceae and Lamiaceae were underestimated.

4. Discussion

Based on the botanical diversity of bumble bee foraged pollens we confirmed that the heterogeneous agricultural region of southern Estonia supports bumble bees well in spring time. We saw the highly variable food availability over the large study territory and no particular preference for any single plant family. At the same time, the data of individual forage flights suggest sufficient availability of the bumble bees' most preferred plant species.

Many bumble bee species are generalists and may forage on almost any entomophilous plant species available. The quality and availability of spring time forage might be limiting factors for the persistence of rich bumble bee fauna [40]. The *Bombus terrestris* forms large colonies naturally and the foragers are evolved to perform flights, starting with hundreds of meters from the nest, allowing to search a large territory around [30]. We used fully developed *B. terrestris* colonies to achieve the proper sampling of flower resources at the time. Natural bumble bee colonies of any species are at their early development phase and are not so numerous to outcompete our test bumble bees.

Plant taxonomic composition and proportional content of pollens collected by bumble bees can give an overview of the forage plant richness of particular environments. Although generalist bumble bees can forage on a large variety of plant species. Each individual bee learns and remembers few rewarding flower types to diminish the energetic costs of foraging [41,42]. Here, a number of workers, as a feature of the developmental stage of the colony, can affect at some point not only food preferences [43] but also foraging intensity. Parmentier et al. [25] show the relatedness of bumble bee preferred pollens to the availability of these plants in bee foraging areas and suggested this phenomenon to be ever-changing and dependent on the season. In case of abundant and diverse forage availability, the flower constancy develops to increase the foraging effort [44] and the poorer the resources, the stronger the pressure to forage on a higher number of plant taxa. The diversity of plant species in bumble bee pollen pellets may indicate the quality of anthropogenic landscapes to bumble bees [25]. Hence, our results here indicate sufficient food source variability over all the study sites. While Fabaceae and Rosaceae were the most preferred plant families, as common, the accompanying plants in bumble bee pollen pellets were different for the two main plant families. Fabaceae were more often accompanied by other plants than Rosaceae, while in more than half of cases the Rosaceae were accompanied with Fabaceae. This might come because of differences in nectar and pollen nutritional value [45]. As Fabaceae plants live in symbiosis with nitrogen-fixing bacteria, their nectar and pollen can be with higher nitrogen content compared to other plant species through a better (self)-fertilisation rate [46]. Flower resources, like pollen and nectar, can benefit from specific fertilisation also in other plant species [47]. Fabaceae plants are often seen to be very attractive to many bumble bee species [48–50], including *B. terrestris* which was used in this study. Pollen nutritional value of Rosaceae plants has been estimated twice as low as that of Fabaceae when estimated by the protein:lipid ratio [51]. Despite the lower nutritional value, the availability of Rosaceae was higher (the strawberry!). Moreover, in Estonia, many spring-flowering fruit trees and berries, including strawberries, belong to the Rosaceae family.

The spring time floral richness offers plenty of flowering plants and as bumble bees prefer to switch between plants, they do not rely on massive crops. Kallioniemi et al. [52] showed that the flowering crops have a negative effect on bumble bees in the spring time but a positive effect in summer. Our results confirm that bumble bees were not interested in crops at this particular time of year. Among Brassicaceae plants, the oilseed rape and turnip rape are commonly grown, but our data indicated that rather the weeds were prevailing in bumble bee forage. Wild species of Brassicaceae prefer open landscapes and are flowering on wastelands, grasslands and inside cereal fields, where these may also form massive flower resources. In our case, while in the study region, also beekeeping is common, the honey bees may have outcompeted bumble bees from larger patches of Brassicaceae, either weeds or crops. Raimets et al. [14] sampled honey bee hives during the flowering of winter oilseed rape, which falls pretty much into the same time frame as our study, and saw that Brassicaceae formed more than a half of foraged pollens.

Lamiaceae are often highly attractive for several species of bumble bees in spring time. In our study, we saw that in some study sites bumble bees foraged on it, but the proportion stayed low. It might come from the unsuitable weather conditions the study year was with rather cool and moist weather, which suppresses the nectar production of this plant [53].

Many studies point to the importance of different land-use types. Flower-rich field margins are shown to be a great food source for bumble bees [54], but this was not confirmed in our study. We saw that proportion of arable land was positively related only with Apiaceae, Lamiaceae and Brassicaceae. The last one belonged to the three most commonly foraged plant families, but was clearly less attractive compared to Fabaceae and Rosaceae. Lye et al. [55] argued that flower-rich grasslands support bumble bee colony initialisation and early development, due to suitability for nesting and foraging and. In our study region, however, the proportion of grasslands supported only forage on Brassicaceae and some minor importance families. Mola et al. [56] demonstrated the importance of herbal plants flowering in forest underwoods at the period of young queens establishing their nests. Osborne et al. [57] found the yards being valuable nesting sites, which simultaneously provide variable forage resources. We saw the relationship of yards and Fabaceae and some other less attractive plant families. The yards might be of higher importance when natural forage is scarce. Surprisingly, in spring time, fruit and berry gardens did not support the forage of Rosaceae pollens. Instead, the Fabaceae was prevailing there like everywhere else. We are not opposed to all aforementioned studies, but we show that every study region, plant communities and particular time windows design their datasets and outcomes. Different habitat types are important at the different time points and bumble bee development stages. Indeed, our study should be continued to estimate the value of this environment throughout the season.

Further improvements are still needed for technical gaps in the quantification of bee-collected pollen samples. In this study, we found at least partial overlap in the results. Although light microscopy and DNA metabarcoding are powerful methods for insect collected pollen identification applying to biological, ecological and even agricultural studies [58–60], both have their limitations. Light microscopy provides more accurate quantitative results, but it takes a lot of time to confirm identification and needs to use different supporting materials and techniques (such as reference samples and an electron microscope). While it is not realistic to identify every single pollen grain, less represented species could be non-detected. DNA metabarcoding produces qualitative information about all plant species found in the pollen sample. Many samples could be processed with a little effort, but thereafter a wide range of the information needs to be analysed to exclude DNA reads, pointing on plants not available in the region or not flowering at the research period. Bell et al. [23] also showed the misinterpretation of related species because of similar nucleotide sequences in the target site of the genome. Based on our results, the DNA metabarcoding did not separate crops like strawberry nor oilseed rape, which is feasible with microscopy (see strawberry pollen illustration in Supplementary materials). Instead, DNA metabarcoding suggested the wild relatives of these plants. As an improvement, the DNA metabarcoding allows to use of plant genera at a taxonomic level in the analyses, this would be hard to achieve with microscopy by large datasets. The DNA metabarcoding is not well calibrated yet to allow a quantitative approach in studies [18,23] Just like Bell et al. [23] described, our data also show strong over and underestimation of some particular plant species. Bell et al. [23] reviewed that the difficulties in quantification in DNA metabarcoding studies might occur due to a number of copies of the DNA regions, preservation, DNA isolation and amplification biases. This might be affected by several factors, and even by chloroplast inheritance modes maternal, paternal and biparental between different plants. Still, we claim that in the most important bumble bee forage plants, the proportional quantification of the reads produces a fairly good result. Bänsch et al. [61] also used the DNA metabarcoding with reads from ITS2 region sequences to quantify the bumble bee-collected pollen but reminded that the interpretation must take the specific limitations of the outcome into account.

5. Conclusions

The availability of nesting sites and food plants can be scarce in modern landscapes. Urbanization gathers people to cities and suburban areas, and thus, populated rural areas are declining. The landscape in southern Estonia represents rural municipalities where the agricultural activity is present but with lower intensity and scattered small settlements are alternated with forest patches, meadows and fields. This region is also rich in traditional villages with privately owned small vegetable and fruit gardens. The disappearance of dairy cattle and unification of smaller agricultural enterprises to larger ones are occurring here like elsewhere alongside modernisation in both agriculture and forestry. These processes usually bring the homogenisation of landscapes and the decrease in the availability of food plants of pollinators.

In order to achieve stable pollinator fauna, we need to understand the parameters supporting them. The needs of bees are described, but the ever-changing conditions are to be overlooked. Until recent times, environmental protection has been focused mainly on preserving certain best suitable areas, often represented by small habitat islands with or without connections between them. Instead of creating a dense network of small various habitats, we should look at larger regions entirely supporting pollinators. As the study of Mola et al. [55] pointed, some habitat types have gained most of the attention and some others have remained disregarded. Based on the Estonian bumble bee monitoring dataset, it is shown that agri-environmental schemes determining the frequency of legume crops in crop rotations and size of field margins [13] most probably support general bumble bee fauna; however, perhaps not for all species. Our study focuses on a small time frame in spring which is most important to support bumble bee colony initialization. If the environment in southern Estonia also supported late-season forage, this region could serve as the bumble bee's best habitat and national rural development plans should consider this.

Supplementary Materials: The following are available online at https://www.mdpi.com/article/10.3390/insects12100922/s1, Figure S1: Examples of landscapes belonging to different categories. Red circles indicate 1 km radius around the bumble bee hives (center of the circles), Table S1: The strength of correlations based on *r*-values between land-use parameters and proportions of plant families, proportions of strawberry pollen and numbers of plant families in bumble bee corbicular pollen, Figure S2: Example of pollen grains (strawberry) seen under light microscopy, maginification x1000, Table S2: Kruskal-Wallis H test and relevant DF values of variation of proportions of plant families on both microscopy and DNA barcoding data at site and landscape level, Figure S3: The correlation matrixes of proportions of pollen grains counted with microscopy and relative abundance of reads of the results of DNA barcoding.

Author Contributions: Conceptualization, A.B., R.K., M.M. and G.S.; methodology, A.B. and G.S.; validation, A.B. and R.K.; formal analysis, A.B. and R.K.; investigation, A.B.; resources, M.M. and G.S.; writing—original draft preparation, A.B. and R.K.; writing—review and editing, M.M. and G.S.; visualization, A.B. and R.K.; supervision, M.M. and G.S.; project administration, M.M. and G.S.; funding acquisition, M.M. and G.S. All authors have read and agreed to the published version of the manuscript.

Funding: This research was funded by the Estonian University of Life Sciences project P200192PKTE, and Estonian Research Council, grant number RITA1/02-10 and the Research Foundation Flanders (FWO Vlaanderen).

Institutional Review Board Statement: Not applicable.

Informed Consent Statement: Not applicable.

Data Availability Statement: The databes was used for Bontšutšnaja, A.; Karise, R.; Mänd, M.; Smagghe, G. Bumble Bee Foraged Pollen Analyses in Spring Time in Southern Estonia Shows Abundant Food Sources. Insects 2021, 12, 922. https://doi.org/10.3390/insects12100922. The database contains the results of DNA metabarcoding (sequences, relative importances) and microscopy pollen count data.

Acknowledgments: The authors want to thank Niels Piot (UGent) for assistance with DNA metabarcoding and Kaire Loit (EMÜ) with further bioanalyses, Matti Pisman (UGent) for QGIS software exploitation and Tanel Kaart (EMÜ) for help with statistical analyses. Ministry of Finance of the Republic of Estonia (Archimedes Foundation) for possibility conduct the part of the study in UGent. We also are very grateful to farmers for their permission to work in their fields.

Conflicts of Interest: The funders had no role in the design of the study; in the collection, analyses, or interpretation of data; in the writing of the manuscript, or in the decision to publish the results.

References

1. Underwood, E.; Darwin, G.; Gerritsen, E. *Pollinator Initiatives in EU Member States Success Factors and Gaps. Report for European Commission under Contract for Provision of Technical Support Related to Target 2 of the EU Biodiversity Strategy to 2020—Maintaining and Restoring Ecosystems and Their Services ENV.B.2/SER/2016/0018*; Institute for European Environmental Policy: Brussels, Belgium, 2017; p. 70.
2. Science for Environment Policy. *Pollinators: Importance for Nature and Human Well-Being, Drivers of Decline and the Need for Monitoring. Future Brief 23. Brief Produced for the European Commission DG Environment*; Science Communication Unit, UWE Bristol: Bristol, UK, 2020.
3. Potts, S.G.; Imperatriz-Fonseca, V.; Ngo, H.T.; Aizen, M.A.; Biesmeijer, J.C.; Breeze, T.; Dicks, L.; Garibaldi, L.A.; Hill, R.; Settele, J.; et al. Safeguarding pollinators and their values to human well-being. *Nature* **2016**, *540*, 220–229. [CrossRef]
4. Avarguès-Weber, A.; Lachlan, R.; Chittka, L. Bumblebee social learning can lead to suboptimal foraging choices. *Anim. Behav.* **2018**, *135*, 209–214. [CrossRef]
5. Harmon-Threatt, A.N.; de Valpine, P.; Kremen, C. Estimating resource preferences of a native bumblebee: The effects of availability and use–availability models on preference estimates. *Oikos* **2017**, *126*, 633–641. [CrossRef]
6. Ogilvie, J.E.; Forrest, J.R. Interactions between bee foraging and floral resource phenology shape bee populations and communities. *Curr. Opin. Insect. Sci.* **2017**, *21*, 75–82. [CrossRef] [PubMed]
7. Vaudo, A.D.; Patch, H.M.; Mortensen, D.A.; Tooker, J.F.; Grozinger, C.M. Macronutrient ratios in pollen shape bumble bee (Bombus impatiens) foraging strategies and floral preferences. *Proc. Natl. Acad. Sci. USA* **2016**, *113*, 4035–4042. [CrossRef]
8. Goulson, D.; Nicholls, E.; Botías, C.; Rotheray, E.L. Bee declines driven by combined stress from parasites, pesticides, and lack of flowers. *Science* **2015**, *347*, 1255957. [CrossRef]
9. Bonoan, R.E.; Gonzalez, J.; Starks, P.T. The perils of forcing a generalist to be a specialist: Lack of dietary essential amino acids impacts honey bee pollen foraging and colony growth. *J. Apic. Res.* **2020**, *59*, 95–103. [CrossRef]
10. Smart, M.D.; Otto, C.R.V.; Lundgren, J.G. Nutritional status of honey bee (Apis mellifera L.) workers across an agricultural land-use gradient. *Sci. Rep.* **2019**, *9*, 1–10. [CrossRef]
11. Park, M.G.; Blitzer, E.J.; Gibbs, J.; Losey, J.E.; Danforth, B.N. Negative effects of pesticides on wild bee communities can be buffered by landscape context. *Proc. R. Soc. B* **2015**, *282*, 20150299. [CrossRef]
12. Marja, R.; Herzon, I.; Viik, E.; Elts, J.; Maend, M.; Tscharntke, T.; Batary, P. Environmentally friendly management as an intermediate strategy between organic and conventional agriculture to support biodiversity. *Biol. Conserv.* **2014**, *178*, 146–154. [CrossRef]
13. Marja, R.; Kleijn, D.; Tscharntke, T.; Klein, A.-M.; Frank, T.; Batáry, P. Effectiveness of agri-environmental management on pollinators is moderated more by ecological contrast than by landscape structure or land-use intensity. *Ecol. Lett.* **2019**, *22*, 1493–1500. [CrossRef]
14. Raimets, R.; Bontsutsnaja, A.; Bartkevics, V.; Pugajeva, I.; Kaart, T.; Puusepp, L.; Pihlik, P.; Keres, I.; Viinalass, H.; Mänd, M.; et al. Pesticide residues in beehive matrices are dependent on collection time and matrix type but independent of proportion of foraged oilseed rape and agricultural land in foraging territory. *Chemosphere* **2020**, *238*, 124555. [CrossRef] [PubMed]
15. Westphal, C.; Steffan-Dewenter, I.; Tscharntke, T. Foraging trip duration of bumblebees in relation to landscape-wide resource availability. *Ecol. Entomol.* **2006**, *31*, 389–394. [CrossRef]
16. Goulson, D.; Lepais, O.; O'Connor, S.; Osborne, J.L.; Sanderson, R.A.; Cussans, J.; Goffe, L.; Darvill, B. Effects of land use at a landscape scale on bumblebee nest density and survival. *J. Appl. Ecol.* **2010**, *47*, 1207–1215. [CrossRef]
17. Rahl, M. Microscopic identification and purity determination of pollen grains. *Methods Mol. Med.* **2008**, *138*, 263–269. [PubMed]
18. Keller, A.; Danner, N.; Grimmer, G.; Ankenbrand, M.; von der Ohe, K.; Rost, S.; Härtel, S.; Steffan-Dewenter, I. Evaluating multiplexed next-generation sequencing as a method in palynology for mixed pollen samples. *Plant. Biol. Stuttg. Ger.* **2015**, *17*, 558–566. [CrossRef]
19. Richardson, R.T.; Lin, C.-H.; Quijia, J.O.; Riusech, N.S.; Goodell, K.; Johnson, R. Rank-based characterization of pollen assemblages collected by honey bees using a multi-locus metabarcoding approach1. *Appl. Plant Sci.* **2015**, *3*, 1500043. [CrossRef]

20. Sickel, W.; Ankenbrand, M.J.; Grimmer, G.; Holzschuh, A.; Härtel, S.; Lanzen, J.; Steffan-Dewenter, I.; Keller, A. Increased efficiency in identifying mixed pollen samples by meta-barcoding with a dual-indexing approach. *BMC Ecol.* **2015**, *15*, 20. [CrossRef]
21. Smart, M.D.; Cornman, R.S.; Iwanowicz, D.D.; McDermott-Kubeczko, M.; Pettis, J.S.; Spivak, M.S.; Otto, C. A comparison of honey bee-collected pollen from working agricultural lands using light microscopy and ITS metabarcoding. *Environ. Entomol.* **2017**, *46*, 38–49. [CrossRef]
22. Richardson, R.T.; Lin, C.-H.; Sponsler, D.B.; Quijia, J.O.; Goodell, K.; Johnson, R.M. Application of ITS2 metabarcoding to determine the provenance of pollen collected by honey bees in an agroecosystem. *Appl. Plant Sci.* **2015**, *3*, apps.1400066. [CrossRef]
23. Bell, K.L.; Burgess, K.S.; Botsch, J.C.; Dobbs, E.K.; Read, T.D.; Brosi, B.J. Quantitative and qualitative assessment of pollen DNA metabarcoding using constructed species mixtures. *Mol. Ecol.* **2019**, *28*, 431–455. [CrossRef] [PubMed]
24. Loit, K.; Soonvald, L.; Astover, A.; Runno-Paurson, E.; Öpik, M.; Tedersoo, L. Temporal and cultivar-specific effects on potato root and soil fungal diversity. *Agronomy* **2020**, *10*, 1535. [CrossRef]
25. Parmentier, L.; Meeus, I.; Cheroutre, L.; Mommaerts, V.; Louwye, S.; Smagghe, G. Commercial bumblebee hives to assess an anthropogenic environment for pollinator support: A case study in the region of Ghent (Belgium). *Environ. Monit. Assess.* **2014**, *186*, 2357–2367. [CrossRef] [PubMed]
26. Statistical Office of Estonia (Ed.). *2000 Population and Housing Census, I. Population De Facto and Usual Resident Population, Location of the Population, Population Sex and Age Structure*; Statistikaamet: Tallinn, Estonia, 2001.
27. Statistical Office of Estonia (Ed). *2000 Population and Housing Census. Population Settlements of Rural Municipalities*; Statistikaamet: Tallinn, Estonia, 2001.
28. Karise, R.; Dreyersdorff, G.; Jahani, M.; Veromann, E.; Runno-Paurson, E.; Kaart, T.; Smagghe, G.; Mänd, M. Reliability of the entomovector technology using Prestop-Mix and Bombus terrestris L. as a fungal disease biocontrol method in open field. *Sci. Rep.* **2016**, *6*, 31650. [CrossRef] [PubMed]
29. Foulis, E.S.J.; Goulson, D. Commercial bumble bees on soft fruit farms collect pollen mainly from wildflowers rather than the target crops. *J. Apic. Res.* **2014**, *53*, 404–407. [CrossRef]
30. Osborne, J.L.; Martin, A.P.; Carreck, N.L.; Swain, J.L.; Knight, M.E.; Goulson, D.; Hale, R.J.; Sanderson, R.A. Bumblebee flight distances in relation to the forage landscape. *J. Anim. Ecol.* **2008**, *77*, 406–415. [CrossRef]
31. Diaz-Forero, I.; Kuusemets, V.; Maend, M.; Liivamaegi, A.; Kaart, T.; Luig, J. Effects of Forest Habitats on the Local Abundance of Bumblebee Species: A Landscape-scale Study. *Balt. For.* **2011**, *17*, 235–242.
32. Walther-Hellwig, K.; Frankl, R. Foraging Distances of Bombus muscorum, Bombus lapidarius, and Bombus terrestris (Hymenoptera, Apidae). *J. Insect Behav.* **2000**, *13*, 239–246. [CrossRef]
33. Walther-Hellwig, K.; Frankl, R. Foraging habitats and foraging distances of bumblebees, Bombus spp. (Hym., Apidae), in an agricultural landscape. *J. Appl. Entomol.* **2000**, *124*, 299–306. [CrossRef]
34. Kreyer, D.; Oed, A.; Walther-Hellwig, K.; Frankl, R. Are forests potential landscape barriers for foraging bumblebees? Landscape scale experiments with Bombus terrestris agg. and Bombus pascuorum (Hymenoptera, Apidae). *Biol. Conserv.* **2004**, *116*, 111–118. [CrossRef]
35. Ruedenauer, F.; Raubenheimer, D.; Kessner-Beierlein, D.; Grund-Mueller, N.; Noack, L.; Spaethe, J.; Leonhardt, S. Best be(e) on low fat: Linking nutrient perception, regulation and fitness. *Ecol. Lett.* **2020**, *23*, 545–554. [CrossRef] [PubMed]
36. Hesse, M.; Halbritter, H.; Weber, M.; Buchner, R.; Frosch-Radivo, A.; Ulrich, S.; Zetter, R. *Pollen Terminology*; Springer: Wien, Austria, 2009; Volume VII, p. 266. [CrossRef]
37. D'Albore, R.G. *An Illustrated Handbook. Mediterranean Melissopalynology*; Università degli studi di Perugia, Facoltà di agraria, Istituto di entomologia agrarian: Perugia, Italy, 1998.
38. Schloss, P.D.; Westcott, S.L.; Ryabin, T.; Hall, J.R.; Hartmann, M.; Hollister, E.B.; Lesniewski, R.A.; Oakley, B.B.; Parks, D.H.; Robinson, C.J.; et al. Introducing mothur: Open-source, platform-independent, community-supported software for describing and comparing microbial communities. *Appl. Environ. Microbiol.* **2009**, *75*, 7537–7541. [CrossRef]
39. Koetschan, C.; Förster, F.; Keller, A.; Schleicher, T.; Ruderisch, B.; Schwarz, R.; Müller, T.; Wolf, M.; Schultz, J. The ITS2 Database III—sequences and structures for phylogeny. *Nucleic Acids Res.* **2010**, *38*, D275–D279. [CrossRef] [PubMed]
40. Goulson, D. The conservation of bumble bees. *Bee World* **2003**, *84*, 105–106. [CrossRef]
41. Karise, R.; Mänd, M.; Ivask, M.; Koskor, E.; Bender, A. The effect of pollen amount and its caloric value in hybrid lucerne (Medicago x varia) on its attractiveness to bumble bees (Bombus terrestris). *Agron. Res.* **2006**, *4*, 211–216.
42. Somme, L.; Vanderplanck, M.; Michez, D.; Lombaerde, I.; Moerman, R.; Wathelet, B.; Wattiez, R.; Lognay, G.; Jacquemart, A.-L. Pollen and nectar quality drive the major and minor floral choices of bumble bees. *Apidologie* **2014**, *46*, 92–106. [CrossRef]
43. Evans, L.J.; Raine, N.E. Changes in learning and foraging behaviour within developing bumble bee (bombus terrestris) colonies. *PLoS ONE* **2014**, *9*, e90556. [CrossRef]
44. Grüter, C.; Moore, H.; Firmin, N.; Helanterä, H.; Ratnieks, F.L.W. Flower constancy in honey bee workers (Apis mellifera) depends on ecologically realistic rewards. *J. Exp. Biol.* **2011**, *214*, 1397–1402. [CrossRef]
45. Ghosh, S.; Jeon, H.; Jung, C. Foraging behaviour and preference of pollen sources by honey bee (Apis mellifera) relative to protein contents. *J. Ecol. Environ.* **2020**, *44*, 4. [CrossRef]

46. Gijbels, P.; Ceulemans, T.; Van den Ende, W.; Honnay, O. Experimental fertilization increases amino acid content in floral nectar, fruit set and degree of selfing in the orchid Gymnadenia conopsea. *Oecologia* **2015**, *179*, 785–795. [CrossRef]
47. Viik, E.; Mänd, M.; Karise, R.; Lääniste, P.; Williams, I.H.; Luik, A. The impact of foliar fertilization on the number of bees (Apoidea) on spring oilseed rape. *Agriculture* **2012**, *99*, 41–46.
48. Hülsmann, M.; von Wehrden, H.; Klein, A.-M.; Leonhardt, S.D. Plant diversity and composition compensate for negative effects of urbanization on foraging bumble bees. *Apidologie* **2015**, *46*, 760–770. [CrossRef]
49. Kleijn, D.; Raemakers, I. A retrospective analysis of pollen host plant use by stable and declining bumble bee species. *Ecology* **2008**, *89*, 1811–1823. [CrossRef] [PubMed]
50. Wood, T.J.; Gibbs, J.; Graham, K.K.; Isaacs, R. Narrow pollen diets are associated with declining Midwestern bumble bee species. *Ecology* **2019**, *100*, e02697. [CrossRef]
51. Vaudo, A.D.; Tooker, J.F.; Patch, H.M.; Biddinger, D.J.; Coccia, M.; Crone, M.K.; Fiely, M.; Francis, J.S.; Hines, H.M.; Hodges, M.; et al. Pollen protein: Lipid macronutrient ratios may guide broad patterns of bee species floral preferences. *Insects* **2020**, *11*, 132. [CrossRef] [PubMed]
52. Kallioniemi, E.; Åström, J.; Rusch, G.M.; Dahle, S.; Åström, S.; Gjershaug, J.O. Local resources, linear elements and mass-flowering crops determine bumblebee occurrences in moderately intensified farmlands. *Agric. Ecosyst. Environ.* **2017**, *239*, 90–100. [CrossRef]
53. Macukanovic, M.; Blazencic, Z. Diurnal and seasonal dynamics of nectar secretion of some species in the family Lamiaceae. *Acta Veterinaria.* 1999, 48, pp. 173–184. Available online: https://agris.fao.org/agris-search/search.do?recordID=YU1998000854 (accessed on 8 June 2021).
54. Marshall, E.J.P.; Moonen, A.C. Field margins in northern Europe: Their functions and interactions with agriculture. *Agric. Ecosyst. Environ.* **2002**, *89*, 5–21. [CrossRef]
55. Lye, G.; Park, K.; Osborne, J.; Holland, J.; Goulson, D. Assessing the value of Rural Stewardship schemes for providing foraging resources and nesting habitat for bumblebee queens (Hymenoptera: Apidae). *Biol. Conserv.* **2009**, *142*, 2023–2032. [CrossRef]
56. Mola, J.M.; Richardson, L.L.; Spyreas, G.; Zaya, D.N.; Pearse, I.S. Long-term surveys support declines in early season forest plants used by bumblebees. *J. Appl. Ecol.* **2021**, *58*, 1431–1441. [CrossRef]
57. Osborne, J.L.; Martin, A.P.; Shortall, C.R.; Todd, A.D.; Goulson, D.; Knight, M.E.; Hale, R.J.; Sanderson, R.A. Quantifying and comparing bumblebee nest densities in gardens and countryside habitats. *J. Appl. Ecol.* **2008**, *45*, 784–792. [CrossRef]
58. Gous, A.; Swanevelder, D.Z.H.; Eardley, C.D.; Willows-Munro, S. Plant–pollinator interactions over time: Pollen metabarcoding from bees in a historic collection. *Evol. Appl.* **2019**, *12*, 187–197. [CrossRef] [PubMed]
59. Gresty, C.E.A.; Clare, E.; Devey, D.S.; Cowan, R.S.; Csiba, L.; Malakasi, P.; Lewis, O.T.; Willis, K. Flower preferences and pollen transport networks for cavity-nesting solitary bees: Implications for the design of agri-environment schemes. *Ecol. Evol.* **2018**, *8*, 7574–7587. [CrossRef]
60. Suchan, T.; Talavera, G.; Sáez, L.; Ronikier, M.; Vila, R. Pollen metabarcoding as a tool for tracking long-distance insect migrations. *Mol. Ecol. Resour.* **2019**, *19*, 149–162. [CrossRef] [PubMed]
61. Bänsch, S.; Tscharntke, T.; Wünschiers, R.; Netter, L.; Brenig, B.; Gabriel, D.; Westphal, C. Using ITS2 metabarcoding and microscopy to analyse shifts in pollen diets of honey bees and bumble bees along a mass-flowering crop gradient. *Mol. Ecol.* **2020**, *29*, 5003–5018. [CrossRef] [PubMed]

insects

MDPI

Article

Intraspecific Variability in Proteomic Profiles and Biological Activities of the Honey Bee Hemolymph

Salma A. Elfar [1], Iman M. Bahgat [1], Mohamed A. Shebl [2], Mathieu Lihoreau [3,*] and Mohamed M. Tawfik [1,*]

1 Zoology Department, Faculty of Science, Port Said University, Port Said 42526, Egypt
2 Department of Plant Protection, Faculty of Agriculture, Suez Canal University, Ismailia 41522, Egypt
3 Research Center on Animal Cognition, Center for Integrative Biology, Centre National de la Recherche Scientifique, University Paul Sabatier, 31062 Toulouse, France
* Correspondence: mathieu.lihoreau@univ-tlse3.fr (M.L.); tawfik@sci.psu.edu.eg (M.M.T.); Tel.: +33-5-61-55-64-71 (M.L.)

Simple Summary: Insect hemolymph is equivalent to blood in higher vertebrates. It is the main site for immune responses, mediates nutrient transportation to organs and tissues, and has antimicrobial and antioxidant properties. Hemolymph can thus provide information about the health status of an insect. Here we report intraspecific variation in hemolymph properties of Western honey bees *Apis mellifera* sampled in four locations providing different diets across Egypt. Bees that had access to a rich and varied diet had higher protein concentrations and levels of biological activities in their hemolymph than bees that were only fed sucrose solution. This suggests hemolymph analyses could be used as a powerful indicator for monitoring bee populations, with the aim of improving their health and pollination efficiency.

Abstract: Pollinator declines have raised major concerns for the maintenance of biodiversity and food security, calling for a better understanding of environmental factors that affect their health. Here we used hemolymph analysis to monitor the health status of Western honey bees *Apis mellifera*. We evaluated the intraspecific proteomic variations and key biological activities of the hemolymph of bees collected from four Egyptian localities characterized by different food diversities and abundances. Overall, the lowest protein concentrations and the weakest biological activities (cytotoxicity, antimicrobial and antioxidant properties) were recorded in the hemolymph of bees artificially fed sucrose solution and no pollen. By contrast, the highest protein concentrations and biological activities were recorded in bees that had the opportunity to feed on various natural resources. While future studies should expand comparisons to honey bee populations exposed to more different diets and localities, our results suggest hemolymph samples can be used as reliable indicators of bee nutrition.

Keywords: honeybee; proteome; anticancer; antimicrobial; antioxidant

Citation: Elfar, S.A.; Bahgat, I.M.; Shebl, M.A.; Lihoreau, M.; Tawfik, M.M. Intraspecific Variability in Proteomic Profiles and Biological Activities of the Honey Bee Hemolymph. *Insects* **2023**, *14*, 365. https://doi.org/10.3390/insects14040365

Academic Editors: Ivana Tlak Gajger and Franco Mutinelli

Received: 15 February 2023
Revised: 29 March 2023
Accepted: 3 April 2023
Published: 6 April 2023

1. Introduction

Bees are key pollinators [1,2] whose widespread declines have raised major concerns for the maintenance of terrestrial biodiversity and food security [3]. Over the past decades, their accelerated losses have raised the need for a better understanding of the environmental stressors that affect population growth and their mechanisms of action, for instance, through large-scale monitoring of bee colony health statuses across different habitats [4].

Like all insects, bees have an open circulatory system composed of hemolymph, which is equivalent to blood in higher vertebrates [5]. Hemolymph is vital to bees because it mediates the distribution of nutrients through the body by supplying tissues and organs [6]. It is the main site for defense against infections [7] and possesses antimicrobial, antioxidant and anticancer activities [8]. The principal components of hemolymph are water, carbohydrates, proteins, inorganic salts, lipids, hormones and immune cells. Variations in protein contents, which are strongly linked to nutrition [9–11], can inform about the physiological and immune status of bees [5,12]. For instance, high hemolymph protein levels minimize the susceptibility of honey bees to pathogens [13,14]. By contrast, low hemolymph protein levels are a signature of poor health status. Therefore, hemolymph analysis can serve as a practical and powerful means for monitoring bee health status. Ultimately, a better understanding of hemolymph variations across individuals and populations may also help better assess its value as a potential therapeutic compound, as recently suggested (i.e., anticancer and hemolytic activities) [15].

Previous research has shown that the hemolymph of the honey bee *Apis mellifera* varies in protein composition across individuals of the same population, especially within castes and among different developmental stages [16]. However, little is known about the variation in hemolymph protein composition among workers with consideration of diet variations. Moreover, these studies have often used different solvents to extract hemolymph samples, which makes it difficult to compare since the nature of solvents affects the solubility and structure of proteins and, thus, the content and biological activities of hemolymph samples [17].

Here we evaluated intraspecific variation in proteomic contents and biological activities of honey bees (*A. mellifera*) hemolymph collected from four localities across Egypt, characterized by contrasted diets (natural vs. artificial diets). We extracted hemolymph samples from 400 honey bees (100 per location) using two different solvents and quantified their proteomic content and potential anticancer, antibacterial, antioxidant, and hemolytic activities.

2. Materials and Methods

2.1. Chemicals

Chemicals were purchased from Sigma Chemical Co. (St. Louis, MO, USA). Hydrophilic Phosphate buffer saline (PBS) and hydrophobic Dimethyl sulfoxide (DMSO) were used as solvents for the extraction of honey bee hemolymph samples. These two complementary solvents were used in order to extract as many molecules as possible.

2.2. Sample Collection and Preparation

We exclusively worked on forager honey bees (hybrids of the subspecies *A. mellifera lamarckii*, *A. mellifera carnica* and *A. mellifera ligustica*). Sweep nets were used to collect bees of unknown age in the different study sites (ca. 100 bees per location) from May to July 2021. Foragers were caught while flying around food sources. All the foragers from each site were collected on the same day. Bees from the faculty of science of Port Said Governorate originated from colonies enclosed in a big flight tent and maintained in an artificial beehive with access to sugar syrup only for the duration of the experiment (from the beginning of May to the end of July). Bees from Ismailia, Suez governorates and Saint Catherine had access to different cultivated plants (see details in Table 1), thus providing diverse nutritional resources. The bees were cold-anesthetized (kept at −20 °C) for 5 min to facilitate hemolymph extraction in live individuals [18].

Table 1. Localities, feeding diet and types of plantations in the study areas.

Symbol	Locality	Main Feeding Diet	Main Plantation
A	Port said Governorate 31.259 N 32.27 E	Artificial (Sucrose syrup)	N.A
B	Ismailia Governorate 30.61 N 32.25 E	Natural	Cotton
C	Suez Governorate 30.02 N 32.34 E	Natural	Corn, cucumber, okra and sesame
D	Saint Catherine 28.56 N 33.94 E	Natural	Medicinal plants (thyme and sial tree)

2.3. Hemolymph Collection and Preparation

Hemolymph was extracted according to Łoś and Strachecka [19] and Basseri [20]. Here, we sampled near the coxal membrane using sterile insulin syringes and pressing the abdomen. Clear and slightly yellow hemolymph was drawn out from the wound. If cloudy yellow intestinal contents were taken, the sample was discarded. The hemolymph samples were kept in sterile Eppendorf tubes and stored at −20 °C until lyophilization. About 9–14 µL of hemolymph were extracted per bee. All the samples from a given site were then pooled, yielding about 1000–1200 µL of hemolymph. After lyophilization, this resulted in about 100 mg per location. Equal weights of lyophilized hemolymph were dissolved in DMSO or PBS.

2.4. Protein Estimation

Protein concentration was measured in mg/mL using a Thermo Scientific™ (Waltham, MA, USA) Nano Drop™ One Micro volume UV-Vis Spectrophotometer. Bovine serum albumin (BSA) was added as a standard in each sample (2 µL, 1 mg/mL) [21]. Three samples were analyzed per site.

2.5. SDS-Polyacrylamide Gel Electrophoresis

Protein gel profiles were obtained to investigate variations in the types of proteins among samples by using sodium dodecyl sulfate-polyacrylamide gel electrophoresis (SDS-PAGE) for separating proteins depending on their molecular weights according to Laemmli [22]. SDS-PAGE was performed at a total acrylamide concentration of 12%. The samples were mixed with the solubilizing buffer that contains 62.5 mM Tris-HCl (pH 6.8), 20% glycerol, 25 (*w*/*v*) SDS, 0.5% 2-mercaptoethanol and 0.01% bromophenol blue, then heated for 4 min at 95 °C. The samples were then immediately loaded into wells (15 µL of the sample with a protein concentration of 100 µg/mL per well). Electrophoresis was carried out at a constant 35 mA for 2 h using Consort N.V. (Turnhout, Belgium) mini vertical system with a running buffer. The gel was stained with 0.1% Coomassie Brilliant Blue (R-250) for the protein bands to be visualized.

2.6. HPLC (High-Performance Liquid Chromatography) Analysis

Reversed-phase HPLC chromatograms of hemolymph extracts were used to separate the protein peaks based on their retention time. Each of the lyophilized hemolymph samples was dissolved in PBS or DMSO at the same concentration (5 mg/mL). 70 µL samples from each location were analyzed using YL9100 HPLC System under the following conditions: C18 column (Promosil C18 Column 5 µm, 150 mm × 4.6 mm) as stationary phase and acetonitrile (ACN) gradients in the range of 10% to 100% acetonitrile in water for 50 min as mobile phase at flow rate 1 mL/min and by using UV detector at wavelength 280 nm [20].

2.7. Anticancer Activity (MTT-Assay)

The anticancer activities of the hemolymph extracted in PBS and DMSO were measured in vitro. Human hepatocellular carcinoma (HepG2) and human cervical carcinoma (HeLa) cells were obtained from the Holding company for biological products and vaccines (VACSERA), Giza, Egypt. Cells were seeded in 96-well plate for 24 h (5×10^3 cells/well). After incubation, the cells were treated with serial concentrations of the 100 µL hemolymph extracts in each solvent (31.25, 62.5, 125, 250, 500 and 1000 µg/mL) and incubated at 37 °C in a 5% CO_2 atmosphere for 48 h.

The cytotoxic effect of extracts on the proliferation of cells was evaluated using the 3-[4,5-methylthiazol-2-yl]-2,5-diphenyl-tetrazolium bromide (MTT) assay [23]. After 48 h of treatment, the medium, including MTT dye, was added to cells and incubated at 37 °C for 4 h to enable the production of formazan crystals in the viable cells only. DMSO was used to solubilize the formazan crystals. The absorbance was then measured at 540 nm using a Bio-Tek microplate reader ELISA. Each experiment was performed in triplicate, and the percentage of cell viability was calculated as follows:

$$\text{Cell viability (\%)} = (\mathbf{A_T}/\mathbf{A_C}) \times 100$$

where $\mathbf{A_T}$ is the absorbance of treated cells, and $\mathbf{A_C}$ is the absorbance of the control cells (untreated cells).

The concentration that inhibits the growth of 50% of the cells (IC_{50}) values was determined for each sample from a dose-response curve between dose concentration (x-axis) and cell inhibition percentage (y-axis).

2.8. *Antimicrobial Activity Screening*

The antimicrobial activity of hemolymph extracts was assessed using agar well diffusion [24]. Antimicrobial activity was determined against two Gram-positive bacteria (*Bacillus subtilis* and *Staphylococcus aureus*) and two Gram-negative bacteria (*Escherichia coli* and *Salmonella typhimurium*). Briefly, the agar plate surface was inoculated by spreading a volume of the microbial inoculum over the entire agar surface, then making a hole with a diameter of 6 to 8 mm aseptically using a sterile tip. A volume of about 20–100 µL of each extract (10 mg/mL) or control at a known concentration was introduced into the well. The agar plates were then incubated at suitable conditions depending on the microorganism used for the test. Gentamycin was used as a positive control, while PBS and DMSO were used as negative controls. The antibacterial activities of the extracts were expressed as inhibition zones in millimeters [25,26].

2.9. *DPPH Radical Scavenging Assay*

The antioxidant properties of hemolymph were assessed in vitro. The radical scavenging capacity of the hemolymph extracts on the stable free radical 1,1-Diphenyl-2-picrylhydrazyl (DPPH) was tested according to Braca [27]. Briefly, 100 µL of DPPH methanolic solution (0.004% in methanol) were mixed with 0.3 mL of sample extracts (1 mg/mL) or standard and incubated in the dark for 30–60 min at 25 °C. Ascorbic acid (Vitamin C) was used as a standard (positive control), and the absorbance was measured at 515 nm by using a spectrophotometer. A negative control was prepared by adding 2.7 mL of DPPH in 0.3 mL of the solvent used in the extract (AC). The antioxidant activity of the samples was calculated as follows [28,29]:

$$\text{Antioxidant activity (\%)} = [(\mathbf{A_C} - \mathbf{A_E})/\mathbf{A_C}] \times 100$$

where $\mathbf{A_C}$ is the mean absorbance of negative control, and $\mathbf{A_E}$ is the mean absorbance of the extract or standard.

2.10. *Hemolytic Activity Assay*

We assessed the hemolytic activity of bee hemolymph against human red blood cells to elucidate its selectivity and safety as a potential therapeutic agent. We compared the ability of extracts of 100 µL at concentrations 125, 250, 500, 1000, 2000 and 4000 µg/mL to lyse human erythrocytes [30]. Fresh blood samples were collected in test tubes containing anticoagulant (EDTA) and then centrifuged for 5 min at 2000 rpm. Samples were washed several times with sterile PBS. Various concentrations of the extracts were added to RBCs and incubated for an hour at room temperature. The samples were then centrifuged for 5 min at 10,000 rpm, and the absorbance of the released hemoglobin was measured at 570 nm. 10% Triton X-100 was used as a positive control (100% hemolysis), and sterile PBS and DMSO as negative controls (0% hemolysis). This experiment was carried out three times, and the hemolysis percentage was calculated for each sample by using the equation:

$$\text{Hemolysis (\%)} = 100 \times [(\mathbf{A_S} - \mathbf{A_{N.C}})/(\mathbf{A_{P.C}} - \mathbf{A_{N.C}}]$$

where $\mathbf{A_S}$ is the absorbance of the sample, $\mathbf{A_{N.C}}$ is the absorbance of the negative control, and $\mathbf{A_{P.C}}$ is the absorbance of the positive control.

2.11. Statistical Analysis

All analyses were done in SPSS 22.0. Pairwise comparisons between parameters of hemolymphs extracted in PBS and DMSO were made with Student's *t*-tests. Multiple comparisons between parameters of hemolymphs from bees of the four study sites were made with One-way ANOVAs followed by a Tukey's test. When the *p*-value was lower than 0.05, the difference in the means of the samples was considered statistically significant. Means are reported with standard errors (mean ± SE).

3. Results

3.1. Naturally Fed Honey Bees Had Higher Protein Concentrations in Hemolymph

The amount of proteins found in each hemolymph extract was measured using spectrophotometry (Figure 1). Among the DMSO dissolved samples, extracts from site **B** had higher protein concentrations (0.12 ± 0.0006 mg/mL, n = 3 samples) than extracts from all the other sites. Among the PBS dissolved samples, extracts from site **D** had the highest protein concentrations (0.19 ± 0.03mg/mL, n = 3 samples). Extracts from site **A** had a significantly lower protein concentration than extracts from all the other sites, being dissolved in PBS (0.006 ± 0.0005 mg/mL, n = 3 samples) or DMSO (0.02 ± 0.0003 mg/mL, n = 3 samples). Protein concentrations thus varied among the hemolymph extracts of bees fed different diets. The lowest concentrations were consistently recorded for the artificially fed bees in site **A**.

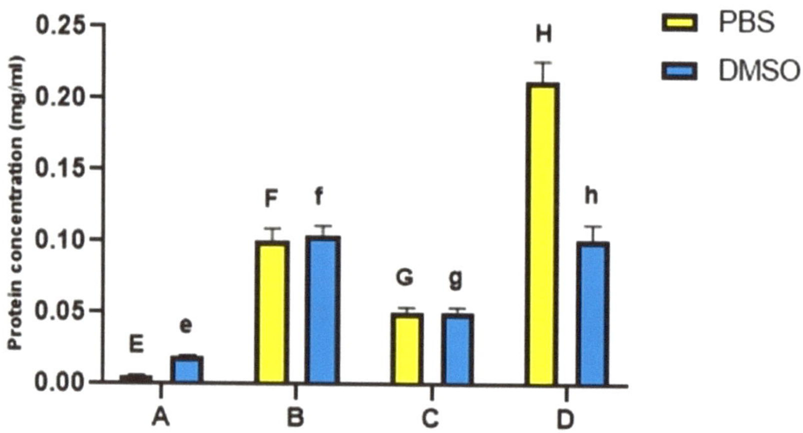

Figure 1. Protein concentrations in mg/mL for honey bee hemolymph extracted in PBS or DMSO where (**A**): bees from Port Said, (**B**): bees from Ismailia governorate, (**C**): bees from Suez governorate, (**D**): bees from Saint Catherine. Bars and error bars represent the mean values ±SE obtained from triplicate measurements. Different letters above bars indicate significant differences (Tukey's test, $p < 0.05$): uppercase letters show comparisons across PBS dissolved samples; lowercase letters show comparisons across DMSO dissolved samples.

3.2. Honey Bee Hemolymph Showed Different Protein Composition across Sites

Quantitative analysis of proteins in the hemolymph extracts revealed a variety of protein bands ranging in mass from 5 to ~250 kDa (Figure 2).

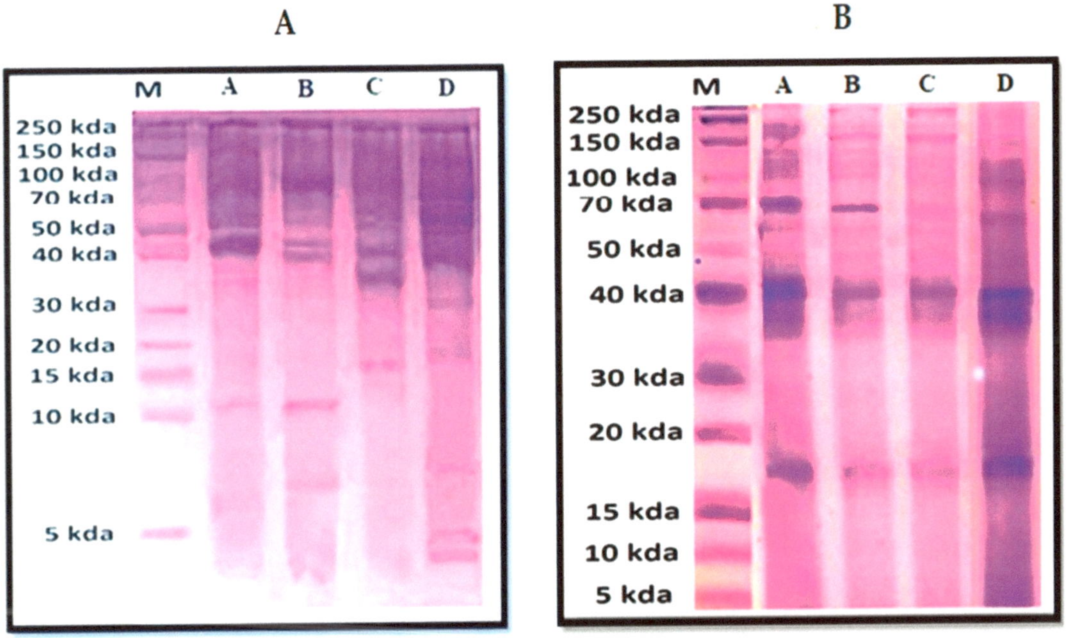

Figure 2. Photos of SDS-PAGE gels for honey bee hemolymph extracts dissolved in PBS (**A**) or DMSO (**B**) where (**A**): bees from Port Said, (**B**): bees from Ismailia governorate, (**C**): bees from Suez governorate, (**D**): bees from Saint Catherine. **M**: marker. The molecular weights of the proteins and their amounts are expressed by the color intensity of the protein bands.

The SDS profile of the samples dissolved in PBS (Figure 2A) revealed six bands with molecular weights ~40, ~50, ~60, ~100, ~180 and ~250 kDa common to extracts from all sites. A band with a molecular weight of ~10 kDa was only recorded in extracts from sites A and B. A band with a molecular weight of ~17 kDa was only recorded in extracts from sites C and D. Three bands with molecular weights ~5, ~30, and ~70 kDa were specific to extracts from site D.

Among the samples dissolved in DMSO (Figure 2B), bands with molecular weights ranging from ~17 to ~250 kDa were observed in all sites. A band with a molecular weight of ~5 kDa was only observed in extracts from site D.

3.3. HPLC Analyses Were Then Used to Separate the Proteins of the Hemolymph Extracts According to Their Peaks Expressed in Retention Time

RP-HPLC chromatograms are supplemented in the supplementary Figures S1 and S2. The chromatograms of the hemolymph extracts dissolved in PBS revealed five peaks that were common to extracts from sites B, C and D at retention times of 7.6, 12.4, 19.9, 25.6 and 43.0 min. The chromatograms of extracts dissolved in DMSO revealed another five peaks common to all sites at retention times of 22.1, 28.7, 30.4, 33.2 and 38.7 min (Table 2).

3.4. Honey Bee Hemolymph Suppresses the Growth of HepG2 and HeLa Cells

The cytotoxic effects of hemolymph were evaluated against the viability of HepG2 and HeLa. The viability of the two cancer cell lines was inhibited in a dose-dependent manner after being treated with different concentrations of the hemolymph extracts in either PBS or DMSO after 48 h of treatment (Figure 3). Overall, DMSO-dissolved extracts possessed higher anti-proliferative activity than PBS-dissolved extracts (Figure 4).

Table 2. RP-HPLC chromatogram peaks profiles of honey bee hemolymph extracts obtained in PBS or DMSO.

Retention Time (min)	Peak Area ($\times 10^2$ mV.s)							
	A		B		C		D	
	PBS	DMSO	PBS	DMSO	PBS	DMSO	PBS	DMSO
2.9	N.A	N.A	1.8	N.A	12.3	N.A	N.A	N.A
4.0	N.A	N.A	N.A	27.7	N.A	1167	N.A	0.2
4.3	N.A	N.A	N.A	N.A	14.4	N.A	3386	N.A
5.0	N.A	N.A	N.A	838.7	N.A	N.A	N.A	397
6.7	36.4	N.A	3.9	N.A	1.0	N.A	N.A	N.A
7.6	N.A	N.A	4.6	N.A	0.8	N.A	163.5	N.A
10.3	N.A	N.A	N.A	N.A	4.8	N.A	247.5	N.A
11.3	N.A	N.A	2.3	N.A	3.9	N.A	N.A	N.A
12.4	N.A	N.A	10.9	N.A	2.5	N.A	1265	N.A
14.4	N.A	N.A	5.9	N.A	N.A	N.A	105.3	N.A
19.9	N.A	N.A	4.1	N.A	5.0	N.A	363.7	N.A
22.1	N.A	0.3	N.A	1.5	N.A	0.6	N.A	2.6
23.2	N.A	N.A	N.A	3.6	N.A	1.0	N.A	N.A
23.5	N.A	0.9	N.A	N.A	N.A	N.A	N.A	1.2
25.6	N.A	N.A	9.1	N.A	18.0	N.A	770.9	N.A
28.7	N.A	0.6	N.A	2.6	N.A	0.3	N.A	1.8
28.8	43.2	N.A	2.3	N.A	4.7	N.A	N.A	N.A
30.3	94.8	N.A	1.8	N.A	5.2	N.A	N.A	N.A
30.4	N.A	1.3	N.A	0.5	N.A	0.2	N.A	0.4
33.2	N.A	0.2	N.A	0.3	N.A	0.3	N.A	0.2
34.8	N.A	N.A	N.A	0.8	N.A	N.A	N.A	1.4
36.5	N.A	1.0	N.A	0.7	N.A	N.A	N.A	0.2
38.1	N.A	N.A	9.1	N.A	3.0	N.A	N.A	N.A
38.7	N.A	10.5	N.A	0.2	N.A	0.1	N.A	0.1
40.7	N.A	N.A	N.A	0.1	N.A	N.A	N.A	0.1
43.0	N.A	N.A	4.4	N.A	2.6	N.A	99.1	N.A
47.7	N.A	N.A	14.6	N.A	10.7	N.A	N.A	N.A

Where, (**A**): bees from Port Said, (**B**): bees from Ismailia governorate, (**C**): bees from Suez governorate, (**D**): bees from Saint Catherine, (**N.A**: not applicable).

Regarding extracts dissolved in DMSO, hemolymph from site C showed the highest cytotoxic effects against HepG2 and HeLa cells (IC_{50} = 52.03 & 97.95 µg/mL, respectively). By contrast, the lowest cytotoxic effect was recorded for hemolymph from site A (IC_{50} = 463.36 & 454.02 µg/mL) (Figure 4).

When considering extracts dissolved in PBS, the highest cytotoxic effect was recorded for hemolymph from site D (IC_{50} = 164.5 & 169.7 µg/mL against HepG2 and HeLa, respectively) and the lowest cytotoxic effect was observed in hemolymph from sites A and B (IC_{50} = 649.36 & 533.08 µg/mL against HepG2 and HeLa respectively) (Figure 4).

3.5. *Honey Bee Hemolymph Inhibits the Growth of Gram-Positive and Gram-Negative Bacteria*

The antimicrobial activity of the hemolymph extracts from sites A, B, C and D was analyzed against two Gram-positive bacteria and two Gram-negative bacteria using the well diffusion method (Table 3). Overall, PBS samples showed higher antibacterial activity than DMSO samples ($p < 0.05$). For both solvents (Table 3), the highest antibacterial activities were observed for the extracts from site B against *E. coli* with inhibition zones 40 mm and 38 mm, respectively. However, extracts from site A showed the lowest antibacterial activities against all tested bacteria, with an inhibition zone ranging from 19 to 28 mm (Figure 5).

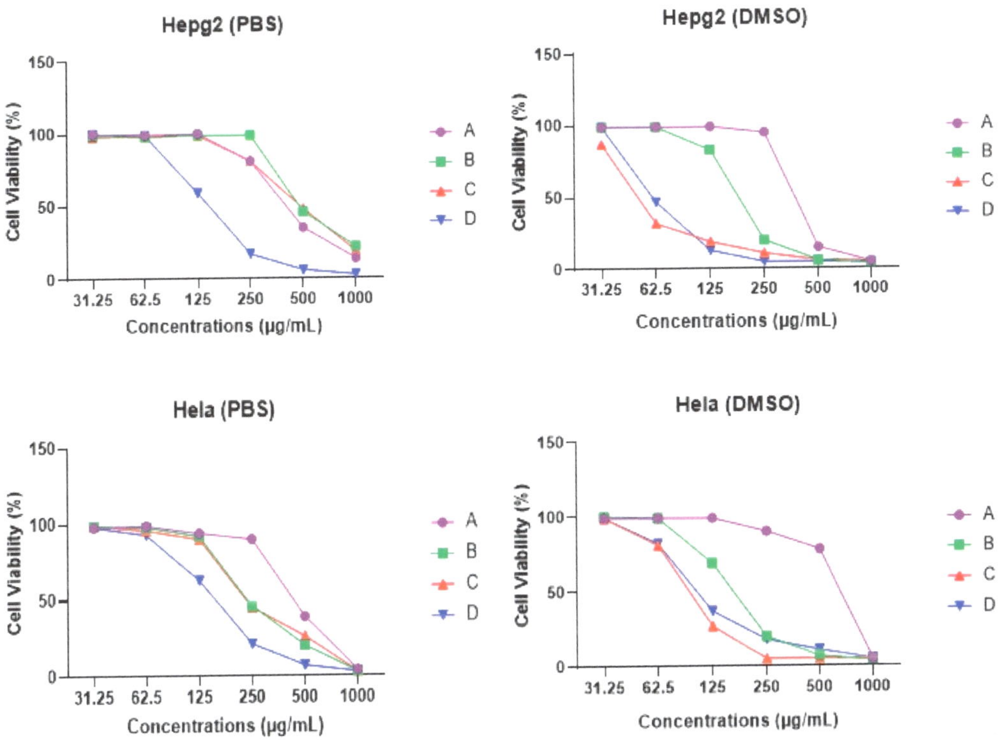

Figure 3. Effects of honey bee hemolymph extracts dissolved at different concentrations in PBS or DMSO on human cancer cell proliferation (HepG2 and HeLa), where (**A**): bees from Port Said, (**B**): bees from Ismailia governorate, (**C**): bees from Suez governorate, (**D**): bees from Saint Catherine.

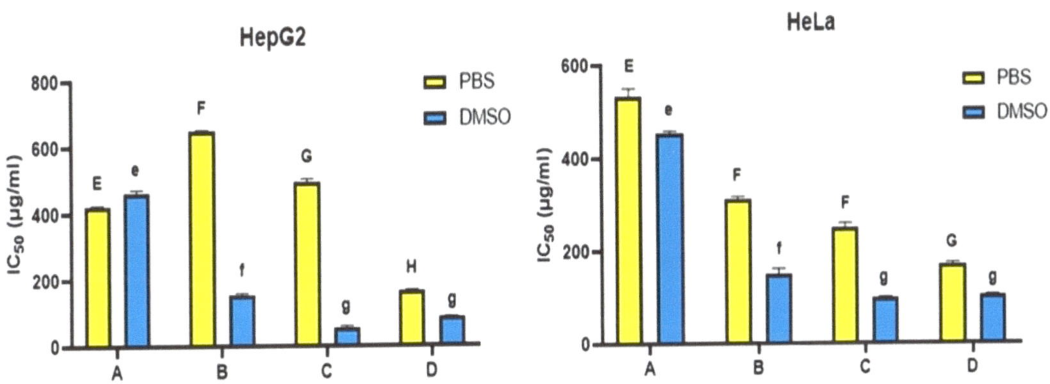

Figure 4. IC_{50} in microgram/mL of hemolymph extracts of honey bees dissolved in PBS or DMSO against Hepg2 and HeLa cell lines, where (**A**): bees from Port Said, (**B**): bees from Ismailia governorate, (**C**): bees from Suez governorate, (**D**): bees from Saint Catherine. Bars and error bars represent the mean values ± SE obtained from triplicate measurements. Different letters above bars indicate significant differences (Tukey's test, $p < 0.05$): uppercase letters show comparisons across PBS dissolved samples; lowercase letters show comparisons across DMSO dissolved samples.

Table 3. Inhibition zone (mm) of honey bee hemolymph extracted in PBS or DMSO on various types of bacteria (*Bacillus subtilis*, *Staphylococcus aureus*, *Escherichia coli* and *Salmonella typhimurium*).

Pathogen \ Sample	PBS				DMSO				CON.
	A	B	C	D	A	B	C	D	
Bacillus subtilis	28 ± 1.2 [a]	34 ± 1.0 [b]	30 ± 0.9 [a]	27 ± 0.9 [a]	26 ± 0.6 [a]	32 ± 1.6 [b]	31 ± 1.5 [ab]	25 ± 1.0 [ac]	22 ± 0.6
Staphylococcus aureus	23 ± 1.0	25 ± 2.0	28 ± 0.9	28 ± 1.0	21 ± 1.6	24 ± 0.9	23 ± 1.0	25 ± 1.9	15 ± 0.9
Escherichia coli	21 ± 1.3 [a]	40 ± 1.5 [b]	32 ± 0.5 [c]	25 ± 1.3 [a]	19 ± 1.3 [a]	38 ± 1.0 [b]	27 ± 1.7 [c]	23 ± 0.7 [ac]	17 ± 1.0
Salmonella typhimurium	22 ± 1.3	N.A	26 ± 1.3	N.A	20 ± 1.5	N.A	25 ± 2.2	N.A	23 ± 0.7

Where **NA** = no activity, (**A**): bees from Port Said, (**B**): bees from Ismailia governorate, (**C**): bees from Suez governorate, (**D**): bees from Saint Catherine, **CON**: Gentamycin as a positive control. Mean values ± SE were obtained from triplicate measurements. Values with different superscript letters (a–c) in the same raw and for a given solvent (PBS or DMSO) are significantly different across locations (Tukey test, $p < 0.05$).

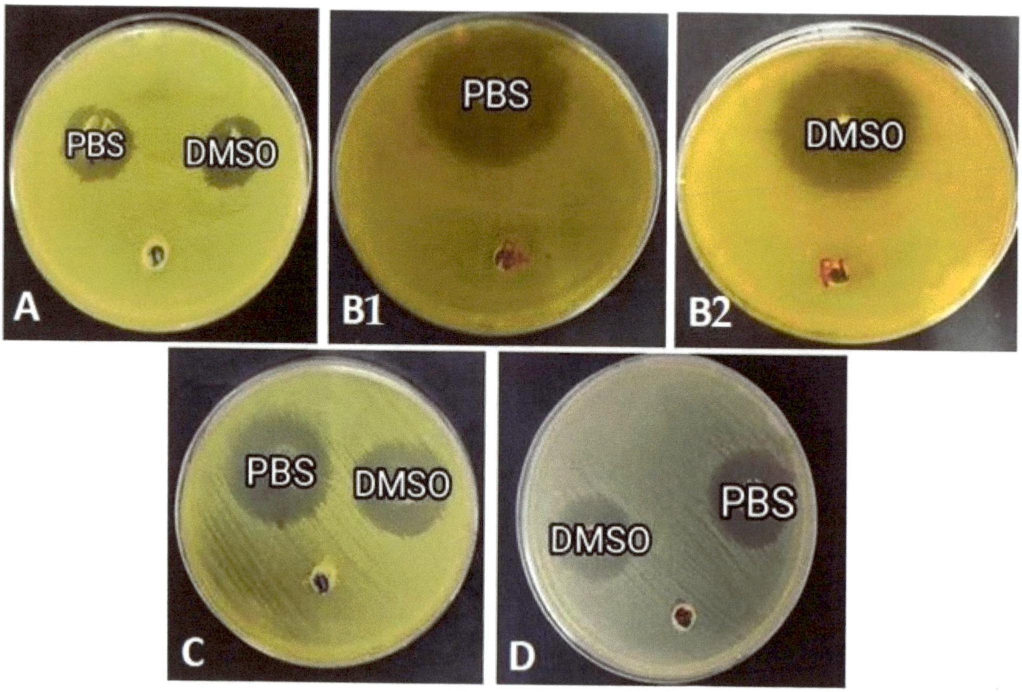

Figure 5. Antimicrobial activity of honey bee hemolymph extracted in PBS or DMSO against *E. coli*, where (**A**): bees from Port Said, (**B1**,**B2**): bees from Ismailia governorate, (**C**): bees from Suez governorate, (**D**): bees from Saint Catherine.

3.6. Honey Bee Hemolymph Scavenges DPPH Free Radical

The antioxidative activity of the hemolymph extracts on DPPH free radicals showed a potential radical scavenging activity in both PBS and DMSO solvents. Overall, PBS-dissolved extracts displayed higher antioxidant properties than DMSO-dissolved extracts (Figure 6) ($p < 0.05$).

Considering PBS- and DMSO-dissolved extracts independently, there were significant differences between sites ($p < 0.001$). Extracts from site D consistently possessed the highest antioxidant activity in both solvents ($p < 0.001$), while extracts from site A possessed the lowest scavenging activity (Figure 6).

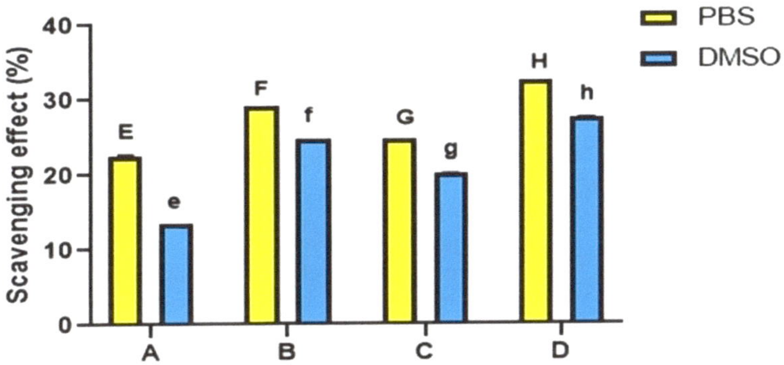

Figure 6. DPPH radical scavenging activity of honey bee hemolymph extracted in PBS or DMSO where (**A**): Bees from Port Said, (**B**): bees from Ismailia governorate, (**C**): bees from Suez governorate, (**D**): bees from Saint Catherine. Bars and error bars represent the mean values ± SEM obtained from triplicate measurements. Different letters above bars indicate significant differences (Tukey's test, $p < 0.05$): uppercase letters show comparisons across PBS dissolved samples; lowercase letters show comparisons across DMSO dissolved samples.

3.7. Honey Bees Possess Low Hemolytic Activity against Human Erythrocytes

Hemolytic activities of the honey bee hemolymph extracts at different concentrations were evaluated on human erythrocytes. None of the extracts displayed hemolytic activity against erythrocytes when compared to negative and positive controls (Figure 7).

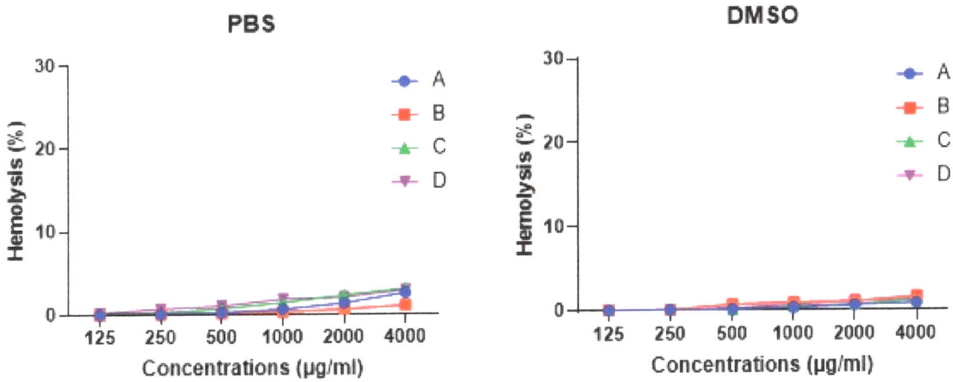

Figure 7. Hemolytic activity of honey bee hemolymph extracted in PBS or DMSO at different concentrations against erythrocytes where (**A**): bees from Port Said, (**B**): bees from Ismailia governorate, (**C**): bees from Suez governorate, (**D**): bees from Saint Catherine.

4. Discussion

Our study describes intraspecific variations in the protein profiles, concentrations and biological activities of the hemolymph of honey bees (*A. mellifera*) collected from different localities in Egypt chosen to offer contrasted diets. The highest protein concentrations, diversity and bioactivity levels were recorded in the hemolymph of bees that had the opportunity to feed on various natural resources when compared to bees fed sucrose solution and no pollen.

The fact that high protein concentrations were recorded in the hemolymph of bees fed natural resources is consistent with the observation of a negative impact of a sugar solution diet on the protein concentration of honey bee hemolymph [31]. Other previous studies have reported the influence of different diets of honey bees on their hemolymph structure [32,33], indicating that the protein content of hemolymph can be used to assess the efficacy of protein diets and pollen quality [10,34–36]. Our study further shows hemolymph proteomic profile can also be used as an indicator of bee health since high protein concentrations extend bee life span and improve immunity and resistance to pathogens [37–39].

Our analysis also revealed important variations in protein concentration depending on the solvent used for extraction. PBS and DMSO were chosen for their complementarity: PBS extracts water-soluble proteins (hydrophilic) [40], while DMSO extracts lipid-soluble proteins (hydrophobic) [41]. Proteins structure, stability and solubility are markedly affected by the polarity of solvents [17], which may explain why hemolymph biological activities greatly varied between PBS and DMSO dissolved samples. We, therefore, recommend future detailed analyses of hemolymph protein contents should use these two complementary solvents for comparative purposes.

The SDS-PAGE gel of the hemolymph of our extracts also showed variations in their protein bands. In general, more protein bands were recorded using DMSO as compared to PBS. Bands with molecular weight ~17 kDa were detected in most extracts dissolved in PBS and DMSO. These bands are in the same range as a class of lectins, which is a common group of proteins in arthropods [42]. Insect lectins have several important roles in immune response, homeostasis and binding for the storage or transport of carbohydrates [43,44]. Insect lectins may also possess antimicrobial, antioxidant and anticancer activities and can be considered as a potential natural bioactive agent [45–49]. Protein bands with molecular weight ~60 kDa were also found in all the tested PBS- and DMSO-dissolved extracts. Similar antimicrobial proteins have been fractionated from cockroach hemolymph [20].

A hemocyanin-like protein band ~70 kDa was also detected in all DMSO dissolved extracts but also some of those dissolved in PBS. Hemocyanin is an oxygen transport protein that has a wide range of biological activities, including antioxidant, antiparasitic, antimicrobial and anticancer activities [48,50,51]. Protein bands with molecular weight ~110 kDa were found in all PBS- and DMSO-dissolved extracts. These bands evoke hexamerins, which are storage proteins that are derived from hemocyanin [52]. The protein bands with molecular weight ~180 kDa were detected in all PBS- and DMSO-dissolved extracts. These bands are in the same range of vitellogenin in honey bee hemolymph [16]. Vitellogenin is a large female-specific gluco-lipoprotein that can act as female nutrient storage and also lipid carrier protein [53,54]. Interestingly, honey bee vitellogenin was shown to have antimicrobial, antioxidant and immunological activities [55].

Protein bands with molecular weight > 250 kDa have also been observed in the proteomic profiles of all PBS and DMSO hemolymph extracts. These bands likely correspond to apolipoprotein, a lipid transporter lipoprotein [56]. These findings agree with Ba et al. (1987) [57] and Chan et al. (2016) [16], who revealed similar bands in honey bee hemolymph.

Interestingly, chromatograms of honey bee hemolymph collected from Saint Catherine showed the highest peak areas among all sites, which is an indicator of the amount of protein. This reinforces the results of the SDS-PAGE gel, which reveals that the protein bands of extract from this location had the highest protein concentrations based on the color intensity of the bands. Saint Catherine is the most floristically diverse region of the Middle East. It is a naturally protected area that accounts for nearly one-fourth of the total flora of Egypt. Such vegetation diversity lets bees in this region have a foraging advantage which may probably enrich the protein content of honey bee hemolymph [58–61].

Both PBS and DMSO dissolved extracts revealed a relative degree of cytotoxicity against HepG2 and HeLa in a dose-dependent manner. There was variation in the cytotoxicity between the hemolymph samples in the two solvents against the two tested cell lines. This variation may be related to the significant differences in their protein profiles and antioxidant activities. The higher cytotoxicity of DMSO-dissolved extracts over PBS-dissolved extracts against both cell lines may be attributed to the role of DMSO in dissolving lipoproteins [41].

The diversity and prevalence of the bioactive proteins such as lectins, hemocyanin, vitellogenin, hexamerins and apolipophorin in PBS and DMSO extracts may elucidate variations in the biological activities of hemolymph described in our results. Hemolymph showed prominent antimicrobial activities against different species of Gram-positive and Gram-negative bacteria. Honey bees, like other invertebrates, possess a very effective innate immune mechanism to defend themselves against microorganisms and pathogens [62], and deficiencies in hemolymph proteins affect their ability to resist diseases [63]. Consequently, feeding on natural resources likely has a great impact on hemolymph bioactivities [64,65].

None of the hemolymph extracts, be they dissolved in PBS or DMSO, showed hemolytic activity when experimented against human erythrocytes at different tested concentrations. This observation agrees with Mokarramat-Yazdi et al. (2021) [15], who reported an absence of hemolytic activity in honey bee hemolymph. Interestingly, in this study, the authors suggested a potential anticancer effect of hemolymph mixed with herbals in mice models. In this regard, honey bee hemolymph is a potential natural selective therapeutic agent in cancer treatment.

5. Conclusions

Our comparative analysis reveals significant intraspecific variability in the composition and concentration of hemolymph of honey bees sampled in Egypt. The hemolymph of the bees that could feed on cultivated plants and wild medicinal plants possessed relatively higher protein concentrations, as well as higher antimicrobial, antioxidant and anticancer properties than that of bees reared only on sugar solutions in artificial conditions. Future studies should expand our approach to more localities and diets in order to clarify the link between diet and hemolymph properties, as well as explore the potential therapeutic usage of honey bee hemolymph.

Supplementary Materials: The following supporting information can be downloaded at: https://www.mdpi.com/article/10.3390/insects14040365/s1, Figure S1: RP-HPLC chromatograms of the PBS dissolved honey bee hemolymph extracts where, (A): Bees from Port Said, (B): bees from Ismailia governorate, (C): bees from Suez governorate, (D): bees from Saint Catherine; Figure S2: RP-HPLC chromatograms of the DMSO dissolved honey bee hemolymph extracts where, (A): Bees from Port Said, (B): bees from Ismailia governorate, (C): bees from Suez governorate, (D): bees from Saint Catherine.

Author Contributions: Conceptualization, M.M.T., I.M.B. and M.A.S.; Methodology, S.A.E., M.M.T., I.M.B. and M.A.S.; Software, validation, formal analysis, investigation, resources and data curation, S.A.E., M.M.T., I.M.B. and M.A.S.; Writing—Original Draft Preparation, S.A.E., M.M.T., I.M.B., M.A.S. and M.L.; Writing—Review & Editing, visualization S.A.E., M.M.T., I.M.B., M.A.S. and M.L.; Supervision, M.M.T., I.M.B. and M.A.S. All authors have read and agreed to the published version of the manuscript.

Funding: This research received no external funding.

Data Availability Statement: Data available on request due to restrictions, e.g., privacy or ethics.

Acknowledgments: We are so grateful to Soliman Kamel and Saied Aboud for their contribution to the honey bee collection.

Conflicts of Interest: The authors declare no conflict of interest.

References

1. Park, M.G.; Orr, M.C.; Danforth, B.N. The Role of Native Bees in Apple Pollination. *Number* **2010**, *18*, 5.
2. Naggar, Y.A.; Codling, G.; Giesy, J.P.; Safer, A. Beekeeping and the Need for Pollination from an Agricultural Perspective in Egypt. *Bee World* **2018**, *95*, 107–112. [CrossRef]
3. Winfree, R. The conservation and restoration of wild bees: Wild bee conservation. *Ann. N. Y. Acad. Sci.* **2010**, *1195*, 169–197. [CrossRef]
4. Marchal, P.; Buatois, A.; Kraus, S.; Klein, S.; Gomez-Moracho, T.; Lihoreau, M. Automated monitoring of bee behaviour using connected hives: Towards a computational apidology. *Apidologie* **2020**, *51*, 356–368. [CrossRef]
5. Horvatinec, J.; Svečnjak, L. Infrared (FTIR) spectral features of honey bee (*Apis mellifera* L.) hemolymph. *J. Cent. Eur. Agric.* **2020**, *21*, 37–41. [CrossRef]

6. Hou, Y.; Zhang, Y.; Gong, J.; Tian, S.; Li, J.; Dong, Z.; Guo, C.; Peng, L.; Zhao, P.; Xia, Q. Comparative proteomics analysis of silkworm hemolymph during the stages of metamorphosis via liquid chromatography and mass spectrometry. *Proteomics* **2016**, *16*, 1421–1431. [CrossRef]
7. Ashrafuzzaman, M.; Razu, M.H.; Showva, N.-N.; Bondhon, T.A.; Moniruzzaman, M.; Rahman, S.A.R.; Rabby, M.R.I.; Akter, F.; Khan, M. Biomolecules of the Horseshoe Crab's Hemolymph: Components of an Ancient Defensive Mechanism and Its Impact on the Pharmaceutical and Biomedical Industry. *Cell. Microbiol.* **2022**, *2022*, e3381162. [CrossRef]
8. Strachecka, A.; Krauze, M.; Olszewski, K.; Borsuk, G.; Paleolog, J.; Merska, M.; Chobotow, J.; Bajda, M.; Grzywnowicz, K. Unexpectedly strong effect of caffeine on the vitality of western honeybees (*Apis mellifera*). *Biochem. Biokhimiia* **2014**, *79*, 1192–1201. [CrossRef]
9. Cremonz, T.M.; De Jong, D.; Bitondi, M.M.G. Quantification of Hemolymph Proteins as a Fast Method for Testing Protein Diets for Honey Bees (Hymenoptera: Apidae). *J. Econ. Entomol.* **1998**, *91*, 1284–1289. [CrossRef]
10. De Jong, D.; da Silva, E.J.; Kevan, P.G.; Atkinson, J.L. Pollen substitutes increase honey bee haemolymph protein levels as much as or more than does pollen. *J. Apic. Res.* **2009**, *48*, 34–37. [CrossRef]
11. DeGrandi-Hoffman, G.; Eckholm, B.J.; Huang, M.H. A comparison of bee bread made by Africanized and European honey bees (*Apis mellifera*) and its effects on hemolymph protein titers. *Apidologie* **2013**, *44*, 52–63. [CrossRef]
12. Feng, M.; Ramadan, H.; Han, B.; Fang, Y.; Li, J. Hemolymph proteome changes during worker brood development match the biological divergences between western honey bees (*Apis mellifera*) and eastern honey bees (*Apis cerana*). *BMC Genomics* **2014**, *15*, 563. [CrossRef] [PubMed]
13. Basualdo, M.; Barragán, S.; Antúnez, K. Bee bread increases honey bee haemolymph protein and promote better survival despite of causing higher *Nosema ceranae* abundance in honey bees. *Environ. Microbiol. Rep.* **2014**, *6*, 396–400. [CrossRef] [PubMed]
14. Cabbri, R.; Ferlizza, E.; Nanetti, A.; Monari, E.; Andreani, G.; Galuppi, R.; Isani, G. Biomarkers of nutritional status in honeybee haemolymph: Effects of different biotechnical approaches for Varroa destructor treatment and wintering phase. *Apidologie* **2018**, *49*, 606–618. [CrossRef]
15. Mokarramat-Yazdi, A.; Soltaninejad, H.; Zardini, H.; Shishehbor, F.; Alemi, A.; Fesahat, F.; Sadeghian, F. Investigating the anticancer effect of a new drug originating from plant and animal: In vitro and in vivo study. *J. Adv. Pharm. Educ. Res.* **2021**, *10*, 72–78.
16. Chan, Q.W.T.; Howes, C.G.; Foster, L.J. Quantitative Comparison of Caste Differences in Honeybee Hemolymph. *Mol. Cell. Proteom.* **2006**, *5*, 2252–2262. [CrossRef] [PubMed]
17. Pace, C.N.; Treviño, S.; Prabhakaran, E.; Scholtz, J.M. Protein structure, stability and solubility in water and other solvents. *Philos. Trans. R. Soc. Lond. B Biol. Sci.* **2004**, *359*, 1225–1234, discussion 1234–1235. [CrossRef] [PubMed]
18. Tutun, H.; Sevin, S.; Çetintav, B. Effects of different chilling procedures on honey bees (*Apis mellifera*) for anesthesia. *Ank. Üniversitesi Vet. Fakültesi Derg.* **2020**, *67*, 289–294. [CrossRef]
19. Łoś, A.; Strachecka, A. Fast and Cost-Effective Biochemical Spectrophotometric Analysis of Solution of Insect "Blood" and Body Surface Elution. *Sensors* **2018**, *18*, 1494. [CrossRef]
20. Basseri, H.R.; Dadi-Khoeni, A.; Bakhtiari, R.; Abolhassani, M.; Hajihosseini-Baghdadabadi, R. Isolation and Purification of an Antibacterial Protein from Immune Induced Haemolymph of American Cockroach, Periplaneta americana. *J. Arthropod-Borne Dis.* **2016**, *10*, 519–527.
21. Desjardins, P.; Hansen, J.B.; Allen, M. Microvolume Protein Concentration Determination Using the NanoDrop 2000c Spectrophotometer. *J. Vis. Exp.* **2009**, *33*, e1610. [CrossRef]
22. Laemmli, U.K. Cleavage of Structural Proteins during the Assembly of the Head of Bacteriophage T4. *Nature* **1970**, *227*, 680–685. [CrossRef] [PubMed]
23. Mosmann, T. Rapid colorimetric assay for cellular growth and survival: Application to proliferation and cytotoxicity assays. *J. Immunol. Methods* **1983**, *65*, 55–63. [CrossRef] [PubMed]
24. Magaldi, S.; Mata-Essayag, S.; Hartung de Capriles, C.; Perez, C.; Colella, M.T.; Olaizola, C.; Ontiveros, Y. Well diffusion for antifungal susceptibility testing. *Int. J. Infect. Dis.* **2004**, *8*, 39–45. [CrossRef]
25. Espinel-Ingroff, A.; Canton, E.; Fothergill, A.; Ghannoum, M.; Johnson, E.; Jones, R.N.; Ostrosky-Zeichner, L.; Schell, W.; Gibbs, D.L.; Wang, A.; et al. Quality Control Guidelines for Amphotericin B, Itraconazole, Posaconazole, and Voriconazole Disk Diffusion Susceptibility Tests with Nonsupplemented Mueller-Hinton Agar (CLSI M51-A Document) for Nondermatophyte Filamentous Fungi. *J. Clin. Microbiol.* **2011**, *49*, 2568–2571. [CrossRef]
26. Espinel-Ingroff, A.; Arthington-Skaggs, B.; Iqbal, N.; Ellis, D.; Pfaller, M.A.; Messer, S.; Rinaldi, M.; Fothergill, A.; Gibbs, D.L.; Wang, A. Multicenter evaluation of a new disk agar diffusion method for susceptibility testing of filamentous fungi with voriconazole, posaconazole, itraconazole, amphotericin B, and caspofungin. *J. Clin. Microbiol.* **2007**, *45*, 1811–1820. [CrossRef]
27. Braca, A.; De Tommasi, N.; Di Bari, L.; Pizza, C.; Politi, M.; Morelli, I. Antioxidant Principles from *Bauhinia t arapotensis*. *J. Nat. Prod.* **2001**, *64*, 892–895. [CrossRef]
28. Hasan, S.M.R.; Jamila, M.; Majumder, M.M.; Akter, R.; Hossain, M.M.; Mazumder, M.E.H.; Alam, M.A.; Jahangir, R.; Rana, M.S.; Arif, M.; et al. Analgesic and Antioxidant Activity of the Hydromethanolic Extract of *Mikania scandens* (L.) Willd. Leaves. *Am. J. Pharmacol. Toxicol.* **2009**, *4*, 1–7. [CrossRef]
29. Sathishkumar, T.; Baskar, R.; Shanmugam, S.K.; Rajasekaran, P.; Sadasivam, S.; Manikandan, V. Optimization of Flavonoids Extraction from the Leaves of *Tabernaemontana heyneana* Wall Using L16 Orthogonal Design. *Nat. Sci.* **2008**, *6*, 10–21. [CrossRef]

30. Malagoli, D. A full-length protocol to test hemolytic activity of palytoxin on human erythrocytes. *Invertebr. Surviv. J.* **2007**, *4*, 92–94.
31. DeGrandi-Hoffman, G.; Chen, Y.; Huang, E.; Huang, M.H. The effect of diet on protein concentration, hypopharyngeal gland development and virus load in worker honey bees (*Apis mellifera* L.). *J. Insect Physiol.* **2010**, *56*, 1184–1191. [CrossRef] [PubMed]
32. El Mohandes, S.; Nafea, E.; Fawzy, A. Effect of different feeding diets on the haemolymph of the newly emerged honeybee workers *Apis mellifera* L. *Egypt. Acad. J. Biol. Sci. Entomol.* **2010**, *3*, 113–220. [CrossRef]
33. Leta, M.A.; Gilbert, C.; Morse, R.A. Levels of hemolymph sugars and body glycogen of honeybees (*Apis mellifera* L.) from colonies preparing to swarm. *J. Insect Physiol.* **1996**, *42*, 239–245. [CrossRef]
34. Kunc, M.; Dobeš, P.; Hurychová, J.; Vojtek, L.; Poiani, S.B.; Danihlík, J.; Havlík, J.; Titěra, D.; Hyršl, P. The Year of the Honey Bee (*Apis mellifera* L.) with Respect to Its Physiology and Immunity: A Search for Biochemical Markers of Longevity. *Insects* **2019**, *10*, 244. [CrossRef]
35. Noordyke, E.; Ellis, J. Reviewing the Efficacy of Pollen Substitutes as a Management Tool for Improving the Health and Productivity of Western Honey Bee (*Apis mellifera*) Colonies. *Front. Sustain. Food Syst.* **2021**, *5*, 772897. [CrossRef]
36. Smart, M.D.; Otto, C.R.V.; Lundgren, J.G. Nutritional status of honey bee (*Apis mellifera* L.) workers across an agricultural land-use gradient. *Sci. Rep.* **2019**, *9*, 16252. [CrossRef]
37. Schulz, M.; Łoś, A.; Grzybek, M.; Ścibior, R.; Strachecka, A. Piperine as a new natural supplement with beneficial effects on the life-span and defence system of honeybees. *J. Agric. Sci.* **2019**, *157*, 140–149. [CrossRef]
38. Skowronek, P.; Wójcik, Ł.; Strachecka, A. Cannabis Extract Has a Positive–Immunostimulating Effect through Proteolytic System and Metabolic Compounds of Honey Bee (*Apis mellifera*) Workers. *Animals* **2021**, *11*, 2190. [CrossRef]
39. Strachecka, A.J.; Olszewski, K.; Paleolog, J. Curcumin Stimulates Biochemical Mechanisms of *Apis mellifera* Resistance and Extends the Apian Life-Span. *J. Apic. Sci.* **2015**, *59*, 129–141. [CrossRef]
40. Nugraha, R.; Ruethers, T.; Johnston, E.B.; Rolland, J.M.; O'Hehir, R.E.; Kamath, S.D.; Lopata, A.L. Effects of Extraction Buffer on the Solubility and Immunoreactivity of the Pacific Oyster Allergens. *Foods* **2021**, *10*, 409. [CrossRef]
41. Arakawa, T. Acetonitrile as solvent for protein interaction analysis. *Int. J. Biol. Macromol.* **2018**, *114*, 728–732. [CrossRef] [PubMed]
42. Viswambari Devi, R.; Basilrose, M.R.; Mercy, P.D. Prospect for lectins in arthropods. *Ital. J. Zool.* **2010**, *77*, 254–260. [CrossRef]
43. Pathak, J.P.N. Haemagglutinins (Lectins) in Insects. In *Insect Immunity*; Pathak, J.P.N., Ed.; Series Entomologica; Springer: Dordrecht, The Netherlands, 1993; pp. 149–169. [CrossRef]
44. Xia, X.; You, M.; Rao, X.-J.; Yu, X.-Q. Insect C-type lectins in innate immunity. *Dev. Comp. Immunol.* **2018**, *83*, 70–79. [CrossRef] [PubMed]
45. Cheung, R.C.F.; Wong, J.H.; Pan, W.; Chan, Y.S.; Yin, C.; Dan, X.; Ng, T.B. Marine lectins and their medicinal applications. *Appl. Microbiol. Biotechnol.* **2015**, *99*, 3755–3773. [CrossRef]
46. Eleftherianos, I.; Zhang, W.; Heryanto, C.; Mohamed, A.; Contreras, G.; Tettamanti, G.; Wink, M.; Bassal, T. Diversity of insect antimicrobial peptides and proteins—A functional perspective: A review. *Int. J. Biol. Macromol.* **2021**, *191*, 277–287. [CrossRef]
47. Fujii, Y.; Dohmae, N.; Takio, K.; Kawsar, S.M.A.; Matsumoto, R.; Hasan, I.; Koide, Y.; Kanaly, R.A.; Yasumitsu, H.; Ogawa, Y.; et al. A Lectin from the Mussel *Mytilus galloprovincialis* Has a Highly Novel Primary Structure and Induces Glycan-mediated Cytotoxicity of Globotriaosylceramide-expressing Lymphoma Cells. *J. Biol. Chem.* **2012**, *287*, 44772–44783. [CrossRef] [PubMed]
48. M. Zakzok Et al., S. Antiproliferative and antioxidant activities of the edible crab *Callinectes sapidus* hepatopancreas and hemolymph extracts. *Egypt. J. Aquat. Biol. Fish.* **2021**, *25*, 531–550. [CrossRef]
49. Pinto, I.R.; Chaves, H.V.; Vasconcelos, A.S.; de Sousa, F.C.F.; Santi-Gadelha, T.; de Lacerda, J.T.J.G.; Ribeiro, K.A.; Freitas, R.S.; Maciel, L.M.; Filho, S.M.P.; et al. Antiulcer and Antioxidant Activity of a Lectin from *Mucuna pruriens* Seeds on Ethanol- induced Gastropathy: Involvement of Alpha-2 Adrenoceptors and Prostaglandins. *Curr. Pharm. Des.* **2019**, *25*, 1430–1439. [CrossRef]
50. Zakzok, S.M.; El-Ghany, A.G.A.; Anas, A.Y.; Dahshan, S.K. Biometric study, sex ratio and potential biological activities of the edible mantis shrimp *Erugosquilla massavensis*. *Egypt. J. Aquat. Biol. Fish.* **2022**, *26*, 229–253. [CrossRef]
51. Guo, D.; Wang, H.; Zeng, D.; Li, X.; Fan, X.; Li, Y. Vaccine potential of hemocyanin from *Oncomelania hupensis* against Schistosoma Japonicum. *Parasitol. Int.* **2011**, *60*, 242–246. [CrossRef]
52. Danty, E.; Arnold, G.; Burmester, T.; Huet, J.C.; Huet, D.; Pernollet, J.C.; Masson, C. Identification and developmental profiles of hexamerins in antenna and hemolymph of the honeybee, *Apis mellifera*. *Insect Biochem. Mol. Biol.* **1998**, *28*, 387–397. [CrossRef] [PubMed]
53. Barchuk, A.R.; Bitondi, M.M.G.; Simões, Z.L.P. Effects of juvenile hormone and ecdysone on the timing of vitellogenin appearance in hemolymph of queen and worker pupae of *Apis mellifera*. *J. Insect Sci.* **2002**, *2*, 1. [CrossRef] [PubMed]
54. Wheeler, D.E.; Kawooya, J.K. Purification and characterization of honey bee vitellogenin. *Arch. Insect Biochem. Physiol.* **1990**, *14*, 253–267. [CrossRef] [PubMed]
55. Park, H.G.; Lee, K.S.; Kim, B.Y.; Yoon, H.J.; Choi, Y.S.; Lee, K.Y.; Wan, H.; Li, J.; Jin, B.R. Honeybee (*Apis cerana*) vitellogenin acts as an antimicrobial and antioxidant agent in the body and venom. *Dev. Comp. Immunol.* **2018**, *85*, 51–60. [CrossRef]
56. Robbs, S.L.; Ryan, R.O.; Schmidt, J.O.; Keim, P.S.; Law, J.H. Lipophorin of the larval honeybee, *Apis mellifera* L. *J. Lipid Res.* **1985**, *26*, 241–247. [CrossRef]
57. Ba, S.; Ro, R.; Jo, S.; Jh, L. Purification and properties of a very high density lipoprotein from the hemolymph of the honeybee *Apis mellifera*. *Biochemistry* **1987**, *26*, 1885–1889. [CrossRef]

58. Ayyad, M.A.; Fakhry, A.M.; Moustafa, A.-R.A. Plant biodiversity in the Saint Catherine area of the Sinai peninsula, Egypt. *Biodivers. Conserv.* **2000**, *9*, 265–281. [CrossRef]
59. Boulos, L. Flora and Vegetation of the Deserts of Egypt. *Flora Mediterr.* **2008**, *18*, 341–359.
60. El-Seedi, H.R.; Burman, R.; Mansour, A.; Turki, Z.; Boulos, L.; Gullbo, J.; Göransson, U. The traditional medical uses and cytotoxic activities of sixty-one Egyptian plants: Discovery of an active cardiac glycoside from *Urginea maritima*. *J. Ethnopharmacol.* **2013**, *145*, 746–757. [CrossRef]
61. Rabei, S.; Khedr, A.; Elgamal, I. Floristic study of Saint Katherine protectorate, Sinai: With one new record to flora of Egypt. *Taeckholmia* **2021**, *41*, 32–55. [CrossRef]
62. Bogaerts, A.; Baggerman, G.; Vierstraete, E.; Schoofs, L.; Verleyen, P. The hemolymph proteome of the honeybee: Gel-based or gel-free? *Proteomics* **2009**, *9*, 3201–3208. [CrossRef] [PubMed]
63. Mattila, H.R.; Otis, G.W. Effects of Pollen Availability and Nosema Infection during the Spring on Division of Labor and Survival of Worker Honey Bees (Hymenoptera: Apidae). *Environ. Entomol.* **2006**, *35*, 708–717. [CrossRef]
64. Foley, K.; Fazio, G.; Jensen, A.B.; Hughes, W.O.H. Nutritional limitation and resistance to opportunistic Aspergillus parasites in honey bee larvae. *J. Invertebr. Pathol.* **2012**, *111*, 68–73. [CrossRef] [PubMed]
65. Omar, E.M.; Darwish, H.Y.A.; Othman, A.A.; El-Seedi, H.R.; Al Naggar, Y. Crushing corn pollen grains increased diet digestibility and hemolymph protein content while decreasing honey bee consumption. *Apidologie* **2022**, *53*, 52. [CrossRef]

MDPI AG
Grosspeteranlage 5
4052 Basel
Switzerland
Tel.: +41 61 683 77 34

Insects Editorial Office
E-mail: insects@mdpi.com
www.mdpi.com/journal/insects

www.ingramcontent.com/pod-product-compliance
Lightning Source LLC
Chambersburg PA
CBHW041426210926
51850CB00013B/3366
* 9 7 8 3 7 2 5 8 3 2 7 7 4 *